MATRIX ALGEBRA

MATRIX ALGEBRA

Using MINImal MATlab™

Joel. W. Robbin
University of Wisconsin - Madison

A K Peters
Wellesley, Massachusetts

Editorial, Sales, and Customer Service Office

A K Peters, Ltd.
289 Linden Street
Wellesley, MA 02181

Library of Congress Cataloging-in-Publication Data

Robbin, Joel, W.
    Matrix algebra using MINImal MATlab / Joel W. Robbin
        p. cm.
    Includes indexes.
    ISBN 1-56881-024-5
    1. Matrices -- Data processing. 2. MINIMAT. I. Title
    QA188.R62   1994B   93-39372
    512.9'434--dc20   CIP

Matlab is a registered trademark of the Mathworks, Inc., of Natick, Massachusetts.

Printed in the United States of America
98 97 96 95 94   10 9 8 7 6 5 4 3 2 1

To Alice and Rachel.

PREFACE

There is, I am told, a small town in Middle America populated entirely by people engaged in the writing of textbooks of elementary linear algebra. I should like to explain what motivates those tormented souls.

In most colleges the first course in linear algebra provides the student with her/his first contact with material that requires *mathematical literacy*. Linear algebra cannot be understood without a partial mastery of the lingo mathematicians use to describe their subject. The student must master hard words like *if*, *unique*, *for all*, and (hardest of all) *there exists*. To one who knows these words, the subject is almost trivial; to the uninitiated, it is impossibly arcane. The course in linear algebra represents a kind of watershed. Its survivors will love mathematics and will exhibit a kind of clarity of thought that will serve them well in almost any future scientific endeavor. Moreover, mathematical literacy isn't much different from the other kind. The first course in linear algebra prepares one for the modern world better than any other course in the college curriculum. So this is why there are so many textbooks in linear algebra: the subject is important.

The book you are holding has ambitious objectives. It aims to teach mathematical literacy by providing a careful treatment of set theoretic notions and elementary mathematical proofs. It aims to develop geometric intuition via low dimensional examples. It gives a complete treatment of the fundamental normal form theorems of matrix algebra. It fully integrates the computer into the course by describing the basic algorithms in the computer language *Matlab* and by providing computer exercises that utilize new pedagogical techniques. It contains enough material for a year long course such as the 340-443 sequence at the University of Wisconsin. The first six chapters of the book can be used in a course which follows the recommendations of NSF's Linear Algebra Curriculum Study Group[1]

The computer program MINIMAT is provided free with this book.[2] This program is upward compatible with commercial programs such as $PC\text{-}Matlab^{TM}$ published by the Mathworks and *Student Edition of Matlab*^{TM} published by Prentice-Hall. MINIMAT contains only those features which are useful in teaching an elementary course. These include a scrolling transcript which enables the user to review earlier output, the capability of saving the transcript at any time, a built-in editor for writing and listing small .M files, and carefully designed menus and help files. All the programs listed in the text are duplicated in .M files supplied on the distribution diskette. These .M files work with most versions of Matlab^{TM} as well. In particular, they have been tested with *The Student Edition of Matlab*^{TM} .

Conversations with many colleagues, too numerous to mention, influenced this book. Typically, one of them would say something like, "Oh, the students just don't understand X", and I would make X an example or exercise in the book. The "Warmup" on Page 2 is based on the first lecture David Fowler gives whenever he teaches this subject. The idea for numbering the theorems (Theorem 99B is the second theorem on Page 99) is due to Anatole Beck. I would like to thank Anatole Beck, Sufian Huseinni, Jerry Keisler, Arnie Miller, Rod Smart, and Dietrich Uhlenbrock, for using preliminary versions of these materials in their Math 340 classes at UW. Their experiences proved invaluable. I also benefitted from conversations with Carl de Boor. Al Letarte found many typographical errors while preparing his study guide for the UW extension course in matrix algebra.

My family did more than provide emotional support and tolerance. My wife Alice Robbin did a magnificent job of proof reading, and my daughter Rachel Robbin provided significant insight into how students learn by allowing me to help her with her math homework.

Thanks also go to IBM's *Project Trochos*, which provided the IBM-AT on which MINIMAT was originally developed.

March 28, 1994   J.W.R

Department of Mathematics
University of Wisconsin
Madison, Wisconsin 53706
e-mail: robbin@math.wisc.edu

---

[1] D. Carlson, C. R. Johnson, D. C. Lay, A. D. Porter: *The linear algebra curriculum study group recommendations for the first course in linear algebra*, College Math Journal **24** Jan. 1993, pp 41-46.

[2] The DOS version is supplied. Versions which run on other platforms (such as the Macintosh or Windows) can be obtained directly from the author for a nominal charge.

Contents

1

WARMUP

First, a pep talk. Matrix algebra is a beautiful subject with many important applications both within mathematics and in other areas of science and industry. To master it you will have to learn to think clearly, read carefully, and write correctly. The computer exercises in this book should help you learn to write correctly. Computers are rather stupid beasts that produce error messages in response to vague or imprecise input. This book will help you read carefully. It contains many questions answered at the back of the book[1] which test your understanding of the text immediately preceding. To learn to think clearly, you should ask questions. Often it is enough to ask yourself the question. A student who can formulate a question correctly can usually answer it.

This short preliminary chapter gathers in one place a few things – about logic, proofs, and notation – that maybe no one ever told you before. There is nothing very difficult here, but students are sometimes afraid to admit they don't understand these things. That's a mistake. In my experience, the best students frequently think they are the worst, because they are confronting their confusion rather than avoiding it.

The elementary set theory introduced in this chapter may be postponed until Chapter 4 is studied.

---

[1]The symbol ▷ indicates that the answer to a question or exercise appears in the answers section at the back of the book.

1.1   Clear Thinking

Sometime in your childhood you learned to solve the equation $2x = 6$.
Divide both sides by 2:

$$2x = 6 \implies x = 6/2 = 3.$$

Perhaps you concluded that, if numbers a and y are given, the solution of
the equation $ax = y$ is $x = y/a$. This isn't quite right; the truth is slightly
more complicated.

Theorem 2A *Assume numbers a, y are given. Then the equation*

$$ax = y$$

- *has a unique solution $x = y/a$ if $a \neq 0$;*

- *has no solution if $a = 0$ and $y \neq 0$;*

- *has any number x as a solution if $a = y = 0$.*

Do you think this degree of precision constitutes nit-picking? Usually
the answer is $x = y/a$, and on a test you would know how to respond to
questions like *solve $2x = 6$* or *solve $0x = 6$*. If this is your attitude, *I beseech
you to think it possible you may be mistaken*, for it is likely that you will
not understand this book.

If you imagine writing a computer program, you will see why such care
is important. A programmer might take $x = y/a$ as the solution to $ax = y$
without first testing whether $a = 0$. Each time the program runs, the
variables a and y receive different values, values beyond the control of the
program. Usually the program works correctly (for the variable a will be
nonzero), but every once in a while the program fails. A good programmer
analyzes all the possibilities in a program and proves that it works in all
circumstances. This activity is not unlike what mathematicians do when
they prove theorems.

1.2   Logic

A statement of form

   if P, then Q

means that Q is true whenever P is true. The **converse** of this statement
is the related statement

if Q, then P.

A statement and its converse do not have the same meaning. For example, the statement

if $x = 2$, then $x^2 = 4$

is true while its converse

if $x^2 = 4$, then $x = 2$

is not generally true (maybe $x = -2$). The following phrases are all synonymous:

- *if P, then Q;*
- *P implies Q;*
- *Q, if P;*
- *P only if Q;*

The mathematical symbol $\implies$ is also used to mean *implies* as in

$$x = 2 \implies x^2 = 4.$$

The **contrapositive** of the statement *if P, then Q* is the statement *if not Q, then not P.* Unlike the converse, an implication and its contrapositive have the same meaning. For example, the two assertions

$$x = 2 \implies x^2 = 4 \quad \text{and} \quad x^2 \neq 4 \implies x \neq 2$$

have exactly the same meaning.

The statement

P if and only if Q

has the same meaning as the statement

if P, then Q and if Q, then P.

This statement asserts a kind of equality – that P and Q have the same meaning: P is true exactly when Q is. The phrase *if and only if* is frequently abbreviated *iff*, especially in definitions. The mathematical symbol $\iff$ is also used to mean *if and only if* as in

$$x^2 = 4 \iff x = 2 \text{ or } x = -2.$$

This equation asserts that a number x satisfies the condition $x^2 = 4$ exactly when it satisfies the condition $x = \pm 2$: the two conditions are "equal".

Sometimes we say

> *the conditions P and Q are equivalent*

when we mean

> *P if and only if Q.*

This is particularly the case when we have more than two conditions as in the following

Example 4A For any number x the following conditions are equivalent:

(1) $x^2 - 5x + 6 = 0$.

(2) Either $x = 2$ or $x = 3$.

(3) The number x is an integer between 1.5 and 3.5.

What this means is that if any one of the three conditions is true, then all of them are.

The notations

$$P \implies Q \qquad (P \textbf{ implies } Q)$$
$$P \iff Q \qquad (P \textbf{ if and only if } Q)$$

are used throughout this book. The word **iff** is an abbreviation for the phrase *if and only if.*

1.3   Proofs

One of the principal aims of this book is to teach the student how to read and, to a lesser extent, write proofs. A proof is an argument intended to convince the reader that a general principle is true in all situations. The amount of detail that an author supplies in a proof should depend on the audience. Too little detail leaves the reader in doubt; too much detail may leave the reader unable to see the forest for the trees. As a general principle, the author of a proof should be able to supply the reader with additional detail on demand. When a student writes a proof for a teacher, the aim is usually not to convince the teacher of the truth of some general principle (the teacher already knows that), but to convince the teacher that the student understands the proof and can write it clearly.

The "theorems" below show the proper format for writing a proof. In each of them you are supposed to imagine that the theorem to be proved has the indicated form. Notice how the key words *choose, assume, let,* and *therefore* are used in the proof. In these sample formats, the phrase "Blah Blah Blah" indicates a sequence of steps, each one justified by earlier steps. The symbol □ is used to indicate the end of the proof.

Theorem *If P, then Q.*

Proof: Assume P. Blah Blah Blah. Therefore Q.   □

Theorem *P if and only if Q.*

Proof: Assume P. Blah Blah Blah. Therefore Q. Conversely, assume Q. Blah Blah Blah. Therefore P.   □

Theorem *P(x) for all x.*

Proof: Choose x. Blah Blah Blah. Therefore P(x).   □

Theorem *There is an x such that P(x).*

Proof: Let $x = \ldots$. Blah Blah Blah. Therefore P(x).   □

Usually P and Q themselves involve the logical phrases *if, for all, there is.* In this case, the proof reflects that structure by using the corresponding key word *assume, choose, let.* For example, consider the following

Theorem 5A *For all a and b, if $a \neq 0$, then there is an x with $ax = b$.*

Proof: Choose a and b. Assume $a \neq 0$. Let $x = b/a$. Then $ax = a(b/a) = b$. Therefore $ax = b$.   □

Of course, this proof is quite trivial and is given here only to illustrate the proper use of the key words *choose, assume, let,* and *therefore.* In general, every step in a proof is either an assumption (based on the structure of the theorem to be proved), an abbreviation (used to introduce notation to make the proof easier to read), or follows from earlier statements by the application of previously justified principles.

1.4   Sets

A set V divides the mathematical universe into two parts: those objects x that **belong** to V and those that don't. The notation $x \in V$ means x

belongs to V. The notation $x \notin V$ means that x does not belong to V. The objects that belong to V are called the **elements** of V or the **members** of V. Other words roughly synonymous with the word *set* are *class*, *collection*, and *aggregate*. These longer words are generally used to avoid using the word *set* twice in one sentence. The situation typically arises when an author wants to talk about sets whose elements are themselves sets. One might write " the collection of all finite sets of integers", rather than "the set of all finite sets of integers".

Defining Sets by Enumeration

The simplest sets are **finite** and these are often defined by simply listing (**enumerating**) their elements between curly brackets. Thus if $V = \{2, 3, 8\}$ then $3 \in V$ and $7 \notin V$. Often an author uses dots as a notational device to mean "et cetera" and indicate that the pattern continues. Thus if

$$V = \{x_1, x_2, \ldots, x_n\}, \tag{1}$$

then for any object y, the phrase "$y \in V$" and the phrase "$y = x_i$ for some $i = 1, 2, \ldots, n$" have the same meaning; that is, one is true if and only if the other is. Having defined V by (1), we have

$$y \in V \iff y = x_1 \text{ or } y = x_2 \text{ or } \ldots \text{ or } y = x_n.$$

In other words, the shorter phrase "$y \in V$" has the same meaning as the more cumbersome phrase "$y = x_1$ or $y = x_2$ or ... $y = x_n$".

The device of listing some of the elements with dots between curly brackets can also be used to define infinite sets provided that the context makes it clear what the dots stand for. For example, we can define the set of **natural numbers** by

$$\mathbf{N} = \{0, 1, 2, 3, \ldots\}$$

and the set of **integers** by

$$\mathbf{Z} = \{\ldots, -2, -1, 0, 1, 2, \ldots\},$$

and hope that the reader understands that $0 \in \mathbf{N}$, $5 \in \mathbf{N}$, $-5 \notin N$, $\frac{3}{5} \notin \mathbf{N}$, $0 \in \mathbf{Z}$, $5 \in \mathbf{Z}$, $-5 \in \mathbf{Z}$, $\frac{3}{5} \notin \mathbf{Z}$, etc.

Common Sets

Certain sets are so important that they have names:

$\emptyset$	(the empty set)
N	(the natural numbers)
Z	(the integers)
Q	(the rational numbers)
R	(the real numbers)
C	(the complex numbers)

These names are almost universally used by mathematicians today, but in older books one may find other notations. Here are some true assertions: $0 \notin \emptyset$, $\frac{3}{5} \in \mathbf{Q}$, $\sqrt{2} \notin \mathbf{Q}$, $\sqrt{2} \in \mathbf{R}$, $x^2 \neq -1$ for all $x \in \mathbf{R}$, and $x^2 = -1$ for some $x \in \mathbf{C}$ (namely $x = \pm i$).

Sets and Properties

If V is a set and $P(x)$ is a property that either holds or fails for each element $x \in V$, then we may form a new set W consisting of all $x \in V$ for which $P(x)$ is true. This set W is denoted by

$$W = \{x \in V : P(x)\} \tag{2}$$

and called "the set of all $x \in V$ such that $P(x)$". Some authors write "$|$" instead of "$:$" as in

$$W = \{x \in V \mid P(x)\}.$$

This is a very handy notation. Having defined W by (2), we may assert that for all x

$$x \in W \iff x \in V \text{ and } P(x)$$

and that for all $x \in V$

$$x \in W \iff P(x).$$

Since the property $P(x)$ may be quite cumbersome to state, the notation $x \in W$ is both shorter and easier to understand.

Example 7A If $W = \{x \in \mathbf{N} : x^2 < 6 + x\}$, then $2 \in W$ (as $2^2 < 6 + 2$), $3 \notin W$ (as $3^2 \not< 6 + 3$), and $-1 \notin W$ (as $-1 \notin \mathbf{N}$).

Another notation that is used to define sets is

$$W = \{f(x) : x \in V\}.$$

This is to be understood as an abbreviation for

$$W = \{y : y = f(x) \text{ for some } x \in V\}$$

so that for any y

$$y \in W \iff y = f(x) \text{ for some } x \in V.$$

It may be difficult to decide if $y \in W$: the definition requires us to examine all solutions x of the equation $y = f(x)$.

Example 8A Using these notations the set W of even natural numbers may be denoted by any of the following three notations:

$$
\begin{aligned}
W & = \{0, 2, 4, \ldots\} \\
& = \{m \in \mathbf{N} : m \text{ is divisible by 2}\} \\
& = \{2n : n \in \mathbf{N}\}
\end{aligned}
$$

Example 8B $\{x^2 : 2 < x < 3\} = \{y : 4 < y < 9\}$.

Example 8C $\{x : 2 < x < 3\} \subset \{x : 4 < x^2 < 9\}$, but these are not equal: the latter set contains negative numbers. The subset symbol $\subset$ is explained below on Page 9.

Crude graphs can be used to get a rough idea of what a set of real numbers is. For example, to graph the set $V = \{x : y_0 < f(x) < y_1\}$ draw the two horizontal lines $y = y_0$ and $y = y_1$, plot the portion of the graph between those lines and project to the x-axis.

Example 8D $\{x : 1 < x^2 < 4\} = \{x : -2 < x < -1\} \cup \{x : 1 < x < 2\}$. (See Figure 1. The union symbol $\cup$ is explained below on Page 11.)

To graph the set $W = \{f(x) : x_0 < x < x_1\}$ draw the two vertical lines lines $x = x_0$ and $x = x_1$, plot the portion of the graph between those lines, and project to the y-axis.

Example 8E $\{x^2 : -1 < x < 2\} = \{y : 0 \leq y < 4\}$. (See Figure 1.)

▷ **Exercise 8F** Simplify $\{x^2 : -2 < x < 3\}$.

▷ **Exercise 8G** For each of the numbers $x = 0, -1, 3, 7/9, 9/7$ and each of the following sets V_i say whether $x \in V_i$. (There are $5 \times 4 = 20$ questions here.)

$$V_1 = \{1, 2, \ldots, 9\} \qquad V_2 = \{x \in \mathbf{Z} : x^2 < 9\}$$

$$V_3 = \{x \in \mathbf{R} : x^2 < 9\} \qquad V_4 = \{x^2 : x \in \mathbf{R}, \ x < 9\}$$

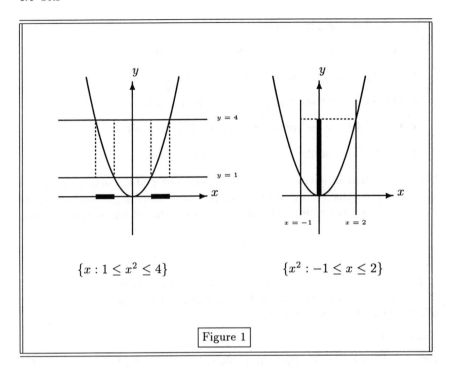

$$\{x : 1 \le x^2 \le 4\} \qquad\qquad \{x^2 : -1 \le x \le 2\}$$

Figure 1

Subsets

Definition 9A A set W is a **subset** of a set V, written

$$W \subset V,$$

iff every element of W is an element of V. The notation $W \not\subset V$ signifies that W is not a subset of V, that is, that there is at least one element of W which is not an element of V.

For example,

$$\{1, 3, 4, 7\} \subset \{0, 1, 2, 3, 4, 7, 9\}$$

since every element on the left appears on the right. However,

$$\{1, 3, 4, 7\} \not\subset \{0, 1, 2, 4, 7, 9\}$$

since $3 \in \{1, 3, 4, 7\}$ but $3 \notin \{0, 1, 2, 4, 7, 9\}$.

Note the following inclusions:

- $\mathbf{N} \subset \mathbf{Z}$ (every natural number is an integer),
- $\mathbf{Z} \subset \mathbf{Q}$ (every integer is a rational number),
- $\mathbf{Q} \subset \mathbf{R}$ (every rational number is a real number),
- $\mathbf{R} \subset \mathbf{C}$ (every real number is a complex number),
- $\emptyset \subset V$ (The empty set is a subset of every set V).

The last statement is true because *every* element x of the empty set lies in V – or indeed satisfies any other property – since there are no such elements x. However, do not confuse the empty set with the set whose only element is 0;

$$\emptyset \neq \{0\}$$

since $0 \in \{0\}$ but $0 \notin \emptyset$.

The following illustrates the proper format for proving a set inclusion. The phrase 'Blah Blah Blah' indicates a sequence of steps each of which follows from the previous ones.

Theorem $V \subset W$.

Proof: Choose $x \in V$. Blah Blah Blah. Therefore $x \in W$.   □

Example 10A We prove $V \subset W$ where

$$V = \{y \in \mathbf{R} : 0 < y < 4\}, \qquad W = \{x^2 : -2 < x < 3\}.$$

Proof: Choose $y \in V$. Then $0 < y < 4$. Let $x = \sqrt{y}$. Then $-2 < x < 3$ and $y = x^2$. Hence, $y \in W$ as required.   □

Example 10B Let $V = \{x^2 : x \in \mathbf{N}, \; -2 < x < 3\}$ and $W = \{0, 1, 4, 9\}$. Then $W \not\subset V$ as $9 \in W$ but $9 \notin V$.

▷ **Exercise 10C** For each of the following pairs of sets (V_i, V_j) decide if $V_i \subset V_j$. If so, prove it; if not, exhibit an x with $x \in V_i$ but $x \notin V_j$. (There are $4 \times (4 - 1) = 12$ problems here.)

$$
\begin{aligned}
V_1 &= \{x \in \mathbf{R} : 1 < 3x + 7 < 20\} \\
V_2 &= \{x \in \mathbf{Z} : 1 < 3x + 7 < 20\} \\
V_3 &= \{3x + 7 : x \in \mathbf{R}, \; 2 < x < 13/3\} \\
V_4 &= \{3x + 7 : x \in \mathbf{R}, \; -2 < x < 13/3\}
\end{aligned}
$$

Boolean Operations

The **intersection**, $V \cap W$, of two sets V and W is the set of objects in both of them:

$$V \cap W = \{z : z \in V \text{ and } z \in W\}.$$

The **union**, $V \cup W$, of two sets V and W is the set of objects in one or the other of them:

$$V \cup W = \{z : z \in V \text{ or } z \in W\}.$$

The **difference**, $V \setminus W$, of two sets V and W is the set of objects in the first and not in the second:

$$V \setminus W = \{z : z \in V \text{ and } z \notin W\}.$$

For example, if

$$V = \{1, 3, 5\}, \quad W = \{2, 3\},$$

then

$$V \cap W = \{3\}, \quad V \cup W = \{1, 2, 3, 5\}, \quad V \setminus W = \{1, 5\}.$$

Equality of Sets

Definition 11A (Equality of Sets) Two sets V and W are **equal**, written $V = W$, iff $V \subset W$ and $W \subset V$, that is, iff every element of V is an element of W and every element of W is an element of V.

The following illustrates the proper format for proving a set equality. The phrase 'Blah Blah Blah' indicates a sequence of steps each of which follows from the previous ones.

Theorem $V = W$.

Proof: We show $V \subset W$. Choose $x \in V$. Blah Blah Blah. Therefore $x \in W$. This proves $V \subset W$.

We show $W \subset V$. Choose $x \in W$. Blah Blah Blah. Therefore $x \in V$. This proves $W \subset V$.   □

To prove that $V \subset W$ we prove

$$x \in V \implies x \in W.$$

To prove that $V = W$ we prove

$$x \in V \iff x \in W.$$

Example 12A Let $f(x) = x^2 - 2x$. We show that $V = W$ where

$$V = \{y \in \mathbf{R} : -1 \le y < 8\}, \quad W = \{f(x) : -1 < x < 4\}.$$

Proof: *We show $V \subset W$. Choose $y \in V$. Then $-1 \le y < 8$. Let* $x = 1 + \sqrt{1 + y}$. *Then* $y = f(x)$ *and* $-1 < 1 + \sqrt{0} \le x < 1 + \sqrt{9} = 4$. *Hence $y \in W$. This proves $V \subset W$.*

   We show $W \subset V$. Choose $y \in W$. Then $y = f(x) = x^2 - 2x = (x-1)^2 - 1$ *for some x with* $-1 < x < 4$. *There are two possibilities:*

case (1) $-1 < x < 1$. *In this case,* $-2 < x - 1 < 0$ *so* $0 < (x-1)^2 < 4$ *so* $-1 < y = (x-1)^2 - 1 = f(x) < 3 < 8$

case (2) $1 \le x < 4$. *In this case,* $0 \le x - 1 < 3$ *so* $0 \le (x-1)^2 < 9$ *so* $-1 \le y = (x-1)^2 - 1 = f(x) < 8$

In either case $y \in V$. This proves $W \subset V$.   □

If you draw the graph of $y = x^2 - 2x$, you'll see how I picked the numbers, but the logic of the proof has nothing to do with the graph.

▷ **Exercise 12B** Let $V = \{n \in \mathbf{N} : n^2 + 7 < 6n\}$ and $W = \{2, 3, 4\}$. Prove that $V = W$.

1.5   Scalars

Throughout this book, $\mathbf{F}$ denotes either the set $\mathbf{R}$ of real numbers or the set $\mathbf{C}$ of complex numbers.[2]   Most of what we do works equally well for real numbers or complex numbers, and so we use the ambiguous symbol $\mathbf{F}$. Occasionally, we must make statements that are true of one but not the other. In such situations we use the symbol $\mathbf{R}$ or $\mathbf{C}$ (as appropriate) in place of the symbol $\mathbf{F}$. In mathematics (though not in physics) the words **scalar** and **number** are synonymous. Some authors call $\mathbf{F}$ the *field of scalars*.

   The notation

$$x \in S$$

means *x is an element of the set S.* For example, the notation

$$x \in \mathbf{R}$$

is an abbreviation for the phrase

---

[2]Complex numbers will not be needed before Chapter 8.

$$x \text{ is a real number}$$

and the notation

$$z \in \mathbf{C}$$

is an abbreviation for the phrase

$$z \text{ is a complex number.}$$

Of course, when we say that z is a complex number we do not mean that it is necessarily nonreal: every real number is a complex number.

A complex number is any number z of form

$$z = x + iy$$

where x and y are real and $i^2 = -1$; the number z is real iff $y = 0$. The real number x is called the **real part** of z and the real number y is called the **imaginary part** of z.   We occasionally use the notations

$$\Re(z) = x, \qquad \Im(z) = y,$$

for the real and imaginary parts of z. Thus z is real iff $\Im(z) = 0$. The **conjugate** of the complex number z is the complex number $\bar{z}$ obtained from z by reversing the sign of the imaginary part:

$$\bar{z} = x - iy.$$

Thus z is real iff $z = \bar{z}$.

The **absolute value** of the complex number z is the nonnegative real number

$$|z| = \sqrt{z\bar{z}} = \sqrt{x^2 + y^2}$$

where $z = x + iy$ and $x, y \in \mathbf{R}$. If z is real, then $\bar{z} = z$, so $z\bar{z} = z^2$, and $|z|$ agrees with the usual notion of absolute value, since $\sqrt{z^2}$ denotes the nonnegative square root of z^2. For example,

$$\sqrt{(-3)^2} = \sqrt{9} = 3 = |-3|.$$

For $z \neq 0$ we can find $1/z$ by

$$\frac{1}{z} = \frac{\bar{z}}{z\bar{z}} = \frac{\bar{z}}{|z|^2}.$$

▷ **Exercise 13A** Let $z = 3 + 4i$. Compute $|z|$ and $1/z$

▷ **Exercise 13B** Prove the laws in the box on Page 25.

▷ **Exercise 13C** What does the symbol ▷ mean?

1.6   Sigma Notation

Mathematicians frequently use the symbol $\sum$ to indicate summation as follows:

$$\sum_{k=m}^{n} a_k = a_m + a_{m+1} + \cdots + a_n.$$

This handy notation is called **sigma notation**. The symbol $\sum$ is the letter capital *sigma* of the Greek alphabet that corresponds to the letter S (for *sum*) of the Latin alphabet. [3]

To evaluate the expression

$$\sum_{k=m}^{n} a_k,$$

you evaluate the expressions a_k for all integer values k between $k = m$ and $k = n$ and add the results. For example,

$$\sum_{k=3}^{5} (k^2 + 1) = 10 + 17 + 26 = 53.$$

The summation index k that appears under the summation sign $\sum$ is called a *dummy variable* because changing it systematically does not change the meaning of the expression. Thus

$$\sum_{k=3}^{5} (k^2 + 1) = \sum_{j=3}^{5} (j^2 + 1).$$

Frequently, the summation index ranges over a set as in the equation

$$S = \sum_{i \in J} a_i.$$

This means that the expression a_i is to be evaluated for each value i in the set J and the results are added to obtain S. For example, if $J = \{3, 5, 9\}$, then

$$\sum_{i \in J} i^2 = 9 + 25 + 81 = 115.$$

Finally, sometimes an author will use words to amplify the meaning of an expression using the summation sign as in the following:

---

[3] Incidentally, the integral sign $\int$ used in calculus also indicates a kind of summation, and, in fact, derives from the old English S.

$$\sum_i i^2 = 115$$

where i ranges over the three-element set $\{3, 5, 9\}$.

Here's an example from high school algebra illustrating the use of the sigma notation.

Theorem 15A (Binomial Formula) *If a and b are numbers and n is a positive integer, then*

$$(a + b)^n = \sum_{p+q=n} \frac{n!}{p!q!} a^p b^q$$

where the sum is over all pairs (p, q) of nonnegative integers that satisfy $p + q = n$.

This formula is proven in Appendix B as Theorem 446A.[4] Here's what it says for $n = 2, 3, 4$:

$$
\begin{aligned}
(a + b)^2 &= a^2 + 2ab + b^2 \\
(a + b)^3 &= a^3 + 3a^2b + 3ab^2 + b^3 \\
(a + b)^4 &= a^4 + 4a^3b + 6a^2b^2 + 4ab^3 + b^4.
\end{aligned}
$$

There is a **pi notation** for products that is analogous to the sigma notation for sums:

$$\prod_{k=m}^n a_k = a_m a_{m+1} \cdots a_n.$$

For example, $n!$ (read n **factorial**) is defined by

$$n! = \prod_{k=1}^n k.$$

Here the symbol $\prod$ is the Greek letter *pi* that corresponds to the Latin letter P for *product*.

---

[4]Numbering in this book is such that Theorem 446A appears on Page 446.

1.7 The Geometry of Linear Systems

In high school you learned that a single equation

$$a_1 x_1 + a_2 x_2 = y$$

determines a line in the (x_1, x_2) plane. Here a_1, a_2, and y are given and x_1, x_2 are the unknowns. (We're using different letters here than you used in high school.) Of course, this equation doesn't always define a line:

- If $a_1 \neq 0$ or $a_2 \neq 0$ (or both), the set of solutions (x_1, x_2) is a line;

- If $a_1 = a_2 = y = 0$, every pair (x_1, x_2) is a solution;

- If $a_1 = a_2 = 0$ but $y \neq 0$, the set of solutions is empty.

In the study of two equations

$$\begin{aligned} a_{11} x_1 + a_{12} x_2 &= y_1 \\ a_{21} x_1 + a_{22} x_2 &= y_2 \end{aligned}$$

in two unknowns x_1, x_2 there are (assuming each equation determines a line) three possibilities:

- *The two lines are distinct and not parallel.* In this case they intersect in a point and there is a unique solution. (See Figure 2.)

- *The two lines are parallel.* In this case there is no solution.

- *The two lines coincide.* In this case any solution of one equation is also a solution of the other, so there are infinitely many solutions.

In three variables (x_1, x_2, x_3) a single equation

$$a_1 x_1 + a_2 x_2 + a_3 x_3 = y$$

determines a (two-dimensional) plane in the three-dimensional space of all triples (x_1, x_2, x_3). (To have a plane at least one of the three coefficients a_1, a_2, a_3 must be nonzero.) For two such equations

$$\begin{aligned} a_{11} x_1 + a_{12} x_2 + a_{13} x_3 &= y_1 \\ a_{21} x_1 + a_{22} x_2 + a_{23} x_3 &= y_2 \end{aligned}$$

we have three possibilities:

- *The two planes are distinct and not parallel.* In this case they intersect in a line.

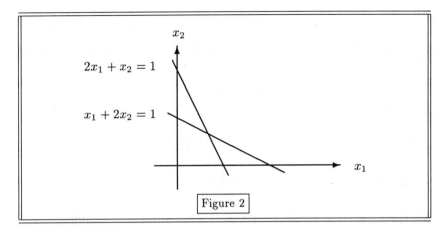

Figure 2

- *The two planes are parallel.* In this case the intersection is empty.

- *The two planes coincide.* In this case the intersection is this common plane.

For three such equations

$$a_{11}x_1 + a_{12}x_2 + a_{13}x_3 = y_1$$
$$a_{21}x_1 + a_{22}x_2 + a_{23}x_3 = y_2$$
$$a_{31}x_1 + a_{32}x_2 + a_{33}x_3 = y_2$$

we have five possibilities:

- *The three planes intersect in a point.* This happens when the first two planes intersect in a line that is not parallel to the third plane. In this case the system of equations has a unique solution.

- *Two of the three planes are parallel.* In this case the system of equations has no solutions.

- *Two of the three planes coincide.* In this case the analysis continues as for two equations in three unknowns.

- *The first two planes intersect in a line parallel to the third.* The pairwise intersections form three parallel lines. In this case the system of equations has no solutions.

- *The three planes intersect in a common line.* In this case, the system of equations has infinitely many solutions.

▷ **Exercise 18A** Match the system on the left with the description on the right.

(1) $\begin{cases} 2x_1 + x_2 = 5 \\ x_1 + x_2 = 3 \end{cases}$ (a) unique solution

(2) $\begin{cases} 2x_1 + 2x_2 = 5 \\ x_1 + x_2 = 3 \end{cases}$ (b) infinitely many solutions

(3) $\begin{cases} 2x_1 + 2x_2 = 6 \\ x_1 + x_2 = 3 \end{cases}$ (c) no solution

▷ **Exercise 18B** Match the system on the left with the description on the right.

(1) $\begin{cases} x_1 + x_2 + x_3 = 3 \\ x_1 + 2x_2 + x_3 = 5 \end{cases}$ (a) one-dimensional intersection

(2) $\begin{cases} x_1 + x_2 + x_3 = 3 \\ 2x_1 + 2x_2 + 2x_3 = 5 \end{cases}$ (b) two-dimensional intersection

(3) $\begin{cases} x_1 + x_2 + x_3 = 3 \\ 2x_1 + 2x_2 + 2x_3 = 6 \end{cases}$ (c) empty intersection

▷ **Exercise 18C** Match the system on the left with the description on the right. (In Chapter 3 you will learn a systematic way for solving problems like this, but for now just solve the problem any way you can.)

(1) $\begin{cases} x_1 + x_2 + x_3 = 3 \\ x_1 + 2x_2 + x_3 = 5 \\ x_1 + x_2 + 2x_3 = 6 \end{cases}$ (a) line of solutions

(2) $\begin{cases} x_1 + x_2 + x_3 = 3 \\ x_1 + 2x_2 + x_3 = 5 \\ 2x_1 + 3x_2 + 2x_3 = 1 \end{cases}$ (b) unique solution

(3) $\begin{cases} x_1 + x_2 + x_3 = 3 \\ x_1 + 2x_2 + x_3 = 5 \\ 2x_1 + 3x_2 + 2x_3 = 8 \end{cases}$ (c) no solution

1.8   Using MINIMAT

Sections with this title appear throughout this book. These sections explain how to instruct a computer to do relevant computations.

MINIMAT is a programming language designed especially for use with this book. It is a subset of a more powerful language called *Matlab*.[5] In fact, the name *MINIMAT* stands for *MINImal MATlab*.

The diskette supplied with this book contains an implementation of MIN-IMAT together with all the auxiliary files (called .M files) needed to use it with this book. Several implementations of Matlab exist on both mainframe computers and PC's and you can use one of these (together with the aforementioned .M files) if you prefer. If you are going to use MINIMAT, you should now try the **MINIMAT TUTORIAL**, supplied with the distribution diskette. If you want to use some other version of Matlab, you will have some other way of becoming acquainted with it.

In the rest of this section we'll learn some of the features of the programming language MINIMAT. More features are explained in each chapter as they are needed. A summary of the language is contained in Appendix C, entitled **Summary of MINIMAT**.

Assignment Statements

The most commonly used commands in MINIMAT are of the form

```
#> variable = expression
```

They are called **assignment statements**. This command causes MINI-MAT to evaluate the expression, display the result on the screen, and assign the result to the variable for future use. If the statement is followed by a semicolon (;) the answer is not printed. Several statements may appear on one line if they are separated by commas and/or semicolons. In the simplest situation the variable is a name: a string of letters and/or digits beginning with a letter. We can omit the part 'variable =" in which case MINIMAT displays the result and assign it to the variable ans. (This is called a **null assignment**.)

The notation #> is MINIMAT's **prompt**. MINIMAT prints this symbol when it is waiting for input from the user. In this book the prompt appears before a command that the user has typed. Lines without the prompt represent MINIMAT's response. For example, the display

```
#>  a=2+3
    a=5
```

indicates that the user typed a=2+3 and that MINIMAT responded with a=5.

---

[5]In the rare instance where we need to stress a difference between MINIMAT and Matlab, we refer to the latter as **classical Matlab**.

Complex Numbers

MINIMAT handles complex numbers as well as real numbers. There is no notation for $\sqrt{-1}$ in MINIMAT, but you can always create one with the command

```
#> i=sqrt(-1)
```

You don't have to use the letter i here, but that's the letter MINIMAT uses when it prints a complex number. MINIMAT's built-in functions `real`, `imag`, `conj`, and `abs`, return the real part, the imaginary part, the complex conjugate, and the absolute value of their inputs.

▷ **Exercise 20A** What happens after the following?

```
#> i=sqrt(-1), x=3, y=4, z=x+i*y
#> real(z), imag(z), conj(z), abs(z)
```

Exercise 20B If $z = 3 + 4i$ what is $1/z$? Confirm your answer by typing the following at MINIMAT:

```
#> z=3+4*sqrt(-1), w=1/z, w*z
```

Expressions

As far as possible, MINIMAT's notations are those of ordinary mathematics. The major differences are imposed by the limitations of the computer keyboard. Since superscripts are hard to handle, we need a special notation for raising to a power. Since the number of symbols is limited, we need names of more than one character: we cannot denote the product of a and b by `ab`. (MINIMAT uses the asterisk * for multiplication.) But usually the conventional notations work. For example, the command

```
#> s=a+b,   d=a-b,  p=a*b,  q=a/b
```

assigns to s, d, p, and q the sum, difference, product, and quotient of a and b respectively. The notation a^p is used for powers. For example,

```
#> a=5^3, b=8^(1/3), c=10^(-2)
```

assigns to a, b, and c, the values 125, 2, and 0.01 respectively. Parentheses ( ) may be used to indicate the order of the operations in a complicated expression as in mathematics. Don't try to use square brackets [ ] or curly brackets { } for this purpose because they have other meanings.

WARNING

In mathematics, one often writes AB as an abbreviation for $A \cdot B$: in the absence of any explicit symbol for an operation, multiplication is understood. In computer languages like MINIMAT this is *never* the case. MINIMAT will regard AB as the name of another quantity having nothing to do with A or B. The product of A and B is denoted by A*B in MINIMAT.

Scalar Built-in Functions

MINIMAT has the usual numerical functions

```
sin,  cos,  tan,  exp
```

and their inverses

```
asin,  acos,  atan,  log
```

found on most hand-held calculators[6] and a few more besides. The table in Appendix C Page 452 lists all of MINIMAT's built-in functions. The square root is indicated with the notation sqrt so that sqrt(9) evaluates to 3. Parentheses are required when applying a built-in function.

MINIMAT differs from calculators in that the built-in functions return complex values in situations where a calculator would indicate an error. For example, the command

```
#> b=sqrt(-5)
```

is legal and assigns the value $i\sqrt{5}$ to b. The question of what is being calculated when MINIMAT computes something like asin(5) (the angle whose sin is 5!?) is interesting but beyond the scope of this book.

Exercise 21A Type the following commands at MINIMAT:

```
#> a=4,   a1=sqrt(a*a)
#> b=-3,  b1=sqrt(b*b)
```

Why does a = a1 but b ≠ b1?

---

[6]The exponential and logarithm are base $e = 2.71\ldots$ and the trigonometric functions use radians, not degrees.

This exercise is easy. The expression `sqrt(c)` is MINIMAT's name for $\sqrt{c}$ which means the *positive* square root of c. Here are some harder exercises.

Exercise 22A Type the following commands at MINIMAT:

```
#>  exp(log(6)),  log(exp(3))
#>  sin(asin(0.456)),  asin(sin(0.75)), asin(sin(8.45))
```

Why did `asin(sin(0.75))` return 0.75 but `asin(sin(8.45))` not return 8.45?

▷ **Exercise 22B** Type the following commands at MINIMAT:

```
#>  t= rand(1,1), i=sqrt(-1)
#>  exp(i*t), cos(t)+i*sin(t)
```

Explain why these are equal.

▷ **Exercise 22C** Type the following commands at MINIMAT:

```
#>  t= rand(1,1),s= rand(1,1), i=sqrt(-1)
#>  u=cos(t)+i*sin(t), v=cos(s)+i*sin(s)
#>  w=cos(t+s)+i*sin(t+s)
#>  u*v, w
```

Explain why these are equal.

Exercise 22D Type the following commands at MINIMAT:

```
#>  t=2*pi/5,  i=sqrt(-1)
#>  w= cos(t)+i*sin(t)
#>  w^5
#>  1+w+w^2+w^3+w^4
```

Explain what happened. Hint: Factor $1 - w^5$.

.M Functions

We have described some of MINIMAT's "built-in" functions. It is possible to create your own functions as programs stored on the computer's disk and executed when invoked. These programs are called **.M functions** and the disk files that hold these programs are called **.M files**. Several .M files are provided on the MINIMAT distribution diskette. These are the only .M files used in this book. Their names are listed in Appendix C on Page 460.

MINIMAT always responds correctly to a command involving a built-in function, but for a .M function it will complain of an *unrecognized identifier* if the corresponding .M file is not present in the default directory. A .M function might use other .M functions and these other functions also need to be present on the disk for the function to work.

In this book, the .M functions are used primarily to save typing. Typically, when students are first taught an algorithm for solving a certain problem, they should type in the steps one by one to reinforce the lesson. Later, the students are encouraged to use the .M function that automates the algorithm.

The what command lists the names of all the .M files present on the disk. The type command can be used to list a particular .M function on the screen so that the user can study how it works. For example, the command

```
#> type randsoln
```

lists the file randsoln.m on the screen so that the user can study it.

Format Command

The command format ww.dd changes the way MINIMAT prints numbers to a field of width ww with dd digits of accuracy. For example, after the command

```
#> format 6.01
    Width =   6  Digits =   1
```

MINIMAT prints numbers in a field of width 6 and with one digit after the decimal point. The zero after the decimal point must be present; format 15.1 produces a field of width 15 with 10 digits of accuracy; format 15.01 produces one digit of accuracy. The format command does not change the way MINIMAT does arithmetic, only the way it prints the answers.

In classical Matlab, the format command is more limited; only format xxx where xxx is one of long, short, longe, shorte are permitted. These work in MINIMAT as well. The default (the format used when MINIMAT starts) is format short. It corresponds to a width of 8 with 3 digits of accuracy.

Control Structures

MINIMAT has the usual control structures

```
if elseif else    for    while    break    return
```

found in most computing languages. These are explained in Appendix C on Page 457. It also has the traditional Boolean operations and the equality and order relations (see Section C.7). For the most part you don't need to worry about these since the meaning is usually clear from (or explained in) the context. However, if you want to write your own .M functions you will have to learn exactly how these control structures work.

Chapter Summary

Complex Numbers

A complex number is a number of form

$$z = x + iy$$

where x and y are real. The following notations are used:

$$\Re(z) = x \qquad \textit{Real part}$$
$$\Im(z) = y \qquad \textit{Imaginary part}$$
$$\bar{z} = x - iy \qquad \textit{Complex conjugate}$$
$$|z| = \sqrt{x^2 + y^2} \quad \textit{Absolute value}$$

The following laws hold for $z, w \in \mathbf{C}$:

$$|z|^2 = z\bar{z}, \qquad z^{-1} = \frac{\bar{z}}{|z|^2},$$

$$\overline{z + w} = \bar{z} + \bar{w}, \qquad \overline{zw} = \bar{z}\bar{w},$$

$$z \in \mathbf{R} \iff z = \bar{z}.$$

$$\Re(z) = \frac{z + \bar{z}}{2}, \qquad \Im(z) = \frac{z - \bar{z}}{2i}.$$

Set Theory

notation	meaning
$V = \{x : P(x)\}$	$x \in V \iff P(x).$
$W = \{f(x) : x \in V\}$	$y \in W \iff y = f(x)$ *for some $x \in V$.*
$V \subset W$	$x \in V \implies x \in W$;
$V = W$	$x \in V \iff x \in W$;

2

MATRIX OPERATIONS

In this chapter we define the basic operations of matrix algebra. We will see that these operations obey most —but not all— of the laws of ordinary algebra. That's the whole idea. Matrix notation enables us to write complicated systems of linear equations, including differential equations, in a compact form. Matrix notations, suitably modified to handle the exigencies of the computer keyboard, provide the ideal way of instructing the computer to solve complicated problems. At the end of this chapter we learn how to use MINIMAT to instruct the computer to perform matrix operations.

The main respects in which the laws of matrix algebra differ from those of ordinary algebra are as follows:

- The operations are only defined when the sizes of the inputs are related in a certain way (which depends on the particular operation).

- The commutative law $AB = BA$ generally fails.

- It is *not* true (as it *is* for numbers) that a matrix A has an inverse A^{-1} if it is nonzero.

Because of the failure of the commutative law, we do not define matrix division B/A. We use $A^{-1}B$ and BA^{-1} instead.

> Throughout the book, the symbol **F** denotes either the set **R** of real numbers or the set **C** of complex numbers.

2.1   Matrices Defined

Fix positive integers m and n. An $m \times n$ **matrix** with entries from **F** is an **array**

$$A = \begin{bmatrix} a_{11} & a_{12} & \dots & a_{1n} \\ a_{21} & a_{22} & \dots & a_{2n} \\ \vdots & \vdots & \ddots & \vdots \\ a_{m1} & a_{m2} & \dots & a_{mn} \end{bmatrix}$$

where the **entry** a_{ij} is a number from **F**. There are m rows and n columns in an $m \times n$ matrix. We say that an $m \times n$ matrix is a matrix of **size** $m \times n$ when we want to call attention to the number of rows and columns.[1] A **square matrix** is one having the same number of rows as columns. We call a matrix **rectangular** when we wish to emphasize that we are not assuming that it is square. A **real matrix** is one whose entries are real (i.e. from **R**). We call a matrix **complex** when we wish to emphasize that its entries might not be real.

The set of all $m \times n$ matrices with entries from **F** is denoted by $\mathbf{F}^{m \times n}$. Thus the notation

$$A \in \mathbf{F}^{m \times n}$$

is an abbreviation for the phrase

A is an $m \times n$ matrix with entries from **F**.

Example 28A The matrix

$$A = \begin{bmatrix} 1 & 2 & 3 \\ 0 & 1 & 0 \end{bmatrix}$$

is a real matrix of size 2×3 so we write $A \in \mathbf{R}^{2 \times 3}$. It is also true that $A \in \mathbf{C}^{2 \times 3}$ since every real number is a complex number. The matrix

$$B = \begin{bmatrix} 3 + 4i & 7 \\ i & 0 \end{bmatrix}$$

is a complex matrix 2×2 matrix which is not real. Thus we write $B \in \mathbf{C}^{2 \times 2}$ and – if we want to emphasize the fact that B is not real – $B \notin \mathbf{R}^{2 \times 2}$.

---

[1]The $\times$ here does *not* indicate multiplication.

We can think of a matrix $A \in \mathbf{F}^{m \times n}$ as a function that assigns to each pair (i, j) of indices (where $i = 1, 2, \ldots, m$ and $j = 1, 2, \ldots, n$) a number a_{ij}. In case we want to specify this entry without displaying A in array form, we denote it by $\text{entry}_{ij}(A)$:

$$\text{entry}_{ij}(A) = a_{ij}.$$

We denote by $\text{row}_i(A)$ the ith row of A and by $\text{col}_j(A)$ the jth column of A:

$$\text{row}_i(A) = \begin{bmatrix} a_{i1} & a_{i2} & \cdots & a_{in} \end{bmatrix}$$

$$\text{col}_j(A) = \begin{bmatrix} a_{1j} \\ a_{2j} \\ \vdots \\ a_{mj} \end{bmatrix}.$$

Note that $\text{row}_i(A)$ is a $1 \times n$ matrix, $\text{col}_j(A)$ is a $m \times 1$ matrix, and $\text{entry}_{ij}(A)$ is a 1×1 matrix.

A **row vector** is a matrix $R \in \mathbf{F}^{1 \times n}$ with only one row. A **column vector** is a matrix $C \in \mathbf{F}^{m \times 1}$ with only one column. For these matrices the redundant subscript 1 is often omitted:

$$\text{entry}_j(R) = \text{entry}_{1j}(R),$$

$$\text{entry}_i(C) = \text{entry}_{i1}(C).$$

In this book the term **vector** is synonymous with the term *column vector*.

Example 29A To illustrate these notations let

$$A = \begin{bmatrix} 7 & -2 & 21 \\ 0 & 4 & -72 \end{bmatrix}.$$

Then

$$\text{row}_1(A) = \begin{bmatrix} 7 & -2 & 21 \end{bmatrix}$$

$$\text{col}_2(A) = \begin{bmatrix} -2 \\ 4 \end{bmatrix}$$

$$\text{entry}_{23}(A) = -72.$$

Remark 29B In mathematics one does not distinguish between a 1×1 matrix and the number that is its only entry:

$$\mathbf{F}^{1 \times 1} = \mathbf{F}.$$

Thus we do not distinguish between the 1×1 matrix $\begin{bmatrix} a \end{bmatrix}$ (which is an element of $\mathbf{F}^{1 \times 1}$) and the number a (which is an element of $\mathbf{F}$).

Definition 30A (Matrix Equality) Two matrices are **equal** iff they have the same size and corresponding entries are equal. In other words, for two matrices $A, B \in \mathbf{F}^{m \times n}$ of the sames size we have

$$A = B \iff \text{entry}_{ij}(A) = \text{entry}_{ij}(B)$$

for $i = 1, 2, \ldots, m, j = 1, 2, \ldots, n$; matrices of different sizes are never equal.

▷ **Question 30B** Does $\begin{bmatrix} a \\ b \end{bmatrix} = \begin{bmatrix} a & b \end{bmatrix}$?

2.2   Additive Operations on Matrices

The operations defined in this section are **entrywise** operations. This means that the (i, j)-entry of the output is determined by the (i, j)-entries of the inputs. This can be contrasted with the operation of matrix multiplication defined in Definition 32C.

Definition 30C (Matrix Addition) The **sum** of two matrices of the same size is defined by entrywise addition. More precisely, two matrices $A \in \mathbf{F}^{m \times n}$ and $B \in \mathbf{F}^{m \times n}$ determine a third matrix $A + B \in \mathbf{F}^{m \times n}$ by the rule

$$\text{entry}_{ij}(A + B) = \text{entry}_{ij}(A) + \text{entry}_{ij}(B)$$

for $i = 1, 2, \ldots, m, j = 1, 2, \ldots, n$.

Example 30D To add two 3×2 matrices you add the corresponding entries:

$$\begin{bmatrix} a_{11} & a_{12} \\ a_{21} & a_{22} \\ a_{31} & a_{32} \end{bmatrix} + \begin{bmatrix} b_{11} & b_{12} \\ b_{21} & b_{22} \\ b_{31} & b_{32} \end{bmatrix} = \begin{bmatrix} a_{11} + b_{11} & a_{12} + b_{12} \\ a_{21} + b_{21} & a_{22} + b_{22} \\ a_{31} + b_{31} & a_{32} + b_{32} \end{bmatrix}.$$

Definition 30E (Matrix Subtraction) The **difference** of two matrices of the same size is defined by entrywise subtraction. More precisely, two matrices $A \in \mathbf{F}^{m \times n}$ and $B \in \mathbf{F}^{m \times n}$ determine a third matrix $A - B \in \mathbf{F}^{m \times n}$ by the rule

$$\text{entry}_{ij}(A - B) = \text{entry}_{ij}(A) - \text{entry}_{ij}(B)$$

for $i = 1, 2, \ldots, m, j = 1, 2, \ldots, n$.

Example 31A To subtract two 3×2 matrices you subtract the corresponding entries:

$$\begin{bmatrix} a_{11} & a_{12} \\ a_{21} & a_{22} \\ a_{31} & a_{32} \end{bmatrix} - \begin{bmatrix} b_{11} & b_{12} \\ b_{21} & b_{22} \\ b_{31} & b_{32} \end{bmatrix} = \begin{bmatrix} a_{11} - b_{11} & a_{12} - b_{12} \\ a_{21} - b_{21} & a_{22} - b_{22} \\ a_{31} - b_{31} & a_{32} - b_{32} \end{bmatrix}.$$

Definition 31B The $m \times n$ **zero matrix** is the element of $\mathbf{F}^{m \times n}$ whose entries are all 0. It is denoted by $0_{m \times n}$ or simply 0 if the size can be understood from the context:

$$\text{entry}_{ij}(0) = \text{entry}_{ij}(0_{m \times n}) = 0$$

for $i = 1, 2, \ldots, m$, $j = 1, 2, \ldots, n$.

Example 31C Here's the 2×3 zero matrix;

$$0_{2 \times 3} = \begin{bmatrix} 0 & 0 & 0 \\ 0 & 0 & 0 \end{bmatrix}.$$

▷ **Question 31D** Does $0_{2 \times 3} = 0_{3 \times 2}$?

Definition 31E (Scalar Multiplication) The **scalar product** of a number and a matrix is defined by entrywise multiplication. More precisely, a number $c \in \mathbf{F}$ and a matrix $A \in \mathbf{F}^{m \times n}$ determine a matrix $cA \in \mathbf{F}^{m \times n}$ by the rule

$$\text{entry}_{ij}(cA) = c\,\text{entry}_{ij}(A)$$

for $i = 1, 2, \ldots, m$, $j = 1, 2, \ldots, n$.

Example 31F To multiply a 2×3 matrix by the scalar c you multiply each entry by c:

$$c \begin{bmatrix} a_{11} & a_{12} & a_{13} \\ a_{21} & a_{22} & a_{23} \end{bmatrix} = \begin{bmatrix} ca_{11} & ca_{12} & ca_{13} \\ ca_{21} & ca_{22} & ca_{23} \end{bmatrix}.$$

Definition 31G The **negative** of a matrix is determined by replacing each entry by its negative. More precisely, a matrix $A \in \mathbf{F}^{m \times n}$ determines another matrix $-A \in \mathbf{F}^{m \times n}$ by the rule

$$\text{entry}_{ij}(-A) = -\,\text{entry}_{ij}(A)$$

for $i = 1, 2, \ldots, m$, $j = 1, 2, \ldots, n$.

Example 31H Of course, $-A = (-1)A$. Thus

$$- \begin{bmatrix} a_{11} & a_{12} & a_{13} \\ a_{21} & a_{22} & a_{23} \end{bmatrix} = \begin{bmatrix} -a_{11} & -a_{12} & -a_{13} \\ -a_{21} & -a_{22} & -a_{23} \end{bmatrix}.$$

▷ **Question 32A** If $A = \begin{bmatrix} 1 & 2 & 3 \end{bmatrix}$ and $A + B = \begin{bmatrix} 3 & 2 & 1 \end{bmatrix}$, what is B?

Addition Laws

Theorem 32B *The additive operations satisfy the following:*

$(A + B) + C = A + (B + C)$ *(Additive Associative Law)*
$A + B = B + A$ *(Additive Commutative Law)*
$c(A + B) = cA + cB$ *(Scalar Left Distributive Law)*
$(a + b)C = aC + bC$ *(Scalar Right Distributive Law)*
$a(bC) = (ab)C$ *(Multiplicative Associative Law)*
$A + 0_{m \times n} = A$ *(Additive Identity)*
$1A = A$ *(One Law)*
$0A = 0_{m \times n}$ *(Zero Law)*

In these laws A, B, and C are matrices of the same size $m \times n$ and a, b, and c are numbers.

Proof: See Appendix A Page 437.

2.3   Multiplicative Operations

We have emphasized that matrices may be added and subtracted only when they are the same size. Two matrices may be multiplied only when the number of columns in the first one is the same as the number of rows in the second one. In other words, the product AB is meaningful when A has size $m \times n$ and B has size $n \times p$. (The result AB has size $m \times p$.)

Definition 32C Matrices $A \in \mathbf{F}^{m \times n}$ and $B \in \mathbf{F}^{n \times p}$ determine a matrix $AB \in \mathbf{F}^{m \times p}$ by the rule

$$\text{entry}_{ik}(AB) = \sum_{j=1}^{n} \text{entry}_{ij}(A)\,\text{entry}_{jk}(B)$$

for $i = 1, 2, \ldots, m$, $k = 1, 2, \ldots, p$. The matrix AB is called the **matrix product** of A and B.

It is important to remember that the product AB of two matrices is only defined when then number of columns of A is the same as the number of rows of B; it has the same number of rows as A and the same number of columns as B.

Example 33A If A and B are 2×2 so is AB.

$$\begin{bmatrix} a_{11} & a_{12} \\ a_{21} & a_{22} \end{bmatrix} \begin{bmatrix} b_{11} & b_{12} \\ b_{21} & b_{22} \end{bmatrix} = \begin{bmatrix} a_{11}b_{11} + a_{12}b_{21} & a_{11}b_{12} + a_{12}b_{22} \\ a_{21}b_{11} + a_{22}b_{21} & a_{21}b_{12} + a_{22}b_{22} \end{bmatrix}.$$

Remark 33B (Row-Column Formula) The formula

$$\text{entry}_{ik}(AB) = \text{row}_i(A)\,\text{col}_k(B).$$

says that to find the (i, k)-entry in the product, we take the ith row of A (it has n entries), the kth column of B (it also has n entries), and form the sum of the products. For example,

$$\begin{bmatrix} c_{11} & c_{12} & c_{13} \\ c_{21} & c_{22} & c_{23} \\ c_{31} & \boxed{c_{32}} & c_{33} \\ c_{41} & c_{42} & c_{43} \\ c_{51} & c_{52} & c_{53} \end{bmatrix} = \begin{bmatrix} a_{11} & a_{12} & a_{13} & a_{14} \\ a_{21} & a_{22} & a_{23} & a_{24} \\ \boxed{a_{31} \quad a_{32} \quad a_{33} \quad a_{34}} \\ a_{41} & a_{42} & a_{43} & a_{44} \\ a_{51} & a_{52} & a_{53} & a_{54} \end{bmatrix} \begin{bmatrix} b_{11} & \boxed{b_{12}} & b_{13} \\ b_{21} & b_{22} & b_{23} \\ b_{31} & b_{32} & b_{33} \\ b_{41} & \boxed{b_{42}} & b_{43} \end{bmatrix}$$

where (for example) $c_{32} = \text{row}_3(A)\,\text{col}_2(B)$ is given by

$$c_{32} = a_{31}b_{12} + a_{32}b_{22} + a_{33}b_{32} + a_{34}b_{42}.$$

Example 33C A row times a column is a number:

$$\begin{bmatrix} a_1 & a_2 \end{bmatrix} \begin{bmatrix} b_1 \\ b_2 \end{bmatrix} = a_1b_1 + a_2b_2,$$

A column times a row is a square matrix:

$$\begin{bmatrix} b_1 \\ b_2 \end{bmatrix} \begin{bmatrix} a_1 & a_2 \end{bmatrix} = \begin{bmatrix} b_1a_1 & b_1a_2 \\ b_2a_1 & b_2a_2 \end{bmatrix}.$$

Definition 33D The $m \times m$ **identity matrix** is the square matrix with 1's on the diagonal and 0's elsewhere. It is denoted by I_m, or simply I if the size can be understood from the context:

$$\text{entry}_{ij}(I) = \text{entry}_{ij}(I_m) = \begin{cases} 1 & \text{for } i = j \\ 0 & \text{otherwise.} \end{cases}$$

Example 34A Here's the 3×3 identity matrix:

$$I_3 = \begin{bmatrix} 1 & 0 & 0 \\ 0 & 1 & 0 \\ 0 & 0 & 1 \end{bmatrix}$$

Definition 34B For a positive integer p and a square matrix A we define the pth **power** A^p of A by

$$A^p = \underbrace{AA \cdots A}_{p}$$

where the notation indicates a product of p copies of A. The 0th power is defined by

$$A^0 = I$$

the identity matrix.This notation assures that the law $A^{p+q} = A^p A^q$ holds for nonnegative integers p and q and not just for positive integers.

▷ **Question 34C** What is the size of A^p?

▷ **Question 34D** Is the notation A^p meaningful if A is not square?

Matrix Multiplication Laws

Theorem 34E *Matrix multiplication satisfies the following:*

$$
\begin{aligned}
&(AB)C = A(BC) && \text{(Multiplicative Associative Law)} \\
&C(A + B) = CA + CB && \text{(Matrix Left Distributive Law)} \\
&(A + B)C = AC + BC && \text{(Matrix Right Distributive Law)} \\
&IA = A && \text{(Left Multiplicative Identity)} \\
&A = AI && \text{(Right Multiplicative Identity)} \\
&A(cB) = c(AB) = (cA)B && \text{(Scalar Commutative Law)} \\
&A^{p+q} = A^p A^q && \text{(Power Law)}
\end{aligned}
$$

In these laws A, B, and C denote matrices and I denotes an identity matrix. In the scalar commutative law c is a number (1×1 matrix) and the meaning of cB is that of scalar multiplication. In the power law, p and q are nonnegative integers.

Proof: See Appendix A Page 438.

The **commutative law** $AB = BA$ generally fails. In place of the commutative law we have the following

Definition 35A Matrices A and B are said to **commute** iff $AB = BA$.

Remark 35B The equation $AB = BA$ is meaningful only when A and B are square and of the same size. When $A \in \mathbf{F}^{m \times n}$ and $B \in \mathbf{F}^{n \times m}$ both products AB and BA are defined and $AB \in \mathbf{F}^{m \times m}$ and $B \in \mathbf{F}^{n \times n}$. Certainly $AB \neq BA$ when $m \neq n$. For example, if

$$A = \begin{bmatrix} a_1 & a_2 & a_3 \end{bmatrix}, \qquad B = \begin{bmatrix} b_1 \\ b_2 \\ b_3 \end{bmatrix},$$

then $A \in \mathbf{F}^{1 \times 3}$, $B \in \mathbf{F}^{3 \times 1}$, $AB \in \mathbf{F}^{1 \times 1} = \mathbf{F}$, $BA \in \mathbf{F}^{3 \times 3}$, and

$$AB = a_1 b_1 + a_2 b_2 + a_3 b_3, \qquad BA = \begin{bmatrix} b_1 a_1 & b_1 a_2 & b_1 a_3 \\ b_2 a_1 & b_2 a_2 & b_2 a_3 \\ b_3 a_1 & b_3 a_2 & b_3 a_3 \end{bmatrix}.$$

Example 35C However even when A and B are square matrices of the same size, they usually do not commute. For example, if

$$A = \begin{bmatrix} 0 & 1 \\ 0 & 0 \end{bmatrix}, \qquad B = \begin{bmatrix} 0 & 0 \\ 1 & 0 \end{bmatrix},$$

then

$$AB = \begin{bmatrix} 1 & 0 \\ 0 & 0 \end{bmatrix}, \qquad BA = \begin{bmatrix} 0 & 0 \\ 0 & 1 \end{bmatrix}.$$

Example 35D Diagonal matrices of the same size always commute:

$$\begin{bmatrix} a_1 & 0 \\ 0 & a_2 \end{bmatrix} \begin{bmatrix} b_1 & 0 \\ 0 & b_2 \end{bmatrix} = \begin{bmatrix} a_1 b_1 & 0 \\ 0 & a_2 b_2 \end{bmatrix} = \begin{bmatrix} b_1 & 0 \\ 0 & b_2 \end{bmatrix} \begin{bmatrix} a_1 & 0 \\ 0 & a_2 \end{bmatrix}.$$

See Corollary 47B below.

Example 35E Here's another example of commuting matrices:

$$\begin{bmatrix} 1 & a \\ 0 & 1 \end{bmatrix} \begin{bmatrix} 1 & b \\ 0 & 1 \end{bmatrix} = \begin{bmatrix} 1 & a+b \\ 0 & 1 \end{bmatrix} = \begin{bmatrix} 1 & b \\ 0 & 1 \end{bmatrix} \begin{bmatrix} 1 & a \\ 0 & 1 \end{bmatrix}$$

▷ **Exercise 36A** Find AX, HA, and HAX where

$$H = \begin{bmatrix} 1 & 1 & 1 \end{bmatrix}, \qquad A = \begin{bmatrix} 1 & 2 & 3 \\ 4 & 5 & 6 \\ 7 & 8 & 9 \end{bmatrix}, \qquad X = \begin{bmatrix} 1 \\ 1 \\ 1 \end{bmatrix}.$$

▷ **Exercise 36B** Compute $\begin{bmatrix} a & b \\ c & d \end{bmatrix} \begin{bmatrix} d & -b \\ -c & a \end{bmatrix}$

▷ **Exercise 36C** Let $X = \begin{bmatrix} x_1 \\ x_2 \end{bmatrix}$, $Y = \begin{bmatrix} y_1 & y_2 \end{bmatrix}$. Which of the following expressions are meaningful? For those that are, evaluate them.

$$(1) \ \ X + Y \qquad\qquad (2) \ \ XY \qquad\qquad (3) \ \ YX.$$

▷ **Exercise 36D** Let $A = \begin{bmatrix} a_1 & a_3 \\ a_5 & a_2 \end{bmatrix}$ and $B = \begin{bmatrix} a_5 & a_3 \\ a_4 & a_6 \end{bmatrix}$. What is the $(2,1)$-entry of the sum $A + B$? of the product AB?

▷ **Exercise 36E** Let $A = \begin{bmatrix} a_1 & a_2 & a_3 \\ b_1 & b_2 & b_3 \end{bmatrix}$, $X = \begin{bmatrix} x_1 \\ x_2 \\ x_3 \end{bmatrix}$, $F = \begin{bmatrix} 0 \\ 1 \\ 0 \end{bmatrix}$.
What is AX? AF?

▷ **Exercise 36F** Let $D_1 = \begin{bmatrix} 1 & 0 & 0 \\ 0 & 1 & 0 \end{bmatrix}$, $D_2 = \begin{bmatrix} 1 & 0 \\ 0 & 1 \\ 0 & 0 \end{bmatrix}$. Compute $D_1 D_2$ and $D_2 D_1$.

▷ **Exercise 36G** Assume $A \in \mathbf{F}^{2\times3}$, $B \in \mathbf{F}^{3\times4}$. True or false?
  (1) **T F** $AB \in \mathbf{F}^{2\times4}$.
  (2) **T F** If A has a row of zeros, so does AB.
  (3) **T F** If A has a column of zeros, so does AB.
  (4) **T F** If B has a row of zeros, so does AB.
  (5) **T F** If B has a column of zeros, so does AB.

Exercise 36H Let

$$A = \begin{bmatrix} 1 & -2 \\ 2 & -4 \end{bmatrix}, \qquad X = \begin{bmatrix} 2 \\ 1 \end{bmatrix}, \qquad Y = \begin{bmatrix} 0 \\ 0 \end{bmatrix}.$$

Show that $AX = AY$ but $X \neq Y$. (Moral: You can't always cancel.)

▷ **Exercise 36I** Which of the following pairs of matrices commute?

$$(1) \qquad\qquad A = \begin{bmatrix} 2 & 2 \\ -1 & 5 \end{bmatrix}, \qquad B = \begin{bmatrix} 3 & 2 \\ 1 & 0 \end{bmatrix}.$$

(2)
$$A = \begin{bmatrix} 3 & 0 \\ 0 & 4 \end{bmatrix}, \qquad B = \begin{bmatrix} 2 & 0 \\ 0 & 1 \end{bmatrix}.$$

(3)
$$A = \begin{bmatrix} 2 & 2 \\ -1 & 5 \end{bmatrix}, \qquad B = \begin{bmatrix} 3 & -2 \\ 1 & 0 \end{bmatrix}.$$

Exercise 37A Give an example of two matrices $H \in \mathbf{F}^{1 \times 3}$, $X \in \mathbf{F}^{3 \times 1}$ so that $HX = 0$ although no entry of either H or X is zero.

▷ **Exercise 37B** How must the sizes of the matrices A, B, C be related in order that the expressions in the right distributive law $(A+B)C = AC+BC$ be meaningful?

▷ **Exercise 37C** Show that the negative $-A$ of a matrix A and the difference $A - B$ of matrices A and B can be defined in terms of matrix addition and scalar multiplication.

▷ **Exercise 37D** Compute AD and DA where

$$A = \begin{bmatrix} a & b & c \\ d & e & f \\ g & h & i \end{bmatrix}, \qquad D = \begin{bmatrix} x & 0 & 0 \\ 0 & y & 0 \\ 0 & 0 & z \end{bmatrix}.$$

▷ **Exercise 37E** Compute AP and PA where

$$A = \begin{bmatrix} a & b & c \\ d & e & f \\ g & h & i \end{bmatrix}, \qquad P = \begin{bmatrix} 0 & 1 & 0 \\ 0 & 0 & 1 \\ 1 & 0 & 0 \end{bmatrix}.$$

▷ **Exercise 37F** Find AB where

$$A = \begin{bmatrix} a_{11} & a_{12} & a_{13} \\ 0 & a_{22} & a_{23} \\ 0 & 0 & a_{33} \end{bmatrix}, \qquad B = \begin{bmatrix} b_{11} & b_{12} & b_{13} \\ 0 & b_{22} & b_{23} \\ 0 & 0 & b_{33} \end{bmatrix}.$$

▷ **Exercise 37G** Calculate AB if

$$A = \begin{bmatrix} 0 & a_{12} & a_{13} & a_{14} \\ 0 & 0 & a_{23} & a_{24} \\ 0 & 0 & 0 & a_{34} \\ 0 & 0 & 0 & 0 \end{bmatrix}, \qquad B = \begin{bmatrix} 0 & 0 & b_{13} & b_{14} \\ 0 & 0 & 0 & b_{24} \\ 0 & 0 & 0 & 0 \\ 0 & 0 & 0 & 0 \end{bmatrix}.$$

2.4   Inverses

In arithmetic the number 1 is a multiplicative identity $(1a = a)$, just as
the number 0 is an additive identity $(0 + a = a)$. The number a^{-1} is
a multiplicative inverse to a $(a^{-1}a = 1)$ just as the number $(-a)$ is an
additive inverse $(-a + a = 0)$. We have generalized three of these four
notions to matrices; we now generalize the fourth.

Definition of the Inverse

Definition 38A A matrix A is called **invertible** iff it is a square
matrix and there is a (necessarily unique) matrix A^{-1} called the
inverse of A which satisfies

$$AA^{-1} = A^{-1}A = I$$

where I is the identity matrix.

Proof of uniqueness: The following calculation shows that if A is invertible, then the only matrix B satisfying $AB = BA = I$ is $B = A^{-1}$:

$$B = BI = B(AA^{-1}) = (BA)A^{-1} = IA^{-1} = A^{-1}. \qquad \square$$

Example 38B The zero matrix is not invertible for there cannot be a
matrix B with $0B = I$ since $0B = 0$ and $I \neq 0$.

Example 38C A 1×1 matrix is invertible iff it is not zero, but this is
false for a $n \times n$ matrix with $n > 1$. For example, let

$$A = \begin{bmatrix} 1 & 0 \\ 0 & 0 \end{bmatrix}.$$

Then $A \neq 0$ since $\text{entry}_{11}(A) = 1 \neq 0$. However,

$$\begin{bmatrix} 1 & 0 \\ 0 & 0 \end{bmatrix} \begin{bmatrix} b_{11} & b_{12} \\ b_{21} & b_{22} \end{bmatrix} = \begin{bmatrix} b_{11} & b_{12} \\ 0 & 0 \end{bmatrix}.$$

We can never have $AB = I$, since $\text{row}_2(AB) = 0$ while $\text{row}_2(I) \neq 0$.

Remark 39A In fact, a matrix (of any size) having a row of zeros is *never* invertible. If A has a row of zeros, then so does AB (no matter what the matrix B). The formula

$$\text{entry}_{ij}(AB) = \text{row}_i(A)\,\text{col}_j(B)$$

shows that $\text{row}_i(AB) = 0$ if $\text{row}_i(A) = 0$. Hence, A cannot be invertible; if it were, we could take $B = A^{-1}$ and obtain $AA^{-1} = I$, which does *not* have a row of zeros.

▷ **Question 39B** Can an invertible matrix have a column of zeros?

▷ **Question 39C** Must a noninvertible matrix have a column of zeros? A row of zeros?

In Theorem 93A we will learn (1) how to tell when a square matrix is invertible, and (2) how to find the inverse of an invertible matrix. For now, let us do this for 2×2 matrices.

Theorem 39D (Inverting a 2×2 Matrix) *Let $A \in \mathbf{F}^{2 \times 2}$ be given by*

$$A = \begin{bmatrix} a & b \\ c & d \end{bmatrix}.$$

Define the **determinant** *of A by*

$$\det(A) = ad - bc.$$

Then

(1) *the matrix A is invertible if and only if its determinant is not zero, and*

(2) *when A is invertible, its inverse is given by the formula*

$$A^{-1} = \det(A)^{-1} \begin{bmatrix} d & -b \\ -c & a \end{bmatrix}.$$

Proof: Let $B = \begin{bmatrix} d & -b \\ -c & a \end{bmatrix}$. Then

$$AB = BA = \begin{bmatrix} ad - bc & 0 \\ 0 & ad - bc \end{bmatrix} = \det(A)I.$$

Assume $\det(A) \neq 0$. Divide by it and obtain

$$A\big(\det(A)^{-1}B\big) = \big(\det(A)^{-1}B\big)A = I.$$

This proves that A is invertible and that $A^{-1} = \det(A)^{-1}B$.

Conversely, assume $\det(A) = 0$; we must show that A is not invertible. According to the Noninvertibility Remark 40B below, it is enough to find a nonzero X with $AX = 0$. If $A = 0$, any nonzero X will do. In the contrary case, at least one of the two columns

$$X_1 = \begin{bmatrix} d \\ -c \end{bmatrix}, \qquad X_2 = \begin{bmatrix} -b \\ a \end{bmatrix},$$

is nonzero and

$$AX_1 = AX_2 = \begin{bmatrix} 0 \\ 0 \end{bmatrix}. \qquad\qquad \square$$

Examples 40A

$$\begin{bmatrix} 1 & 3 \\ 2 & 8 \end{bmatrix}^{-1} = \begin{bmatrix} 4 & -3/2 \\ -1 & 1/2 \end{bmatrix}, \qquad\qquad \begin{bmatrix} 2 & 1 \\ 1 & 1 \end{bmatrix}^{-1} = \begin{bmatrix} 1 & -1 \\ -1 & 2 \end{bmatrix}.$$

Noninvertibility Remark

Remark 40B If $X \neq 0$ and $AX = 0$, then the matrix A is not invertible.

Proof: We prove the contrapositive, that is, if A is invertible and $AX = 0$, then $X = 0$:

$$X = IX = (A^{-1}A)X = A^{-1}(AX) = A^{-1}0 = 0. \qquad\qquad \square$$

Example 40C For the matrices $A = \begin{bmatrix} 1 & 2 \\ 2 & 4 \end{bmatrix}$, $X = \begin{bmatrix} 2 \\ -1 \end{bmatrix}$, we have $X \neq 0$ and $AX = 0$. Thus A is not invertible. Notice that $\det(A) = 0$.

Theorem 40D *The invertible matrices satisfy the following:*

(1) *the identity matrix is invertible;*

(2) *the inverse of an invertible matrix is invertible;*

(3) *a product of invertible matrices is invertible.*

Proof:

(1) Since $I^2 = I$, the identity matrix is invertible and is its own inverse.

(2) To say that A^{-1} is invertible is to say that there is a matrix B such that $A^{-1}B = BA^{-1} = I$; when so, $B = (A^{-1})^{-1}$. But the matrix $B = A$ satisfies these conditions, since A is invertible. Thus the inverse of A^{-1} is A: in other words,

$$(A^{-1})^{-1} = A.$$

(3) By definition, AB is invertible if and only if there is a (necessarily unique) matrix $C = (AB)^{-1}$ satisfying $(AB)C = C(AB) = I$. But the matrix $C = B^{-1}A^{-1}$ satisfies these conditions since

$$(AB)(B^{-1}A^{-1}) = A(BB^{-1})A^{-1} = AIA^{-1} = AA^{-1} = I$$

and

$$(B^{-1}A^{-1})(AB) = B^{-1}(A^{-1}A)B = B^{-1}IB = B^{-1}B = I.$$

Thus

$$(AB)^{-1} = B^{-1}A^{-1}. \qquad \square$$

Remark 41A The law

$$(AB)^{-1} = B^{-1}A^{-1}.$$

is called the **antimultiplicative law** for inverses. It says that *the inverse of the product is the product of the inverses in reverse order.*

Definition 41B (Negative Powers) If A is invertible, define negative powers by

$$A^{-p} = (A^{-1})^p = (A^p)^{-1}$$

We justify the assertion that $(A^{-1})^p = (A^p)^{-1}$: The matrix $B = (A^{-1})^p$ satisfies $A^pB = BA^p = I$. Therefore, $B = (A^p)^{-1}$, by the uniqueness of the inverse.

Invertible Matrix Multiplication Laws

Theorem 42A *Invertible matrices satisfy the following:*

$$A^{-1}A = AA^{-1} = I \qquad \textit{(Definition of Inverse)}$$
$$(AB)C = A(BC) \qquad \textit{(Associative Law)}$$
$$IA = A \qquad \textit{(Left Multiplicative Identity)}$$
$$AI = A \qquad \textit{(Right Multiplicative Identity)}$$
$$I^{-1} = I \qquad \textit{(Inverse of the Identity)}$$
$$(A^{-1})^{-1} = A \qquad \textit{(Double Inverse Law)}$$
$$(AB)^{-1} = B^{-1}A^{-1} \qquad \textit{(Inverse Antimultiplicative Law)}$$
$$A^{p+q} = A^p A^q \qquad \textit{(Power Law)}$$

In these laws, A, B, and C denote invertible matrices of the same size, and I denotes the identity matrix of that size. In the power law, p and q are arbitrary integers.

▷ **Exercise 42B** Prove the inverse antimultiplicative law.

▷ **Exercise 42C** If $AX = \begin{bmatrix} 3 \\ 5 \end{bmatrix}$ and $A^{-1} = \begin{bmatrix} 2 & 1 \\ 1 & 1 \end{bmatrix}$, what is X?

▷ **Exercise 42D** Find the inverse of $A = \begin{bmatrix} 3 & 1 \\ 1 & 1 \end{bmatrix}$ and confirm your answer by computing AA^{-1} and $A^{-1}A$.

▷ **Exercise 42E** Give an example of a 2×2 matrix which is not invertible even though all its entries are nonzero.

▷ **Exercise 42F** Give an example, other than $A = \pm I$, of an invertible 2×2 matrix A with integer entries whose inverse A^{-1} also has integer entries. (The result is surprising since the only integers a whose reciprocal a^{-1} is also an integer are $a = \pm 1$.)

▷ **Exercise 42G** Show that a matrix $A \in \mathbf{F}^{2\times 2}$ is invertible if and only if the only solution $X \in \mathbf{F}^{2\times 1}$ of $AX = 0$ is $X = 0$. (In Theorem 94A we shall learn that this is true for square matrices of any size, not just 2×2.)

Exercise 43A Which of the following matrices $A \in \mathbf{F}^{2 \times 2}$ are invertible? If A is invertible, find A^{-1}; if not, find a nonzero[2] $X \in \mathbf{F}^{2 \times 1}$ with $AX = 0$.

(1) $\quad A = \begin{bmatrix} 1 & 1 \\ 0 & 1 \end{bmatrix}$
 (2) $\quad A = \begin{bmatrix} 2 & 1 \\ 1 & 1 \end{bmatrix}$
 (3) $\quad A = \begin{bmatrix} 1 & 3 \\ 2 & 6 \end{bmatrix}$

(4) $\quad A = \begin{bmatrix} 1 & 1 \\ 0 & 0 \end{bmatrix}$
 (5) $\quad A = \begin{bmatrix} 1 & 0 \\ 2 & 0 \end{bmatrix}$
 (6) $\quad A = \begin{bmatrix} 1 & 0 \\ 0 & 1 \end{bmatrix}$

Exercise 43B If $MA = \begin{bmatrix} 1 & 2 & 3 \\ 4 & 5 & 6 \end{bmatrix}$ and $M = \begin{bmatrix} 3 & 2 \\ 7 & 5 \end{bmatrix}$, what is A?

2.5 Transpose

Now we introduce two similar operations called the *transpose* and the *conjugate transpose*. The transpose $A^\top$ of a matrix A has the crucial property that the columns of A are the rows of $A^\top$. Thus this operation has the virtue of translating properties of the rows of a matrix into properties of the columns and vice versa. The conjugate transpose is the complex conjugate of the transpose. For real matrices the transpose and the conjugate transpose agree, but for complex matrices the conjugate transpose is more important.

Definition 43C The **transpose** of a matrix $A \in \mathbf{F}^{m \times n}$ is the matrix $A^\top \in \mathbf{F}^{n \times m}$ whose columns are the rows of A. More precisely:

$$\text{entry}_{ji}(A^\top) = \text{entry}_{ij}(A)$$

for $i = 1, 2, \ldots, m$, $j = 1, 2, \ldots, n$. (There are many other notations in common use for the transpose of a matrix. Some examples are A', A^*, A^{tr}, A^t, $^\top A$, etc.

Example 43D The transpose of a 2×3 matrix is 3×2:

$$\begin{bmatrix} a & b & c \\ d & e & f \end{bmatrix}^\top = \begin{bmatrix} a & d \\ b & e \\ c & f \end{bmatrix}.$$

---

[2]Remember that a matrix is nonzero iff it has at least one nonzero entry.

Transpose Laws

Theorem 44A *The transpose operation satisfies the following laws:*

$$(aA + bB)^\top = aA^\top + bB^\top \quad \textit{(Linearity)}$$
$$I^\top = I \quad \textit{(Transpose of Identity)}$$
$$(A^\top)^\top = A \quad \textit{(Double Transpose)}$$
$$(AB)^\top = B^\top A^\top \quad \textit{(Transpose Antimultiplicative)}$$
$$(A^\top)^{-1} = (A^{-1})^\top \quad \textit{(Transpose Inverse)}$$

Example 44B We prove that $I^\top = I$. For the identity matrix I we have

$$\text{entry}_{ij}(I^\top) = \text{entry}_{ji}(I) = \text{entry}_{ij}(I)$$

since $\text{entry}_{ij}(I) = \text{entry}_{ji}(I) = 0$ if $i \neq j$. As $I^\top$ and I have equal entries, they are equal.

Example 44C We prove the Transpose Inverse Law. Assume that the matrix A is invertible (and hence square). Let $B = (A^{-1})^\top$. Then

$$
\begin{aligned}
A^\top B \quad &= A^\top (A^{-1})^\top \quad && \text{definition of } B; \\
&= (A^{-1}A)^\top \quad && \text{transpose antimultiplicative law;} \\
&= I^\top \quad && \text{definition of } A^{-1}; \\
&= I \quad && \text{transpose of identity.}
\end{aligned}
$$

Similarly, $BA^\top = I$. Thus B is the inverse of $A^\top$.

Because of the obvious laws

$$\text{row}_i(A)^\top = \text{col}_i(A^\top), \qquad \text{col}_j(A)^\top = \text{row}_j(A^\top),$$

we can use the transpose operation to convert theorems about rows into theorems about columns and vice versa. These laws also hold for the conjugate transpose.

Definition 45A The matrix $A^* \in \mathbf{F}^{n \times m}$ defined by the formula

$$\text{entry}_{ji}(A^*) = \overline{\text{entry}_{ij}(A)}$$

for $i = 1, 2, \ldots, m$, $j = 1, 2, \ldots, n$ is called the **conjugate transpose** of the matrix $A \in \mathbf{F}^{m \times n}$. The bar on the right denotes **complex conjugation**: $\bar{z} = x - iy$ for the complex number $z = x + iy$ (where x and y are real).

Example 45B The conjugate transpose of a 2×3 matrix is 3×2:

$$\begin{bmatrix} a & b & c \\ d & e & f \end{bmatrix}^* = \begin{bmatrix} \bar{a} & \bar{d} \\ \bar{b} & \bar{e} \\ \bar{c} & \bar{f} \end{bmatrix}.$$

Remark 45C For real matrices A the transpose $A^\top$ and conjugate transpose A^* are the same:

$$A^\top = A^* \text{ for } A \in \mathbf{R}^{m \times n}.$$

This is because $\bar{z} = z$ if (and only if) z is real.

The following theorem emphasizes that the two operations of *transpose* and *conjugate transpose* have similar properties.

Conjugate Transpose Laws

Theorem 45D *The conjugate transpose satisfies the following laws:*

$(aA + bB)^* = \bar{a}A^* + \bar{b}B^*$ *(Antilinearity)*
$I^* = I$ *(Conjugate Transpose of Identity)*
$A^{**} = A$ *(Double Conjugate Transpose)*
$(AB)^* = B^*A^*$ *(Antimultiplicative)*
$(A^*)^{-1} = (A^{-1})^*$ *(Conjugate Transpose Inverse)*

▷ **Exercise 45E** Let $H = \begin{bmatrix} x & y & z \end{bmatrix}$, $A = \begin{bmatrix} a & b \\ c & d \\ e & f \end{bmatrix}$. Calculate $A^\top$, $H^\top$, HA, $(HA)^\top$, $A^\top H^\top$, $\text{row}_1(A^\top)$, $\text{col}_1(A)^\top$.

2.6   Diagonal Matrices

The **diagonal entries** of a matrix are those whose column index is the same as the row index. For example, the diagonal entries of the following matrix are a, e, and i:

$$A = \begin{bmatrix} a & b & c \\ d & e & f \\ g & h & i \end{bmatrix}.$$

Diagonal Matrix

Definition 46A A matrix D is called **diagonal** iff it is square and all its nondiagonal entries are zero, that is, iff

$$\text{entry}_{ij}(D) = 0 \qquad \text{for } i \neq j.$$

Let $d_1, d_2, \ldots, d_n$ be numbers. We denote by

$$D = \text{diag}(d_1, d_2, \ldots, d_n)$$

the $n \times n$ diagonal matrix with the indicated entries on the diagonal:

$$\text{entry}_{ii}(D) = d_i$$

for $i = 1, 2, \ldots, n$. For example,

$$\text{diag}(a, b, c) = \begin{bmatrix} a & 0 & 0 \\ 0 & b & 0 \\ 0 & 0 & c \end{bmatrix}.$$

Theorem 46B *The product of two square diagonal matrices of the same size is again diagonal and may be computed by multiplying corresponding diagonal entries. More precisely, if*

$$D = \text{diag}(d_1, d_2, \ldots, d_n)$$

and

$$C = \text{diag}(c_1, c_2, \ldots, c_n)$$

then

$$DC = \text{diag}(d_1 c_1, d_2 c_2, \ldots, d_n c_n).$$

Proof:

$$\text{entry}_{ij}(DC) = \sum_{k=1}^{n} \text{entry}_{ik}(D)\,\text{entry}_{kj}(C)$$

and $\text{entry}_{ik}(D) = 0$ if $i \neq k$ and $\text{entry}_{kj}(C) = 0$ if $k \neq j$. Hence, if $i \neq j$ all the terms on the right are zero, and if $i = j$ then the only nonzero term is the one with $k = i = j$. But $\text{entry}_{ii}(D) = d_i$ and $\text{entry}_{jj}(C) = c_j$.   □

Corollary 47A *A diagonal matrix is invertible if and only if its diagonal entries are nonzero. If*

$$D = \text{diag}(d_1, d_2, \ldots, d_n)$$

is invertible, then

$$D^{-1} = \text{diag}(d_1^{-1}, d_2^{-1}, \ldots, d_n^{-1}).$$

Corollary 47B *Any two diagonal $n \times n$ matrices commute; that is, if D and C are diagonal (as in the theorem), then $DC = CD$.*

Proof: $d_i c_i = c_i d_i$.   □

Theorem 47C *If the matrices A and B commute, then so do the matrices PAP^{-1} and PBP^{-1} for any invertible matrix P.*

Proof: Exercise 48C.

Theorem 47D (Diagonal Multiplication)   *The effect of multiplying an $m \times n$ matrix A on the right by the diagonal matrix*

$$D = \text{diag}(d_1, d_2, \ldots, d_n)$$

is to multiply the jth column of A by d_j (for each $j = 1, 2, \ldots, n$):

$$\text{col}_j(AD) = d_j\,\text{col}_j(A).$$

Proof: The ith entry in the jth column of AD is $\text{entry}_{ij}(AD)$. By the definition of matrix multiplication,

$$\text{entry}_{ij}(AD) = \sum_{k=1}^{n} \text{entry}_{ik}(A)\,\text{entry}_{kj}(D).$$

But $\text{entry}_{kj}(D) = 0$ for $k \neq j$ since D is diagonal so the sum on the right has only one nonzero term. Hence

$$\text{entry}_{ij}(AD) = \text{entry}_{ij}(A)\,\text{entry}_{jj}(D) = \text{entry}_{ij}(A)d_j.$$

As this holds for each $i = 1, 2, \ldots, m$ we obtain $\text{col}_j(AD) = d_j\,\text{col}_j(A)$.   □

▷ **Exercise 48A** Compute PD and DP where

$$P = \begin{bmatrix} x_1 & y_1 & z_1 \\ x_2 & y_2 & z_2 \\ x_3 & y_3 & z_3 \end{bmatrix}, \qquad D = \begin{bmatrix} a & 0 & 0 \\ 0 & b & 0 \\ 0 & 0 & c \end{bmatrix}.$$

▷ **Exercise 48B** Let

$$P = \begin{bmatrix} 2 & 1 \\ 1 & 1 \end{bmatrix}, \qquad C = \begin{bmatrix} 3 & 0 \\ 0 & 4 \end{bmatrix}, \qquad D = \begin{bmatrix} 2 & 0 \\ 0 & 1 \end{bmatrix}.$$

Compute $A = PCP^{-1}$, $B = PDP^{-1}$, AB, and BA.

▷ **Exercise 48C** Prove Theorem 47C.

▷ **Exercise 48D** Prove that the effect of multiplying an $m \times n$ matrix A on the left by the diagonal matrix

$$D = \operatorname{diag}(d_1, d_2, \ldots, d_m)$$

is to multiply the ith row of A by d_i (for $i = 1, 2, \ldots m$):

$$\operatorname{row}_i(DA) = d_i \operatorname{row}_i(A).$$

2.7   Triangular Matrices

The entries of a matrix whose column index is greater than the row index are said to be **above the diagonal**; those whose column index is less than the row index are said to be **below the diagonal**; the remaining entries (the diagonal entries) are said to be **on the diagonal**. For example, if

$$A = \begin{bmatrix} a & b & c \\ d & e & f \\ g & h & i \end{bmatrix},$$

then the entries b, c, f are above the diagonal, the entries a, e, i are on the diagonal, and the entries d, g, h are below the diagonal.

Definition 48E A square matrix is **triangular** iff all the entries below the diagonal vanish and **strictly triangular** iff all the entries on or below the diagonal vanish.

Example 49A The matrix R is triangular and the matrix N is strictly triangular:

$$R = \begin{bmatrix} * & * & * \\ 0 & * & * \\ 0 & 0 & * \end{bmatrix}, \qquad N = \begin{bmatrix} 0 & * & * \\ 0 & 0 & * \\ 0 & 0 & 0 \end{bmatrix}.$$

Here the asterisks represent any numbers, zero or not.

Theorem 49B (Triangular Invertibility) *A triangular matrix is invertible if and only if all its diagonal entries are nonzero.*

This theorem will be proved in Exercise 96D, again in Exercise 248B, and yet again in Exercise 348F. The 2×2 case is an immediate consequence of Theorem 39D. Exercises 49E and 49C below treat the 3×3 case.

Exercise 49C Suppose that $A = \begin{bmatrix} a & b & c \\ 0 & d & e \\ 0 & 0 & f \end{bmatrix}$ and

$$X_1 = \begin{bmatrix} 1 \\ 0 \\ 0 \end{bmatrix}, \qquad X_2 = \begin{bmatrix} -b \\ a \\ 0 \end{bmatrix}, \qquad X_3 = \begin{bmatrix} be - cd \\ -ae \\ ad \end{bmatrix}.$$

Show that $AX_1 = 0$ if $a = 0$, that $AX_2 = 0$ if $d = 0$, and that $AX_3 = 0$ if $f = 0$. Conclude that R is not invertible if one of a, d, f is zero. (See the Noninvertibility Remark 40B.)

Exercise 49D Suppose that

$$U = \begin{bmatrix} 1 & b & c \\ 0 & 1 & e \\ 0 & 0 & 1 \end{bmatrix}, \qquad V = \begin{bmatrix} 1 & x & y \\ 0 & 1 & z \\ 0 & 0 & 1 \end{bmatrix}.$$

Solve $UV = I_3$ for x, y, z in terms of b, c, e. Show that the solution satisfies $VU = I_3$. Conclude that U is invertible.

Exercise 49E Suppose that A is 3×3 triangular with nonzero diagonal entries. Show that A may be written in the form $A = DU$ where D is invertible and diagonal (see Corollary 47A) and U is triangular with 1's on the diagonal. Conclude that A is invertible.

2.8   Matrices of Matrices

We shall sometimes make "matrices of matrices". Thus suppose we are given a q-tuple $(m_1, m_2, \ldots, m_q)$ and a r-tuple $(n_1, n_2, \ldots, n_r)$ of positive integers, and that for each $u = 1, 2, \ldots, q$ and each $v = 1, 2, \ldots, r$ we are given a matrix A_{uv} of size $m_u \times n_v$. We define

$$m = m_1 + m_2 + \cdots + m_q, \qquad n = n_1 + n_2 + \cdots + n_r,$$

and then define a matrix $A \in \mathbf{F}^{m \times n}$ by

$$A = \begin{bmatrix} A_{11} & A_{12} & \cdots & A_{1r} \\ A_{21} & A_{22} & \cdots & A_{2r} \\ \vdots & \vdots & \ddots & \vdots \\ A_{q1} & A_{q2} & \cdots & A_{qr} \end{bmatrix}.$$

The entries of the matrix A can be defined precisely by the equation

$$\text{entry}_{ij}(A) = \text{entry}_{xy}(A_{uv})$$

for $i = 1, 2, \ldots, m$, $j = 1, 2, \ldots, n$ where $i = m_1 + m_2 + \cdots + m_{u-1} + x$, $j = n_1 + n_2 + \cdots + n_{v-1} + y$, $1 \le x \le m_u$, and $1 \le y \le n_v$.

Example 50A Any matrix can be viewed as a matrix of matrices in many ways. For example, let

$$A = \begin{bmatrix} a_{11} & a_{12} & a_{13} & a_{14} & a_{15} \\ a_{21} & a_{22} & a_{23} & a_{24} & a_{25} \\ a_{31} & a_{32} & a_{33} & a_{34} & a_{35} \end{bmatrix}.$$

Then

$$A = \begin{bmatrix} A_{11} & A_{12} \\ A_{21} & A_{22} \end{bmatrix}$$

where

$$A_{11} = \begin{bmatrix} a_{11} & a_{12} \\ a_{21} & a_{22} \end{bmatrix}, \qquad A_{12} = \begin{bmatrix} a_{13} & a_{14} & a_{15} \\ a_{23} & a_{24} & a_{25} \end{bmatrix},$$

$$A_{21} = \begin{bmatrix} a_{31} & a_{32} \end{bmatrix}, \qquad A_{22} = \begin{bmatrix} a_{33} & a_{34} & a_{35} \end{bmatrix}.$$

The notion of a matrix of matrices is very handy for doing calculations. The point is that *most of the formulas for the various matrix operations hold when the entries are themselves matrices (not just numbers) provided that they are of the correct size*. The most general formula of this kind is the following

Theorem 51A (Block Multiplication Law)   *Let A and B be given by*

$$A = \begin{bmatrix} A_{11} & A_{12} & \cdots & A_{1r} \\ A_{21} & A_{22} & \cdots & A_{2r} \\ \vdots & \vdots & \ddots & \vdots \\ A_{q1} & A_{q2} & \cdots & A_{qr} \end{bmatrix}, \qquad B = \begin{bmatrix} B_{11} & B_{12} & \cdots & B_{1s} \\ B_{21} & B_{22} & \cdots & B_{2s} \\ \vdots & \vdots & \ddots & \vdots \\ B_{r1} & B_{r2} & \cdots & B_{rs} \end{bmatrix},$$

where A_{wv} of size $m_w \times n_v$ and B_{vu} of size $n_v \times p_u$. Then the product AB is given by

$$AB = \begin{bmatrix} C_{11} & C_{12} & \cdots & C_{1s} \\ C_{21} & C_{22} & \cdots & C_{2s} \\ \vdots & \vdots & \ddots & \vdots \\ C_{q1} & C_{q2} & \cdots & C_{qs} \end{bmatrix}$$

where C_{wu} is of size $m_w \times p_u$ and

$$C_{wu} = \sum_{v=1}^{r} A_{wv} B_{vu}$$

for $w = 1, 2, \ldots, q$, $u = 1, 2, \ldots s$. In other words, the rule is that the matrices multiply exactly as if all the submatrices are 1×1.

This is proved in Appendix A on Page 439. In this section we'll list the most useful special cases and give some examples.[3]

First let's consider the case $q = r = s = 2$. Then the block multiplication law asserts that the formula

$$\begin{bmatrix} A_{11} & A_{12} \\ A_{21} & A_{22} \end{bmatrix} \begin{bmatrix} B_{11} & B_{12} \\ B_{21} & B_{22} \end{bmatrix} = \begin{bmatrix} A_{11}B_{11} + A_{12}B_{21} & A_{11}B_{12} + A_{12}B_{22} \\ A_{21}B_{11} + A_{22}B_{21} & A_{21}B_{12} + A_{22}B_{22} \end{bmatrix}$$

holds when each of the multiplications $A_{uv}B_{vw}$ on the right side is meaningful, that is, when the number of n_v columns in A_{uv} is the same as the number of rows in B_{vw} for $u, v, w = 1, 2$.

Theorem 51B (Block Shear Formula)   *If $R \in \mathbf{F}^{m \times n}$ and $P \in \mathbf{F}^{n \times n}$ are given by*

$$R = \begin{bmatrix} I_r & C \\ 0_{(m-r) \times r} & 0_{(m-r) \times (n-r)} \end{bmatrix}, \qquad P = \begin{bmatrix} I_r & K \\ 0_{(n-r) \times r} & I_{n-r} \end{bmatrix}.$$

---

[3]We'll give each special case a suggestive name, but these names are not standard and there is no need to memorize them.

where $C, K \in \mathbf{F}^{(n-r) \times r}$, then

$$RP = \begin{bmatrix} I_r & C + K \\ 0_{(m-r) \times r} & 0_{(m-r) \times (n-r)} \end{bmatrix}.$$

With $r = 2$, $m = 3$, $n = 4$, this says

$$\left[\begin{array}{cc|cc} 1 & 0 & c_1 & c_2 \\ 0 & 1 & c_3 & c_4 \\ \hline 0 & 0 & 0 & 0 \end{array}\right] \left[\begin{array}{cc|cc} 1 & 0 & k_1 & k_2 \\ 0 & 1 & k_3 & k_4 \\ \hline 0 & 0 & 1 & 0 \\ 0 & 0 & 0 & 1 \end{array}\right] = \left[\begin{array}{cc|cc} 1 & 0 & c_1 + k_1 & c_2 + k_2 \\ 0 & 1 & c_3 + k_3 & c_4 + k_4 \\ \hline 0 & 0 & 0 & 0 \end{array}\right].$$

Theorem 52A (Corner Formula)   *Suppose that $P_1, P_2 \in \mathbf{F}^{n \times n}$ are given by*

$$P_1 = \begin{bmatrix} I_p & C_1 \\ 0_{q \times p} & I_q \end{bmatrix}, \qquad P_2 = \begin{bmatrix} I_p & C_2 \\ 0_{q \times p} & I_q \end{bmatrix}.$$

where $C_1, C_2 \in \mathbf{F}^{p \times q}$ and $n = p + q$. Then

$$P_1 P_2 = \begin{bmatrix} I_p & C_1 + C_2 \\ 0_{q \times p} & I_q \end{bmatrix}.$$

With $p = 2$ and $q = 1$, this says

$$\begin{bmatrix} 1 & 0 & a_1 \\ 0 & 1 & b_1 \\ 0 & 0 & 1 \end{bmatrix} \begin{bmatrix} 1 & 0 & a_2 \\ 0 & 1 & b_2 \\ 0 & 0 & 1 \end{bmatrix} = \begin{bmatrix} 1 & 0 & a_1 + a_2 \\ 0 & 1 & b_1 + b_2 \\ 0 & 0 & 1 \end{bmatrix}.$$

Corollary 52B (Block Inverse Formula)   *If $C \in \mathbf{F}^{p \times q}$ and $n = p + q$, then the $n \times n$ matrix P is invertible with the indicated inverse:*

$$P = \begin{bmatrix} I_p & C \\ 0_{q \times p} & I_q \end{bmatrix}, \qquad P^{-1} = \begin{bmatrix} I_p & -C \\ 0_{q \times p} & I_q \end{bmatrix}.$$

With $p = 2$ and $q = 1$, this says

$$\begin{bmatrix} 1 & 0 & a \\ 0 & 1 & b \\ 0 & 0 & 1 \end{bmatrix}^{-1} = \begin{bmatrix} 1 & 0 & -a \\ 0 & 1 & -b \\ 0 & 0 & 1 \end{bmatrix}.$$

In the case $q = r = 2$ and $s = 1$ the block multiplication law asserts that the formula

$$\begin{bmatrix} A_{11} & A_{12} \\ A_{21} & A_{22} \end{bmatrix} \begin{bmatrix} X_1 \\ X_2 \end{bmatrix} = \begin{bmatrix} A_{11}X_1 + A_{12}X_2 \\ A_{21}X_1 + A_{22}X_2 \end{bmatrix}$$

holds when A_{uv} has size $m_u \times n_v$ and X_v has size $n_v \times 1$ for $u, v = 1, 2$. The following special case will be used to parameterize the space of solutions of a linear system. (Theorem 136B)

Theorem 53A (Solution Formula)   *If $R \in \mathbf{F}^{m \times n}$ and $X \in \mathbf{F}^{n \times k}$ have the forms*

$$R = \begin{bmatrix} I_r & C \\ 0_{(m-r) \times r} & 0_{(m-r) \times (n-r)} \end{bmatrix}, \qquad X = \begin{bmatrix} W \\ Z \end{bmatrix},$$

where $C \in \mathbf{F}^{r \times (n-r)}$, $W \in \mathbf{F}^{r \times k}$, $Z \in \mathbf{F}^{(n-r) \times k}$, then

$$RX = \begin{bmatrix} W + CZ \\ 0_{(m-r) \times k} \end{bmatrix}.$$

For example,

$$\begin{bmatrix} 1 & 0 & c_1 & c_2 \\ 0 & 1 & c_3 & c_4 \\ 0 & 0 & 0 & 0 \end{bmatrix} \begin{bmatrix} w_1 \\ w_2 \\ z_2 \\ z_2 \end{bmatrix} = \begin{bmatrix} w_1 + c_1 z_1 + c_2 z_2 \\ w_2 + c_3 z_1 + c_4 z_2 \\ 0 \end{bmatrix}.$$

In the case $q = s = 1$ and $r = n$, the block multiplication law asserts that

$$AB = \sum_{j=1}^{n} A_j B_j$$

where

$$A = \begin{bmatrix} A_1 & A_2 & \cdots & A_n \end{bmatrix}, \qquad B = \begin{bmatrix} B_1 \\ B_2 \\ \vdots \\ B_n \end{bmatrix}.$$

If we read X for B, x_j for B_j and drop the sigma notation, we get

$$AX = x_1 A_1 + x_2 A_2 + \cdots x_n A_n.$$

We restate this as

Theorem 53B (Column Formula)   *Let $A \in \mathbf{F}^{m \times n}$ and $X \in \mathbf{F}^{n \times 1}$ so $AX \in \mathbf{F}^{m \times 1}$. Then AX is a linear combination of the columns of A. More precisely:*

$$AX = x_1 \operatorname{col}_1(A) + x_2 \operatorname{col}_2(A) + \cdots + x_n \operatorname{col}_n(A)$$

where $x_j = \operatorname{entry}_j(X)$ is the jth entry of X.

When $n = 2$ and $m = 3$, the formula says that

$$\begin{bmatrix} a_1 & a_2 \\ b_1 & b_2 \\ c_1 & c_2 \end{bmatrix} \begin{bmatrix} x_1 \\ x_2 \end{bmatrix} = x_1 \begin{bmatrix} a_1 \\ b_1 \\ c_1 \end{bmatrix} + x_2 \begin{bmatrix} a_2 \\ b_2 \\ c_2 \end{bmatrix}.$$

Corollary 54A (Selection Formula) *For $A \in \mathbf{F}^{m \times n}$ we have*

$$AE_j = \mathrm{col}_j(A).$$

where $E_j = \mathrm{col}_j(I_n)$ is the jth column of the identity matrix.

Proof: This is Theorem 53B with $x_j = 1$ and $x_k = 0$ for $k \neq j$.   □

When $n = 2$, $m = 3$, and $j = 2$, the selection formula says that

$$\begin{bmatrix} a_1 & a_2 \\ b_1 & b_2 \\ c_1 & c_2 \end{bmatrix} \begin{bmatrix} 0 \\ 1 \end{bmatrix} = \begin{bmatrix} a_2 \\ b_2 \\ c_2 \end{bmatrix}.$$

Finally, we consider the special case of the block multiplication law where $q = r = 1$ and $s = p$. In this case, the formula asserts that

$$A \begin{bmatrix} B_1 & B_2 & \cdots & B_p \end{bmatrix} = \begin{bmatrix} AB_1 & AB_2 & \cdots & AB_p \end{bmatrix}.$$

In other words, $\mathrm{col}_k(AB) = A\,\mathrm{col}_k(B)$. For future reference we restate this formula with X_k in place of B_k.

Theorem 54B (Catenation Formula) *Suppose that $A \in \mathbf{F}^{m \times n}$ and that $X_1, X_2, \ldots, X_p \in \mathbf{F}^{n \times 1}$. Then*

$$A \begin{bmatrix} X_1 & X_2 & \cdots & X_p \end{bmatrix} = \begin{bmatrix} AX_1 & AX_2 & \cdots & AX_p \end{bmatrix}.$$

▷ **Exercise 54C** Suppose $A \in \mathbf{F}^{3 \times 5}$ and $E = \mathrm{col}_2(I_5)$ the second column of the 5×5 identity matrix. What is AE?

▷ **Exercise 54D** Suppose $A \in \mathbf{F}^{4 \times 3}$. Find a matrix $H \in \mathbf{F}^{1 \times 4}$ so that $HA = \mathrm{row}_2(A)$.

▷ **Exercise 54E** Suppose $A \in \mathbf{F}^{5 \times 9}$ and $B \in \mathbf{F}^{9 \times 13}$ are given by

$$A = \begin{bmatrix} A_{11} & A_{12} \\ A_{21} & A_{22} \end{bmatrix}, \quad B = \begin{bmatrix} B_{11} & B_{12} \\ B_{21} & B_{22} \end{bmatrix},$$

where $A_{11} \in \mathbf{F}^{3 \times 4}$, $A_{12} \in \mathbf{F}^{3 \times 5}$, $A_{21} \in \mathbf{F}^{2 \times 4}$, $A_{22} \in \mathbf{F}^{2 \times 5}$, $B_{11} \in \mathbf{F}^{4 \times 7}$, $B_{12} \in \mathbf{F}^{4 \times 6}$, $B_{21} \in \mathbf{F}^{5 \times 7}$, $B_{22} \in \mathbf{F}^{5 \times 6}$. By Theorem 51A, $AB \in \mathbf{F}^{5 \times 13}$ is given by

$$AB = \begin{bmatrix} A_{11}B_{11} + A_{12}B_{21} & A_{11}B_{12} + A_{12}B_{22} \\ A_{21}B_{11} + A_{22}B_{21} & A_{21}B_{12} + A_{22}B_{22} \end{bmatrix}.$$

What are the sizes of the matrices in this decomposition? What is the fourth row of A? The eighth column of B? The $(4, 8)$-entry of AB?

Block Multiplication Laws

Block Shear Formula.

$$\begin{bmatrix} I & C \\ 0 & 0 \end{bmatrix} \begin{bmatrix} I & K \\ 0 & I \end{bmatrix} = \begin{bmatrix} I & C + K \\ 0 & 0 \end{bmatrix}$$

Corner Formula.

$$\begin{bmatrix} I & C_1 \\ 0 & I \end{bmatrix} \begin{bmatrix} I & C_2 \\ 0 & I \end{bmatrix} = \begin{bmatrix} I & C_1 + C_2 \\ 0 & I \end{bmatrix}$$

Solution Formula.

$$\begin{bmatrix} I & C \\ 0 & 0 \end{bmatrix} \begin{bmatrix} W \\ Z \end{bmatrix} = \begin{bmatrix} W + CZ \\ 0 \end{bmatrix}$$

Column Formula.

$$AX = \sum_{j=1}^{n} x_j \operatorname{col}_j(A)$$

Selection Formula.

$$AE_j = \operatorname{col}_j(A).$$

Catenation Formula.

$$A \begin{bmatrix} X_1 & X_2 & \ldots & X_p \end{bmatrix} = \begin{bmatrix} AX_1 & AX_2 & \ldots & AX_p \end{bmatrix}.$$

This box contains the most important special cases of the block multiplication law. Refer to the text for precise statements.

▷ **Exercise 56A** Let $M = \begin{bmatrix} A & B \\ C & D \end{bmatrix}$, $Z = \begin{bmatrix} X \\ Y \end{bmatrix}$, where

$$A = \begin{bmatrix} a_{11} & a_{12} \\ a_{21} & a_{22} \end{bmatrix}, \quad B = \begin{bmatrix} b_1 \\ b_2 \end{bmatrix}, \quad X = \begin{bmatrix} x_1 \\ x_2 \end{bmatrix},$$

$$C = \begin{bmatrix} c_1 & c_2 \end{bmatrix}, \quad D = d, \quad Y = y,$$

where all the lower-case letters are numbers. (1) Write out M and Z displaying all their entries. (2) Calculate MZ using the definition of matrix multiplication. (3) Compare your answer with the result predicted by the block multiplication law.

Exercise 56B Suppose

$$L = \begin{bmatrix} P & Q \end{bmatrix}, \quad M = \begin{bmatrix} A & B \\ C & D \end{bmatrix}, \quad W = \begin{bmatrix} U \\ V \end{bmatrix},$$

where $A \in \mathbf{F}^{3 \times 3}$, $B \in \mathbf{F}^{3 \times 2}$, $C \in \mathbf{F}^{1 \times 3}$, $D \in \mathbf{F}^{1 \times 2}$, $U \in \mathbf{F}^{3 \times 5}$, $V \in \mathbf{F}^{2 \times 5}$, $P \in \mathbf{F}^{1 \times 3}$, $Q \in \mathbf{F}^{1 \times 1}$. 1. What size is M? W? L? MW? LM? LMW? 2. Express MW in terms of A, B, C, D, U, V. 3. Express LM in terms of P, Q, A, B, C, D. 4. Express LMW in terms of P, Q, A, B, C, D, U, V.

Exercise 56C Suppose $A_1, B_1 \in \mathbf{F}^{p \times q}$, that $n = p + q$, and that $A, B \in \mathbf{F}^{n \times n}$ are given by

$$A = \begin{bmatrix} I_p & A_1 \\ 0_{q \times p} & I_q \end{bmatrix}, \quad B = \begin{bmatrix} I_p & B_1 \\ 0_{q \times p} & I_q \end{bmatrix}.$$

Find AB. Do A and B commute?

▷ **Exercise 56D** Let $A \in \mathbf{F}^{m \times n}$ and $H \in \mathbf{F}^{1 \times m}$ so $HA \in \mathbf{F}^{1 \times n}$. Express HA as a linear combination of the rows of A.

Exercise 56E Find AB where

$$A = \begin{bmatrix} 1 & 0 & 0 & 2 & 3 \\ 0 & 1 & 0 & 4 & 5 \\ 0 & 0 & 1 & 6 & 7 \\ 0 & 0 & 0 & 1 & 0 \\ 0 & 0 & 0 & 0 & 1 \end{bmatrix}, \quad B = \begin{bmatrix} 1 & 0 & 0 & 6 & 5 \\ 0 & 1 & 0 & 4 & 3 \\ 0 & 0 & 1 & 2 & 1 \\ 0 & 0 & 0 & 1 & 0 \\ 0 & 0 & 0 & 0 & 1 \end{bmatrix}.$$

▷ **Exercise 56F** Suppose $A \in \mathbf{F}^{4 \times 4}$ satisfies

$$\mathrm{col}_4(A) = 2\,\mathrm{col}_1(A) + 3\,\mathrm{col}_2(A).$$

Find a nonzero $X \in \mathbf{F}^{4 \times 1}$ with $AX = 0$. Is A invertible?

▷ **Exercise 57A** Suppose that $A \in \mathbf{F}^{n \times 2}$ and $B \in \mathbf{F}^{n \times 3}$ satisfy

$$A = \begin{bmatrix} A_1 & A_2 \end{bmatrix}, \qquad B = \begin{bmatrix} A_1 & A_2 & A_1 + A_2 \end{bmatrix}.$$

Find a matrix $C \in \mathbf{F}^{2 \times 3}$ such that $AC = B$.

Recall Theorem 39D, the formula for inverting a 2×2 matrix:

$$A = \begin{bmatrix} a_{11} & a_{12} \\ a_{21} & a_{22} \end{bmatrix}, \qquad A^{-1} = \begin{bmatrix} d^{-1}a_{22} & -d^{-1}a_{12} \\ -d^{-1}a_{21} & d^{-1}a_{11} \end{bmatrix},$$

where $d = a_{11}a_{22} - a_{12}a_{21}$ is the determinant. When $a_{12} = 0$ the formula simplifies considerably:

$$A = \begin{bmatrix} a_{11} & a_{12} \\ 0 & a_{22} \end{bmatrix}, \qquad A^{-1} = \begin{bmatrix} a_{11}^{-1} & -a_{11}^{-1}a_{22}^{-1}a_{12} \\ 0 & a_{22}^{-1} \end{bmatrix}.$$

If we are careful about the order in which we write the multiplication in the $(1, 2)$-position, this formula works for blocks as well.

▷ **Exercise 57B (Improved Block Inverse Formula)** Suppose that $A \in \mathbf{F}^{n \times n}$ has the form

$$A = \begin{bmatrix} A_{11} & A_{12} \\ 0_{q \times p} & A_{22} \end{bmatrix}$$

where $p + q = n$, $A_{11} \in \mathbf{F}^{p \times p}$, $A_{12} \in \mathbf{F}^{p \times q}$, $A_{22} \in \mathbf{F}^{q \times q}$. Suppose that both of the square matrices A_{11} and A_{22} are invertible. Then A is invertible and its inverse is given by the formula

$$A^{-1} = \begin{bmatrix} A_{11}^{-1} & -A_{11}^{-1}A_{12}A_{22}^{-1} \\ 0_{q \times p} & A_{22}^{-1} \end{bmatrix}.$$

Exercise 57C Find A^{-1} and A^2 where

$$A = \begin{bmatrix} 1 & 0 & 0 & 2 & 3 \\ 0 & 1 & 0 & 4 & 5 \\ 0 & 0 & 1 & 6 & 7 \\ 0 & 0 & 0 & 1 & 0 \\ 0 & 0 & 0 & 0 & 1 \end{bmatrix}$$

▷ **Exercise 57D** Suppose that $n = p + q$ and that $A \in \mathbf{F}^{n \times n}$ has form

$$A = \begin{bmatrix} I_p & 0_{p \times q} \\ C & I_q \end{bmatrix}$$

where $C \in \mathbf{F}^{q \times p}$. What is A^{-1}?

Exercise 58A Using the block multiplication law 52B find the inverses of

$$A = \begin{bmatrix} 1 & 4 & 1 & 2 \\ 2 & 7 & 3 & 4 \\ 0 & 0 & 2 & 1 \\ 0 & 0 & 1 & 2 \end{bmatrix}, \quad B = \begin{bmatrix} 1 & 0 & a & b \\ 0 & 1 & c & d \\ 0 & 0 & 1 & 0 \\ 0 & 0 & 0 & 1 \end{bmatrix}.$$

Check your answers by computing AA^{-1} and BB^{-1}.

Exercise 58B Find A^{-1} if

$$A = \begin{bmatrix} A_1 & 0 & 0 \\ 0 & A_2 & 0 \\ 0 & 0 & A_3 \end{bmatrix}$$

where A_1, A_2, A_3 are invertible. What is

$$\begin{bmatrix} 1 & 2 & 0 & 0 \\ 1 & 1 & 0 & 0 \\ 0 & 0 & 3 & 0 \\ 0 & 0 & 0 & 7 \end{bmatrix}^{-1} ?$$

2.9   Using MINIMAT

In this section we explain the notations used by MINIMAT for the operations defined in this chapter. First note that the notations explained in Section 1.8 apply to matrices as well as numbers. The operations of addition, subtraction, multiplication, and division work as for numbers, but MINIMAT will produce an error message if the sizes of the operands aren't compatible with the operation. For example, the command

```
#> C = A*B
```

assigns the matrix product of A and B to C, provided that A and B are already defined and the number of columns in A is the same as the number of rows in B. If b is 1×1, then either of

```
#> b*A,    A*b
```

is accepted as the result of multiplying the matrix A by the scalar b and the two notations

```
#> (1/b)*A,    A/b
```

produce the same result.

Creating New Matrices

In MINIMAT, matrices can be created by entering a sequence of expressions between square brackets. The entries in a row are separated by spaces (you can also use commas) and the rows are separated by semicolons (you can also use carriage returns).   The (i,j)-entry of A is denoted by A(i,j) since there is no way to type a_{ij} into the computer. The ith row is denoted by A(i,:) and the jth column is denoted A(:,j). You can type A(i) instead of A(i,1) if A has only one column, and A(j) instead of A(1,j) if A has only one row.

For example, the MINIMAT statement

```
#>    A = [ 7  -2  21; 0  4  -7/2 ]
```

causes MINIMAT to respond with

```
A =  [  7.000       -2.000       21.000
        0.000        4.000       -3.500  ]
```

and MINIMAT now remembers that A denotes this matrix (until you change the meaning of A with another similar command).   We can recover the entries of A using functional notation and the rows and columns can be obtained using the *colon* (:).

```
#>   e = A(2,3), r = A(1,:),    c = A(:,2),   r(2)

e =   -3.500

r =  [  7.000       -2.000       21.000  ]

c =  [  -2.000
         4.000  ]

ans = -2.000
```

White Space

For the most part MINIMAT ignores white space (blanks). We have already remarked that a space can be used in place of a comma to separate columns in a matrix. However, spaces do not act as separators when they occur within parentheses.

▷ **Exercise 59A** Predict the result of the following:

```
#> x = 3
#> A = [ 1  x-1  3; (2*x-1)  4  6]
#> A = [ 1  x-1  3; (2*x - 1)  4  6]
#> A = [ 1  x  -1  3; (2*x-1)  4  6]
```

Zero Matrices

The built-in function `zeros(m,n)` returns a zero matrix with m rows and n columns. For example,

```
#> Z = zeros(2,3)
   Z = [0.000 0.000 0.000
        0.000 0.000 0.000]
```

Random Matrices

MINIMAT has the capability to generate random matrices; the built-in function `rand(m,n)` returns a random matrix with m rows and n columns. The entries are always between zero and one. For example,

```
#> A = rand(2,3)
   A = [0.392  0.277  0.824
        0.481  0.935  0.618]
```

Exercise 60A Enter the command `A=rand(3,4)` into MINIMAT a few times and note that you get a different result each time.

The capability to generate random matrices can be used to confirm matrix laws "experimentally". We can confirm the true laws by plugging in randomly chosen matrices and observing that the two sides of the law give the same answer. If a putative law is in fact false, we can discover this quickly by plugging in random matrices.

Identity Matrix

In MINIMAT the identity matrix is denoted by `eye(m)`.

```
#> I=eye(3)

I = [ 1.000      0.000      0.000
      0.000      1.000      0.000
      0.000      0.000      1.000 ]
```

The funny name is chosen since the letter I is popular for other purposes and computers can't handle subscripts and superscripts easily.

▷ **Exercise 61A** Confirm that multiplication by the identity matrix produces no change.

Confirming Laws

Many computer exercises in this book ask you to 'confirm' some general principle. These means that you are asked to choose random matrices of the appropriate size, compute both sides of the equation, and confirm that they are equal. For example, you confirm the commutative law for addition as follows:

```
#> A=rand(2,3), B=rand(2,3)
#> A+B, B+A  % (the answers agree.)
```

Example 61B We confirm the associative law for matrix multiplication using MINIMAT. We generate three random matrices of the appropriate size.

```
#> A=rand(2,4),  B=rand(4,3),  C=rand(3,5)
```

```
A = [ 0.9      0.5      1.0      0.2
      0.7      0.2      0.6      0.4 ]

B = [ 0.9      0.2      0.5
      0.5      0.3      0.4
      0.3      0.6      0.9
      0.2      0.9      0.6 ]

C = [ 0.3      0.1      0.3      1.0      0.5
      0.0      0.6      0.9      0.9      0.2
      0.6      0.1      0.8      0.4      0.8 ]
```

We then evaluate both sides of the equation $(AB)C = A(BC)$, displaying the intermediate results so that the content of the law is clear.

```
#> AB = A*B,  BC = B*C   %intermediate results
```

```
AB = [ 1.4      1.1      1.7
       1.1      0.9      1.3]

BC = [ 0.5      0.3      0.8      1.2      0.9
       0.6      0.5      1.3      1.1      0.9
       0.5      0.6      1.4      1.3      0.8 ]
```

Now we compare the two sides of the equation.

```
#> AB*C, A*BC              % check that (AB)C=A(BC)

   ans = [ 1.4       1.0       2.7       3.0       2.2
           1.1       0.9       2.2       2.4       1.7
   ans = [ 1.4       1.0       2.7       3.0       2.2
           1.1       0.9       2.2       2.4       1.7 ]
```

The answers agree as expected.

▷ **Question 62A** Why did MINIMAT apparently make an error of 0.1 in calculating the (1,1) entry of BC?

▷ **Exercise 62B** Confirm some of the other multiplication laws in Theorem 34E analogously.

Transpose

MINIMAT uses the notation A' for the **conjugate transpose** of A and the notation A.' for the **transpose** of A. For real matrices the conjugate transpose and the transpose are the same so you might as well use the shorter notation A'. For complex matrices MINIMAT reserves the shorter notation for the conjugate transpose because it is more important.

This calculation confirms the double transpose law 44A:

```
#> A=[1 2 3; 4 5 6]
   A =  [ 1.000      2.000      3.000
          4.000      5.000      6.000 ]
#> A'
   ans = [ 1.000      4.000
           2.000      5.000
           3.000      6.000 ]
#> A''
   ans = [ 1.000      2.000      3.000
           4.000      5.000      6.000 ]
```

Inverses and Powers

MINIMAT accepts two notations for the inverse of the matrix A: either of the commands B=A^(-1) or B=inv(A) assigns the inverse of the matrix A to B.

▷ **Exercise 62C** Most square matrices are invertible. To convince yourself of this have MINIMAT pick one at random and invert it.

Example 63A There are lots of matrices that are not invertible. To see this, try the following:

```
#>   A=rand(3,2), p=rand(1,1), q=rand(1,1)
#>   A=[A, p*A(:,1)+q*A(:,2)]
#>   B=inv(A)
#>   A*B, B*A
```

MINIMAT may or may not report an error here. However, the 3×3 matrix A constructed here is definitely not invertible since

$$\mathrm{col}_3(A) = p\,\mathrm{col}_1(A) + q\,\mathrm{col}_2(A)$$

so that $AX = 0$ where

$$X = \begin{bmatrix} -p \\ -q \\ 1 \end{bmatrix}.$$

If A were invertible, we would have

$$X = IX = (A^{-1}A)X = A^{-1}(AX) = A^{-1}0 = 0$$

which is not the case. (See the Noninvertibility Remark 40B.) Even if the command B=inv(A) fails to report an error, the subsequent commands B*A and A*B do not produce the identity matrix as they would if A were invertible. MINIMAT will sometimes produce the error message "vanishing determinant" when you ask it to compute the inverse of a noninvertible matrix. However, due to round-off error, it may instead produce an (incorrect) answer without warning. Always check that a candidate B for the inverse A^{-1} is in fact the inverse of A by computing BA and AB.

▷ **Exercise 63B** Use MINIMAT to confirm the antimultiplicative law for inverses (41A).

Since computers don't handle superscripts well MINIMAT uses the notation A^p for the pth power A^p of A. MINIMAT only computes the power A^p when A is square and p is an integer or when both A and p are both scalars (1×1 matrices).

▷ **Exercise 63C** Use MINIMAT to confirm some of the multiplication laws of Theorem 42A. The exponents must be integers.

Diagonal Matrices

MINIMAT uses the built-in function `diag` to convert a row or column matrix to a diagonal matrix. If X is $1 \times n$ or $n \times 1$, then the MINIMAT command

```
#> D=diag(X)
```

assigns to D a diagonal matrix whose diagonal entries agree with the entries of X. Hence, the command

```
#> D=diag(rand(1,4))
```

assigns to D a random 4×4 diagonal matrix. The function that undoes this has the same name; the command

```
#> R = diag(A)
```

assigns to R a row matrix whose entries are the diagonal entries of A. The diag function returns a square matrix if its input is a row (or column) and returns a row if its input is square. To create $D = \mathrm{diag}(1, 2, 3, 4)$ in MINIMAT, use D=diag([1 2 3 4]) not D=diag(1,2,3,4).

▷ **Exercise 64A** Use MINIMAT to verify empirically that diagonal matrices commute.

▷ **Exercise 64B** Use MINIMAT to construct two 4×4 nondiagonal matrices which commute.

Recall that matrix multiplication is not performed elementwise; the MINMAT expressions A*B and A.*B are quite different as are A^p and A.^p. It is important to note, however, that these elementwise operations agree with the usual operations for diagonal matrices. Theorem 46B states that the product of two diagonal matrices can be computed by elementwise multiplication. In particular, if D is diagonal and p is an integer, then the matrices D^p and D.^p agree.

▷ **Exercise 64C** Confirm Theorem 46B using MINIMAT's diag function and its notation L.*M for elementwise multiplication.

▷ **Exercise 64D** Use MINIMAT to confirm Theorem 47D empirically .

Exercise 64E In Section 9.3 we will study the problem of solving the matrix equation $B^p = A$ for B. It is easy to find solutions when A is diagonal for we may take B diagonal with pth roots of the entries of A as entries. Using MINIMAT, find a cube root of the matrix $D = \mathrm{diag}(1, 8, 27, 64)$.

Triangular Matrices

Function 64F The .M function randsutr returns a random strictly upper triangular matrix. The MINIMAT command

```
#>  N=randsutr(n)
```

assigns to the variable N a random strictly triangular matrix of size n × n. The command

```
#>  A=randsutr(n)+diag(rand(1,n))
```

generates a random triangular matrix. (The built-in function diag converts a row matrix R to a square matrix with the entries of R on the diagonal and zeros off the diagonal.)

▷ **Exercise 65A** Generate a random 5 × 5 strictly upper triangular matrix and compute its powers.

▷ **Exercise 65B** Generate a random 5 × 5 strictly upper triangular matrix N and use the formula

$$I - N^5 = (I - N)(I + N + N^2 + N^3 + N^4)$$

to find $(I - N)^{-1}$. (Do Exercise 65A first.)

▷ **Exercise 65C** Generate a random 5 × 5 triangular matrix and compute its inverse. (Hint: D=diag(diag(A)) will create a diagonal matrix D having the same diagonal entries as A.)

Matrices of Matrices

MINIMAT handles "matrices of matrices" routinely. It reports an error if the constituent parts don't fit together correctly. Thus the command

```
#>  M = [A B; C D]
```

works iff A and B have the same number of rows, C and D have the same number of rows, A and C have the same number of columns, and B and D have the same number of columns.

▷ **Exercise 65D** Use MINIMAT to construct a "matrix of matrices" with the following commands:

```
#>  Z=[1;2;3], X=Z(1:2), Y=Z(3:3)
#>  A=[1 2; 3 4],    B=[  -1;-2]
#>  C=[5 9],         D=89
#>  M=[A B; C D]
#>  M(1:2,1:2), A
```

Then confirm the block multiplication law.

Submatrices, Colon Notation

If I and J are row matrices with integer entries in the ranges $1, 2, ..., m$ and $1, , 2, ..., n$ respectively and if A is a matrix with m rows and n columns, then A(I,J) returns the submatrix of A with rows indexed by I and columns indexed by J. For example,

```
#>   A = [11 12 13 14 15
          21 22 23 24 25
          31 32 33 34 35];

#>   B = A([2 1], [1 3 4]),

     B = [ 21 23 24
           11 13 14  ]
```

These expressions can also be used on the left side of an assignment statement provided both sides denote matrices of the same size. For example,

```
#>   B(1,[2 3]) = B(2,[1 2]),

     B = [ 21 11 13
           11 13 14  ]
```

It is unreasonable to use the submatrix notation A(I,J) if the entries of I or J are not distinct, but MINIMAT does not report an error.

The expression p:q evaluates to the row matrix [p p+1 ... q]. A single colon can be used as a subscript to denote all the rows or all the columns. For example, with A as above,

```
#>   C = A(:,3:5)

     C = [ 13 14 15
           23 24 25
           33 34 35 ]
```

Example 66A We confirm that the (1,2)-entry of the product A*B is obtained by multiplying row 1 of A by column 2 of B.

```
#>   A=[1 2 3; 4 5 6]; B=[1 2; 3 4; 5 6]; C=A*B;
#>   A(1,:)*B(:,2), C(1,2)

     ans = 28.000
     ans = 28.000
```

Entrywise Operations

We have already emphasized that the product AB of two matrices A and B is *not* defined entrywise: the (i,j)-entry of AB is usually not the (i,j)-entry of A multiplied by the (i,j)-entry of B. Similarly the (i,j)-entry of a power A^p is not the pth power of the (i,j)-entry of A. MINIMAT provides notations for certain entrywise operations even though these operations are of little interest mathematically. Generally, this notation is obtained by preceding the less important (entrywise) operation by a period.[4] For example, the entrywise product of A and B is defined when A and B are the same size and denoted in MINIMAT by A.*B. MINIMAT also computes the entrywise power denoted by A.^p when either p is a scalar (in which case the (i,j)-entry of A.^p is the (i,j)-entry of A raised to the power p), or when p and A are matrices of the same size (in which case the (i,j)-entry of A.^p is the (i,j)-entry of A raised to the power the (i,j)-entry of p). In either case, the matrix A may be of any size, and A.^p has the same size. (By contrast, the ordinary power is defined only if A is square.)

The built-in functions (sin, cos etc.) accept matrices as arguments with the result that they are applied entrywise. For example,

```
#> A=[1 4 9 25], B=sqrt(A),
   B=[1 2 3 5]
```

WARNING: This is *not* the customary notation used in mathematics. Note in particular that sqrt(A*A) usually is not the same as A (unlike the 1×1 case when sqrt(A*A) equals A if A is positive).

▷ **Exercise 67A** What output will the following produce?

```
#> A=[1 8 27], B=[1 2 3]
#> A.*B,       A*B
#> A.^(1/3),   A^(1/3)
```

Some Computer Exercises

▷ **Exercise 67B** Use MINIMAT to confirm the antimultiplicative law for transposes in Theorem 44A.

▷ **Exercise 67C** Type the following at MINIMAT:

```
#> A=rand(2,3)+sqrt(-1)*rand(2,3)
#> A' ,   A.' ,   conj(A.')
```

---

[4] However, recall (2.9) that MINIMAT's notation for the ordinary transpose (which is *not* an entrywise operation) also involves a period.

What is the conjugate transpose of a matrix?

▷ **Exercise 68A** Confirm the antimultiplicative law for inverses.

▷ **Exercise 68B** What does the following do?

```
#> I=eye(4), E=I(:,3), I=eye(3), F=I(3,:)
#> A=rand(3,4), A*E, F*A
```

▷ **Exercise 68C** A matrix is symmetric if it is equal to its transpose. How can we generate a random symmetric matrix in MINIMAT?

▷ **Exercise 68D** Try the following:

```
#> H=rand(3,3)/4,  I=eye(3)
#> A=I-H, B=I+H+H^2+H^3+H^4
#> A*B, B*A
#> H^5
```

Why is B a good approximation to A^{-1}?

▷ **Exercise 68E** For small $H \in \mathbf{F}^{n \times n}$ the formula

$$(I - H)^{-1} = \sum_{j=0}^{\infty} H^j$$

is valid. Confirm this using MINIMAT.

The .M function **randsoln** computes a random solution to a homogeneous system. More precisely, the MINIMAT command

```
#> X=randsoln(A)
```

assigns to X a random solution of $AX = 0$.

▷ **Exercise 68F** Using MINIMAT, produce a nonzero solution for the homogeneous linear system

$$
\begin{array}{rrrrcl}
x_1 & +2x_2 & +3x_3 - x_4 & = & 0 \\
4x_1 & +5x_2 & +6x_3 - x_4 & = & 0 \\
7x_1 & +8x_2 & +9x_3 + x_4 & = & 0.
\end{array}
$$

▷ **Exercise 68G** Produce a random 5×3 matrix A and then find a random solution H of H*A=0.

2.10   More Exercises

▷ **Exercise 69A** Suppose $A = BP_1$ and $C = BP_2$ where P_1, P_2 are invertible. Find P such that $A = CP$.

▷ **Exercise 69B** A matrix is called **symmetric** iff it equals its transpose. Show that any square matrix A of form $A = B + B^\top$ is symmetric.

▷ **Exercise 69C** Let $A = \begin{bmatrix} 1 & 2 & 3 \\ 4 & 5 & 6 \\ 7 & 8 & 9 \end{bmatrix}$. Find a nonzero vector X such that $AX = 0$. Is A invertible?

Exercise 69D Let

$$P = \begin{bmatrix} 2 & 1 \\ 1 & 1 \end{bmatrix}, \qquad A = \begin{bmatrix} 1 & 0 \\ 0 & 0 \end{bmatrix}, \qquad B = PAP^{-1}.$$

Compute B. Compute A^2 and B^2. What's the point?

Preview of the Exponential

▷ **Exercise 69E** What is $\begin{bmatrix} 1 & 1 \\ 0 & 1 \end{bmatrix}^n$?

Exercise 69F Show that $U(t + s) = U(t)U(s)$ if

$$U(t) = \begin{bmatrix} \cos t & \sin t \\ -\sin t & \cos t \end{bmatrix}$$

What is $U(0)$? $U(t)^{-1}$? $U(t)^p$?

Exercise 69G Show that $U(t + s) = U(t)U(s)$ if

$$U(t) = \begin{bmatrix} \cosh t & \sinh t \\ \sinh t & \cosh t \end{bmatrix}$$

What is $U(0)$? $U(t)^{-1}$? $U(t)^p$?

Exercise 69H [Hard]   For $t \in \mathbf{F}$ define a matrix $U(t) \in \mathbf{F}^{n \times n}$ by

$$\text{entry}_{ij}\left(U(t)\right) = \begin{cases} t^{j-i}/(j-i)! & \text{if } i \leq j, \\ 0 & \text{if } j < i. \end{cases}$$

For example, if $n = 5$ then

$$U(t) = \begin{bmatrix} 1 & t & t^2/2 & t^3/6 & t^4/24 \\ 0 & 1 & t & t^2/2 & t^3/6 \\ 0 & 0 & 1 & t & t^2/2 \\ 0 & 0 & 0 & 1 & t \\ 0 & 0 & 0 & 0 & 1 \end{bmatrix}.$$

Show that $U(t + s) = U(t)U(s)$. (If you can't do the general case, try the cases $n = 2$ and $n = 3$.) What is $U(0)$? $U(t)^{-1}$? $U(t)^p$?

Geometric Series

The formula

$$(I - H)(I + H + H^2 + H^3 + \cdots + H^{p-1}) = I - H^p \qquad (*)$$

holds in matrix theory just as in high school algebra. It has already been used in Exercise 22D.

Exercise 70A Show that if $H \in \mathbf{F}^{n \times n}$ is strictly upper triangular, then $H^n = 0$. Deduce that $I - H$ is invertible and

$$(I - H)^{-1} = I + H + H^2 + H^3 + \cdots + H^{n-1}.$$

(See Exercises 65A and 65B.)

Exercise 70B Suppose that $N, M \in \mathbf{F}^{n \times n}$ are strictly upper triangular and have integer entries. Show that the matrix

$$A = (I + N)(I + M)^{\top}$$

is invertible and both A and A^{-1} have integer entries.

Exercise 70C [Hard]   The formula for the **geometric series**

$$(I - H)^{-1} = I + H + H^2 + H^3 + \cdots$$

holds for matrices H with small entries just as it hold for small numbers. (See Exercise 68E.) Prove this as follows:

(1) Show that if $H \in \mathbf{F}^{n \times n}$ and each entry of H has absolute value less than h/n, then each entry of H^p has absolute value less than h^p.

(2) Conclude that if $h < 1$, then $\lim_{p \to \infty} H^p = 0_{n \times n}$.

(3) Now use $(*)$.

Chapter Summary

Matrix Algebra

Matrix algebra satisfies the same laws as ordinary algebra with a few notable exceptions:

- *Matrices may be added only when they have the same size.*

- *Matrices may be multiplied only when the number of columns in the first factor equals the number of rows in the second,*

- *Matrix multiplication is usually not commutative.*

- *A matrix can be both nonzero and noninvertible.*

Matrix Multiplication

The **Row-Column Formula**

$$\text{entry}_{ik}(AB) = \text{row}_i(A)\ \text{col}_k(B).$$

is used to compute the product.

Block Multiplication

Matrices of matrices can be manipulated by the same laws as matrices of numbers provided the sizes of the constituent matrices match correctly.

3

INVERTIBLE MATRICES

In this chapter we answer two fundamental questions:

- How can we tell if a matrix is invertible?

- If it is, how do we find its inverse?

We saw in Theorem 39D how to do this for a 2×2 matrix and now we will learn how to do it in the $n \times n$ case. The algorithm we employ is called **Gauss-Jordan Elimination**. It is quite easy to implement on a computer. This algorithm is also quite useful for solving linear systems $AX = Y$ where A might not be a square matrix. We will use it for this purpose in Chapter 4. That's why we will also learn how to apply the algorithm to matrices that aren't square.

Here's the idea. In Section 3.2 we will define a certain class of matrices called the **elementary matrices**. These matrices are easily seen to be invertible. Moreover, if E is elementary, it's easy to compute EA from A. Since E is invertible, it follows that A is invertible if and only if EA is invertible. The Gauss-Jordan Elimination algorithm is explained in Theorem 84C. It constructs from A a sequence $E_1, E_2, \ldots, E_k$ of elementary matrices so that

$$E_k \cdots E_2 E_1 A = R$$

where R is particularly easy to understand. If A is invertible, then R will be the identity matrix and

$$A^{-1} = E_1^{-1} E_2^{-2} \cdots E_k^{-1};$$

if A is not invertible, then R will not be invertible and, *because of the special form of R*, this will be obvious.

3.1   Elementary Row Operations

Before we define the *elementary matrices*, we define the *elementary row operations*. Let $A \in \mathbf{F}^{m \times n}$ be any matrix (not necessarily square). There are three kinds of **elementary row operations** that may be applied to the matrix A:

- **Scale.** For $p = 1, 2, \ldots, m$ and $c \neq 0$, the matrix $\mathrm{Scale}(A, p, c)$ is the matrix that results from A by multiplying the pth row by c.

- **Swap.** For $p, q = 1, 2, \ldots, m$, the matrix $\mathrm{Swap}(A, p, q)$ is the matrix that results from A by exchanging the pth and qth rows.

- **Shear.** For $p, q = 1, 2, \ldots, m$ with $p \neq q$ and any scalar c, the matrix $\mathrm{Shear}(A, p, q, c)$ is the matrix that results from A by adding c times the qth row to the pth row.

Example 74A Take $m = 4$, $n = 3$, and let

$$A = \begin{bmatrix} a_{11} & a_{12} & a_{13} \\ a_{21} & a_{22} & a_{23} \\ a_{31} & a_{32} & a_{33} \\ a_{41} & a_{42} & a_{43} \end{bmatrix}.$$

- The result of rescaling the first row of A by c is

$$\mathrm{Scale}(A, 1, c) = \begin{bmatrix} ca_{11} & ca_{12} & ca_{13} \\ a_{21} & a_{22} & a_{23} \\ a_{31} & a_{32} & a_{33} \\ a_{41} & a_{42} & a_{43} \end{bmatrix},$$

- The result of swapping rows two and three is

$$\mathrm{Swap}(A, 2, 3) = \begin{bmatrix} a_{11} & a_{12} & a_{13} \\ a_{31} & a_{32} & a_{33} \\ a_{21} & a_{22} & a_{23} \\ a_{41} & a_{42} & a_{43} \end{bmatrix},$$

- The result of adding c times the third row to the second is

$$\text{Shear}(A, 2, 3, c) = \begin{bmatrix} a_{11} & a_{12} & a_{13} \\ a_{21} + ca_{31} & a_{22} + ca_{32} & a_{23} + ca_{33} \\ a_{31} & a_{32} & a_{33} \\ a_{41} & a_{42} & a_{43} \end{bmatrix}.$$

Theorem 75A *Each elementary row operation is reversible in the sense that it can be undone by another operation of the same kind. Specifically:*

- *If $B = \text{Scale}(A, p, c)$, then $A = \text{Scale}(B, p, c^{-1})$.*

- *If $B = \text{Swap}(A, p, q)$, then $A = \text{Swap}(B, p, q)$.*

- *If $B = \text{Shear}(A, p, q, c)$, then $A = \text{Shear}(B, p, q, -c)$.*

Proof:

- If B is obtained from A by multiplying the pth row by c, then A is obtained from B by multiplying the pth row by c^{-1}.

- If B is obtained from A by swapping rows p and q, then A is obtained from B by swapping again.

- If B is obtained from A by adding c times the qth row to the pth row, then A is obtained from B by subtracting c times this row.   □

3.2   Elementary Matrices

The elementary matrices are the simplest of all invertible matrices, except for the identity matrix. We shall see that they are the building blocks from which the invertible matrices are constructed. Here's the definition.

Definition 75B A matrix that results from the identity matrix by applying a single elementary row operation is called an **elementary matrix**.

An elementary matrix is always a square matrix. There are three kinds.

- **Scale.** The matrix $E = \text{Scale}(I, p, c)$ is an elementary matrix for $p = 1, 2, \ldots, m$ and $c \neq 0$. It differs from the $m \times m$ identity matrix $I = I_m$ in that $\text{entry}_{pp}(E) = c$ rather than 1.

- **Swap.** The matrix $E = \text{Swap}(I, p, q)$ is an elementary matrix for $p, q = 1, 2, \ldots, m$, $p \neq q$. It differs from the identity matrix in that

$$\text{entry}_{pp}(E) = 0 \quad \text{entry}_{pq}(E) = 1$$
$$\text{entry}_{qp}(E) = 1 \quad \text{entry}_{qq}(E) = 0.$$

- **Shear.** The matrix $E = \text{Shear}(I, p, q, c)$ is an elementary matrix for $p, q = 1, 2, \ldots, m$, $p \neq q$. It differs from the identity matrix in that

$$\text{entry}_{pq}(E) = c.$$

Example 76A Let I be the 4×4 identity matrix. Then,

- $\text{Scale}(I, 1, c) = \begin{bmatrix} c & 0 & 0 & 0 \\ 0 & 1 & 0 & 0 \\ 0 & 0 & 1 & 0 \\ 0 & 0 & 0 & 1 \end{bmatrix}$,

- $\text{Swap}(I, 2, 3) = \begin{bmatrix} 1 & 0 & 0 & 0 \\ 0 & 0 & 1 & 0 \\ 0 & 1 & 0 & 0 \\ 0 & 0 & 0 & 1 \end{bmatrix}$,

- $\text{Shear}(I, 2, 3, c) = \begin{bmatrix} 1 & 0 & 0 & 0 \\ 0 & 1 & c & 0 \\ 0 & 0 & 1 & 0 \\ 0 & 0 & 0 & 1 \end{bmatrix}$.

Fundamental Theorem on Row Operations

Theorem 76B *The matrix EA that results by multiplying a matrix A on the left by an elementary matrix E is the same as the matrix that results by applying the corresponding elementary row operation to A.*

Let the elementary matrix $E \in \mathbf{F}^{m \times m}$ result by applying some elementary row operation to the identity matrix $I = I_m \in \mathbf{F}^{m \times m}$. The Fundamental Theorem says that for any matrix $A \in \mathbf{F}^{m \times n}$, the matrix EA results by applying that same row operation to A. More precisely,

- $E = \text{Scale}(I_m, p, c) \implies EA = \text{Scale}(A, p, c)$.

- $E = \text{Swap}(I_m, p, q) \implies EA = \text{Swap}(A, p, q)$.

- $E = \text{Shear}(I_m, p, q, c) \implies EA = \text{Shear}(A, p, q, c)$.

For a careful proof see Appendix A on Page 441. The following three examples, one for each type of elementary matrix, should be more convincing than a formal proof.

- $E = \text{Scale}(I_4, 1, c)$, $EA = \text{Scale}(A, 1, c)$,

$$
\begin{bmatrix}
c & 0 & 0 & 0 \\
0 & 1 & 0 & 0 \\
0 & 0 & 1 & 0 \\
0 & 0 & 0 & 1
\end{bmatrix}
\begin{bmatrix}
a_{11} & a_{12} & a_{13} \\
a_{21} & a_{22} & a_{23} \\
a_{31} & a_{32} & a_{33} \\
a_{41} & a_{42} & a_{43}
\end{bmatrix}
=
\begin{bmatrix}
ca_{11} & ca_{12} & ca_{13} \\
a_{21} & a_{22} & a_{23} \\
a_{31} & a_{32} & a_{33} \\
a_{41} & a_{42} & a_{43}
\end{bmatrix}.
$$

- $E = \text{Swap}(I_4, 2, 3)$, $EA = \text{Swap}(A, 2, 3)$,

$$
\begin{bmatrix}
1 & 0 & 0 & 0 \\
0 & 0 & 1 & 0 \\
0 & 1 & 0 & 0 \\
0 & 0 & 0 & 1
\end{bmatrix}
\begin{bmatrix}
a_{11} & a_{12} & a_{13} \\
a_{21} & a_{22} & a_{23} \\
a_{31} & a_{32} & a_{33} \\
a_{41} & a_{42} & a_{43}
\end{bmatrix}
=
\begin{bmatrix}
a_{11} & a_{12} & a_{13} \\
a_{31} & a_{32} & a_{33} \\
a_{21} & a_{22} & a_{23} \\
a_{41} & a_{42} & a_{43}
\end{bmatrix}.
$$

- $E = \text{Shear}(I_4, 2, 3, c)$, $EA = \text{Shear}(A, 2, 3, c)$,

$$
\begin{bmatrix}
1 & 0 & 0 & 0 \\
0 & 1 & c & 0 \\
0 & 0 & 1 & 0 \\
0 & 0 & 0 & 1
\end{bmatrix}
\begin{bmatrix}
a_{11} & a_{12} & a_{13} \\
a_{21} & a_{22} & a_{23} \\
a_{31} & a_{32} & a_{33} \\
a_{41} & a_{42} & a_{43}
\end{bmatrix}
=
$$

$$
=
\begin{bmatrix}
a_{11} & a_{12} & a_{13} \\
a_{21} + ca_{31} & a_{22} + ca_{32} & a_{23} + ca_{33} \\
a_{31} & a_{32} & a_{33} \\
a_{41} & a_{42} & a_{43}
\end{bmatrix}.
$$

Theorem 77A *Elementary matrices are invertible. In fact,*

- $\text{Scale}(I, p, c)^{-1} = \text{Scale}(I, p, c^{-1})$.

- $\text{Swap}(I, p, q)^{-1} = \text{Swap}(I, p, q)$.

- *Shear*$(I, p, q, c)^{-1} = \text{Shear}(I, p, q, -c)$.

Proof: Let E be an elementary matrix and F be its putative inverse. In other words,

- if $E = \text{Scale}(I, p, c)$, then $F = \text{Scale}(I, p, c^{-1})$;

- if $E = \text{Swap}(I, p, q)$, then $F = \text{Swap}(I, p, q)$;

- if $E = \text{Shear}(I, p, q, c)$, then $F = \text{Shear}(I, p, q, -c)$.

Now E is the result of applying an elementary row operation to I. According to Theorem 75A, applying to E the elementary row operation corresponding to F restores I. But by the Fundamental Theorem, this is FE. In other words, $FE = I$. Similarly, $EF = I$. Thus E is invertible and $F = E^{-1}$ as required.   □

Corollary 78A *A product of any number of elementary matrices is invertible.*

Proof: According to Theorem 40D, a product of invertible matrices is invertible. The antimultiplicative law shows that the inverse of such a product is again a product of elementary matrices:

$$M = E_1 E_2 \cdots E_k \implies M^{-1} = E_k^{-1} \cdots E_2^{-1} E_1^{-1}.$$ □

▷ **Exercise 78B** Verify that each of the elementary matrices

$$E = \text{Scale}(I_4, 1, c), \ \text{Swap}(I_4, 2, 3), \ \text{Shear}(I_4, 2, 3, c)$$

is invertible by finding an elementary matrix F such that $EF = FE = I$.

▷ **Exercise 78C** Let

$$E_1 = \begin{bmatrix} 1 & 0 & 0 \\ 0 & 1 & c \\ 0 & 0 & 1 \end{bmatrix}, \qquad E_2 = \begin{bmatrix} 0 & 1 & 0 \\ 1 & 0 & 0 \\ 0 & 0 & 1 \end{bmatrix}, \qquad E_3 = \begin{bmatrix} 1 & 0 & 0 \\ 0 & 0 & 1 \\ 0 & 1 & 0 \end{bmatrix}.$$

Find $E_2 E_3$, $E_1 E_2 E_3$, $E_1 E_2 E_1$ and their inverses.

▷ **Exercise 78D** Let

$$M = \begin{bmatrix} 3 & 0 & 0 \\ 0 & 1 & 0 \\ 0 & 0 & 1 \end{bmatrix} \begin{bmatrix} 1 & 0 & 0 \\ 5 & 1 & 0 \\ 0 & 0 & 1 \end{bmatrix} \begin{bmatrix} 1 & 0 & 0 \\ 0 & 1 & 4 \\ 0 & 0 & 1 \end{bmatrix} \begin{bmatrix} 1 & 0 & 3 \\ 0 & 1 & 0 \\ 0 & 0 & 1 \end{bmatrix}.$$

Find M^{-1}.

Using MINIMAT

Using MINIMAT's submatrix and column notations, it is easy to implement the elementary row operations. For example, consider the following dialog:

```
#>   A = [1 2; 3 4; 5 6]

A = [ 1    2
      3    4
      4    5]

#>   A(1,:)= 4*A(1, :)

A = [ 4    8
      3    4
      4    5]

#>   A([1 2],:)=A([2 1], :)

A = [ 3    4
      4    8
      4    5]

#>   A(3,:)=A(3, :)-2*A(1,:)

A = [ 3    4
      4    8
     -2   -3]
```

Each command changed the value of A. If we want MINIMAT to remember
the original value, we must make a copy as in

```
#>  B=A;   B(3,:)=B(3, :)+B(1,:)

B = [ 3    4
      4    8
      1    1]
```

The semicolon in the command suppressed the printing that would normally follow the first assignment statement. Make an elementary matrix
as follows:

```
#>  E=eye(3);   E(3,:)=E(3, :)+E(1,:)

E = [ 1    0   0
      0    1   0
      1    0   1]
```

Elementary Row Operations in MINIMAT

MINIMAT's Notation	Mathematical Notation
B=A, B(p,:) = c*B(p,:) B=A, B([p q],:) = B([q p],:) B=A, B(p,:) = B(p,:)+c*B(q,:)	$B = \text{Scale}(A, p, c)$ $B = \text{Swap}(A, p, q)$ $B = \text{Shear}(A, p, q, c)$

Elementary Matrices in MINIMAT

MINIMAT's Notation	Mathematical Notation
E=eye(m), E(p,:)=c*E(p,:) E=eye(m), E([p q],:)=E([q p],:) E=eye(m), E(p,:)=E(p,:)+c*E(q,:)	$E = \text{Scale}(I_m, p, c)$ $E = \text{Swap}(I_m, p, q)$ $E = \text{Shear}(I_m, p, q, c)$

▷ **Exercise 80A** Verify instances of the Fundamental Theorem empirically as follows:

(1) Generate a random 3×4 matrix A0.

(2) Choose an elementary row operation and generate the corresponding elementary matrix E.

(3) Generate the matrix A that results from A0 by performing this operation and compare the result with E*A0.

Do this for a scale, swap, and a shear.

▷ **Exercise 80B** Use MINIMAT to verify empirically that the elementary matrices are invertible by generating a random elementary matrix E, computing the elementary matrix F which is its inverse, and computing E*F and F*E.

Elementary Row Operations as .M Functions

```
function B = scale(A,p,c)

B = A; B(p,:) = c*A(p,:);


function B = swap(A,p,q)

B = A; B([p q],:) = A([q p],:);


function B = shear(A,p,q,c)

B = A; B(p,:) =A(p,:) + c*A(q,:);
```

Listing 1

Function 81A The .M functions shown in Listing 1 perform the elementary row operations. The commands

```
#> B = scale(A,p,c)
#> B = swap(A,p,q)
#> B = shear(A,p,q,c)
```

assign to B the result of performing the corresponding elementary operation on A. The commands

```
#> E = scale(eye(m),p,c)
#> E = swap(eye(m),p,q)
#> E = shear(eye(m),p,q,c)
```

create the corresponding elementary matrices.

Exercise 81B Generate a matrix A with the command

```
#> A = round(4*rand(3,5))
```

After generating each of the following matrices B, find an elementary matrix E so that B and E*A are the same.

```
#> B=A; B(2,:)=5*A(2,:)
#> B=A; B([1 3],:) = A([3 1],:)
#> B=A; B(3,:) = A(3,:)+7*A(1,:)
#> B=scale(A,2,5)
#> B=swap(A,1,3)
#> B=shear(A,3,1,7)
```

3.3   Reduced Row Echelon Form

As indicated at the beginning of this chapter, our strategy for inverting a matrix involves transforming it to a certain normal form. This form is called *reduced row echelon form* and is defined in this section. In the next section we give the algorithm for transforming a given matrix to this normal form.

Reduced Row Echelon Form (RREF)

Definition 82A An $m \times n$ matrix R is in **reduced row echelon form**, abbreviated **RREF**, iff

- all the rows that vanish identically (if any) appear below the other (nonzero) rows;

- the leading entry in any row appears to the left of the leading entry of any nonzero row below;

- the leading entry in any nonzero row is 1;

- all other entries in the column of a leading entry are 0.

The **leading entry** of a row is the first nonzero entry in that row. The columns which hold the leading entries are called the **leading columns**; the other columns are called the **free columns**. (The reason for the term *free* will become apparent when we study linear systems in Chapter 4. See Remark 121A.)

Examples 83A The following matrix is in RREF:

$$R = \begin{bmatrix} 1 & c_{11} & 0 & c_{12} & 0 & c_{13} & c_{14} \\ 0 & 0 & 1 & c_{22} & 0 & c_{23} & c_{24} \\ 0 & 0 & 0 & 0 & 1 & c_{33} & c_{34} \\ 0 & 0 & 0 & 0 & 0 & 0 & 0 \\ 0 & 0 & 0 & 0 & 0 & 0 & 0 \end{bmatrix}$$

The leading entries are in the $(1,1)$, $(2,3)$, and $(3,5)$ positions as the entries to the left of them vanish. The zero rows appear at the bottom, each leading entry is to the right of the leading entry in the row above, and the other entries in columns 1, 3, and 5 vanish. Columns 1, 3, and 5 are leading; columns 2, 4, 6, and 7 are free.

The following matrix is in RREF:

$$R = \begin{bmatrix} 1 & 0 & 0 & c_{11} & c_{12} & c_{13} & c_{14} \\ 0 & 1 & 0 & c_{21} & c_{22} & c_{23} & c_{24} \\ 0 & 0 & 1 & c_{31} & c_{32} & c_{33} & c_{34} \\ 0 & 0 & 0 & 0 & 0 & 0 & 0 \\ 0 & 0 & 0 & 0 & 0 & 0 & 0 \end{bmatrix}$$

The leading entries are in the $(1,1)$, $(2,2)$, and $(3,3)$ positions as the entries to the left of them vanish. Columns 1, 2, and 3 are leading; columns 4, 5, 6, and 7 are free. The zero rows appear at the bottom. Each leading entry is to the right of the leading entry in the row above. The nonleading entries in columns 1, 2, and 3 vanish.

More generally, any matrix $R \in \mathbf{F}^{m \times n}$ of form

$$R = \begin{bmatrix} I_r & C \\ 0_{(m-r) \times r} & 0_{(m-r) \times (n-r)} \end{bmatrix}$$

where I_r is the $r \times r$ identity matrix and $C \in \mathbf{F}^{r \times (n-r)}$ is in RREF; the first r columns are leading, the last $n - r$ are free. Moreover, any matrix in RREF has this form after a possible rearrangement of its columns. Either of the equalities $r = m$ or $r = n$ is allowed so that the following are all in RREF:

$$R = \begin{bmatrix} I_m & C \end{bmatrix} \qquad (r = m \le n)$$

$$R = \begin{bmatrix} I_n \\ 0_{(m-n) \times n} \end{bmatrix} \qquad (r = n \le m)$$

$$R = I_n \qquad (r = n = m).$$

Remark 84A A square matrix in RREF is either the identity matrix (in which case it is invertible) or has a row of zeros (in which case it is not invertible). For example, the 3×3 matrix

$$R = \begin{bmatrix} 1 & 0 & c_{13} \\ 0 & 1 & c_{23} \\ 0 & 0 & 0 \end{bmatrix}$$

is a square matrix in RREF and it is not the identity matrix.

▷ **Question 84B** Why is a matrix with a zero row not invertible?

Gauss-Jordan Elimination

The algorithm for transforming a given matrix A to a matrix R in RREF is called **Gauss-Jordan Elimination**. This algorithm is nothing more than a set of rules prescribing the order of applying elementary row operations to achieve the transformation. It is very easy to understand because each step brings us closer to the desired form. It is also very easy to implement on the computer. The algorithm here is described in the natural language English. For a description in the programming language MINIMAT, see Listing 2 on Page 91.

Theorem 84C (Gauss-Jordan Elimination) *A matrix may be transformed to a matrix in RREF by applying a sequence of elementary row operations.*

Proof: Our proof is a description of the algorithm that describes how to choose the elementary row operations to transform a given matrix A to a matrix R in RREF. The algorithm processes the columns one at a time, in order from left to right. It maintains a count r of the number of leading columns found so far.

When the algorithm begins, no leading columns have been found ($r = 0$). The following steps process the kth column.

- First decide if the kth column is free or leading. If all the entries in the positions at or below the $(r + 1, k)$ entry are zero, the kth column is free, so we proceed to the next column. In the contrary case the kth column is leading, so we increase r.

- Swap rows so that the (r, k) entry is nonzero. (For reasons of accuracy, a computer program chooses the swap so that the nonzero entry is as large as possible, but this is not required.)

- Scale so that (r, k) entry becomes 1.

- Subtract appropriate multiples of the rth row from the other rows, so that all other entries in the kth column are 0.

These steps are to be repeated until the number r of leading columns is the number m of rows or until all the columns have been processed. □

Example 85A We transform the matrix

$$A = \begin{bmatrix} 0 & 0 & 2 & 2 & -2 \\ 2 & 2 & 6 & 14 & 4 \end{bmatrix}$$

to RREF via elementary row operations.

$$A_1 = \mathrm{Swap}(A, 1, 2) = \begin{bmatrix} 2 & 2 & 6 & 14 & 4 \\ 0 & 0 & 2 & 2 & -2 \end{bmatrix}$$

gets a nonzero entry in the $(1,1)$-position.

$$A_2 = \mathrm{Scale}(A_1, 1, 1/2) = \begin{bmatrix} 1 & 1 & 3 & 7 & 2 \\ 0 & 0 & 2 & 2 & -2 \end{bmatrix}$$

makes the $(1,1)$-entry 1. Since the $(2,1)$-entry is already zero, it is not necessary to shear.

$$A_3 = \mathrm{Scale}(A_2, 2, 1/2) = \begin{bmatrix} 1 & 1 & 3 & 7 & 2 \\ 0 & 0 & 1 & 1 & -1 \end{bmatrix}$$

makes the $(2,3)$-entry 1.

$$A_4 = \mathrm{Shear}(A_3, 1, 2, -3) = \begin{bmatrix} 1 & 1 & 0 & 4 & 5 \\ 0 & 0 & 1 & 1 & -1 \end{bmatrix}$$

gets a 0 in the $(1,3)$-position. Now, $R = A_4$ is in RREF with leading 1s in the $(1,1)$ and $(2,3)$ positions. Columns 1 and 3 form a 2×2 identity matrix. Columns 2, 4, and 5 are free.

Computing the Multiplier

It is convenient to express Theorem 84C as a theorem about matrix multiplication. We can do this using the Fundamental Theorem on Row Operations. Here's how it goes.

Theorem 85B (Multiplier Theorem)   *For any $A \in \mathbf{F}^{m \times n}$ there is an invertible matrix M such that the matrix $R = MA$ is in RREF.*

Proof: According to Theorem 84C, there is a sequence $A_0, A_1, \ldots, A_h$ of matrices such that

(1) The first matrix A_0 of the sequence is the original matrix A;

(2) The last matrix A_h of the sequence is in RREF;

(3) Each matrix of the sequence is obtained from the preceding one by applying an elementary row operation.

By the Fundamental Theorem 76B on Row Operations, we may express condition (3) with an equation

$$A_k = E_k A_{k-1}$$

for $k = 1, 2, \ldots, h$ where E_k is an elementary matrix. Now define $R = A_h$ (condition (2) says that R is in RREF) and let M be the product of the elementary matrices E_k in reverse order:

$$M = E_h E_{h-1} \cdots E_2 E_1.$$

Condition (1) says that $A_0 = A$ so

$$
\begin{aligned}
R &= A_h \\
&= E_h A_{h-1} \\
&= E_h E_{h-1} A_{h-2} \\
&\quad \cdots \\
&= E_h E_{h-1} \cdots E_2 E_1 A_0 \\
&= M A
\end{aligned}
$$

as required. The multiplier M is invertible by Corollary 78A.   □

Remark 86A In Chapter 5 (Theorem 182D) it is proved that the reduced row echelon form R is unique: If $MA = R$ and $\widetilde{M} A = \widetilde{R}$ where M and $\widetilde{M}$ are invertible and R and $\widetilde{R}$ are in RREF, then $R = \widetilde{R}$.

The following handy trick, called the *Multiplier Trick*, enables us to compute the multiplier matrix M as we perform elementary row operations. Form the $m \times (n + m)$ matrix $\begin{bmatrix} A & I_m \end{bmatrix}$. Using the block multiplication law 54B, we have

$$M \begin{bmatrix} A & I \end{bmatrix} = \begin{bmatrix} MA & MI \end{bmatrix} = \begin{bmatrix} MA & M \end{bmatrix}$$

where $I = I_m$ is the $m \times m$ identity matrix.

Theorem 87A (Multiplier Trick) *If we apply the same row operations to the matrix $\begin{bmatrix} A & I_m \end{bmatrix}$ that we applied to A to transform it to MA, the last m columns of the result contain the matrix M.*

Proof: Computer programmers like this trick because the algorithm has in it something called an **invariant relation**. The algorithm begins with

$$\begin{bmatrix} A_0 & M_0 \end{bmatrix} = \begin{bmatrix} A & I \end{bmatrix}$$

and ends with

$$\begin{bmatrix} A_h & M_h \end{bmatrix} = \begin{bmatrix} R & M \end{bmatrix}$$

with successive steps related by

$$A_k = E_k A_{k-1}, \qquad M_k = E_k M_{k-1},$$

for $k = 1, 2, \ldots, h$ where E_k is an elementary matrix. The aforementioned invariant relation is the equation

$$M_k A = A_k.$$

The relation can be proved by induction. It holds for $k = 0$ since $M_0 = I$ and $A_0 = A$. If $M_{k-1} A = A_{k-1}$ (that is, if it holds at some stage of the process), then $M_k A = E_k M_{k-1} A = E_k A_{k-1} = A_k$ (it holds at the next stage of the process). This algorithm is implemented in the MINIMAT programming language in Listing 3 on Page 92. □

Example 87B We apply the Multipier Trick to the matrix

$$A = \begin{bmatrix} 0 & 0 & 2 & 2 & -2 \\ 2 & 2 & 6 & 14 & 4 \end{bmatrix}$$

of Example 85A. We do elementary row operations on $\begin{bmatrix} A & I \end{bmatrix}$:

$$\begin{bmatrix} A_0 & M_0 \end{bmatrix} = \left[\begin{array}{ccccc|cc} 0 & 0 & 2 & 2 & -2 & 1 & 0 \\ 2 & 2 & 6 & 14 & 4 & 0 & 1 \end{array} \right]$$

$$\begin{bmatrix} A_1 & M_1 \end{bmatrix} = \left[\begin{array}{ccccc|cc} 2 & 2 & 6 & 14 & 4 & 0 & 1 \\ 0 & 0 & 2 & 2 & -2 & 1 & 0 \end{array} \right]$$

$$\begin{bmatrix} A_2 & M_2 \end{bmatrix} = \left[\begin{array}{ccccc|cc} 1 & 1 & 3 & 7 & 2 & 0 & 1/2 \\ 0 & 0 & 2 & 2 & -2 & 1 & 0 \end{array} \right]$$

$$\begin{bmatrix} A_3 & M_3 \end{bmatrix} = \left[\begin{array}{ccccc|cc} 1 & 1 & 3 & 7 & 2 & 0 & 1/2 \\ 0 & 0 & 1 & 1 & -1 & 1/2 & 0 \end{array} \right]$$

$$\begin{bmatrix} A_4 & M_4 \end{bmatrix} = \left[\begin{array}{ccccc|cc} 1 & 1 & 0 & 4 & 5 & -3/2 & 1/2 \\ 0 & 0 & 1 & 1 & -1 & 1/2 & 0 \end{array} \right].$$

so

$$MA = \begin{bmatrix} 1 & 1 & 0 & 4 & 5 \\ 0 & 0 & 1 & 1 & -1 \end{bmatrix}, \qquad M = \begin{bmatrix} -3/2 & 1/2 \\ 1/2 & 0 \end{bmatrix}.$$

▷ **Exercise 88A** For each of the matrices A_1, A_2, A_3, A_4 of Example 85A, find the elementary matrix E_k such that $A_k = E_k A_{k-1}$. Compute

$$M = E_4 E_3 E_2 E_1$$

and check that $A_4 = MA$.

Exercise 88B Find an elementary matrix E so that EA is in RREF:

(1) $A = \begin{bmatrix} 1 & 2 & 3 \\ 0 & 1 & 5 \end{bmatrix}$ (2) $A = \begin{bmatrix} 1 & a & b \\ 0 & 0 & 1 \end{bmatrix}$

(3) $A = \begin{bmatrix} 0 & 0 & 1 \\ 1 & 2 & 0 \end{bmatrix}$ (4) $A = \begin{bmatrix} 2 & 6 & 0 \\ 0 & 0 & 1 \end{bmatrix}$

Exercise 88C For each A find an invertible matrix M and a matrix R in RREF such that $R = MA$.

(1) $A = \begin{bmatrix} 1 & 2 \\ 3 & 7 \end{bmatrix}$ (2) $A = \begin{bmatrix} 1 & 0 \\ 0 & 1 \end{bmatrix}$ (3) $A = \begin{bmatrix} 1 & 2 & 2 \\ 3 & 7 & 6 \end{bmatrix}$

(4) $A = \begin{bmatrix} 1 & 2 \\ 3 & 7 \\ 2 & 4 \end{bmatrix}$ (5) $A = \begin{bmatrix} 1 & 2 & 2 \\ 3 & 6 & 7 \end{bmatrix}$ (6) $A = \begin{bmatrix} 1 & 2 \\ 2 & 4 \\ 3 & 7 \end{bmatrix}$

(7) $A = \begin{bmatrix} 1 & 2 & 2 \\ 3 & 7 & 6 \\ 4 & 9 & 8 \\ 2 & 5 & 4 \end{bmatrix}$ (8) $A = \begin{bmatrix} 1 & 2 & 3 & 1 \\ 3 & 7 & 10 & 4 \\ 2 & 4 & 6 & 2 \end{bmatrix}$

Exercise 88D Confirm that the following matrices are inverses of one another by multiplying them.

$$M = \begin{bmatrix} 1 & -1 & 4 \\ 1 & 0 & 2 \\ 0 & -2 & 5 \end{bmatrix}, \qquad M^{-1} = \begin{bmatrix} 4 & -3 & -2 \\ -5 & 5 & 2 \\ -2 & 2 & 1 \end{bmatrix}.$$

Exercise 88E Confirm that $MA = R$ for each of the following pairs of matrices A, R. (Here M is the invertible matrix of the previous exercise.)

(1) $A = \begin{bmatrix} 4 & -3 & -2 & -8 \\ -5 & 5 & 2 & 11 \\ -2 & 2 & 1 & 5 \end{bmatrix}, \qquad R = \begin{bmatrix} 1 & 0 & 0 & 1 \\ 0 & 1 & 0 & 2 \\ 0 & 0 & 1 & 3 \end{bmatrix}.$

(2) $A = \begin{bmatrix} 4 & -3 & -2 & 0 \\ -5 & 5 & 5 & 5 \\ -2 & 2 & 2 & 2 \end{bmatrix}$, $R = \begin{bmatrix} 1 & 0 & 1 & 3 \\ 0 & 1 & 2 & 4 \\ 0 & 0 & 0 & 0 \end{bmatrix}$.

(3) $A = \begin{bmatrix} 0 & 4 & 8 & -3 \\ 0 & -5 & -10 & 5 \\ 0 & -2 & -4 & 2 \end{bmatrix}$, $R = \begin{bmatrix} 0 & 1 & 2 & 0 \\ 0 & 0 & 0 & 1 \\ 0 & 0 & 0 & 0 \end{bmatrix}$.

(4) $A = \begin{bmatrix} 4 & -3 & -2 & -2 \\ -5 & 5 & 5 & 2 \\ -2 & 2 & 2 & 1 \end{bmatrix}$, $R = \begin{bmatrix} 1 & 0 & 1 & 0 \\ 0 & 1 & 2 & 0 \\ 0 & 0 & 0 & 1 \end{bmatrix}$.

(5) $A = \begin{bmatrix} 4 & -3 \\ -5 & 5 \\ -2 & 2 \end{bmatrix}$, $R = \begin{bmatrix} 1 & 0 \\ 0 & 1 \\ 0 & 0 \end{bmatrix}$.

(6) $A = \begin{bmatrix} 4 & 8 \\ -5 & -10 \\ -2 & -4 \end{bmatrix}$, $R = \begin{bmatrix} 1 & 2 \\ 0 & 0 \\ 0 & 0 \end{bmatrix}$.

(7) $A = \begin{bmatrix} 0 & 4 \\ 0 & -5 \\ 0 & -2 \end{bmatrix}$, $R = \begin{bmatrix} 0 & 1 \\ 0 & 0 \\ 0 & 0 \end{bmatrix}$.

Using MINIMAT

In Section 3.2, we learned how to perform the elementary row operations using MINIMAT; so now we can compute the RREF. This involves a certain amount of tedious typing; the MINIMAT command for implementing a single elementary row operation is rather long. After you have learned the algorithm of Gauss-Jordan elimination described in Theorem 84C, you can use the .M functions gj and gjm described in this section to avoid this typing. Study the listings of these functions and compare them with the descriptions of the algorithms given the text.

Example 89A We transform the matrix of Example 85A to RREF without using MINIMAT'S .M function gj.

```
#> A  = [ 0  0  2  2  -2;  2  2  6  14 4]
#> A1 = A,  A1([1 2],:) = A1([2 1],:)
```

```
#> A2 = A1, A2(1,:) = (1/A2(1,1))*A2(1,:)
#> A3 = A2, A3(2,:) = (1/A3(2,3))*A3(2,:)
#> A4 = A3, A4(1,:) = A4(1,:)-A4(1,3)*A4(2,:)
#> A4, gj(A) % (These agree.)
```

Example 90A Usually, it is not necessary to do any swaps to put a matrix in RREF. We illustrate this by putting a random 3×5 matrix in RREF (without using gj or gjm).

```
#> A=rand(3,5), R=A
#> R(1,:) = R(1,:)/R(1,1)
#> R(2,:) = R(2,:)-R(2,1)*R(1,:)
#> R(3,:) = R(3,:)-R(3,1)*R(1,:)
#> R(2,:) = R(2,:)/R(2,2)
#> R(1,:) = R(1,:)-R(1,2)*R(2,:)
#> R(3,:) = R(3,:)-R(3,2)*R(2,:)
#> R(3,:) = R(3,:)/R(3,3)
#> R(1,:) = R(1,:)-R(1,3)*R(3,:)
#> R(2,:) = R(2,:)-R(2,3)*R(3,:)
```

▷ **Exercise 90B** Using MINIMAT, generate a 3×5 matrix with a zero in the $(1, 1)$-position but which is otherwise random. Put it in RREF. (Your first step is a swap.)

Function 90C The .M function gj, shown in Listing 2 computes the RREF of its input.[1] After the MINIMAT command

```
#> [R, lead, free] = gj(A)
```

the variable R holds RREF of A, and the row matrices lead and free hold the indices of the leading and free columns, respectively. If gj.m is invoked in the form

```
#> R = gj(A)
```

the outputs lead and free are discarded.

The function gj uses MINIMAT's built-in size that returns the size of its input. For example,

```
#> A = rand(3,5); [m n] = size(A)
   m = 3
   n = 5
```

---

[1]This is named gj rather than rref to avoid a name clash with a .M file supplied with PC Matlab,

Reduced Row Echelon Form in MINIMAT

```
function [R, lead, free] = gj(A)

[m n] = size(A);
R=A; lead=zeros(1,0); free=zeros(1,0);
r = 0;  % rank of first k columns
for k=1:n
    if r==m, free=[free, r+1:n]; return; end
    [y,h] = max(abs(R(r+1:m, k)));  h=r+h;  % (*)
    if  (y < 1.0E-9)  % (i.e if y == 0)
        free = [free, k];
    else
        lead = [lead, k];  r=r+1;
        R([r h],:) = R([h r],:);     % swap
        R(r,:) = R(r,:)/R(r,k);      % scale
        for i = [1:r-1,r+1:m]        % shear
            R(i,:) = R(i,:) - R(i,k)*R(r,:);
        end
    end % if
end  % for
```

Listing 2

The variable k indexes the column being processed, and the variable r counts the number[2] of leading columns found thus far. The built-in functions max and abs (see Pages 451 and 452) are used in the line marked (*). After this line is executed, the variable h holds the row index of the largest entry in the kth column that is below the rth row, and the variable y holds the absolute value of that entry. This is because the expression A(r+1:m, k) returns a column matrix holding the entries in question, abs takes the absolute value of each entry, and the built-in function [y h]=max(C) returns the largest element y in the column C together with its position h. At this point h holds a relative index in the range 1 to m-r. The command h=r+h assures that h has the correct value in the range r+1 to m.

---

[2]The comment calls this the *rank*; this will be explained later in Section 5.4.

The Multiplier Trick in MINIMAT

```
function [M, R] = gjm(A)

    [m,n] = size(A);
    RaM = gj([A eye(m)]);
    R = RaM(:,1:n);
    M = RaM(:,n+1:n+m);
```

Listing 3

Example 92A It is easy to implement the Multiplier Trick using MINI-MAT. We'll redo the Example 89A using it. Here are the steps:

```
#> A= [ 0  0  2   2  -2; 2  2  6  14   4 ]
#> RaM = gj([A,  eye(2)])
#> R = RaM(:,  1:5),   M = RaM(:,  6:7)
#> R, M*A  % (These agree.)
```

Function 92B After you understand how this works, you can use the .M function gjm, shown in Listing 3 to save a little typing. This function implements the multiplier trick. The command

```
#>  [M, R]=gjm(A)
```

assigns to the variable R the RREF of A and to the variable M an invertible matrix that transforms A to R, so that M*A = R.

Example 92C We repeat Example 89A using the gjm function.

```
#> A= [ 0  0  2   2  -2
        2  2  6  14   4 ]
#> [M, R]=gjm(A)
#> R, M*A   % (These agree.)
```

3.4   How to Invert

The Multiplier Trick 87A provides an efficient method to

- decide if a square matrix A is invertible, and

- compute the inverse A^{-1} of an invertible matrix A.

We put A in reduced row echelon form R, simultaneously computing the invertible matrix M with $MA = R$. Of course, the matrix M is invertible by Corollary 78A. Now we apply the following

---

How to Invert

Theorem 93A *Let $A, M, R \in \mathbf{F}^{n \times n}$ be square matrices with M invertible, R in RREF, and*

$$MA = R.$$

Then

- *A is invertible if and only if R is the identity matrix.*

- *If A is invertible, then $A^{-1} = M$.*

Proof: Assume that R is the identity matrix. Then $MA = I$, so (since M is invertible) $A = M^{-1}$. Therefore, A is invertible, since the inverse of an invertible matrix is invertible (by Theorem 40D).

Conversely, assume that R is not the identity matrix. Then (since it is in RREF and square), it has at least one row of zeros. Now R cannot be invertible, since it has a row of zeros. But if A were invertible, then (by Theorem 40D again) $R = MA$ would be invertible. Thus A is also not invertible.   □

Factorization Theorem

Corollary 93B *A matrix is invertible if and only if it is a product of elementary matrices.*

Proof: The matrix M (and hence its inverse) is a product of elementary matrices. ("If" was Corollary 78A.)   □

A Criterion for Invertibility

Corollary 94A *Suppose that A is a square matrix. Then either*

- *A is invertible, or else*

- *there is a nonzero X with $AX = 0$.*

Proof: If R is not the identity matrix, it is easy to find a nonzero X with $RX = 0$. It follows that $AX = M^{-1}RX = 0$. We'll be more precise on finding such an X in the next chapter, but for the moment consider the example

$$R = \begin{bmatrix} 1 & 0 & c_{13} \\ 0 & 1 & c_{23} \\ 0 & 0 & 0 \end{bmatrix}, \quad X = \begin{bmatrix} -c_{13} \\ -c_{23} \\ 1 \end{bmatrix}.$$

Then $RX = 0$ but $X \neq 0$ since $\mathrm{entry}_3(X) = 1$.   □

Remark 94B It is impossible that both alternatives occur. If A is invertible and $AX = 0$, then

$$X = A^{-1}AX = A^{-1}0 = 0.$$

(This is a restatement of the Noninvertibility Remark 40B.)

▷ **Exercise 94C** Use the method explained in the proof of Theorem 93A to compute A^{-1} where $A = \begin{bmatrix} 1 & 2 \\ 3 & 4 \end{bmatrix}$.

▷ **Exercise 94D** Let $A = \begin{bmatrix} 2 & 1 \\ 3 & 1 \end{bmatrix}$. Write A^{-1} as a product of elementary matrices. Then write A as a product of elementary matrices.

▷ **Exercise 94E** Prove that the matrix $A = \begin{bmatrix} 1 & 2 & 3 \\ 4 & 5 & 9 \\ 5 & 2 & 7 \end{bmatrix}$ is not invertible.

▷ **Exercise 95A** Decide which of the following matrices A are invertible. If A is invertible, then find its inverse. If A is not invertible, then find a nonzero vector X such that $AX = 0$.

$$(1) \quad A = \begin{bmatrix} 1 & 0 & 3 \\ 0 & 1 & 6 \\ 0 & 0 & 0 \end{bmatrix} \qquad (2) \quad A = \begin{bmatrix} 1 & 0 & 0 \\ 0 & 1 & 0 \\ 0 & 0 & 1 \end{bmatrix}$$

$$(3) \quad A = \begin{bmatrix} 1 & 2 & 3 \\ 4 & 5 & 6 \\ 7 & 8 & 9 \end{bmatrix} \qquad (4) \quad A = \begin{bmatrix} 1 & 2 & 3 \\ 2 & 5 & 6 \\ 3 & 6 & 10 \end{bmatrix}$$

$$(5) \quad A = \begin{bmatrix} 1 & 2 & 3 \\ 4 & 5 & 6 \\ 0 & 0 & 0 \end{bmatrix} \qquad (6) \quad A = \begin{bmatrix} 1 & 2 & 0 \\ 2 & 5 & 0 \\ 3 & 6 & 0 \end{bmatrix}$$

Exercise 95B Find A^{-1} if $A = \begin{bmatrix} 1 & 1 & 2 \\ 2 & 1 & 3 \\ 0 & 1 & 2 \end{bmatrix}$.

▷ **Exercise 95C** Is the matrix $A = \begin{bmatrix} 1 & 2 & 3 \\ 4 & 5 & 9 \\ 1 & 3 & 4 \end{bmatrix}$ invertible? If yes, find its inverse; if no, find a nonzero X with $AX = 0$.

▷ **Exercise 95D** Let A be as in Exercise 95C. Is the square matrix $A^\top$ invertible? If yes, find its inverse; if no, find Y such that the equation $Y = AX$ has no solution X. Hint: If $HA^\top = 0$ and $HY \neq 0$, can there be an X with $A^\top X = Y$?

Exercise 95E Write the matrix

$$A = \begin{bmatrix} 1 & 0 & 0 & 0 & 0 \\ 0 & 1 & 0 & 0 & 0 \\ 0 & 0 & 1 & 0 & 0 \\ 1 & 2 & 3 & 1 & 0 \\ 4 & 5 & 6 & 0 & 1 \end{bmatrix}$$

as a product of elementary matrices. (In fact, this matrix can be written as a product of six shears.)

▷ **Exercise 95F** Find the inverse of the matrix

$$A = \begin{bmatrix} I_p & 0_{p \times q} \\ C & I_q \end{bmatrix}$$

where $C \in \mathbf{F}^{q \times p}$. How many elementary matrices are required to write A as a product of elementary matrices?

Exercise 96A Find the inverse of the matrix

$$A = \begin{bmatrix} 1 & 0 & 0 & 0 & 0 \\ 0 & 1 & 0 & 0 & 0 \\ 2 & 3 & 1 & 0 & 0 \\ 4 & 5 & 0 & 1 & 0 \\ 6 & 6 & 0 & 0 & 1 \end{bmatrix}.$$

How many elementary matrices are required to write A as a product of elementary matrices?

Exercise 96B Find an invertible matrix M so that $R = MA$ is in RREF where

$$A = \begin{bmatrix} 1 & 0 & 0 \\ 0 & 1 & 0 \\ 2 & 3 & 0 \\ 4 & 5 & 0 \\ 6 & 6 & 0 \end{bmatrix}.$$

How many elementary row operations are required?

▷ **Exercise 96C** Let $A \in \mathbf{F}^{m \times n}$ have form

$$A = \begin{bmatrix} I_r & 0_{r \times (n-r)} \\ C & 0_{(m-r) \times (n-r)} \end{bmatrix}$$

where $C \in \mathbf{F}^{(m-r) \times r}$. Find an invertible matrix M so that $R = MA$ is in RREF. How many elementary row operations are required?

▷ **Exercise 96D** Prove the Triangular Invertibility Theorem 49B.

Using MINIMAT

MINIMAT has a built-in function $A\hat{\ }(-1)$ or $inv(A)$ to invert a matrix. (These are different notations for the same operation.) Now we can see how it works.

Example 96E The following commands assign a random 3×3 matrix to A and then compute its inverse using the Multiplier Trick 87A.

```
#> A =  rand(3,3)
#> [B, I] = gjm(A)
#> B*A, A*B, I
#> B, A^(-1)
```

All three of the matrices in the third line evaluate to the 3×3 identity matrix. The last line compares the computed inverse B with MINIMAT's built-in inverse function. It is hardly surprising that the two matrices on the last line agree: this is precisely the algorithm MINIMAT uses for its built-in inverse function.

Example 97A MINIMAT can factor $A = \begin{bmatrix} 2 & 1 \\ 1 & 1 \end{bmatrix}$ into elementary matrices.

```
#> A=[2 1; 1 1],          I=eye(2),           R=A
#> E1=I, E1(1,:)=E1(1,:)/R(1,1),              R=E1*R % Scale
#> E2=I, E2(2,:)=E2(2,:)-R(2,1)*E2(1,:),      R=E2*R % Shear
#> E3=I, E3(2,:)=E3(2,:)/R(2,2),              R=E3*R % Scale
#> E4=I, E4(1,:)=E4(1,:)-R(1,2)*E4(2,:),      R=E4*R % Shear
```

At this point R is the identity matrix and $I = E_4 E_3 E_2 E_1 A$, so we can write $A = F_1 F_2 F_3 F_4$ where $F_i = E_i^{-1}$. The inverse of an elementary matrix is elementary, so this is the answer:

```
#> F1=inv(E1), F2=inv(E2), F3=inv(E3), F4=inv(E4)
#> F1*F2*F3*F4,  A      % (Confirms the answer)
```

Exercise 97B Use MINIMAT to create a random 2×2 matrix. Then write it as a product of elementary matrices.

Exercise 97C Use MINIMAT to create a random 3×3 matrix. Then write it as a product of elementary matrices.

▷ **Exercise 97D** Use the Multiplier Trick and MINIMAT to find the inverse of a random 4×4 matrix A. You may use the gjm function but not MINIMAT's built-in inverse function (A^(-1) or inv(A)). Confirm your answer and also compare it with the result of performing MINIMAT's built-in inverse function..

▷ **Exercise 97E** Write a .M function inverse that acts like MINIMAT's built-in inverse function. If A holds an invertible matrix, the command

```
#> inverse(A), inv(A), A^(-1)
```

should produce three copies of the same matrix.

Exercise 97F What is the result of the following? (Explain.)

```
#> B=rand(3,3), A=[B B], R=gj(A)
```

A Random Integer Matrix with an Integer Inverse

```
function A = intinv(n)

    U = eye(n);   L = eye(n);
    for i = 1:n
       for j = i+1:n
             U(i,j) = 3-round(6*rand(1,1));
             L(j,i) = 3-round(6*rand(1,1));
       end
    end
    A = U*L;
```

Listing 4

▷ **Exercise 98A** Notice that the inverse of an integer elementary matrix is again an integer elementary matrix in the case of shear and swap matrices. (This is not true for a scale matrix.) Also, the transpose of an elementary matrix is elementary. Use these facts to construct a 3×3 integer matrix, having no zero entries, whose inverse is an integer matrix.

Function 98B The command

```
#>   A=intinv(n)
```

assigns to the variable A a randomly chosen $n \times n$ integer matrix whose inverse is also an integer matrix. (See Listing 4.) This is useful for making up problems that avoid fractions. Exercise 98A shows why this .M function works. (See also 70B.) The .M function intinv uses MINIMAT's built-in round function that rounds to the nearest integer.

▷ **Exercise 98C** Use intinv(n) to create a random 4×4 matrix whose inverse has integer entries. Confirm the result.

3.5   Elementary Column Operations

In this section we formulate and prove the **Fundamental Theorem on Column Operations**. We could do this by mimicking the arguments for the Fundamental Theorem on Row Operations, But it is easier to prove the former from the latter using the transpose operation. The key point is that the transpose operation interchanges rows and columns:

$$\text{row}_p(A^\top) = \text{col}_p(A)^\top.$$

There are three kinds of **elementary column operations** that may be applied to a matrix A. They may be defined by applying the corresponding elementary row operation to the transpose of A and then taking the transpose again:

- **Scale** If $B^\top = \text{Scale}(A^\top, p, c)$, then B results from A by multiplying the pth column by c.

- **Swap** If $B^\top = \text{Swap}(A^\top, p, q)$, then B results from A by exchanging the pth and qth columns.

- **Shear** If $B^\top = \text{Shear}(A^\top, p, q, c)$, then B results from A by adding c times the qth column to the pth column.

Fundamental Theorem for Column Operations

Theorem 99A *The matrix AE that results by multiplying a matrix A on the right by an elementary matrix E is the same as the matrix that results by applying the corresponding elementary column operation to A.*

By taking transposes and using the formula $(AE)^\top = E^\top A^\top$, this theorem can be deduced from the Fundamental Theorem on Row Operations (76B). Here are some examples.

Example 99B Scaling the first column.

$$\begin{bmatrix} a_{11} & a_{12} & a_{13} \\ a_{21} & a_{22} & a_{23} \\ a_{31} & a_{32} & a_{33} \\ a_{41} & a_{42} & a_{43} \end{bmatrix} \begin{bmatrix} c & 0 & 0 \\ 0 & 1 & 0 \\ 0 & 0 & 1 \end{bmatrix} = \begin{bmatrix} ca_{11} & a_{12} & a_{13} \\ ca_{21} & a_{22} & a_{23} \\ ca_{31} & a_{32} & a_{33} \\ ca_{41} & a_{42} & a_{43} \end{bmatrix},$$

Example 100A Swapping columns 2 and 3.

$$
\begin{bmatrix} a_{11} & a_{12} & a_{13} \\ a_{21} & a_{22} & a_{23} \\ a_{31} & a_{32} & a_{33} \\ a_{41} & a_{42} & a_{43} \end{bmatrix}
\begin{bmatrix} 1 & 0 & 0 \\ 0 & 0 & 1 \\ 0 & 1 & 0 \end{bmatrix}
=
\begin{bmatrix} a_{11} & a_{13} & a_{12} \\ a_{21} & a_{23} & a_{22} \\ a_{31} & a_{33} & a_{32} \\ a_{41} & a_{43} & a_{42} \end{bmatrix}
$$

Example 100B Adding c times column 2 to column 3.

$$
\begin{bmatrix} a_{11} & a_{12} & a_{13} \\ a_{21} & a_{22} & a_{23} \\ a_{31} & a_{32} & a_{33} \\ a_{41} & a_{42} & a_{43} \end{bmatrix}
\begin{bmatrix} 1 & 0 & 0 \\ 0 & 1 & c \\ 0 & 0 & 1 \end{bmatrix}
=
\begin{bmatrix} a_{11} & a_{12} & a_{13} + ca_{12} \\ a_{21} & a_{22} & a_{23} + ca_{22} \\ a_{31} & a_{32} & a_{33} + ca_{32} \\ a_{41} & a_{42} & a_{43} + ca_{42} \end{bmatrix} .
$$

3.6   Permutation Matrices

Sometimes we need to permute the rows (or columns) of a matrix. This can be accomplished by multiplying it on the left (or right) by a permutation matrix. Here's the definition.

Definition 100C A **permutation matrix** is a square matrix S that satisfies the following three equivalent conditions:

- All the entries of S are either zero or one and there is exactly one nonzero entry in every row and exactly one nonzero entry in every column;

- The matrix S is obtained from the identity matrix by permuting its rows;

- The matrix S is obtained from the identity matrix by permuting its columns.

Example 100D The matrix $S = \begin{bmatrix} 0 & 1 & 0 \\ 0 & 0 & 1 \\ 1 & 0 & 0 \end{bmatrix}$ is a permutation matrix. It can be obtained from the identity matrix by permuting its rows:

$$
\begin{aligned}
\text{row}_1(S) &= \text{row}_2(I_3) \\
\text{row}_2(S) &= \text{row}_3(I_3) \\
\text{row}_3(S) &= \text{row}_1(I_3).
\end{aligned}
$$

It can also be obtained from the identity matrix by permuting its columns:

$$\begin{aligned}
\text{col}_1(S) &= \text{col}_3(I_3) \\
\text{col}_2(S) &= \text{col}_1(I_3) \\
\text{col}_3(S) &= \text{col}_2(I_3).
\end{aligned}$$

Row Permutation Theorem

Theorem 101A *Suppose that $S \in \mathbf{F}^{m \times m}$ is a permutation matrix. Then for any matrix $A \in \mathbf{F}^{m \times n}$ with m rows, the matrix SA is obtained from the matrix A by permuting its rows.*

Column Permutation Theorem

Theorem 101B *Suppose that $T \in \mathbf{F}^{n \times n}$ is a permutation matrix. Then for any matrix $A \in \mathbf{F}^{m \times n}$ with n columns, the matrix AT is obtained from the matrix A by permuting its columns.*

These two permutation theorems are special cases of the Fundamental Theorem. The Fundamental Theorem says that if a matrix B results from a matrix A by performing a sequence of elementary operations on the rows of X, then $B = SA$ where S is the matrix that results by applying those same operations to the identity matrix. When all these elementary operations are swaps, the matrix S is a permutation matrix. (See Theorem 102B below.) Taking transposes we see that this also works for columns.

Example 101C Take

$$S = \begin{bmatrix} 0 & 1 & 0 \\ 0 & 0 & 1 \\ 1 & 0 & 0 \end{bmatrix}, \quad X = \begin{bmatrix} x_1 & x_2 \\ y_1 & y_2 \\ z_1 & z_2 \end{bmatrix}, \quad Y = \begin{bmatrix} y_1 & y_2 \\ z_1 & z_2 \\ x_1 & x_2 \end{bmatrix}.$$

Then Y results from X by permuting its rows, and S results by permuting the rows of the identity matrix in exactly the same way. Hence $SX = Y$.

Example 102A Take

$$H = \left[\begin{array}{ccc} a_1 & b_1 & c_1 \\ a_2 & b_2 & c_3 \end{array} \right], \ T = \left[\begin{array}{ccc} 0 & 0 & 1 \\ 1 & 0 & 0 \\ 0 & 1 & 0 \end{array} \right], \ K = \left[\begin{array}{ccc} b_2 & c_2 & a_2 \\ b_2 & c_2 & a_2 \end{array} \right].$$

Then K results from H by permuting its columns, and T results by permuting the columns of the identity matrix in exactly the same way. Hence $HT = K$.

Factorization Theorem

Corollary 102B *A matrix is a permutation matrix if and only if it is a product of swap matrices.*

Proof: Let S be a permutation matrix. Transform S to the identity using elementary row operations in the proof of the Factorization Theorem 93B. Only swap operations are required. Any product of swap matrices is a permutation matrix since, according to the Fundamental Theorem 76B, it can be obtained from the identity matrix by permuting the rows.   □

Theorem 102C *The inverse of a permutation matrix is its transpose.*

Proof: For a swap matrix E we have

$$E^\top = E = E^{-1},$$

since swapping two rows of the identity matrix produces the same effect as swapping the corresponding columns, and swapping again reproduces the identity matrix. Hence, if

$$S = E_1 E_2 \cdots E_h$$

is a product of swaps, we have

$$
\begin{aligned}
S^{-1} &= (E_1 E_2 \cdots E_h)^{-1} \\
&= E_h^{-1} \cdots E_2^{-1} E_1^{-1} \\
&= E_h^\top \cdots E_2^\top E_1^\top \\
&= (E_1 E_2 \cdots E_h)^\top \\
&= S^\top
\end{aligned}
$$

as required.
□

Exercise 103A How are A, S, and B related?

$$A = \begin{bmatrix} 1 & 0 & c_1 \\ 0 & 1 & c_2 \\ 0 & 0 & 1 \end{bmatrix}, \quad S = \begin{bmatrix} 0 & 0 & 1 \\ 1 & 0 & 0 \\ 0 & 1 & 0 \end{bmatrix}, \quad B = \begin{bmatrix} 0 & c_1 & 1 \\ 1 & c_2 & 0 \\ 0 & 1 & 0 \end{bmatrix}.$$

Find A^{-1}, S^{-1}, B^{-1}.

Exercise 103B Do Exercise 37E.

Exercise 103C Find $\begin{bmatrix} 0 & 0 & 1 \\ 1 & 0 & 0 \\ 0 & 1 & 0 \end{bmatrix}^{-1}$.

▷ **Exercise 103D** Suppose that

$$A = \begin{bmatrix} a & b & c & d & e & f \end{bmatrix},$$
$$B = \begin{bmatrix} e & d & c & f & b & a \end{bmatrix}, \qquad X = \begin{bmatrix} u \\ v \\ w \\ x \\ y \\ z \end{bmatrix},$$

that $X \neq 0$, and that $AX = 0$. Find $Y \neq 0$ such that $BY = 0$.

3.7 Equivalence

A central strategy in mathematics in general, and this book in particular, is the strategy of solving a problem by transforming it into an equivalent, but easier, problem. For example, this is how we solved the problem of deciding whether a matrix A is invertible. We found an invertible matrix M and a matrix R in RREF with $MA = R$. Then A is invertible if and only if R is, and R is invertible if and only if it has no zero row. The first step in applying this strategy is to define the notion of "equivalent problem".

Equivalence

Definition 104A Let A and B be two matrices of the same size, say $A, B \in \mathbf{F}^{m \times n}$. We say that

- A is **left equivalent** to B iff there is an invertible matrix Q with
$$A = QB.$$

- A is **right equivalent** to B iff there is an invertible matrix P with
$$A = BP^{-1}.$$

- A is **biequivalent** to B iff there are invertible matrices Q and P with
$$A = QBP^{-1}.$$

Remark 104B Left equivalent matrices are trivially biequivalent: we may take P to be the identity. Similarly, right equivalent matrices are biequivalent.

Theorem 104C *All three of these relations satisfy the following:*

(Reflexive Law) $A \equiv A$;

(Symmetric Law) *if $A \equiv B$, then $B \equiv A$;*

(Transitive Law) *if $A \equiv B$ and $B \equiv C$, then $A \equiv C$.*

Here $A \equiv B$ means either "A is left equivalent to B", "A is right equivalent to B", or "A is biequivalent to B".

Proof: We shall prove the theorem for left equivalence and leave the other two cases as exercises.

- $A = I_m A$.

- If $A = QB$, then $B = Q^{-1}A$.

- If $A = Q_1 B$ and $B = Q_2 C$, then $A = (Q_1 Q_2)C$.   □

When mathematicians have some sort of equivalence relation, they look for a **normal form**. The idea is that every object should be equivalent to an object in normal form, and the objects in normal form are easy to understand. In the case of matrices, "easy to understand" means "a lot of zeros". Our next theorem says that reduced row echelon form is a normal form under left equivalence.

Gauss-Jordan Decomposition

Theorem 105A *Any matrix is left equivalent to a unique matrix in reduced row echelon form.*

Proof: In other words any matrix $A \in \mathbf{F}^{m \times n}$ can be written in the form

$$A = QR.$$

where $Q \in \mathbf{F}^{m \times m}$ is invertible, and $R \in \mathbf{F}^{m \times n}$ is in RREF. Moreover, in this decomposition the matrix R is unique: if $\widetilde{Q}\widetilde{R} = QR$, then $\widetilde{R} = R$. The Multiplier Theorem 85B showed how to find M with $MA = R$: take $Q = M^{-1}$. This proves existence. The uniqueness of R is trickier and will be proved later in Theorem 182D. The problem is that the matrix Q (as opposed to R) need *not* be unique. For example, the zero matrix is in RREF, but $Q0 = 0$ for *any* Q.

Biequivalence Normal Form

Definition 105B The matrix

$$D_{m,n,r} = \begin{bmatrix} I_r & 0_{r \times (n-r)} \\ 0_{(m-r) \times r} & 0_{(m-r) \times (n-r)} \end{bmatrix}$$

is called the $m \times n$ rank r matrix in biequivalence normal form. A matrix is said to be in **biequivalence normal form**, abbreviated **BENF**, iff it is equal to some $D_{m,n,r}$.

Example 106A The 3×4 matrix

$$D = D_{3,4,2} = \begin{bmatrix} 1 & 0 & 0 & 0 \\ 0 & 1 & 0 & 0 \\ 0 & 0 & 0 & 0 \end{bmatrix}$$

is in biequivalence normal form. The identity matrix

$$D_{n,n,n} = I_n$$

is also in biequivalence normal form.

▷ **Question 106B** When is $D_{m,n,r}$ invertible?

Lemma 106C *Suppose $R \in \mathbf{F}^{m \times n}$ is in RREF with r nonzero rows. Then R is right equivalent to $D_{m,n,r}$.*

Proof: By the Column Permutation Theorem 101B, we have

$$RT = \begin{bmatrix} I_r & C \\ 0_{(m-r) \times r} & 0_{(m-r) \times (n-r)} \end{bmatrix}$$

where T is a permutation matrix and $C \in \mathbf{F}^{r \times (n-r)}$. By the block multiplication law 51B, write $RTP_1 = D$ with $D = D_{m,n,r}$ as above and

$$P_1^{-1} = \begin{bmatrix} I_r & C \\ 0_{(n-r) \times r} & I_{n-r} \end{bmatrix}, \quad P_1 = \begin{bmatrix} I_r & -C \\ 0_{(n-r) \times r} & I_{n-r} \end{bmatrix}.$$

(The formula for the inverse is the block multiplication law 52B.) Now take $P = TP_1$ so $RP = D$ or $R = PD^{-1}$ as required. To test your understanding of this proof, do Exercise 107F below.   □

Biequivalence Decomposition

Theorem 106D *Every matrix is biequivalent to a unique matrix in biequivalence normal form.*

Proof: In other words any matrix $A \in \mathbf{F}^{m \times n}$ may be written in the form

$$A = QDP^{-1}$$

where $Q \in \mathbf{F}^{m \times m}$ and $P \in \mathbf{F}^{n \times n}$ are invertible and $D = D_{m,n,r}$ for some r. Moreover, the D in this decomposition is unique: if $QDP^{-1} = \tilde{Q}\tilde{D}\tilde{P}^{-1}$ with $D = D_{m,n,r}$ and $\tilde{D} = D_{m,n,\tilde{r}}$, then $D = \tilde{D}$. The uniqueness will be proved later. (See Theorem 171D and Example 171B.) Existence is proved as follows: First, write $A = QR$ as in Theorem 105A. Then, write $RP = D$ in Lemma 106C. It follows that $A = QR = QDP^{-1}$ as required.   □

▷ **Exercise 107A** Suppose that $A_2 = Q_1 A_1 = Q_2 A_3$, that $A_4 = Q_3 A_3$, and that Q_1, Q_2, Q_3 are invertible. Find a matrix Q such that $A_4 = QA_1$.

Exercise 107B Write each of the following matrices A in the form $A = QDP^{-1}$ where Q and P are invertible and $D = D_{m,n,r}$. Check your answers by comparing A and QDP^{-1}.

(1) $A = \begin{bmatrix} 1 & 0 & a & b \\ 0 & 1 & c & d \end{bmatrix}$,

(3) $A = \begin{bmatrix} 1 & 0 & a & b \\ 0 & 1 & c & d \\ 1 & 1 & a+c & b+d \end{bmatrix}$,

(2) $A = \begin{bmatrix} 2 & 1 & a & b \\ 1 & 1 & c & d \end{bmatrix}$,

(4) $A = \begin{bmatrix} 2 & 1 & a & b \\ 1 & 1 & c & d \\ 3 & 2 & a+c & b+d \end{bmatrix}$.

Exercise 107C Repeat the previous exercise with $A^\top$ in place of A.

▷ **Exercise 107D** For each of the matrices A in Exercise 88C find invertible matrices Q and P and a matrix $D = D_{m,n,r}$ such that $A = QDP^{-1}$. Check your answers by comparing A and QDP^{-1}.

▷ **Exercise 107E** For each of the matrices A in Exercise 88E find invertible matrices Q^{-1} and P so that $Q^{-1}AP$ is in biequivalence normal form.

▷ **Exercise 107F** Find an invertible matrix P so that RP is in biequivalence normal form where

$$R = \begin{bmatrix} 1 & c_{11} & 0 & c_{12} & 0 & c_{13} & c_{14} \\ 0 & 0 & 1 & c_{22} & 0 & c_{23} & c_{24} \\ 0 & 0 & 0 & 0 & 1 & c_{33} & c_{34} \\ 0 & 0 & 0 & 0 & 0 & 0 & 0 \end{bmatrix}$$

Hint: Imitate the proof of Lemma 106C. First, find a permutation matrix T such that

$$RT = \begin{bmatrix} I & C \\ 0 & 0 \end{bmatrix}.$$

Then, take

$$P_1 = \begin{bmatrix} I & -C \\ 0 & I \end{bmatrix}, \qquad \text{so} \qquad RTP_1 = \begin{bmatrix} I & 0 \\ 0 & 0 \end{bmatrix}.$$

The answer will be $P = TP_1$.

Using MINIMAT

These commands produce a biequivalence decomposition $A = QDP^{-1}$ of the matrix A:

```
[M, R]= gjm(A);    Q= M^(-1);
[N, T]= gjm(R');   P=N'; D=T';
```

The point is that if $MA = R$, then $A = QR$ with $Q = M^{-1}$, so we can find Q with the gjm and inv functions. We can then factor $R = DP^{-1}$ by taking transposes and using the gjm function again. This works because the RREF of R^T is D^T. Thus to find P we find N solving $NR^\mathsf{T} = D^\mathsf{T}$ and then take $P = N^\mathsf{T}$. Remember that the mathematical notation for the transpose of A is A^T; MINIMAT's notation for the transpose of A is A' for real matrices and A.' for complex matrices.

▷ **Exercise 108A** In each of the following problems construct the matrix A as indicated and use MINIMAT to solve $A = QDP^{-1}$ as in the Biequivalence Decomposition. Predict the value of D before you solve the problem.
(1)   A=rand(3,5)
(2)   A=rand(5,3)
(3)   A=rand(3,6), A=[A; 2*A(1,:)+7*A(2,:)]
(4)   B=zeros(3,5), B(1,1)=1, B(2,2)=1
(5)   A=rand(3,3)*B*rand(5,5)

▷ **Exercise 108B** What is the result of the following?

```
#> A=rand(3,5), R=gj(A), B=rand(3,3)*A, S=gj(B),
#> R-S
```

Function 108C The .M function bed, shown in Listing 5, calculates the Biequivalence Decomposition. The command

```
#> [Q, D, P] = bed(A)
```

produces invertible matrices Q and P and a matrix D in biequivalence normal form such that the subsequent command #> Q*D*P^(-1),A   produces two copies of the same matrix. (Try it!)

Biequivalence Decomposition in MINIMAT

```
function [Q, D, P] = bed(A)

[m n] = size(A);  r = 0;
D = A; Q = eye(m); P = eye(n);
% Invariant relation:   A = Q*D*P^(-1)
while (r<m & r<n)
   y = 1.0E-9;
   for i=r+1:m for j=r+1:n
      if (abs(D(i,j)) >  y)
            y = abs(D(i,j)); h=i; k=j;
      end
   end; end
   if  y == 1.0E-9;  return; else r=r+1; end
   E = swap(eye(m),r,h);               % swap rows
   D = E*D;  Q = Q*E^(-1);
   F = swap(eye(n),r,k);               % swap cols
   D = D*F; P = P*F;
   E = scale(eye(m),r,1/D(r,r));       % scale row
   D = E*D; Q = Q*E^(-1);
   for i = r+1:m
      E = shear(eye(m), i, r,-D(i,r)); % shear rows
      D = E*D; Q = Q*E^(-1);
   end
   for j = r+1:n
      F = shear(eye(n),r,j, -D(r,j));  % shear cols
      D = D*F;  P = P*F;
   end
end % while
```

Listing 5

The implementation of **bed** shown in Listing 5 does not use the .M function **gjm**. Instead, it illustrates how to use an **invariant relation** to design a program. At every stage in the program the relation

$$A = QDP^{-1} \tag{$*$}$$

is true. At the beginning of the program this is trivially true since $D = A$ and Q and P are identity matrices. Every time we replace D by ED, we also replace Q by QE^{-1}, so that $(*)$ continues to hold. Similarly, every time we replace D by DF, we also replace P by PF, and again $(*)$ is preserved. The main loop terminates when D is in biequivalence normal form. The program is much like the **gj** function in Listing 2 on Page 91.

Exercise 110A The 24 matrices on Pages 111-112 were generated by MINIMAT. In each of the four sets the matrices Q and P are invertible with the indicated inverses and are related to the matrices A and D via the formula

$$A = QDP^{-1}.$$

Confirm this by computing (by hand) several entries in each of the matrices QQ^{-1}, PP^{-1}, QDP^{-1}. (These matrices will be used in later exercises.)

---

(1)

$$Q = \begin{bmatrix} 4 & -3 & -2 \\ -5 & 5 & 2 \\ -2 & 2 & 1 \end{bmatrix}, \qquad P = \begin{bmatrix} -3 & 3 & -4 & -2 & 0 \\ 0 & 1 & 0 & 2 & 0 \\ 3 & -3 & 3 & 2 & 0 \\ -2 & 2 & 4 & 4 & -3 \\ 1 & -1 & -1 & -1 & 1 \end{bmatrix},$$

$$Q^{-1} = \begin{bmatrix} 1 & -1 & 4 \\ 1 & 0 & 2 \\ 0 & -2 & 5 \end{bmatrix}, \qquad P^{-1} = \begin{bmatrix} 1 & 1 & 4 & -8 & -24 \\ 0 & 1 & 2 & -6 & -18 \\ -1 & 0 & -1 & 0 & 0 \\ 0 & 0 & -1 & 3 & 9 \\ -2 & 0 & -4 & 5 & 16 \end{bmatrix},$$

$$D = \begin{bmatrix} 1 & 0 & 0 & 0 & 0 \\ 0 & 1 & 0 & 0 & 0 \\ 0 & 0 & 1 & 0 & 0 \end{bmatrix}, \qquad A = \begin{bmatrix} 6 & 1 & 12 & -14 & -42 \\ -7 & 0 & -12 & 10 & 30 \\ -3 & 0 & -5 & 4 & 12 \end{bmatrix}.$$

(2)

$$Q = \begin{bmatrix} -10 & -4 & 0 & 4 & 3 \\ 11 & -3 & 1 & -2 & 1 \\ -5 & 6 & -2 & 0 & -3 \\ -4 & -1 & -1 & 3 & 2 \\ -1 & -1 & 0 & 1 & 1 \end{bmatrix}, \qquad P = \begin{bmatrix} -3 & -4 & -1 \\ 2 & 2 & 1 \\ 1 & 1 & 1 \end{bmatrix},$$

$$Q^{-1} = \begin{bmatrix} 1 & 2 & 3 & -4 & 12 \\ -2 & -3 & -4 & 5 & -13 \\ 2 & 4 & 7 & -11 & 33 \\ 6 & 11 & 17 & -23 & 68 \\ -7 & -12 & -18 & 24 & -68 \end{bmatrix}, \qquad P^{-1} = \begin{bmatrix} 1 & 3 & -2 \\ -1 & -2 & 1 \\ 0 & -1 & 2 \end{bmatrix},$$

$$D = \begin{bmatrix} 1 & 0 & 0 \\ 0 & 1 & 0 \\ 0 & 0 & 1 \\ 0 & 0 & 0 \\ 0 & 0 & 0 \end{bmatrix}, \qquad A = \begin{bmatrix} -6 & -22 & 16 \\ 14 & 38 & -23 \\ -11 & -25 & 12 \\ -3 & -9 & 5 \\ 0 & -1 & 1 \end{bmatrix}.$$

(3)

$$Q = \begin{bmatrix} 4 & -3 & -2 \\ -5 & 5 & 2 \\ -2 & 2 & 1 \end{bmatrix}, \qquad P = \begin{bmatrix} -3 & 3 & -4 & -2 & 0 \\ 0 & 1 & 0 & 2 & 0 \\ 3 & -3 & 3 & 2 & 0 \\ -2 & 2 & 4 & 4 & -3 \\ 1 & -1 & -1 & -1 & 1 \end{bmatrix},$$

$$Q^{-1} = \begin{bmatrix} 1 & -1 & 4 \\ 1 & 0 & 2 \\ 0 & -2 & 5 \end{bmatrix}, \qquad P^{-1} = \begin{bmatrix} 1 & 1 & 4 & -8 & -24 \\ 0 & 1 & 2 & -6 & -18 \\ -1 & 0 & -1 & 0 & 0 \\ 0 & 0 & -1 & 3 & 9 \\ -2 & 0 & -4 & 5 & 16 \end{bmatrix},$$

$$D = \begin{bmatrix} 1 & 0 & 0 & 0 & 0 \\ 0 & 1 & 0 & 0 & 0 \\ 0 & 0 & 0 & 0 & 0 \end{bmatrix}, \qquad A = \begin{bmatrix} 4 & 1 & 10 & -14 & -42 \\ -5 & 0 & -10 & 10 & 30 \\ -2 & 0 & -4 & 4 & 12 \end{bmatrix}.$$

(4)

$$Q = \begin{bmatrix} -10 & -4 & 0 & 4 & 3 \\ 11 & -3 & 1 & -2 & 1 \\ -5 & 6 & -2 & 0 & -3 \\ -4 & -1 & -1 & 3 & 2 \\ -1 & -1 & 0 & 1 & 1 \end{bmatrix}, \qquad P = \begin{bmatrix} -3 & -4 & -1 \\ 2 & 2 & 1 \\ 1 & 1 & 1 \end{bmatrix},$$

$$\mathbf{Q}^{-1} = \begin{bmatrix} 1 & 2 & 3 & -4 & 12 \\ -2 & -3 & -4 & 5 & -13 \\ 2 & 4 & 7 & -11 & 33 \\ 6 & 11 & 17 & -23 & 68 \\ -7 & -12 & -18 & 24 & -68 \end{bmatrix}, \qquad P^{-1} = \begin{bmatrix} 1 & 3 & -2 \\ -1 & -2 & 1 \\ 0 & -1 & 2 \end{bmatrix},$$

$$D = \begin{bmatrix} 1 & 0 & 0 \\ 0 & 1 & 0 \\ 0 & 0 & 0 \\ 0 & 0 & 0 \\ 0 & 0 & 0 \end{bmatrix}, \qquad A = \begin{bmatrix} -6 & -22 & 16 \\ 14 & 39 & -25 \\ -11 & -27 & 16 \\ -3 & -10 & 7 \\ 0 & -1 & 1 \end{bmatrix}.$$

Chapter Summary

Row Multiplier Theorem

If a matrix B results from a matrix A by elementary row operations and the matrix M results from the identity matrix by these same operations, then $B = MA$.

Column Multiplier Theorem

If B results from A by elementary column operations and the matrix P results from the identity matrix by these same operations, then $B = AP$.

Gauss-Jordan Decomposition

Every matrix is left equivalent to a unique matrix in reduced row echelon form.

Biequivalence Decomposition

Every matrix is biequivalent to a unique matrix in biequivalence normal form.

4

SUBSPACES

This chapter introduces some concepts that are useful for describing linear systems of equations. A system of m linear equations in n unknowns can be written as a single matrix equation

$$AX = Y$$

where $A \in \mathbf{F}^{m \times n}$, $X \in \mathbf{F}^{n \times 1}$, and $Y \in \mathbf{F}^{m \times 1}$. Here the entries of X are the unknowns, and each row of A corresponds to one equation.

There are two basic problems associated to the system $AX = Y$. They are the problems of

- **Existence.** *Does the system $AX = Y$ have any solutions X?*

- **Parameterization.** *If so, describe the set of all solutions.*

Our first step in understanding these problems will be to reformulate them in the language of set theory. We will define the notion of *subspace*. To each matrix A we will associate two subspaces: its null space $\mathcal{N}(A)$ and its range $\mathcal{R}(A)$. The null space $\mathcal{N}(A)$ of A is the set of all solutions X of the system $AX = 0$. The range $\mathcal{R}(A)$ is the set of all Y for which the system $AX = Y$ has a solution X. The existence problem asks whether Y is an element of $\mathcal{R}(A)$. The parameterization problem, at least in the case $Y = 0$, requires us to describe the subspace $\mathcal{N}(A)$.

This chapter uses some elementary set-theoretic notations, namely

- $X \in V$ (X is an element of the set V),

- $V \subset W$ (the set V is a subset of the set W), and

- $V = W$ (the sets V and W are equal).

Reread Section 1.4 if you are uncomfortable with these notations.

4.1   Linear Systems

Matrix theory provides a convenient notation for describing systems of linear equations in several unknowns. The system of m equations in n unknowns

$$
\begin{aligned}
a_{11}x_1 + a_{12}x_2 + \cdots + a_{1n}x_n &= y_1 \\
a_{21}x_1 + a_{22}x_2 + \cdots + a_{2n}x_n &= y_2 \\
&\vdots \\
a_{m1}x_1 + a_{m2}x_2 + \cdots + a_{mn}x_n &= y_m
\end{aligned}
$$

(here the unknowns are $x_1, x_2, \ldots, x_n$) can be written more succinctly as a single matrix equation

$$ AX = Y $$

where

$$
A = \begin{bmatrix}
a_{11} & a_{12} & \cdots & a_{1n} \\
a_{21} & a_{22} & \cdots & a_{2n} \\
\vdots & \vdots & \ddots & \vdots \\
a_{m1} & a_{m2} & \cdots & a_{mn}
\end{bmatrix}, \qquad
X = \begin{bmatrix} x_1 \\ x_2 \\ \vdots \\ x_n \end{bmatrix}, \qquad
Y = \begin{bmatrix} y_1 \\ y_2 \\ \vdots \\ y_m \end{bmatrix}.
$$

The matrix A is called the **matrix of coefficients**, the vector Y is called the **inhomogeneous term**, and the vector X is called the **column of unknowns**. In a particular problem, the matrix of coefficients A and the inhomogeneous term Y will be *given*; the problem is to find the solutions X. The system is called **homogeneous** when the inhomogeneous term Y is zero, that is, when $X = 0$ is a solution. Otherwise, it is called **inhomogeneous**.

For a system $AX = Y$ where A is invertible, there is exactly one solution X. We can find it by multiplying by A^{-1}:

$$ AX = Y \iff X = A^{-1}AX = A^{-1}Y. $$

This won't always work because there are lots of matrices that are not invertible. In particular, it will *never* work for systems where the coefficient matrix A is not square: such a matrix is not invertible by definition. The coefficient matrix is not square when the number n of unknowns is different from the number m of equations.

When the coefficient matrix is not square *either*

- there are more equations than unknowns $(m > n)$, in which case the system is called **overdetermined** *or else*

- there are are more unknowns than equations $(n > m)$, in which case the system is called **underdetermined**.

In the overdetermined case, it usually (but not always) happens that the system is *inconsistent,* meaning that the equations contradict one another For example, we might solve the first n equations and discover that the result doesn't satisfy the remaining $m-n$ equations. In the underdetermined case, it usually (but not always) happens that we can assign $n - m$ of the unknowns arbitrary values thus producing m equations in the remaining m unknowns. We can then solve these m equations uniquely. The rest of this chapter makes these ideas precise. In this section we give some examples that illustrate what can happen.

Here's the general strategy for solving the linear system. We have learned how to transform the matrix A to RREF (reduced row echelon form): Theorem 85B says there is an invertible matrix $M \in \mathbf{F}^{m \times m}$ and a matrix $R \in \mathbf{F}^{m \times n}$ in RREF such that $MA = R$. (Using the Multiplier Trick 87A we learned how to compute M and R at the same time.) In this situation the inhomogeneous systems $AX = Y$ and $RX = MY$ have exactly the same solutions:

If $MA = R$ where M is invertible, then

$$AX = Y \iff RX = MY.$$

Here's a proof of the statement in the last box:

$$AX = Y \implies RX = MAX = MY$$

(multiply by M) and

$$RX = MY \implies AX = M^{-1}RX = M^{-1}MY = Y$$

(multiply by M^{-1}). In other words.

The process of multiplying by M transforms the system $AX = Y$
to an equivalent system $RX = MY$. This latter system is easy to
understand because R is so simple.

Now we give a few examples. We haven't shown the calculation to find
M and R. You can do that yourself either by hand (using the Multiplier
Trick) or with MINIMAT's gjm function.

Example 118A *An inhomogeneous system with no solution.* The system

$$
\begin{aligned}
2x_1 + x_2 &= 4 \\
x_1 + x_2 &= 3 \\
3x_1 + 2x_2 &= 9
\end{aligned}
$$

is overdetermined since there are 3 equations in 2 unknowns. It has the
matrix form $AX = Y_1$ where

$$
A = \begin{bmatrix} 2 & 1 \\ 1 & 1 \\ 3 & 2 \end{bmatrix}, \qquad
X = \begin{bmatrix} x_1 \\ x_2 \end{bmatrix}, \qquad
Y_1 = \begin{bmatrix} 4 \\ 3 \\ 9 \end{bmatrix}.
$$

Now $MA = R$ where

$$
M = \begin{bmatrix} 1 & -1 & 0 \\ -1 & 2 & 0 \\ -1 & -1 & 1 \end{bmatrix}, \qquad
R = \begin{bmatrix} 1 & 0 \\ 0 & 1 \\ 0 & 0 \end{bmatrix}, \qquad
MY_1 = \begin{bmatrix} 1 \\ 2 \\ 2 \end{bmatrix},
$$

and the equation $RX = MY_1$ is

$$
\begin{aligned}
1x_1 + 0x_2 &= 1 \\
0x_1 + 1x_2 &= 2 \\
0x_1 + 0x_2 &= 2
\end{aligned}
$$

so there is no solution since $0 \neq 2$. The equations contradict one another.

Example 118B *An inhomogeneous system with a unique solution.* The
system

$$
\begin{aligned}
2x_1 + x_2 &= 4 \\
x_1 + x_2 &= 3 \\
3x_1 + 2x_2 &= 7
\end{aligned}
$$

has the form $AX = Y_2$ where A, X, M, R are as in Example 118A, and

$$Y_2 = \begin{bmatrix} 4 \\ 3 \\ 7 \end{bmatrix}, \qquad MY_2 = \begin{bmatrix} 1 \\ 2 \\ 0 \end{bmatrix}.$$

The equation $RX = MY_2$ is

$$
\begin{aligned}
1x_1 + 0x_2 &= 1 \\
0x_1 + 1x_2 &= 2 \\
0x_1 + 0x_2 &= 0
\end{aligned}
$$

so there is exactly one solution $x_1 = 1$ and $x_2 = 2$. (As in Example 118A, this system is overdetermined, but in this case the calculation shows that the equations do not contradict one another.)

Example 119A *Whether or not an overdetermined inhomogeneous system $AX = Y$ has a solution X depends on the inhomogeneous term Y. For example, the system*

$$
\begin{aligned}
2x_1 + x_2 &= y_1 \\
x_1 + x_2 &= y_2 \\
3x_1 + 2x_2 &= y_3
\end{aligned}
$$

has the form $AX = Y$ with A, X, M, R as in Examples 118A and 118B, and

$$Y = \begin{bmatrix} y_1 \\ y_2 \\ y_3 \end{bmatrix}, \qquad MY = \begin{bmatrix} y_1 - y_2 \\ -y_1 + 2y_2 \\ -y_1 - y_2 + y_3 \end{bmatrix}.$$

The equation $RX = MY$ is

$$
\begin{aligned}
1x_1 + 0x_2 &= y_1 - y_2 \\
0x_1 + 1x_2 &= -y_1 + 2y_2 \\
0x_1 + 0x_2 &= -y_1 - y_2 + y_3
\end{aligned}
$$

so there is a solution if and only if Y satisfies

$$-y_1 - y_2 + y_3 = 0. \qquad (*)$$

When equation $(*)$ holds there is exactly one solution X.

Example 120A *An underdetermined homogeneous system has infinitely many solutions.* For example, the system

$$x_1 + x_2 - x_3 - x_4 = 0$$
$$x_1 + 2x_2 + 4x_3 + 5x_4 = 0$$

is underdetermined since there are 2 equations in 4 unknowns. It has the matrix form $AX = 0$ where

$$A = \begin{bmatrix} 1 & 1 & -1 & -1 \\ 1 & 2 & 4 & 5 \end{bmatrix}, \qquad X = \begin{bmatrix} x_1 \\ x_2 \\ x_3 \\ x_4 \end{bmatrix}.$$

Then $MA = R$ where

$$M = \begin{bmatrix} 2 & -1 \\ -1 & 1 \end{bmatrix}, \qquad R = \begin{bmatrix} 1 & 0 & -6 & -7 \\ 0 & 1 & 5 & 6 \end{bmatrix},$$

and the equivalent system $RX = MAX = 0$ has the form

$$x_1 - 6x_3 - 7x_4 = 0$$
$$x_2 + 5x_3 + 6x_4 = 0.$$

Notice how the unknowns x_1 and x_2 that correspond to the leading columns (that is, to the first two columns) each appear once in different equations. This is no accident since R has the form

$$R = \begin{bmatrix} I_2 & C \end{bmatrix}, \qquad C = \begin{bmatrix} -6 & -7 \\ 5 & 6 \end{bmatrix}.$$

This means that the unknowns x_3 and x_4 corresponding to the free columns may be assigned arbitrary values that then determine the values of x_1 and x_2. Thus solutions are not unique: there are infinitely many. Two interesting solutions of $RX = 0$ are

$$X_1 = \begin{bmatrix} 6 \\ -5 \\ 1 \\ 0 \end{bmatrix}, \qquad X_2 = \begin{bmatrix} 7 \\ -6 \\ 0 \\ 1 \end{bmatrix}.$$

Each is obtained by assigning the value 1 to one of the free unknowns and the value 0 to the other. The general solution is a linear combination of these two solutions. It is

$$X = \begin{bmatrix} 6x_3 + 7x_4 \\ -5x_3 - 6x_4 \\ x_3 \\ x_4 \end{bmatrix} = x_3 X_1 + x_4 X_2$$

where the free unknowns x_3 and x_4 are arbitrary.

Remark 121A Recall from Definition 82A that for a matrix R in RREF, the columns that hold the leading entries are called *leading* and the others are called *free*. We also refer to the corresponding unknowns $x_j = \text{entry}_j(X)$ as leading or free. Example 120A shows the origin of this terminology. In the general solution the free unknowns take arbitrary values and these values then determine the values of leading unknowns.

Example 121B *An underdetermined inhomogeneous system need not have solutions.* For example, the system

$$1x_1 + 2x_2 + 3x_3 + 4x_4 = y_1$$
$$2x_1 + 4x_2 + 6x_3 + 8x_4 = y_2$$

has the matrix from $Y = AX$ where

$$A = \begin{bmatrix} 1 & 2 & 3 & 4 \\ 2 & 4 & 6 & 8 \end{bmatrix}, \quad X = \begin{bmatrix} x_1 \\ x_2 \\ x_3 \\ x_4 \end{bmatrix}, \quad Y = \begin{bmatrix} y_1 \\ y_2 \end{bmatrix}.$$

Then $MA = R$ where

$$M = \begin{bmatrix} 1 & 0 \\ -2 & 1 \end{bmatrix}, \quad R = \begin{bmatrix} 1 & 2 & 3 & 4 \\ 0 & 0 & 0 & 0 \end{bmatrix}$$

and the equations $RX = MY$ are

$$1x_1 + 2x_2 + 3x_3 + 4x_4 = y_1$$
$$0x_1 + 0x_2 + 0x_3 + 0x_4 = -2y_1 + y_2.$$

We see that system has solutions X when $-2y_1 + y_2 = 0$, and that in this case, the most general solution is given by

$$X = X_0 + x_2 X_1 + x_3 X_2 + x_4 X_3$$

where x_2, x_3, x_4 are arbitrary and

$$X_0 = \begin{bmatrix} y_1 \\ 0 \\ 0 \\ 0 \end{bmatrix}, \quad X_1 = \begin{bmatrix} -2 \\ 1 \\ 0 \\ 0 \end{bmatrix}, \quad X_2 = \begin{bmatrix} -3 \\ 0 \\ 1 \\ 0 \end{bmatrix}, \quad X_3 = \begin{bmatrix} -4 \\ 0 \\ 0 \\ 1 \end{bmatrix}.$$

Note that X_0 is a solution of the inhomogeneous system $RX = MY$, and that each of the columns of X_1, X_2, X_3 is obtained by finding the solution X of the homogeneous system $RX = 0$ where one of the free unknowns is 1 and the others are 0.

▷ **Exercise 122A** Find a solution X of the system $AX = Y$ if

$$M = \begin{bmatrix} 1 & -1 & 4 \\ 1 & 0 & 2 \\ 0 & -2 & 5 \end{bmatrix}, \quad MA = \begin{bmatrix} 1 & 0 & 2 \\ 0 & 1 & 3 \\ 0 & 0 & 0 \end{bmatrix}, \quad Y = \begin{bmatrix} -2 \\ 5 \\ 2 \end{bmatrix}.$$

Show that there is no solution if the last entry of Y is changed from 2 to 0.

Exercise 122B In each of the examples in this section multiply M and A and verify that $MA = R$.

▷ **Exercise 122C** For each of the matrices A in Exercise 88E and each of the vectors

$$Y_1 = \begin{bmatrix} -2 \\ 5 \\ 2 \end{bmatrix}, \quad Y_2 = \begin{bmatrix} -2 \\ 5 \\ 0 \end{bmatrix}, \quad Y_3 = \begin{bmatrix} -2 \\ 5 \\ 5 \end{bmatrix},$$

say whether or not the system $Y_i = AX$ has a solution X.

▷ **Exercise 122D** Take $Y = 0$ in Example 121B. Find the solution X such that $x_3 = 3$ and $x_4 = 4$.

▷ **Exercise 122E** For the matrix of coefficients A of Example 121B, give an example of an inhomogeneous term Y_1 for which there is no solution X, and an example of another inhomogeneous term Y_2 for which there are infinitely many solutions X. Is there an inhomogeneous term Y_3 for which there is exactly one solution X?

▷ **Exercise 122F** Discuss the system $AX = Y$ given by

$$\begin{aligned} x_1 + x_2 - x_3 - x_4 &= y_1 \\ x_1 + 2x_2 + 4x_3 + 5x_4 &= y_2 \\ x_1 - x_2 - 11x_3 - 13x_4 &= y_3 \end{aligned}$$

following the model of the Example 121B.

Exercise 122G Redo all the exercises in Section 1.7 in the style of this section.

Using MINIMAT

If the matrix A is invertible, then the command

```
#> X=A^(-1)*Y
```

assigns to X the unique solution of the system $Y = AX$. The gjm function (Function 92B) can be used in case A is not invertible. Execute the command

```
#> [M, R]=gjm(A); R, M*Y
```

and examine the values of R and M*Y. If some row of R vanishes identically while the corresponding entry of M*Y does not vanish, then there is no solution X for the system $Y = AX$. In the contrary case, one solves the system $Y = AX$ by solving the equivalent system $MY = RX$.

Exercise 123A Redo the examples in the Section 4.1 using MINIMAT. In any case where the inhomogeneous term Y is arbitrary, use MINIMAT's built-in rand function to choose a value.

4.2   Null Space and Range

A matrix $A \in \mathbf{F}^{m \times n}$ determines two sets, its *null space* $\mathcal{N}(A) \subset \mathbf{F}^{n \times 1}$ and its *range* $\mathcal{R}(A) \subset \mathbf{F}^{m \times 1}$. These sets are useful for discussing properties of linear systems.

Null Space of a Matrix

Definition 123B The **null space** of a matrix $A \in \mathbf{F}^{m \times n}$ is the set

$$\mathcal{N}(A) = \{X \in \mathbf{F}^{n \times 1} : AX = 0\}$$

of all solutions X of the homogeneous system $AX = 0$.

The notation means that to decide if $X \in \mathcal{N}(A)$, we simply check whether $AX = 0$. In other words, for $X \in \mathbf{F}^{n \times 1}$ we have

$$X \in \mathcal{N}(A) \iff AX = 0.$$

Example 124A Let

$$A = \begin{bmatrix} 1 & 2 & 3 \end{bmatrix}, \qquad X_1 = \begin{bmatrix} 2 \\ -1 \\ 0 \end{bmatrix}, \qquad X_2 = \begin{bmatrix} 1 \\ 2 \\ 3 \end{bmatrix},$$

so that $AX_1 = 0$ and $AX_2 = 14$. Thus $X_1 \in \mathcal{N}(A)$ and $X_2 \notin \mathcal{N}(A)$.

▷ **Question 124B** What is the null space of the zero matrix? of the identity matrix?

Range of a Matrix

Definition 124C The **range** of a matrix $A \in \mathbf{F}^{m \times n}$ is the set of all inhomogeneous terms $Y \in \mathbf{F}^{m \times 1}$ for which the inhomogeneous system $Y = AX$ has at least one solution X. In set-theoretic notation

$$\mathcal{R}(A) = \{AX : X \in \mathbf{F}^{n \times 1}\}.$$

The notation means that to decide if $Y \in \mathcal{R}(A)$, we decide whether the inhomogeneous system $AX = Y$ has a solution X. In other words, for $Y \in \mathbf{F}^{m \times 1}$ we have

$$Y \in \mathcal{R}(A) \iff Y = AX \text{ for some } X \in \mathbf{F}^{n \times 1}.$$

Example 124D Let $A = \begin{bmatrix} 2 & 1 \\ 1 & 1 \\ 3 & 2 \end{bmatrix}$, $Y_1 = \begin{bmatrix} 4 \\ 3 \\ 9 \end{bmatrix}$, $Y_2 = \begin{bmatrix} 4 \\ 3 \\ 7 \end{bmatrix}$.

To decide if $Y_1 \in \mathcal{R}(A)$, we attempt to solve the inhomogeneous system $Y_1 = AX$ for X. This leads to a system of three inhomogeneous equations in two unknowns:

$$\begin{aligned} 2x_1 + x_2 &= 4 \\ x_1 + x_2 &= 3 \\ 3x_1 + 2x_2 &= 9. \end{aligned}$$

This is the system studied in Example 118A. There we found that there is no solution X, so we conclude that $Y_1 \notin \mathcal{R}(A)$. On the other hand, the system $Y_2 = AX$ has the solution $X = \begin{bmatrix} x_1 \\ x_2 \end{bmatrix} = \begin{bmatrix} 1 \\ 2 \end{bmatrix}$ (see Example 118B), so we conclude that $Y_2 \in \mathcal{R}(A)$.

▷ **Question 125A** What is the range of the zero matrix? of the identity matrix?

Example 125B If $D = \begin{bmatrix} 1 & 0 & 0 & 0 \\ 0 & 1 & 0 & 0 \\ 0 & 0 & 0 & 0 \end{bmatrix}$, $X = \begin{bmatrix} x_1 \\ x_2 \\ x_3 \\ x_4 \end{bmatrix}$, $Y = \begin{bmatrix} y_1 \\ y_2 \\ y_3 \end{bmatrix}$,

the equation $DX = Y$ takes the form

$$\begin{bmatrix} 1 & 0 & 0 & 0 \\ 0 & 1 & 0 & 0 \\ 0 & 0 & 0 & 0 \end{bmatrix} \begin{bmatrix} x_1 \\ x_2 \\ x_3 \\ x_4 \end{bmatrix} = \begin{bmatrix} x_1 \\ x_2 \\ 0 \end{bmatrix} = \begin{bmatrix} y_1 \\ y_2 \\ y_3 \end{bmatrix},$$

so that $X \in \mathcal{N}(D) \iff x_1 = x_2 = 0$ and $Y \in \mathcal{R}(D) \iff y_3 = 0$. For example, if

$$X_1 = \begin{bmatrix} 0 \\ 0 \\ 2 \\ 0 \end{bmatrix}, \quad X_2 = \begin{bmatrix} 0 \\ 2 \\ 2 \\ 0 \end{bmatrix}, \quad Y_1 = \begin{bmatrix} 0 \\ 2 \\ 0 \end{bmatrix}, \quad Y_2 = \begin{bmatrix} 0 \\ 2 \\ 3 \end{bmatrix},$$

then $X_1 \in \mathcal{N}(D)$, $X_2 \notin \mathcal{N}(D)$, $Y_1 \in \mathcal{R}(D)$, $Y_2 \notin \mathcal{R}(D)$.

Example 125C More generally, for a matrix

$$D = D_{m,n,r} = \begin{bmatrix} I_r & 0_{r \times (n-r)} \\ 0_{(m-r) \times r} & 0_{(m-r) \times (n-r)} \end{bmatrix}$$

in biequivalence normal form, the null space $\mathcal{N}(D)$ consists of all $X \in \mathbf{F}^{n \times 1}$ whose first r entries vanish, and the range $\mathcal{R}(D)$ consists of all $Y \in \mathbf{F}^{m \times 1}$ whose last $m - r$ entries vanish:

$$X \in \mathcal{N}(D) \iff X = \begin{bmatrix} 0_{r \times 1} \\ Z \end{bmatrix} \qquad \text{for some } Z \in \mathbf{F}^{(n-r) \times 1},$$

and

$$Y \in \mathcal{R}(D) \iff Y = \begin{bmatrix} W \\ 0_{(m-r) \times 1} \end{bmatrix} \qquad \text{for some } W \in \mathbf{F}^{r \times 1}.$$

Remark 125D Corollary 94A may be expressed as follows: a square matrix A is invertible if and only if $\mathcal{N}(A) = \{0\}$.

▷ **Question 125E** True or False? If the null space of a square matrix is nonempty, then the matrix is not invertible.

Remark 126A The block multiplication law 53B

$$AX = x_1 \operatorname{col}_1(A) + x_1 \operatorname{col}_2(A) + \cdots + x_n \operatorname{col}_n(A)$$

shows that the elements AX of the range $\mathcal{R}(A)$ are precisely the linear combinations of the columns of A. For this reason some authors call the range the **column space**.

▷ **Exercise 126B** Let $A = \begin{bmatrix} Y_1 & Y_2 & Y_3 & Y_4 \end{bmatrix}$ and $Y = Y_1 - Y_2 + 3Y_4$. Show that $Y \in \mathcal{R}(A)$ by exhibiting a solution $X \in \mathbf{F}^{4 \times 1}$ of $Y = AX$.

▷ **Exercise 126C** Let $A = \begin{bmatrix} 1 & 2 & 3 & 4 \\ 5 & 6 & 7 & 8 \end{bmatrix}$ and $Y = Y_1 - Y_2 + 3Y_4$ where $Y_j = \operatorname{col}_j(A)$. Is Y a column of A? Is Y a linear combination of the columns of A? Is $Y \in \mathcal{R}(A)$?

▷ **Exercise 126D** For each of the matrices A in Exercise 88E and each of the matrices

$$Y_1 = \begin{bmatrix} -2 \\ 5 \\ 2 \end{bmatrix}, \qquad Y_2 = \begin{bmatrix} -2 \\ 5 \\ 0 \end{bmatrix}, \qquad Y_3 = \begin{bmatrix} -2 \\ 5 \\ 5 \end{bmatrix},$$

say whether or not $Y_i \in \mathcal{R}(A)$.

▷ **Exercise 126E** Show that every column of a matrix A is in the range $\mathcal{R}(A)$ of A.

▷ **Exercise 126F** The following satisfy $MA = R$ and M is invertible.

$$M = \begin{bmatrix} 2 & -1 & 0 \\ -1 & 1 & 0 \\ 3 & -2 & -1 \end{bmatrix}, \qquad A = \begin{bmatrix} 1 & 1 & -1 & -1 \\ 1 & 2 & 4 & 5 \\ 1 & -1 & -11 & -13 \end{bmatrix}$$

$$R = \begin{bmatrix} 1 & 0 & -6 & -7 \\ 0 & 1 & 5 & 6 \\ 0 & 0 & 0 & 0 \end{bmatrix}.$$

Which of the following lie in $\mathcal{N}(A)$? In $\mathcal{N}(R)$? In $\mathcal{R}(A)$? In $\mathcal{R}(R)$?

$$X_1 = \begin{bmatrix} 6 \\ -5 \\ 1 \\ 0 \end{bmatrix}, \quad X_2 = \begin{bmatrix} 7 \\ -6 \\ 0 \\ 1 \end{bmatrix}, \quad X_3 = \begin{bmatrix} 13 \\ -11 \\ 1 \\ 1 \end{bmatrix}, \quad X_4 = \begin{bmatrix} 13 \\ 10 \\ 1 \\ 1 \end{bmatrix},$$

$$Y_1 = \begin{bmatrix} 1 \\ 1 \\ 0 \end{bmatrix}, \quad Y_2 = \begin{bmatrix} 2 \\ 3 \\ 0 \end{bmatrix}, \quad Y_3 = \begin{bmatrix} 1 \\ 1 \\ 1 \end{bmatrix}, \quad Y_4 = \begin{bmatrix} 2 \\ 3 \\ 1 \end{bmatrix}.$$

▷ **Exercise 127A** Continue the notation of the previous exercise. Which Y_j satisfy $MY_j \in \mathcal{R}(R)$? $M^{-1}Y_j \in \mathcal{R}(A)$?

4.3 Set Equality

If $\mathcal{V}$ and $\mathcal{W}$ are two subsets of $\mathbf{F}^{n \times 1}$, then to prove $\mathcal{V} \subset \mathcal{W}$ we must prove that

$$X \in \mathcal{V} \implies X \in \mathcal{W}.$$

(That's the definition of $\subset$.) This means the proof must look something like this:

Choose $X \in \mathcal{V}$. BLAH BLAH BLAH *Therefore $X \in \mathcal{W}$.*

Here BLAH BLAH BLAH stands for a sequence of steps, each following from previous steps or definitions. Here's an example.

Theorem 127B *Assume that $A \in \mathbf{F}^{m \times n}$, $B \in \mathbf{F}^{n \times p}$, and that $AB = 0$. Then $\mathcal{R}(B) \subset \mathcal{N}(A)$.*

Proof: Choose $X \in \mathcal{R}(B)$. Then $X = BZ$ for some $Z \in \mathbf{F}^{p \times 1}$ by the definition of $\mathcal{R}(B)$. Hence,

$$AX = A(BZ) = (AB)Z = 0Z = 0$$

so $X \in \mathcal{N}(A)$. □

Two sets are equal when each is a subset of the other:

$$\mathcal{V} = \mathcal{W} \iff \mathcal{V} \subset \mathcal{W} \text{ and } \mathcal{W} \subset \mathcal{V}.$$

Thus a proof that $\mathcal{V} = \mathcal{W}$ looks like a proof of $\mathcal{V} \subset \mathcal{W}$ followed by a proof of $\mathcal{W} \subset \mathcal{V}$.

Example 127C Let $A = \begin{bmatrix} 1 & 2 & 3 \end{bmatrix}$, $B = \begin{bmatrix} -2 & -3 \\ 1 & 0 \\ 0 & 1 \end{bmatrix}$. Then the null space of A equals the range of B:

$$\mathcal{N}(A) = \mathcal{R}(B).$$

Proof: We prove $\mathcal{R}(B) \subset \mathcal{N}(A)$. Choose $X \in \mathcal{R}(B)$. Then

$$X = BZ = \begin{bmatrix} -2z_1 - 3z_2 \\ z_1 \\ z_2 \end{bmatrix}$$

for some $Z = \begin{bmatrix} z_1 \\ z_2 \end{bmatrix} \in \mathbf{F}^{2 \times 1}$. Hence

$$AX = \begin{bmatrix} 1 & 2 & 3 \end{bmatrix} \begin{bmatrix} -2z_1 - 3z_2 \\ z_1 \\ z_2 \end{bmatrix} = 0$$

so $X \in \mathcal{N}(A)$. This proves that $\mathcal{R}(B) \subset \mathcal{N}(A)$.

We prove $\mathcal{N}(A) \subset \mathcal{R}(B)$. Choose $X \in \mathcal{N}(A)$. Then $AX = 0$, that is,

$$x_1 + 2x_2 + 3x_3 = 0 \qquad \text{where} \quad X = \begin{bmatrix} x_1 \\ x_2 \\ x_3 \end{bmatrix}.$$

Thus $X = BZ$ where $Z = \begin{bmatrix} x_2 \\ x_3 \end{bmatrix} \in \mathbf{F}^{2 \times 1}$. Hence $X \in \mathcal{R}(B)$. This proves that $\mathcal{N}(A) \subset \mathcal{R}(B)$.   □

Invariance of the Null Space

Theorem 128A *Left equivalent matrices have the same null space.*

Proof: Assume that $A = QB$ with Q invertible. Then

$$X \in \mathcal{N}(B) \implies BX = 0 \implies AX = QBX = 0 \implies X \in \mathcal{N}(A).$$

This shows that $\mathcal{N}(B) \subset \mathcal{N}(A)$. Since $B = Q^{-1}A$, the same argument shows $\mathcal{N}(A) \subset \mathcal{N}(B)$. Hence $\mathcal{N}(B) = \mathcal{N}(A)$ as required.   □

Invariance of the Range

Theorem 128B *Right equivalent matrices have the same range.*

Proof: Assume that $AP = B$ with P invertible. Choose $Y \in \mathcal{R}(B)$. Then $Y = BX$ for some X. Let $\widetilde{X} = PX$. Then

$$Y = BX = APX = A\widetilde{X}$$

so $Y \in \mathcal{R}(A)$. This shows that $\mathcal{R}(B) \subset \mathcal{R}(A)$. Since $A = BP^{-1}$, the same argument shows $\mathcal{R}(A) \subset \mathcal{R}(B)$. Hence $\mathcal{R}(B) = \mathcal{R}(A)$ as required.   □

▷ **Exercise 129A** . Let $V = \mathcal{R}(A)$ and $W = \mathcal{R}(B)$ where

$$A = \begin{bmatrix} 1 & 0 \\ 0 & 1 \\ 1 & 0 \\ 0 & 0 \end{bmatrix}, \qquad B = \begin{bmatrix} 2 & 1 \\ 1 & 1 \\ 2 & 1 \\ 3 & 3 \end{bmatrix}.$$

Let $A_1 = \mathrm{col}_1(A)$ and $B_1 = \mathrm{col}_1(B)$. Is

$$A_1 \in V? \qquad A_1 \in W? \qquad B_1 \in W?$$

$$B_1 \in V? \qquad W \subset V? \qquad V \subset W?$$

▷ **Exercise 129B** Let $A = \begin{bmatrix} a_1 & a_2 \end{bmatrix}$ and $B = \begin{bmatrix} -a_2 \\ a_1 \end{bmatrix}$. Assume that $A \neq 0$. Show that $\mathcal{R}(B) = \mathcal{N}(A)$.

▷ **Exercise 129C** Suppose that $C \in \mathbf{F}^{n \times k}$ results from $B \in \mathbf{F}^{n \times (k+1)}$ by deleting its ith column. Show that $\mathcal{R}(C) \subset \mathcal{R}(B)$.

▷ **Exercise 129D** Define $A \in \mathbf{F}^{1 \times 3}$ and $B \in \mathbf{F}^{3 \times 3}$ by

$$A = \begin{bmatrix} a_1 & a_2 & a_3 \end{bmatrix}, \qquad B = \begin{bmatrix} 0 & a_3 & -a_2 \\ -a_3 & 0 & a_1 \\ a_2 & -a_1 & 0 \end{bmatrix},$$

and let $B_i \in \mathbf{F}^{3 \times 2}$ result from B by deleting the ith column. Assume that $A \neq 0$. Then $\mathcal{N}(A)$ is a plane through the origin and each column of B lies on this plane. Show that $\mathcal{R}(B) = \mathcal{N}(A)$. Show that $\mathcal{R}(B_i) \subset \mathcal{N}(A)$ for $i = 1, 2, 3$ with equality for at least one value of i. Give an example where $\mathcal{R}(B_1) \neq \mathcal{N}(A)$ even though $A \neq 0$.

▷ **Exercise 129E** Do left equivalent matrices have the same range?

▷ **Exercise 129F** Show that if $A = QB$, then $\mathcal{N}(B) \subset \mathcal{N}(A)$. Give an example (with Q not invertible) where $A = QB$ but $\mathcal{N}(B) \neq \mathcal{N}(A)$.

Exercise 129G Show that if $AP = B$, then $\mathcal{R}(B) \subset \mathcal{N}(B)$. Give an example (with P not invertible) where $AP = B$ but $\mathcal{R}(A) \neq \mathcal{R}(B)$.

▷ **Exercise 129H** Is the converse of Theorem 127B true?

Using MINIMAT

It is quite easy to use MINIMAT to decide if a given column vector X is an element of the null space of a given matrix A: the command

```
#> A*X
```

prints the zero matrix if $X \in \mathcal{N}(A)$ and a nonzero matrix if $X \notin \mathcal{N}(A)$. The .M function randsoln (Function 149B) creates a random element of the null space; the command

```
#> X=randsoln(A)
```

assigns to A a random solution of $AX = 0$. If $\mathcal{N}(A) = \{0\}$, then X will be the zero vector.

It is quite easy to use MINIMAT to generate a random element Y of the range of a matrix A; the command

```
#> Y = A*rand(n,1)
```

does this (if A is $m \times n$). To decide if a given vector Y is in the range $\mathcal{R}(A)$ of a given matrix A one solves $Y = AX$ as in Section 4.1. If this attempt produces a solution X, then $Y \in \mathcal{R}(A)$; if we discover that there is no solution X, then $Y \notin \mathcal{R}(A)$. Another approach is

```
#> X = randsoln(A, Y)
```

If $Y \in \mathcal{R}(A)$, then this command will assign to X a solution of $Y = AX$. If $Y \notin \mathcal{R}(A)$, then the value assigned to X will not satisfy $AX = Y$. *It doesn't matter which method you use to find a solution X of an inhomogeneous system $AX = Y$. No matter which method you use, you should check your answer by comparing AX and Y.*

To confirm that the null space or range of one matrix is subset of the null space or range of another, generate a random element of the former and prove it is in the latter.

Example 130A We confirm that left equivalent matrices have the same null space.

```
#> A = rand(3,5), B = rand(3,3)*A
#> X = randsoln(A), B*X  % (= zero)
#> X = randsoln(B), A*X  % (= zero)
```

The second line confirms $\mathcal{N}(A) \subset \mathcal{N}(B)$; the third confirms $\mathcal{N}(B) \subset \mathcal{N}(A)$.

Exercise 131A In each of the following generate the matrices A and B as indicated. Then confirm that $\mathcal{N}(A) \subset \mathcal{N}(B)$ but $\mathcal{N}(B) \not\subset \mathcal{N}(A)$. Explain why the inclusion holds.

(1)　　#> A=rand(3,5), B=A(1:2,:)
(2)　　#> B=rand(2,5), A=[B; rand(1,5)]
(3)　　#> B=rand(3,5), A= rand(2,3)*B

▷ **Exercise 131B** Confirm that right equivalent matrices have the same range.

▷ **Exercise 131C** In each of the following, generate the matrices A and B as indicated. Then confirm that $\mathcal{R}(B) \subset \mathcal{R}(A)$ but $\mathcal{R}(A) \not\subset \mathcal{R}(B)$. Explain why the inclusion holds.

(1)　　#> A=rand(5,3), B=A(:,1:2)
(2)　　#> B=rand(5,2), A=[B rand(5,1)]
(3)　　#> A=rand(5,3), B = A*rand(3,2)

4.4　Subspaces

A *subspace* of $\mathbf{F}^{n \times 1}$ is a subset[1] $\mathcal{V} \subset \mathbf{F}^{n \times 1}$ which contains $0_{n \times 1}$ and is **closed** under matrix addition and scalar multiplication. The closure property means that $c_1 X_1 + c_2 X_2 \in \mathcal{V}$ whenever $X_1, X_2 \in \mathcal{V}$ and $c_1, c_2 \in \mathbf{F}$. Here's a restatement of the definition:

Subspace

Definition 131D A subset $\mathcal{V} \subset \mathbf{F}^{n \times 1}$ is called a **subspace** iff it satisfies the following three conditions:

(1) $0 \in \mathcal{V}$;

(2) $X_1, X_2 \in \mathcal{V} \implies X_1 + X_2 \in \mathcal{V}$;

(3) $X \in \mathcal{V}, c \in \mathbf{F} \implies cX \in \mathcal{V}$.

Two extreme examples of subspaces are $\mathcal{V} = \{0\}$ the subspace with the single element 0 and $\mathcal{V} = \mathbf{F}^{n \times 1}$ the subspace of all $n \times 1$ matrices. These

---
[1] The definition is meaningful with $\mathbf{F}^{n \times 1}$ replaced by $\mathbf{F}^{p \times q}$, but this is not needed.

are extreme examples in the sense that any subspace $\mathcal{V}$ lies between:

$$\{0\} \subset \mathcal{V} \subset \mathbf{F}^{n \times 1}.$$

We'll discuss many other examples of subspaces in this chapter.

Theorem 132A *The null space $\mathcal{N}(A)$ of a matrix $A \in \mathbf{F}^{m \times n}$ is a subspace of $\mathbf{F}^{n \times 1}$.*

Proof: (1) $0_{n \times 1} \in \mathcal{N}(A)$ since $A0_{n \times 1} = 0_{m \times 1}$. (2) Assume $X_1, X_2 \in \mathcal{N}(A)$. Then $AX_1 = AX_2 = 0$ by the definition of $\mathcal{N}(A)$. Hence,

$$A(X_1 + X_2) = AX_1 + AX_2 = 0,$$

so $X_1 + X_2 \in \mathcal{N}(A)$. (3) Assume $X \in \mathcal{N}(A)$ and $c \in \mathbf{F}$. Then $A(cX) = cAX = 0$ so $cX \in \mathcal{N}(A)$.   □

Theorem 132B *The range $\mathcal{R}(A)$ of a matrix $A \in \mathbf{F}^{m \times n}$ is a subspace of $\mathbf{F}^{m \times 1}$.*

Proof: (1) $0 \in \mathcal{R}(A)$ since $0_{m \times 1} = AX$ for $X = 0_{n \times 1}$. (2) Assume $Y_1, Y_2 \in \mathcal{R}(A)$. Then there exist $X_1, X_2 \in \mathbf{F}^{n \times 1}$ with $Y_1 = AX_1$ and $Y_2 = AX_2$. Hence

$$Y_1 + Y_2 = AX_1 + AX_2 = A(X_1 + X_2)$$

so $Y_1 + Y_2 \in \mathcal{R}(A)$. (3) Assume $Y \in \mathcal{R}(A)$ and $c \in \mathbf{F}$. Then $Y = AX$ for some X. Hence, $cY = cAX = A(cX)$ so $cY \in \mathcal{R}(A)$.   □

Theorem 132C (Subset Criterion)   *Suppose that $\mathcal{W}$ is a subspace of $\mathbf{F}^{m \times 1}$ and that $A \in \mathbf{F}^{m \times n}$. Let $A_j = \mathrm{col}_j(A)$ denote the jth column of A:*

$$A = \begin{bmatrix} A_1 & A_2 & \cdots & A_n \end{bmatrix}.$$

Then $\mathcal{R}(A) \subset \mathcal{W}$ if and only if $A_j \in \mathcal{W}$ for $j = 1, 2, \ldots, m$.

Proof: Assume that $A_j \in \mathcal{W}$ for $j = 1, 2, \ldots, m$. Choose $Y \in \mathcal{R}(A)$. Then $Y = AX$ for some X. By the block multiplication law 53B,

$$Y = x_1 A_1 + x_2 A_2 + \cdots x_n A_n$$

where $x_j = \mathrm{entry}_j(X)$. Each $A_j \in \mathcal{V}$ and subspaces are closed under scalar multiplication and addition; it follows that $Y \in \mathcal{W}$. Conversely, assume $A_j \notin \mathcal{W}$ for some j. But $A_j \in \mathcal{R}(A)$. Hence $\mathcal{R}(A) \not\subset \mathcal{W}$.   □

Corollary 133A (Equality Criterion) *Assume* $A, B \in \mathbf{F}^{m \times n}$. *Then* $\mathcal{R}(A) = \mathcal{R}(B)$ *if and only if every column of A is an element of $\mathcal{R}(B)$ and every column of B is an element of $\mathcal{R}(A)$.*

Proof: $\mathcal{R}(A) = \mathcal{R}(B) \iff \mathcal{R}(A) \subset \mathcal{R}(B)$ and $\mathcal{R}(B) \subset \mathcal{R}(A)$.   □

▷ **Exercise 133B** For each of the following decide if $\mathcal{R}(A_i) \subset \mathcal{R}(A_k)$. There are $12 = 4 \cdot (4 - 1)$ questions here.

(1) $A_1 = \begin{bmatrix} 1 & 1 \\ 1 & 2 \\ 2 & 1 \end{bmatrix}$ (2) $A_2 = \begin{bmatrix} 1 & 0 \\ 1 & 1 \\ 2 & -1 \end{bmatrix}$,

(3) $A_3 = \begin{bmatrix} 1 & 0 \\ 0 & 1 \\ 0 & 0 \end{bmatrix}$, (4) $A_4 = \begin{bmatrix} 1 & 2 \\ 1 & 2 \\ 2 & 4 \end{bmatrix}$.

▷ **Exercise 133C** Give an example where

$$A \in \mathbf{F}^{3 \times 2}, \qquad B \in \mathbf{F}^{3 \times 3}, \qquad \mathcal{R}(A) = \mathcal{R}(B).$$

4.5   New Subspaces from Old

The same subspace $\mathcal{V}$ may be described in several ways. There are many matrices whose null space is $\mathcal{V}$ and many other matrices whose range is $\mathcal{V}$. Subspaces may appear in other guises as well. The following operations construct new subspaces from old.

Definition 133D For $\mathcal{W} \subset \mathbf{F}^{m \times 1}$ and $A \in \mathbf{F}^{m \times n}$ the set

$$A^{-1}\mathcal{W} = \{X \in \mathbf{F}^{n \times 1} : AX \in \mathcal{W}\}$$

is called the **preimage** of $\mathcal{W}$ by A.

Definition 133E For $\mathcal{V} \subset \mathbf{F}^{n \times 1}$ and $A \in \mathbf{F}^{m \times n}$ the set

$$A\mathcal{V} = \{AX \in \mathbf{F}^{m \times 1} : X \in \mathcal{V}\}$$

is called the **image** of $\mathcal{V}$ by A.

▷ **Question 133F** Suppose that $A \in \mathbf{F}^{m \times n}$. What is the image of $\mathbf{F}^{n \times 1}$ by A? The image of $\{0\}$? The preimage of $\mathbf{F}^{m \times 1}$? The preimage of $\{0\}$?

Definition 134A The **intersection** $V_1 \cap V_2$ of two subsets $V_1, V_2 \subset \mathbf{F}^{n \times 1}$ is defined by

$$V_1 \cap V_2 = \{X \in \mathbf{F}^{n \times 1} : X \in V_1 \text{ and } X \in V_2\}.$$

Definition 134B The **sum** $V_1 + V_2$ of two subsets $V_1, V_2 \subset \mathbf{F}^{n \times 1}$ is defined by

$$V_1 + V_2 = \{X_1 + X_2 : X_1 \in V_1 \text{ and } X_2 \in V_2\}.$$

Exercise 134C Let $A = \begin{bmatrix} 2 & 0 \\ 0 & 3 \end{bmatrix}$ and $V = \mathcal{N}(B)$. For each of the following values of B draw a graph showing V, $A^{-1}V$, and AV:

(1) $B = \begin{bmatrix} 1 & 1 \end{bmatrix}$ (2) $B = \begin{bmatrix} 0 & 1 \end{bmatrix}$,

(3) $B = \begin{bmatrix} 1 & 0 \end{bmatrix}$,   (4) $B = \begin{bmatrix} 2 & 3 \end{bmatrix}$.

The subspaces of $\mathbf{F}^{2 \times 1}$, apart from $\{0\}$ and $\mathbf{F}^{2 \times 1}$, are lines through the origin. It suffices to pick a nonzero $X \in V$ and plot X, $A^{-1}X$, and AX and draw the lines connecting these to the origin. Draw four different graphs, one for each B, each showing the coordinate axes and the three lines.

Exercise 134D Repeat the previous exercise with $A = \begin{bmatrix} 1 & 1 \\ 0 & 1 \end{bmatrix}$.

▷ **Exercise 134E** Let $P = \begin{bmatrix} 2 & 1 \\ 1 & 1 \end{bmatrix}$ and $V = \left\{ \begin{bmatrix} x_1 \\ x_2 \end{bmatrix} : x_1 = x_2 \right\}$. Which of the following are elements of PV?

$$X_1 = \begin{bmatrix} 1 \\ 1 \end{bmatrix}, \quad X_2 = \begin{bmatrix} 0 \\ 1 \end{bmatrix}, \quad X_3 = \begin{bmatrix} 1 \\ 0 \end{bmatrix}, \quad X_4 = \begin{bmatrix} 3 \\ 2 \end{bmatrix}.$$

Exercise 134F Show that the preimage of a subspace is a subspace.

Exercise 134G Show that the image of a subspace is a subspace.

Exercise 134H Show that the intersection of two subspaces is a subspace.

Exercise 134I Show that the sum of two subspaces is a subspace.

Exercise 134J Show that $\mathcal{N}(BA) = A^{-1}\mathcal{N}(B)$.

Exercise 134K Show that $\mathcal{R}(AB) = A\mathcal{R}(B)$.

Exercise 135A Let $m = m_1 + m_2$ and

$$A_1 \in \mathbf{F}^{m_1 \times n}, \quad A_2 \in \mathbf{F}^{m_2 \times n}, \quad A = \begin{bmatrix} A_1 \\ A_2 \end{bmatrix} \in \mathbf{F}^{m \times n}.$$

Prove that
$$\mathcal{N}(A) = \mathcal{N}(A_1) \cap \mathcal{N}(A_2).$$

Exercise 135B Let $n = n_1 + n_2$ and

$$A_1 \in \mathbf{F}^{m \times n_1}, \quad A_2 \in \mathbf{F}^{m \times n_1}, \quad A = \begin{bmatrix} A_1 & A_2 \end{bmatrix} \in \mathbf{F}^{m \times n}.$$

Prove that
$$\mathcal{R}(A) = \mathcal{R}(A_1) + \mathcal{R}(A_2).$$

▷ **Exercise 135C** Consider the set

$$A^{-1}\{Y\} = \{X \in \mathbf{F}^{n \times 1} : AX = Y\}$$

of solutions X of the inhomogeneous system $AX = Y$. Is it a subspace?

▷ **Exercise 135D** The preimage of a set is defined even when A is not invertible, i.e. the notation $A^{-1}\mathcal{W}$ is meaningful even when the notation A^{-1} is not. When A *is* invertible the notation $A^{-1}\mathcal{W}$ is apparently ambiguous, denoting both the preimage of $\mathcal{W}$ by A and the image of $\mathcal{W}$ by A^{-1}. How should the ambiguity be resolved?

4.6   Bases

Basis for a Subspace

Definition 135E A basis[2] for a subspace $\mathcal{V}$ is a matrix Φ such that
$$\mathcal{R}(\Phi) = \mathcal{V}, \qquad \mathcal{N}(\Phi) = \{0\}.$$

▷ **Question 135F** When is a matrix a basis for its own range?

---

[2]The plural of *basis* is **bases**.

In Chapter 5 we will prove that every subspace has a basis. The following theorem shows that there are many bases for a nonzero subspace. We'll prove a converse in the next chapter. (See Theorem 156D.)

Change of Basis

Theorem 136A *Any matrix that is right equivalent to a basis for a subspace $\mathcal{V}$ is itself a basis for $\mathcal{V}$.*

Proof: Assume that Φ is a basis for $\mathcal{V}$, and that Ψ is right equivalent to Φ. Then $\mathcal{R}(\Phi) = \mathcal{V}$, $\mathcal{N}(\Phi) = \{0\}$, and $\Psi = \Phi T$ where T is invertible. Thus

$$\mathcal{R}(\Psi) = \mathcal{R}(\Phi T) = \mathcal{R}(\Phi) = \mathcal{V}$$

by Theorem 128B. If $X \in \mathcal{N}(\Psi)$, then $0 = \Psi X = (\Phi T)X = \Phi(TX)$, so $TX \in \mathcal{N}(\Phi) = \{0\}$, so $TX = 0$ so $X = 0$ (as T is invertible). Hence $\mathcal{N}(\Psi) = \{0\}$ as required. $\square$

Basis for the Null Space

Left equivalent matrices have the same null space (see Theorem 128A). Therefore a basis for the null space $\mathcal{N}(A)$ of a matrix A is the same as a basis for the null space $\mathcal{N}(R)$ of its reduced row echelon form R. The following theorem gives a method for finding a basis for this null space.

Theorem 136B *Suppose that $MA = R$ where $A, R \in \mathbf{F}^{m \times n}$, that R is in RREF, and that $M \in \mathbf{F}^{m \times m}$ is invertible. Suppose that all the leading columns of R are at the left so that R has the form*

$$R = \begin{bmatrix} I_r & C \\ 0_{(m-r) \times r} & 0_{(m-r) \times (n-r)} \end{bmatrix},$$

where $C \in \mathbf{F}^{r \times (n-r)}$. Then the matrix

$$\Phi = \begin{bmatrix} -C \\ I_{n-r} \end{bmatrix}$$

is a basis for the common null space $\mathcal{N}(A) = \mathcal{N}(R)$ of the (left equivalent) matrices A and R.

Proof: Choose $X \in \mathbf{F}^{n \times 1}$. To show that $\mathcal{N}(R) = \mathcal{R}(\Phi)$ we must show that

$$RX = 0 \iff X = \Phi Z \quad \text{for some } Z.$$

Let W denote the first r entries of X and Z denote the last $n - r$ entries. By the block multiplication law 53A, we get

$$X = \begin{bmatrix} W \\ Z \end{bmatrix}, \qquad RX = \begin{bmatrix} W + CZ \\ 0_{(m-r) \times 1} \end{bmatrix}, \qquad \Phi Z = \begin{bmatrix} -CZ \\ Z \end{bmatrix}.$$

This shows that if $\Phi Z_1 = X$, then $Z_1 = Z$; in particular,

$$\Phi Z = 0 \iff Z = 0$$

so that $\mathcal{N}(\Phi) = \{0\}$. Also,

$$
\begin{aligned}
X \in \mathcal{N}(R) &\iff RX = 0 \\
&\iff W + CZ = 0 \\
&\iff W = -CZ \\
&\iff X = \Phi Z \\
&\iff X \in \mathcal{R}(\Phi).
\end{aligned}
$$

Thus $\mathcal{N}(R) = \mathcal{R}(\Phi)$ as required. $\qquad\qquad\qquad\qquad\qquad\square$

Example 137A With $m = 3$, $n = 5$, $r = 2$, we have

$$R = \begin{bmatrix} 1 & 0 & c_{11} & c_{12} & c_{13} \\ 0 & 1 & c_{21} & c_{22} & c_{23} \\ 0 & 0 & 0 & 0 & 0 \end{bmatrix}, \qquad \Phi = \begin{bmatrix} -c_{11} & -c_{12} & -c_{13} \\ -c_{21} & -c_{22} & -c_{23} \\ 1 & 0 & 0 \\ 0 & 1 & 0 \\ 0 & 0 & 1 \end{bmatrix}.$$

Let $X \in \mathbf{F}^{5 \times 1}$ be the column of unknowns and $Z \in \mathbf{F}^{3 \times 1}$ be the column of free unknowns:

$$X = \begin{bmatrix} x_1 \\ x_2 \\ x_3 \\ x_4 \\ x_5 \end{bmatrix}, \qquad Z = \begin{bmatrix} x_3 \\ x_4 \\ x_5 \end{bmatrix}.$$

After discarding redundant equations $0 = 0$, $x_3 = x_3$, $x_4 = x_4$, $x_5 = x_5$, the equations $RX = 0$ and $X = \Phi Z$ are the same:

$$
\begin{aligned}
x_1 + c_{11}x_3 + c_{12}x_4 + c_{13}x_5 &= 0 \\
x_2 + c_{21}x_3 + c_{22}x_4 + c_{23}x_5 &= 0.
\end{aligned}
$$

Notice that each column of Φ is the solution of $AX = 0$ obtained by setting one of the free unknowns to 1, the others to 0, and solving for the leading unknowns x_1 and x_2.

Example 138A The same idea works when the leading columns aren't at the left. For example, a basis for the null space of

$$R = \begin{bmatrix} 1 & c_{11} & 0 & c_{12} & 0 & c_{13} & c_{14} \\ 0 & 0 & 1 & c_{22} & 0 & c_{23} & c_{24} \\ 0 & 0 & 0 & 0 & 1 & c_{33} & c_{34} \\ 0 & 0 & 0 & 0 & 0 & 0 & 0 \\ 0 & 0 & 0 & 0 & 0 & 0 & 0 \end{bmatrix}$$

is given by

$$\Phi = \begin{bmatrix} -c_{11} & -c_{12} & -c_{13} & -c_{14} \\ 1 & 0 & 0 & 0 \\ 0 & -c_{22} & -c_{23} & -c_{24} \\ 0 & 1 & 0 & 0 \\ 0 & 0 & -c_{33} & -c_{34} \\ 0 & 0 & 1 & 0 \\ 0 & 0 & 0 & 1 \end{bmatrix}.$$

For $X \in \mathbf{F}^{7 \times 1}$ put $x_j = \text{entry}_j(X)$. The leading unknowns are x_1, x_3, x_5 and the free unknowns are x_2, x_4, x_6, x_7. Each column of Φ is obtained by setting one of the free unknowns to 1, the others to 0, and solving for the leading unknowns.

$$RX = 0 \iff X = \Phi Z \quad \text{where} \quad Z = \begin{bmatrix} x_2 \\ x_4 \\ x_6 \\ x_7 \end{bmatrix}.$$

Basis for the Range

Our next aim is Theorem 139A which gives a method for finding a basis for the range. First we need the following

Theorem 138B *Every matrix A is right equivalent to a matrix B whose transpose $B^\top$ is in reduced row echelon form.*

Proof: According to the Gauss-Jordan Decomposition (Theorem 105A), every matrix is left equivalent to a matrix in RREF. We apply this to the transpose $A^\top$ of A: we can find an invertible matrix M and a matrix R in RREF with $A^\top = MR$. Let $B = R^\top$ and $P = M^\top$. The matrix

P is invertible (because M is), the transpose of B is in RREF (because $B^\top = (R^\top)^\top = R$), and the calculation

$$A = (A^\top)^\top = (MR)^\top = R^\top M^\top = BP$$

shows that the matrices A and B are right equivalent.   □

Theorem 139A *Assume $A = BP$ where P is invertible and $B^\top$ is in RREF. Then the matrix Ψ formed from the nonzero columns of B is a basis for the common range $\mathcal{R}(A) = \mathcal{R}(B)$ of the (right equivalent) matrices A and B.*

Proof: Suppose[3] that the leading columns in $B^\top$ come first so that B and Ψ have the forms

$$B = \begin{bmatrix} I_r & 0_{r \times (n-r)} \\ G & 0_{(m-r) \times (n-r)} \end{bmatrix}, \qquad \Psi = \begin{bmatrix} I_r \\ G \end{bmatrix},$$

where $G \in \mathbf{F}^{(m-r) \times r}$. For $X \in \mathbf{F}^{n \times 1}$ denote by $W \in \mathbf{F}^{r \times 1}$ the matrix formed from the first r entries of X and by Z the matrix formed by the last $n - r$ entries. Then

$$X = \begin{bmatrix} W \\ Z \end{bmatrix}, \qquad BX = \begin{bmatrix} W \\ GW \end{bmatrix} = \Psi W.$$

This shows that $\Psi W = 0$ only if $W = 0$, so that $\mathcal{N}(\Psi) = \{0\}$. Also

$$
\begin{aligned}
Y \in \mathcal{R}(B) \quad &\Longleftrightarrow \quad Y = BX \text{ for some } X \\
&\Longleftrightarrow \quad Y = \Psi W \text{ for some } W \\
&\Longleftrightarrow \quad Y \in \mathcal{R}(\Psi)
\end{aligned}
$$

so $\mathcal{R}(B) = \mathcal{R}(\Psi)$ as required.   □

▷ **Exercise 139B** Let A be the matrix in Example 120A. Find a basis Φ for its null space.

▷ **Exercise 139C** For each of the matrices A in Exercise 88C, decide if $\mathcal{N}(A) = \{0\}$. If not, find a basis for $\mathcal{N}(A)$.

▷ **Exercise 139D** For each of the matrices A in Exercise 88E, decide if $\mathcal{N}(A) = \{0\}$. If not, find a basis for $\mathcal{N}(A)$.

---

[3]The case where the leading columns do not come first is treated in Exercise 145C.

Using MINIMAT

Function 140A The .M function Nbasis1, shown in Listing 6, computes a basis for the null space as in Theorem 136B and Example 138A. The command

```
#>  Phi = Nbasis1(A)
```

assigns to Phi a basis for the null space of A.

Function 140B The .M function Rbasis1, shown in Listing 6, computes a basis for the range as in Theorem 139A. The command

```
#>  Psi = Rbasis1(A)
```

assigns to Psi a basis for the range of A.

Basis for the Null Space using MINIMAT

```
function Phi = Nbasis1(A)

    [m, n] = size(A);
    [R, lead, free] = gj(A);
    [toss, rank] = size(lead);
    [toss, nullity] = size (free);
    Phi = zeros(n,nullity);
    if nullity==0 return; end % for PC Matlab
    Phi(free,:) = eye(nullity);
    Phi(lead,:) = -R(1:rank,free);
```

Basis for the Range using MINIMAT

```
function Psi = Rbasis1(A)

    [R lead] = gj(A'); B = R';
    [toss, rank] = size(lead);
    Psi = B(:,1:rank);
```

Listing 6

Exercise 141A For each of the following matrices, find a basis for its null space.

$$A_1 = \begin{bmatrix} 3 & 6 & -2 & -1 & -4 \\ -10 & -20 & 10 & 3 & 31 \\ -3 & -6 & 3 & 1 & 10 \end{bmatrix}, \quad A_2 = \begin{bmatrix} 12 & 24 & 2 & 3 & 84 \\ 4 & 8 & 1 & 1 & 30 \\ 3 & 6 & 0 & 1 & 19 \end{bmatrix},$$

$$A_3 = \begin{bmatrix} 3 & 6 & -2 & -1 & -4 \\ -1 & -2 & 1 & 0 & 1 \\ -2 & -4 & 2 & 1 & 9 \end{bmatrix}, \quad A_4 = \begin{bmatrix} -1 & -2 & 0 & 0 & -5 \\ -1 & -2 & 5 & 0 & 25 \\ 0 & 0 & -2 & 0 & -12 \end{bmatrix}.$$

Exercise 141B For each of the transposes $B_i = A_i^\top$ of the matrices in the previous exerccise find a basis for its range.

4.7 Bases and Biequivalence

Now we will see how to use biequivalence to find bases. Our first observation, Theorem 141C, is that it is very easy to find bases for the null space and range of a matrix in biequivalence normal form. Then we show in Theorem 142B how the null space and range of biequivalent matrices are related. Corollary 144A then tells us how to find a basis for the null space or range of a matrix from a basis for the null space or range of a biequivalent matrix. Combining Theorem 141C and Corollary 144A tells us how to find a basis for the null space or range of a matrix from its biequivalence normal form decomposition.

Theorem 141C *For a matrix*

$$D = \begin{bmatrix} I_r & 0_{r \times (n-r)} \\ 0_{(m-r) \times r} & 0_{(m-r) \times (n-r)} \end{bmatrix}$$

in biequivalence normal form:

- $\Phi = \begin{bmatrix} 0_{r \times (n-r)} \\ I_{n-r} \end{bmatrix}$ *is a basis for the null space $\mathcal{N}(D)$;*

- $\Psi = \begin{bmatrix} I_r \\ 0_{(m-r) \times r} \end{bmatrix}$ *is a basis for the range $\mathcal{R}(D)$.*

Proof: Take $C = 0$ in 136B and $G = 0$ in 139A. □

Example 141D Let D and Φ be defined by

$$D = \begin{bmatrix} 1 & 0 & 0 \end{bmatrix}, \quad \Phi = \begin{bmatrix} 0 & 0 \\ 1 & 0 \\ 0 & 1 \end{bmatrix}.$$

Then $\mathcal{N}(\Phi) = \{0\}$, and

$$\mathcal{R}(\Phi) = \mathcal{N}(D) = \left\{ \begin{bmatrix} 0 \\ x_2 \\ x_3 \end{bmatrix} : x_2, x_3 \in \mathbf{F} \right\},$$

so Φ is a basis for $\mathcal{N}(D)$.

Example 142A Let D and Ψ be defined by

$$D = \begin{bmatrix} 1 & 0 & 0 \\ 0 & 1 & 0 \\ 0 & 0 & 0 \end{bmatrix}, \qquad \Psi = \begin{bmatrix} 1 & 0 \\ 0 & 1 \\ 0 & 0 \end{bmatrix}.$$

Then $\mathcal{N}(\Psi) = \{0\}$, and

$$\mathcal{R}(\Psi) = \mathcal{R}(D) = \left\{ \begin{bmatrix} x_1 \\ x_2 \\ 0 \end{bmatrix} : x_1, x_2 \in \mathbf{F} \right\},$$

so Ψ is a basis for $\mathcal{R}(D)$. Note that $\mathcal{N}(D) \neq \{0\}$ so that D is not a basis for its own range.

As noted in Theorems 128A and 128B, left equivalent matrices have the same null space and right equivalent matrices have the same range. Usually, however, left equivalent matrices have different ranges and right equivalent matrices have different null spaces. However, there is a relation.

Subspaces and Biequivalence

Theorem 142B *Suppose the matrices A and B are biequivalent, say $A = QBP^{-1}$ where Q and P are invertible. Then*

$$P\mathcal{N}(B) = \mathcal{N}(A), \qquad Q\mathcal{R}(B) = \mathcal{R}(A),$$

Proof: Assume $X, Y, \widetilde{X}, \widetilde{Y}$ are related by $X = P\widetilde{X}$ and $Y = Q\widetilde{Y}$. Then

$$Y = AX \iff \widetilde{Y} = B\widetilde{X}.$$

In particular, taking $Y = 0$ gives

$$
\begin{aligned}
X \in \mathcal{N}(A) \quad &\Longleftrightarrow \quad AX = 0 \\
&\Longleftrightarrow \quad B\widetilde{X} = 0 \\
&\Longleftrightarrow \quad X = P\widetilde{X} \text{ where } \widetilde{X} \in \mathcal{N}(B) \\
&\Longleftrightarrow \quad X \in P\mathcal{N}(B)
\end{aligned}
$$

which says that $\mathcal{N}(A) = P\mathcal{N}(B)$. Similarly,

$$
\begin{aligned}
Y \in \mathcal{R}(A) \quad &\Longleftrightarrow \quad Y = AX \text{ for some } X \\
&\Longleftrightarrow \quad \widetilde{Y} = B\widetilde{X} \text{ for some } \widetilde{X} \\
&\Longleftrightarrow \quad Y = Q\widetilde{Y} \text{ where } \widetilde{Y} \in \mathcal{R}(B) \\
&\Longleftrightarrow \quad Y \in Q\mathcal{R}(B)
\end{aligned}
$$

which says that $\mathcal{R}(A) = Q\mathcal{R}(B)$. $\qquad\qquad\square$

Remark 143A Here's a little trick to help you remember when to use Q and when to use Q^{-1}. Represent the equation $A = QBP^{-1}$ with the diagram

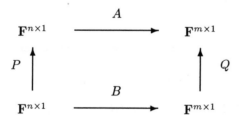

Here's how to understand this diagram. The equation $A = QBP^{-1}$ says that

$$
AX = Q(B(P^{-1}(X)))
$$

for all $X \in \mathbf{F}^{n \times 1}$. Think of X as chosen from the copy of $\mathbf{F}^{n \times 1}$ in the upper left hand corner. We can compute AX in two ways: either (1) by applying A to X (going with the arrow across the top); or (2) by applying P^{-1} (going down on the left against the arrow), then B (going across the bottom with the arrow), then Q (going up on the right with the arrow). The vertical arrows indicate the equations

$$
X = P\widetilde{X}, \qquad Y = Q\widetilde{Y}.
$$

These define correspondences $\widetilde{X} \to X$, $\widetilde{Y} \to Y$. Since the matrices P and Q are invertible, the correspondences are one-one meaning that to each X

corresponds exactly one $\widetilde{X}$ such that $X = P\widetilde{X}$, and similarly, to each Y corresponds exactly one $\widetilde{Y}$ such that $Y = Q\widetilde{Y}$. Each vertical arrow acts as a *change of variables*. Now we extend the above diagram to obtain

$$
\begin{array}{ccc}
\mathcal{N}(A) \subset \mathbf{F}^{n \times 1} & \xrightarrow{\ \ A\ \ } & \mathcal{R}(A) \subset \mathbf{F}^{m \times 1} \\[2pt]
\Big\uparrow{\scriptstyle P} & & \Big\uparrow{\scriptstyle Q} \\[2pt]
\mathcal{N}(B) \subset \mathbf{F}^{n \times 1} & \xrightarrow{\ \ B\ \ } & \mathcal{R}(B) \subset \mathbf{F}^{m \times 1}
\end{array}
$$

The equations

$$\mathcal{N}(A) = P\mathcal{N}(B), \qquad \mathcal{R}(A) = Q\mathcal{R}(B)$$

and

$$P^{-1}\mathcal{N}(A) = \mathcal{N}(B), \qquad Q^{-1}\mathcal{R}(A) = \mathcal{R}(B)$$

can be read from the diagram.

Bases and Biequivalence

Corollary 144A *Suppose $A = QBP^{-1}$ as in Theorem 142B. Then*

(1) *If $\widetilde{\Phi}$ is a basis for $\mathcal{N}(B)$, then $\Phi = P\widetilde{\Phi}$ is a basis for $\mathcal{N}(A)$.*

(2) *If $\widetilde{\Psi}$ is a basis for $\mathcal{R}(B)$, then $\Psi = Q\widetilde{\Psi}$ is a basis for $\mathcal{R}(A)$.*

Proof: (1) Assume $\mathcal{R}(\widetilde{\Phi}) = \mathcal{N}(B)$ and $\mathcal{N}(\widetilde{\Phi}) = \{0\}$. Then

$$\mathcal{R}(\Phi) = \mathcal{R}(P\widetilde{\Phi}) = P\mathcal{R}(\widetilde{\Phi}) = P\mathcal{N}(B) = \mathcal{N}(A)$$

and

$$\mathcal{N}(\Phi) = \mathcal{N}(P\widetilde{\Phi}) = P\mathcal{N}(\widetilde{\Phi}) = P\{0\} = \{0\}$$

as required. Part (2) has a similar proof.   □

To help remember the relationships expressed in Corollary 144A, we use the diagram

$$
\begin{array}{ccc}
& \mathcal{N}(A) \subset \mathbf{F}^{n\times 1} \xrightarrow{\quad A \quad} \mathbf{F}^{m\times 1} \supset \mathcal{R}(A) & \\
\Phi \nearrow \quad P \uparrow \qquad\qquad \uparrow Q \quad \nwarrow \Psi & \\
\mathbf{F}^{(n-r)\times 1} \qquad\qquad\qquad\qquad\qquad\qquad \mathbf{F}^{r\times 1} & \\
\tilde{\Phi} \searrow \qquad\qquad\qquad\qquad\qquad \swarrow \tilde{\Psi} & \\
& \mathcal{N}(B) \subset \mathbf{F}^{n\times 1} \xrightarrow[\quad B \quad]{} \mathbf{F}^{m\times 1} \supset \mathcal{R}(B) &
\end{array}
$$

Combining Theorem 141C and Corollary 144A tells us how to find a basis for the null space or range of a matrix from its biequivalence normal form decomposition. Use this to solve Exercises 145A and 145B.

▷ **Exercise 145A** For each of the four sets of matrices on Pages 111-112, find a basis $\tilde{\Phi}$ for the null space $\mathcal{N}(D)$ and use it to find a basis Φ for the null space $\mathcal{N}(A)$.

▷ **Exercise 145B** For each of the four sets of matrices on Pages 111-112 find a basis $\tilde{\Psi}$ for the range $\mathcal{R}(D)$ and use it to find a basis Ψ for the range $\mathcal{R}(A)$.

▷ **Exercise 145C** In the proof of Theorem 139A it was assumed that B and Ψ have the forms

$$
B = \begin{bmatrix} I & 0 \\ G & 0 \end{bmatrix}, \qquad \Psi = \begin{bmatrix} I \\ G \end{bmatrix}.
$$

Eliminate this assumption.

▷ **Exercise 145D** Suppose that

$$
A = \begin{bmatrix} a_1 & b_1 & c_1 & d_1 & e_1 & f_1 \\ a_2 & b_2 & c_2 & d_2 & e_2 & f_2 \end{bmatrix},
$$

$$
B = \begin{bmatrix} e_1 & d_1 & c_1 & f_1 & b_1 & a_1 \\ e_2 & d_2 & c_2 & f_2 & b_2 & a_2 \end{bmatrix},
$$

$$
\Phi = \begin{bmatrix} u_1 & u_2 & u_3 & u_4 \\ v_1 & v_2 & v_3 & v_4 \\ w_1 & w_2 & w_3 & w_4 \\ x_1 & x_2 & x_3 & x_4 \\ y_1 & y_2 & y_3 & y_4 \\ z_1 & z_2 & z_3 & z_4 \end{bmatrix},
$$

and that the matrix Φ is a basis for $\mathcal{N}(A)$. Find a basis Ψ for $\mathcal{N}(B)$.

Using MINIMAT

Function 146A The .M function Nbasis2, shown in Listing 7, computes
a basis for the null space using Theorem 141C and Corollary 144A. The
command

```
#>  Phi = Nbasis2(A)
```

assigns to Phi a basis for the null space of A. This function uses the bed
Function 108C. It uses the built-in functions sum and round (see Pages 451
and 452 of Appendix C).

Function 146B The .M function Rbasis2, shown in Listing 7, computes
a basis for the range using Theorem 141C and Corollary 144A. The command

```
#>  Psi = Rbasis2(A)
```

assigns to Psi a basis for the range of A.

Basis for the Null Space using MINIMAT

```
function Phi = Nbasis2(A)

    [m, n] = size(A);
    [Q D P] = bed(A);
    r = round(sum(sum(D)));
    if abs(n-r) < 0.5,  return, end
    Phi = P * [zeros(r,n-r); eye(n-r)];
```

Basis for the Range using MINIMAT

```
function Psi = Rbasis2(A)

    [m, n] = size(A);
    [P D Q] = bed(A);
    r = round(sum(sum(D)));
    if abs(r) < 0.5,  return, end
    Psi = P * [eye(r); zeros(m-r,r)];
```

Listing 7

4.8   The Range from the RREF

In general, a matrix and its RREF have different ranges. In Theorem 139A
we computed a basis for $\mathcal{R}(A)$ by transforming the transpose $A^\top$ of A to
RREF. Now we show how to compute a basis for the range $\mathcal{R}(A)$ of a
matrix A directly from the RREF of A itself.

Theorem 147A *Suppose that* $MA = R$ *where* $A, R \in \mathbf{F}^{m \times n}$, M *is in-*
vertible, and R *is in RREF. Then the matrix* $\Psi \in \mathbf{F}^{m \times r}$ *formed from the*
columns of A *that correspond to leading columns of* R *is a basis for the*
range $\mathcal{R}(A)$ *of* A.

Proof: Let $1 \le j_1 < j_2 < \cdots < j_r \le n$ be the indices of the leading
columns of R. These columns are columns of the $m \times m$ identity matrix
$I = I_m$ and the corresponding columns of A are the columns of Ψ:

$$\mathrm{col}_{j_k}(R) = \mathrm{col}_k(I), \qquad \mathrm{col}_{j_k}(A) = \mathrm{col}_k(\Psi), \qquad\qquad 1 \le k \le r.$$

We apply Theorem 144A with $B = R$, $Q = M^{-1}$, and $\widetilde{\Psi}$ formed from the
first r columns of I_m. The matrix $\widetilde{\Psi}$ is a basis for the range $\mathcal{R}(R)$ of the
RREF. (See Exercise 148G below.) Since $A = QR$, we have $\mathrm{col}_j(A) = \mathrm{col}_j(QR) = Q\,\mathrm{col}_j(R)$ for each j. In particular,

$$\mathrm{col}_k(\Psi) = \mathrm{col}_{j_k}(A) = Q\,\mathrm{col}_{j_k}(R) = Q\,\mathrm{col}_j(\widetilde{\Psi}) = \mathrm{col}_j(Q\widetilde{\Psi}),$$

so $\Psi = Q\widetilde{\Psi}$ as required.   □

Example 147B Assume that $MA = R$ where M is invertible and

$$A = \begin{bmatrix} a_{11} & a_{12} & a_{13} & a_{14} \\ a_{21} & a_{22} & a_{23} & a_{24} \\ a_{31} & a_{32} & a_{33} & a_{34} \end{bmatrix}, \qquad R = \begin{bmatrix} c_{11} & 1 & c_{12} & 0 \\ c_{21} & 0 & c_{22} & 1 \\ 0 & 0 & 0 & 0 \end{bmatrix}.$$

Theorem 147A says that the matrices

$$\widetilde{\Psi} = \begin{bmatrix} 1 & 0 \\ 0 & 1 \\ 0 & 0 \end{bmatrix}, \qquad \Psi = \begin{bmatrix} a_{12} & a_{14} \\ a_{22} & a_{24} \\ a_{32} & a_{34} \end{bmatrix},$$

are bases for $\mathcal{R}(R)$ and $\mathcal{R}(A)$ respectively. Note that $M\Psi = \widetilde{\Psi}$. (The
matrix R is in RREF only if $c_{11} = c_{12} = c_{22} = 0$, but the reasoning applies
in the same way.)

▷ **Exercise 148A** Suppose that $MA = R$ where M is invertible and

$$R = \begin{bmatrix} 0 & 1 & * & 0 & * \\ 0 & 0 & * & 1 & * \\ 0 & 0 & 0 & 0 & 0 \end{bmatrix}$$

where the $*$'s are any numbers. Find a basis for $\mathcal{R}(A)$.

▷ **Exercise 148B** For each of the matrices A in Exercise 88E, find a basis for its range.

▷ **Exercise 148C** For each of the matrices A in Exercise 88C, find a basis for its range.

Exercise 148D Show that any two columns of the matrix

$$A = \begin{bmatrix} 1 & 2 & 3 \\ 4 & 5 & 6 \\ 7 & 8 & 9 \end{bmatrix}$$

form a basis for its range. Hint: Use Theorem 136A.

▷ **Exercise 148E** Continue the notation of Example 147B. This exercise and the next give a direct proof that $\mathcal{R}(A) = \mathcal{R}(\Psi)$. By the block multiplication law 53B, a typical element $Y \in \mathcal{R}(\Psi)$ has the form

$$Y = \Psi Z = z_1 A_2 + z_2 A_4$$

where $Z \in \mathbf{F}^{2 \times 1}$, $z_j = \text{entry}_j(Z)$, and $A_j = \text{col}_j(A)$. Show directly that $Y \in \mathcal{R}(A)$ by solving $Y = \Psi X$ for $X \in \mathbf{F}^{4 \times 1}$.

▷ **Exercise 148F** Continue the notation of Example 147B. By the block multiplication law 53B, a typical element $Y \in \mathcal{R}(A)$ has the form

$$Y = AX = x_1 A_1 + x_2 A_2 + x_3 A_3 + x_4 A_4$$

where $X \in \mathbf{F}^{4 \times 1}$ and $x_j = \text{entry}_j(X)$. Show directly that $Y \in \mathcal{R}(\Psi)$ by solving $Y = \Psi Z$ for $Z \in \mathbf{F}^{2 \times 1}$.

▷ **Exercise 148G** The proof of Theorem 147A asserts that the matrix $\widetilde{\Psi}$ formed from the first r columns of the $m \times m$ identity matrix is a basis the range $\mathcal{R}(R)$ of any matrix $\mathbf{F}^{m \times n}$ in RREF with r nonzero rows. Prove this.

▷ **Exercise 148H** Prove that for any matrix A there is a basis Ψ for its range $\mathcal{R}(A)$ with the property that every column of Ψ is a column of A. Must *every* basis for $\mathcal{R}(A)$ satisfy this condition?

Basis for the Range using MINIMAT

```
function Psi = Rbasis3(A)

    [R, lead] = gj(A);
    Psi = A(:,lead);
```

Listing 8

Using MINIMAT

Function 149A The .M function Rbasis3, shown in Listing 8, computes a basis for the range using the method of Theorem 147A. The command

```
#>   Psi = Rbasis3(A)
```

assigns to Psi a basis for the range of A.

4.9 Random Solutions

Function 149B The .M function randsoln, shown in Listing 9, computes a random element of the null space. The MINIMAT statement

```
#> X = randsoln(A)
```

assigns to the variable X a random solution of the homogeneous system $AX = 0$. When invoked with two inputs as in

```
#> X = randsoln(A,Y)
```

the variable X receives a random solution of the inhomogeneous system $AX = Y$, *if this system has a solution.*

Here is how this function works. The .M function randsoln invokes the .M function Nbasis1 explained in Function 140A. This gives a basis Φ for the null space $\mathcal{N}(A)$, and hence a one-one correspondence $X = \Phi Z$ between vectors $Z \in \mathbf{F}^{\nu \times 1}$ vectors $X \in \mathcal{N}(A)$. We simply choose the element of $\mathbf{F}^{\nu \times 1}$ randomly using the built-in rand function. (In the listing, the value of the variable called Phi is Φ and the value of the variable called nullity is the the number ν of columns in Φ.)

The function uses the MINIMAT value nargin, described in Appendix C Page 461, to determine whether the call is randsoln(A) or randsoln(A,Y). In the latter case, it computes a random solution of $AV - wY = 0$ and returns $X = V/w$ if $w \neq 0$.

Random Solutions in MINIMAT

```
function X = randsoln(A, Y)

[m, n] = size(A);
X = zeros(n,1);              % change if possible
if nargin == 1               % homogeneous case
    Phi  =  Nbasis1(A);
    [toss, nullity] = size(Phi);
    if nullity>0
        X = Phi*rand(nullity,1);
    end
elseif nargin == 2           % inhomogeneous case
    W = randsoln([A -Y]);    % recursive call
    if abs(W(n+1)) > 1.0E-9
        X = W(1:n)/W(n+1); % normalize
    end
end
```

Listing 9

4.10   Co-bases(*)

It is easy to decide if $X \in \mathcal{N}(A)$: we just compute AX. It is harder to decide if $Y \in \mathcal{R}(A)$: we must attempt to solve $Y = AX$ for X. A co-basis for a subspace exhibits that subspace as its null space. Once we have a co-basis for a subspace $\mathcal{V}$ it is easy to decide if $X \in \mathcal{V}$.

---

[3](*) This section may be omitted.

Definition 150A Let $\mathcal{V} \subset \mathbf{F}^{n\times 1}$ be a subspace. A **co-basis** for the subspace $\mathcal{V}$ is a matrix $\Lambda \in \mathbf{F}^{k\times n}$ such that

$$\mathcal{R}(\Lambda) = \{0\}, \qquad \mathcal{N}(\Lambda) = \mathcal{V}.$$

▷ **Question 151A** When is a matrix $A \in \mathbf{F}^{m\times n}$ a co-basis for its own null space?

Theorem 151B *Suppose that*

$$R = \begin{bmatrix} I_r & C \\ 0_{(m-r)\times r} & 0_{(m-r)\times(n-r)} \end{bmatrix},$$

and $MA = R$ *as in Theorem 136B and that* Λ *is formed from the last* $(m-r)$ *rows of* M. *Then* Λ *is a co-basis for the range* $\mathcal{R}(A)$ *of* A.

Proof: The matrices M and MY have the form

$$M = \begin{bmatrix} V \\ \Lambda \end{bmatrix}, \qquad MY = \begin{bmatrix} VY \\ \Lambda Y \end{bmatrix},$$

where $V \in \mathbf{F}^{r\times m}$, $\Lambda \in \mathbf{F}^{(m-r)\times m}$, and $Y \in \mathbf{F}^{m\times 1}$. For $X \in \mathbf{F}^{n\times 1}$ denote by $W \in \mathbf{F}^{r\times 1}$ the matrix formed from the first r entries of X and by Z the matrix formed from the last $n - r$ entries. Then

$$X = \begin{bmatrix} W \\ Z \end{bmatrix}, \qquad RX = \begin{bmatrix} W + CZ \\ 0_{(m-r)\times 1} \end{bmatrix}.$$

Hence,

$$\begin{aligned} Y \in \mathcal{R}(A) \quad &\Longleftrightarrow \quad Y = AX \text{ for some } X \\ &\Longleftrightarrow \quad MY = RX \text{ for some } X \\ &\Longleftrightarrow \quad \Lambda Y = 0 \\ &\Longleftrightarrow \quad Y \in \mathcal{N}(\Lambda) \end{aligned}$$

so $\mathcal{R}(A) = \mathcal{N}(\Lambda)$. To see that $\mathcal{R}(\Lambda) = \mathbf{F}^{(m-r)\times 1}$, choose $U \in \mathbf{F}^{(m-r)\times 1}$. Let $Y_1 \in \mathbf{F}^{m\times 1}$ be the column whose last $m - r$ entries are those of U and whose first r entries are zero, and let $Y = M^{-1}Y_1$. Then

$$\begin{bmatrix} 0_{r\times 1} \\ U \end{bmatrix} = Y_1 = MY = \begin{bmatrix} VY \\ \Lambda Y \end{bmatrix}$$

so $U = \Lambda Y \in \mathcal{R}(\Lambda)$ as required. $\square$

Consider the following diagram:

$$\{0\} \to \mathbf{F}^{(n-r)\times 1} \xrightarrow{\Phi} \mathbf{F}^{n\times 1} \xrightarrow{A} \mathbf{F}^{m\times 1} \xrightarrow{\Lambda} \mathbf{F}^{(m-r)\times 1} \to \{0\}$$

Here $\Phi \in \mathbf{F}^{n\times(n-r)}$ is a basis for the null space $\mathcal{N}(A)$ of A, and $\Lambda \in \mathbf{F}^{(m-r)\times m}$ is a co-basis for $\mathcal{R}(A)$. At each space in the diagram, the range of the matrix coming in is equal to the null space of the matrix going out. For example, at $\mathbf{F}^{n\times 1}$ we have $\mathcal{R}(\Phi) = \mathcal{N}(A)$ and at $\mathbf{F}^{m\times 1}$ we have $\mathcal{R}(A) = \mathcal{N}(\Lambda)$. Mathematicians call such a diagram an **exact sequence**, *but this term is not used in this book.*

Example 152A For the matrices A, M, and R of Example 119A the last row of M is $\Lambda = \begin{bmatrix} -1 & -1 & 1 \end{bmatrix}$ and the equation $\Lambda Y = 0$ is just equation $(*)$ on Page 119. The fact that $AX = Y$ has a solution if and only if $\Lambda Y = 0$ is another way of saying that $\mathcal{R}(A) = \mathcal{N}(\Lambda)$.

▷ **Exercise 152B** Let $D = \begin{bmatrix} I_r & 0_{r\times(n-r)} \\ 0_{(m-r)\times r} & 0_{(m-r)\times(n-r)} \end{bmatrix}$ be in biequivalence normal form. Find a co-basis Θ for $\mathcal{N}(D)$ and co-basis Λ for $\mathcal{R}(D)$.

Exercise 152C Suppose $A = QBP^{-1}$ with Q and P invertible. Show that

(1) If $\widetilde{\Lambda}$ is a co-basis for $\mathcal{R}(B)$, then $\Lambda = \widetilde{\Lambda}Q^{-1}$ is a co-basis for $\mathcal{R}(A)$.

(2) If $\widetilde{\Theta}$ is a co-basis for $\mathcal{N}(B)$, then $\Theta = \widetilde{\Theta}P^{-1}$ is a co-basis for $\mathcal{N}(A)$.

(Compare with Theorem 142B.) Here is a diagram to help:

Exercise 152D For each of the four sets of matrices on Pages 111-112 find a co-basis $\widetilde{\Lambda}$ for the range $\mathcal{R}(D)$ and use it to find a basis Λ for the range $\mathcal{R}(A)$.

Exercise 153A For each of the four sets of matrices on Pages 111-112 find a co-basis $\tilde{\Theta}$ for the null space $\mathcal{N}(D)$ and use it to find a co-basis Θ for the null space $\mathcal{N}(A)$.

Exercise 153B The matrices M, A, and R defined by

$$M = \begin{bmatrix} 1 & -1 & 0 \\ -1 & 2 & 0 \\ -1 & -1 & 1 \end{bmatrix}, \qquad A = \begin{bmatrix} 2 & 1 \\ 1 & 1 \\ 3 & 2 \end{bmatrix}, \qquad R = \begin{bmatrix} 1 & 0 \\ 0 & 1 \\ 0 & 0 \end{bmatrix},$$

satisfy $MA = R$. Find a co-basis for $\mathcal{R}(R)$. Find a co-basis for $\mathcal{R}(A)$.

▷ **Exercise 153C** What's the quickest way to find a co-basis for the null space from the RREF?

Exercise 153D Write a .M function Rcobasis such that the command

    #> Lambda = Rcobasis(A)

assigns to Lambda a co-basis for $\mathcal{R}(A)$.

Exercise 153E Write a .M function Ncobasis such that the command

    #> Theta = Rcobasis(A)

assigns to Theta a co-basis for $\mathcal{N}(A)$.

Chapter Summary

Left equivalent matrices have the same null space:

$$\mathcal{N}(QB) = \mathcal{N}(B).$$

Right equivalent matrices have the same range:

$$\mathcal{R}(BP^{-1}) = \mathcal{R}(B).$$

Biequivalent matrices satisfy

$$\mathcal{N}(QBP^{-1}) = P\mathcal{N}(B), \qquad \mathcal{R}(QBP^{-1}) = Q\mathcal{R}(B).$$

A matrix Φ is a basis for a subspace $\mathcal{V}$ iff

$$\mathcal{R}(\Phi) = \mathcal{V}, \qquad \mathcal{N}(\Phi) = \{0\}.$$

A matrix $\Lambda \in \mathbf{F}^{k \times n}$ is a co-basis for a subspace $\mathcal{V}$ iff

$$\mathcal{N}(\Phi) = \mathcal{V}, \qquad \mathcal{R}(\Phi) = \mathbf{F}^{k \times 1}.$$

If $A = QBP^{-1}$, then

- $\widetilde{\Phi}$ *a basis for* $\mathcal{N}(B) \implies \Phi = P\widetilde{\Phi}$ *a basis for* $\mathcal{N}(A)$.

- $\widetilde{\Psi}$ *a basis for* $\mathcal{R}(B) \implies \Psi = Q\widetilde{\Psi}$ *a basis for* $\mathcal{R}(A)$.

- $\widetilde{\Lambda}$ *a co-basis for* $\mathcal{R}(B) \implies \Lambda = \widetilde{\Lambda}Q^{-1}$ *a co-basis for* $\mathcal{R}(A)$.

- $\widetilde{\Theta}$ *a co-basis for* $\mathcal{N}(B) \implies \Theta = \widetilde{\Theta}P^{-1}$ *a co-basis for* $\mathcal{N}(A)$.

5

RANK AND DIMENSION

Associated to every subspace is an integer called its *dimension*. The *rank* of a matrix A is the dimension of its range $\mathcal{R}(A)$. These integers are the most fundamental invariants in the theory of matrix algebra.

5.1 The Definition of Dimension

Definition 155A The number ν of columns in a basis for $\mathcal{V}$ is called the **dimension** of $\mathcal{V}$. In symbols,

$$\nu = \dim(\mathcal{V}).$$

There are two things wrong with this definition. We haven't proved that any two bases for $\mathcal{V}$ have the same number of columns, and we haven't proved that every subspace has a basis. We prove the former assertion in this section and the latter in the next.

The following lemma is crucial. It could have been proved in Chapter 3. In the language of linear systems it says that an underdetermined homogeneous system (more unknowns than equations) has a nonzero solution.

Lemma 156A *If $A \in \mathbf{F}^{m \times n}$ and $m < n$, then $\mathcal{N}(A) \neq \{0\}$.*

Proof: There is a matrix R in RREF that is left equivalent to A and hence has the same null space. Since $m < n$ there must be at least one free column in R, so there is a solution X of $AX = 0$ where the corresponding free unknown takes a nonzero value. (See Chapter 4, e.g., Example 120A.) Hence, $X \neq 0$ and $X \in \mathcal{N}(A)$ as required.   □

Corollary 156B *If $A \in \mathbf{F}^{m \times n}$ and $\mathcal{N}(A) = \{0\}$, then $n \leq m$.*

Proof: This is the contrapositive form of Lemma 156A.   □

▷ **Question 156C** Is the converse to Corollary 156B true?

Change of Basis

Theorem 156D *Any two bases for the same subspace $V \subset \mathbf{F}^{n \times 1}$ are right equivalent. In particular, they have the same size.*

Proof: This is the converse to Theorem 136A. We use the same notation as there. Denote the two bases of the subspace V by

$$\Phi \in \mathbf{F}^{n \times \nu} \quad \text{and} \quad \Psi \in \mathbf{F}^{n \times \mu}.$$

We are required to find an invertible matrix T such that

$$\Psi = \Phi T.$$

Let $E_j = \mathrm{col}_j(I_\mu) \in \mathbf{F}^{\mu \times 1}$ be the jth column of the identity matrix. Then

$$\Psi E_j \in \mathcal{R}(\Psi) = V = \mathcal{R}(\Phi)$$

so $\Psi E_j = \Phi T_j$ for some matrix $T_j \in \mathbf{F}^{\nu \times 1}$. Form the matrix $T \in \mathbf{F}^{\nu \times \mu}$ whose j-column is T_j. By the block multiplication laws 54A and 54B, we have

$$\mathrm{col}_j(\Psi) = \Psi E_j = \Phi T_j = \Phi \, \mathrm{col}_j(T) = \mathrm{col}_j(\Phi T)$$

for $j = 1, 2, \ldots, \mu$, so $\Psi = \Phi T$. Now $\mathcal{N}(T) = \{0\}$: if $TZ = 0$, then $\Psi Z = \Phi T Z = 0$, so $Z \in \mathcal{N}(\Psi) = \{0\}$. From Corollary 156B it follows that $\mu \le \nu$. Reversing the roles of Ψ and Φ we see that $\nu \le \mu$ as well. Hence, $\nu = \mu$: the matrix T is square. But we have shown that $\mathcal{N}(T) = \{0\}$, so T is invertible by Theorem 94A. $\qquad\square$

Definition 157A Assume that the two matrices Φ and Ψ are both bases for the same subspace $\mathcal{V}$. Theorem 156D says that there is a (necessarily unique) invertible matrix T such that

$$\Psi = \Phi T.$$

This matrix T is called the **transition matrix** from Ψ to Φ.

Remark 157B The proof of Theorem 156D gave a method for computing the transition matrix. In Theorem 175C we'll give another method using left inverses.

Remark 157C The inverse matrix T^{-1} is the transition matrix from Φ to Ψ. The following diagram helps us remember which is which.

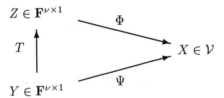

The triangle represents the equation $\Psi Y = \Phi(TY)$. The sloping arrows at the right represent the two parameterizations

$$X = \Phi Z, \quad X = \Psi Y, \qquad X \in \mathcal{V}, \quad Z, Y \in \mathbf{F}^{\nu \times 1}.$$

If $X = \Psi Y$ and $Z = TY$, then $X = \Phi Z$. In other words,

> *Multiplying the representative Y of X in the Ψ parameterization by the transition matrix T from Ψ to Φ produces the representative Z of X in the Φ parameterization.*

Some authors call a basis a **frame** because it gives a *frame of reference* that enables us to *coordinatize*[1] or *name* the elements of a subspace. Choosing another basis gives a different frame of reference and leads to a *change of coordinates* defined by the transition matrix.

---

[1] The words *parameterize* and *coordinatize* are roughly synonymous.

Example 158A Let $V \subset \mathbf{F}^{3 \times 1}$ be the null space of $A = \begin{bmatrix} 1 & 2 & 3 \end{bmatrix}$ so that

$$\Psi = \begin{bmatrix} -7 & -5 \\ 2 & 1 \\ 1 & 1 \end{bmatrix}, \qquad \Phi = \begin{bmatrix} -2 & -3 \\ 1 & 0 \\ 0 & 1 \end{bmatrix},$$

are two bases for V. To find the transition matrix T, we find its columns $T_j = \mathrm{col}_j(T)$ by solving the inhomogeneous system $\mathrm{col}_j(\Psi) = \Phi T_j$. Since

$$\begin{bmatrix} -7 \\ 2 \\ 1 \end{bmatrix} = \begin{bmatrix} -2 & -3 \\ 1 & 0 \\ 0 & 1 \end{bmatrix} \begin{bmatrix} 2 \\ 1 \end{bmatrix}, \qquad \begin{bmatrix} -5 \\ 1 \\ 1 \end{bmatrix} = \begin{bmatrix} -2 & -3 \\ 1 & 0 \\ 0 & 1 \end{bmatrix} \begin{bmatrix} 1 \\ 1 \end{bmatrix},$$

the transition matrix is $T = \begin{bmatrix} 2 & 1 \\ 1 & 1 \end{bmatrix}$.

Theorem 158B *Any subspace V has a unique basis Φ whose transpose $\Phi^\top$ is in reduced row echelon form.*

Proof: In Theorem 163A below, it will be proved that every subspace has a basis Ψ. By the Gauss-Jordan Decomposition (Theorem 105A), the transpose $\Psi^\top$ is left equivalent to a unique matrix R in RREF. Hence, Ψ is right equivalent to a unique matrix $\Phi = R^\top$ whose transpose $R = \Phi^\top$ is in RREF.   □

Remark 158C This provides a method for deciding if two subspaces are equal: find a basis for each, transform the transposes to RREF, and compare. To illustrate, consider the two subspaces $V = \mathcal{N}(A)$ and $W = \mathcal{R}(B)$ where

$$A = \begin{bmatrix} 2 & 4 & 6 \end{bmatrix}, \qquad B = \begin{bmatrix} 1 & -2 & 2 \\ -2 & 4 & -1 \\ 1 & -2 & 0 \end{bmatrix}.$$

Using the methods of Theorem 136B and Theorem 139A, we get bases

$$\Phi = \begin{bmatrix} -2 & -3 \\ 1 & 0 \\ 0 & 1 \end{bmatrix}, \qquad \Psi = \begin{bmatrix} 1 & 2 \\ -2 & -1 \\ 1 & 0 \end{bmatrix},$$

for $\mathcal{V}$ and $\mathcal{W}$, respectively. Transforming $\Phi^\top$ and $\Psi^\top$ to RREF gives

$$\begin{bmatrix} -2 & -3 \\ 1 & 0 \\ 0 & 1 \end{bmatrix} \begin{bmatrix} 0 & 1 \\ -1/3 & -2/3 \end{bmatrix} = \begin{bmatrix} 1 & 0 \\ 0 & 1 \\ -1/3 & -2/3 \end{bmatrix}$$

and

$$\begin{bmatrix} 1 & 2 \\ -2 & -1 \\ 1 & 0 \end{bmatrix} \begin{bmatrix} -1/3 & -2/3 \\ 2/3 & 1/3 \end{bmatrix} = \begin{bmatrix} 1 & 0 \\ 0 & 1 \\ -1/3 & -2/3 \end{bmatrix}.$$

Thus Φ and Ψ are right equivalent. We conclude that $\mathcal{V} = \mathcal{W}$.

▷ **Exercise 159A** Suppose that

$$\Phi = \begin{bmatrix} X_1 & X_2 & \cdots & X_\nu \end{bmatrix} \quad \text{and} \quad \Psi = \begin{bmatrix} Y_1 & Y_2 & \cdots & Y_\nu \end{bmatrix}$$

are bases for the same subspace:

$$\mathcal{R}(\Phi) = \mathcal{R}(\Psi), \qquad \mathcal{N}(\Phi) = \mathcal{N}(\Psi) = \{0\}.$$

(a) Show that each Y_i is a linear combination of $X_1, X_2, \ldots, X_\nu$, that is, that there are numbers b_{ij} such that

$$Y_i = \sum_{j=1}^{\nu} b_{ij} X_j.$$

(b) Let $B \in \mathbf{F}^{\nu \times \nu}$ be the matrix defined by defined by $\text{entry}_{ij}(B) = b_{ij}$, and let T be the transition matrix from Ψ to Φ: $\Psi = \Phi T$. Which of the following is true?

(1) $T = B$ (2) $T = B^{-1}$ (3) $T = B^\top$ (4) $T = \left(B^\top \right)^{-1}$

Exercise 159B Suppose that $\Phi \in \mathbf{F}^{n \times \nu}$ is a basis for $\mathcal{V}$ and $\Psi \in \mathbf{F}^{n \times \mu}$ is a basis for $\mathcal{W}$. Show that $\mathcal{W} \subset \mathcal{V}$ if and only if there is a matrix $T \in \mathbf{F}^{\nu \times \mu}$ such that $\Psi = \Phi T$.

▷ **Exercise 159C** Do not assume that $\mathcal{N}(\Psi) = \{0\}$, $\mathcal{N}(\Phi) = \{0\}$. Is it still true that $\mathcal{R}(\Psi) \subset \mathcal{R}(\Phi)$ if and only if there is a matrix $T \in \mathbf{F}^{\nu \times \mu}$ such that $\Psi = \Phi T$?

Exercise 159D Suppose that Φ is a basis for $\mathcal{V}$ and Ψ is a basis for $\mathcal{W}$. Assume that $\mathcal{V} \not\subset \mathcal{W}$. Show that there is a column X of Φ with $X \notin \mathcal{W}$.

▷ **Exercise 160A** Each of the following matrices Φ_k is a basis for a sub-space $V_k \subset \mathbf{F}^{4 \times 1}$. For each pair (j, k) *either*

- find an invertible matrix T such that $\Phi_k = \Phi_j T$, *or else*

- find a vector $X \in V_j$ with $X \notin V_k$.

(There are $30 = 6 \cdot 5$ problems here.)

$$\Phi_1 = \begin{bmatrix} 2 & 1 \\ 1 & 1 \\ 3 & 4 \\ 0 & 1 \end{bmatrix}, \quad \Phi_2 = \begin{bmatrix} 1 & 2 \\ 1 & 1 \\ 4 & 3 \\ 1 & 0 \end{bmatrix}, \quad \Phi_3 = \begin{bmatrix} 1 & 0 \\ 0 & 1 \\ 0 & 0 \\ 0 & 0 \end{bmatrix},$$

$$\Phi_4 = \begin{bmatrix} 2 & 1 \\ 1 & 1 \\ 0 & 0 \\ 0 & 0 \end{bmatrix}, \quad \Phi_5 = \begin{bmatrix} 1 & 0 \\ 0 & 1 \\ 1 & 0 \\ 0 & 0 \end{bmatrix}, \quad \Phi_6 = \begin{bmatrix} 2 & 1 \\ 1 & 1 \\ 2 & 1 \\ 3 & 3 \end{bmatrix}.$$

Using MINIMAT

Exercise 160B Perform the following exercises for each of the matrices A generated as follows:

(1) `A=rand(4,4)`

(2) `A=rand(3,5)`

(3) `A=zeros(3,5);`
      `for j=1:5, A(:,j)=randsoln([1 2 3]); end`

(4) `A=rand(5,3)`

(5) `A=zeros(5,3);`
      `for j=1:3, A(:,j)=randsoln([1 2 3 4 5]); end`

(6) `A=rand(4,3); A=[A, 2*A(:,1)+A(:,2)-2*A(:,3)]`

(7) `A=rand(3,4); A=[A; 2*A(1,:)+A(2,:)-2*A(3,:)]`

(8) `A=rand(4,3); A=[A, 2*A(:,1)+A(:,2)-2*A(:,3)]`
      `A=[A; 2*A(1,:)+A(2,:)-2*A(3,:)]`

Each exercise of the following exercises asks you to construct two bases Φ and Ψ for a subspace (either null space or range) and to find the transition matrix T. The variables A, Ph, Ps, and T hold the matrices A, Φ, Ψ amd T respectively. You should check your answer with the command

```
#> Ps,  Ph*T
```

(the answers should agree since $\Psi = \Phi T$) and you should check that T is invertible.

▷ **Exercise 161A** For each of the matrices A in Exercise 160B, use the command

```
#> Ph = Nbasis1(A)
```

to generate a basis for the null space. Then use the command

```
#> Ps = Ph*rand(nu,nu)
```

to generate another basis. (You must define the value ν of nu.) The transition matrix T was held in the expression rand(nu,nu), but you have not instructed MINIMAT to remember it. Use the method explained in the text to reconstruct this transition matrix T from Ps and Ph.

Exercise 161B Repeat the previous exercise with the Rbasis1 in place of Nbasis1.

Exercise 161C For each of the matrices A in Exercise 160B, execute

```
#> Ph1=Nbasis1(A), Ph2=Nbasis2(A)
```

Then find the transition matrix from Ph1 to Ph2 and the transition matrix from Ph2 to Ph1.

Exercise 161D For each of the matrices A in Exercise 160B, execute

```
#> Ps1=Rbasis1(A), Ps2=Rbasis2(A), Ps3=Rbasis3(A)
```

Then find all six transition matrices.

5.2   Existence

Now we'll prove that every subspace has a basis. In order to assure that this is true even for the zero subspace, we introduce the convention that the empty matrix[2]

$$0_{n \times 0} \in \mathbf{F}^{n \times 0}$$

with n rows and no columns is a basis for the zero subspace $\{0_{n \times 1}\} \subset \mathbf{F}^{n \times 1}$.

---

[2]See Appendix C Page 462 for a discussion of empty matrices.

Lemma 162A *Suppose that* $\Phi_k \in \mathbf{F}^{n \times k}$ *and that* $\Phi_{k+1} \in \mathbf{F}^{n \times (k+1)}$ *is obtained from* Φ_k *by adjoining a new column:*

$$\Phi_k = \begin{bmatrix} X_1 & X_2 & \cdots & X_k \end{bmatrix}, \qquad \Phi_{k+1} = \begin{bmatrix} \Phi_k & X_{k+1} \end{bmatrix}.$$

Assume that $\mathcal{N}(\Phi_k) = \{0\}$. *Then the following conditions are equivalent:*

(1) $X_{k+1} \in \mathcal{R}(\Phi_k)$;

(2) $\mathcal{R}(\Phi_k) = \mathcal{R}(\Phi_{k+1})$;

(3) $\mathcal{N}(\Phi_{k+1}) \neq \{0\}$.

Proof: First we introduce some notation to be used in the proof. Let $Z_k \in \mathbf{F}^{k \times 1}$ have entries $z_1, \ldots, z_k$, and let $Z_{k+1} \in \mathbf{F}^{(k+1) \times 1}$ be obtained from Z_k by adjoining a new entry z_{k+1} at the bottom.

$$Z_k = \begin{bmatrix} z_1 \\ \vdots \\ z_k \end{bmatrix}, \qquad Z_{k+1} = \begin{bmatrix} Z_k \\ z_{k+1} \end{bmatrix}.$$

By the block multiplication law 53B, we have

$$\Phi_k Z_k = \sum_{j=1}^{k} z_j X_j, \qquad \Phi_{k+1} Z_{k+1} = \sum_{j=1}^{k+1} z_j X_j,$$

so

$$\Phi_{k+1} Z_{k+1} = \Phi_k Z_z + z_{k+1} X_{k+1}. \tag{$*$}$$

The inclusion

$$\mathcal{R}(\Phi_k) \subset \mathcal{R}(\Phi_{k+1}) \tag{$**$}$$

is always true. [Proof: Choose $X \in \mathcal{R}(\Phi_k)$. Then $X = \Phi_k Z_k$ for some Z_k. Let $z_{k+1} = 0$. Then $X = \Phi_k Z_k = \Phi_{k+1} Z_{k+1}$. Therefore $X \in \mathcal{R}(\Phi_{k+1})$.] Now we prove Lemma 162A in four steps.

$(2) \implies (1)$. By Theorem 54A, $X_{k+1} \in \mathcal{R}(\Phi_{k+1})$. Hence, $X_{k+1} \in \mathcal{R}(\Phi_k)$ if (2) holds.

$(1) \implies (2)$. Assume $X_{k+1} \in \mathcal{R}(\Phi_k)$. Choose $\Phi_{k+1} Z_{k+1} \in \mathcal{R}(\Phi_{k+1})$. By the definition of the range, $\Phi_k Z_k \in \mathcal{R}(\Phi_k)$. Since $\mathcal{R}(\Phi_k)$ is a subspace, we conclude that $\Phi_k Z_z + z_{k+1} X_{k+1} \in \mathcal{R}(\Phi_k)$. From $(*)$ we conclude $\Phi_{k+1} Z_{k+1} \in \mathcal{R}(\Phi_k)$. This shows that $\mathcal{R}(\Phi_{k+1}) \subset \mathcal{R}(\Phi_k)$. Hence (2) holds by $(**)$.

(1) $\implies$ (3). Assume (1). Then $X_{k+1} = \Phi_k Z_k$ for some Z_k. Let $z_{k+1} = -1$. By $(*)$, $\Phi_{k+1} Z_{k+1} = 0$, i.e. $Z_{k+1} \in \mathcal{N}(\Phi_{k+1})$. Since $X_{k+1} \neq 0$, (3) holds.

(3) $\implies$ (1). Assume (3). Then $\Phi_{k+1} Z_{k+1} = 0$ for some $Z_{k+1} \neq 0$. By $(*)$, we must have $z_{k+1} \neq 0$, else we would have $\Phi_k Z_k = 0$, contradicting the hypothesis that $\mathcal{N}(\Phi_k) = \{0\}$. Hence, we can divide by z_{k+1} and obtain $X_{k+1} = \Phi_k Y$ where $Y = -Z_k/z_{k+1}$. Thus (1) holds as required. $\square$

Existence of a Basis

Theorem 163A *Every subspace has a basis.*

Proof: Suppose $\mathcal{V} \subset \mathbf{F}^{n \times 1}$ is a subspace. We construct a sequence of matrices $\Phi_k \in \mathbf{F}^{n \times k}$ of form

$$\Phi_k = \begin{bmatrix} X_1 & X_2 & \cdots & X_k \end{bmatrix}$$

where $X_j \in \mathcal{V}$. Define Φ_0 to be the empty matrix (no columns), the only element of $\mathbf{F}^{n \times 0}$. From the block multiplication law 53B and the fact that $\mathcal{V}$ is a subspace, it follows that

$$\mathcal{R}(\Phi_k) \subset \mathcal{V}.$$

Assume that

$$\mathcal{N}(\Phi_k) = \{0\}.$$

(This is true when $k = 0$ by convention.) If $\mathcal{R}(\Phi_k) = \mathcal{V}$, then Φ_k is the desired basis. If not, there is an $X_{k+1} \in \mathcal{V}$ with $X_{k+1} \notin \mathcal{R}(\Phi_k)$. By Lemma 162A, the matrix Φ_{k+1} obtained by adjoining the new column X_{k+1} to Φ_k satisfies the condition that $\mathcal{N}(\Phi_{k+1}) = \{0\}$ and the process continues. This process must stop with $k \leq n$, since a matrix with $k > n$ columns has a nonzero null space by Lemma 156A. $\square$

Extension Theorem

Theorem 164A *Suppose that $\mathcal{V} \subset \mathbf{F}^{n \times 1}$ is a subspace and that $\Psi \in \mathbf{F}^{n \times \mu}$ is a matrix satisfying*

$$\mathcal{R}(\Psi) \subset \mathcal{V}, \qquad \mathcal{N}(\Psi) = \{0\}.$$

Then Ψ may be extended to a basis $\Phi \in \mathbf{F}^{n \times \nu}$ for $\mathcal{V}$ by adjoining additional columns.

Proof: In other words, if $X_1, X_2, \ldots, X_\mu \in \mathcal{V}$ and the matrix

$$\Psi = \begin{bmatrix} X_1 & X_2 & \ldots & X_\mu \end{bmatrix}$$

satisfies $\mathcal{N}(\Psi) = \{0\}$, then we may find columns $X_{\mu+1}, \ldots, X_\nu \in \mathcal{V}$ such that the matrix

$$\Phi = \begin{bmatrix} X_1 & \ldots & X_\mu & X_{\mu+1} & \ldots & X_\nu \end{bmatrix}$$

satisfies

$$\mathcal{N}(\Phi) = \{0\}, \qquad \mathcal{R}(\Phi) = \mathcal{V}.$$

This is the same argument as in Theorem 163A; we start with $\Phi_\mu = \Psi$ instead of with $\Phi_0 =$ the empty matrix.   □

Corollary 164B *A basis for a subspace may be extended to a invertible matrix by adjoining additional columns.*

Extraction Theorem

Theorem 164C *Suppose that $\mathcal{V}$ is a subspace and that $A \in \mathbf{F}^{n \times k}$ is a matrix satisfying $\mathcal{R}(A) = \mathcal{V}$. Then a basis Ψ for $\mathcal{V}$ may be formed from A by deleting some (possibly none) of its columns.*

Proof: This is Theorem 147A. See also Exercise 164D.   □

Exercise 164D Prove the following analog of Lemma 162A. *Suppose that $A \in \mathbf{F}^{n \times (k+1)}$ has a nonzero nullspace: $\mathcal{N}(A) \neq \{0\}$. Then it is possible to delete a column of A to yield a matrix $B \in \mathbf{F}^{n \times k}$ with the same range: $\mathcal{R}(B) = \mathcal{R}(A)$.* Use this to give an alternate proof of Theorem 164C.

Using MINIMAT

▷ **Exercise 165A** To find a basis for the null space of a matrix, we can emulate the proof of Theorem 163A by randomly choosing elements of the null space. Generate matrices with the commands

```
#> A=rand(3,5)
#> Ph1=randsoln(A)
#> Ph2=[Ph1 randsoln(A)]
#> Ph3=[Ph2 randsoln(A)]
```

Which have null space $\{0\}$? Which have range equal to the null space of A? Which is a basis for the null space?

▷ **Exercise 165B** The following commands produce a matrix $A \in \mathbf{F}^{m \times n}$ and matrices $\Phi_k \in \mathbf{F}^{n \times k}$ where $m = 2$, $n = 5$, and $k = 1, 2, 3, 4$. (The variables Ph1, Ph2, Ph3, Ph4, hold the values Φ_1, Φ_2, Φ_3, Φ_4.)

```
#> A=rand(2,5), Ph1=randsoln(A)
#> Ph2 = [Ph1 randsoln(A)]
#> Ph3 = [Ph2 randsoln(A)]
#> Ph4 = [Ph3 randsoln(A)]
```

(1) For which k is $\mathcal{N}(\Phi_k) = \{0\}$?

(2) For which k is $\mathcal{R}(\Phi_k) \subset \mathcal{N}(A)$?

(3) For which k is $\mathcal{R}(\Phi_k) = \mathcal{N}(A)$?

(4) For which k is $\mathcal{R}(\Phi_k) \subset \mathcal{R}(\Phi_{k+1})$?

(5) For which k is $\mathcal{R}(\Phi_k) = \mathcal{R}(\Phi_{k+1})$?

Predict the answer before doing the calculation.

▷ **Exercise 165C** Use MINIMAT to confirm your answers to the previous exercise. The **randsoln** function can be used to decide whether $\mathcal{N}(\Phi) = \{0\}$. To decide whether $\mathcal{R}(B) \subset \mathcal{R}(C)$ choose a random element $X \in \mathcal{R}(B)$ using the command X=B*rand(n,1) and then attempt to solve $X = CZ$ for Z.

Function 165D The .M function Nbasis3, shown in Listing 10, computes a basis for the null space. The command

```
#>  Phi = Nbasis3(A)
```

Basis via Extension

```
function Phi = Nbasis3(A)

  [m,n] = size(A); Phi = zeros(n,0);
  while 0==0 % infinite loop
     X = randsoln(A);
     if sum(abs(randsoln([Phi X])))>0
         return;
     else
         Phi = [Phi X];
     end
  end
```

Basis via Extraction

```
function Psi = Rbasis4(A)

Psi = A; [m k]=size(Psi);
while k>0 % or until return
      X=randsoln(Psi);
      [b j] = max(abs(X'));
      if b==0 return end
      Psi = Psi(:, [1:j-1,j+1,k]);
      k = k-1;
end
```

Listing 10

assigns to Phi a basis for the null space of A. The algorithm mimics the extension argument used in the proof of Theorem 163A. That argument starts out with an empty matrix and adjoins columns successively so long as the null space of the basis being constructed remains {0}. The issue of initializing to an empty matrix is discussed in Appendix C Page 462.

Function 166A The .M function Rbasis4, shown in Listing 10, computes a basis for the range. The command

```
#>  Psi = Rbasis4(A)
```

assigns to Psi a basis for the range of A. The algorithm mimics the proof in Exercise 164D. It initializes Psi to A and then tests whether the null space of Psi is $\{0\}$ by using the randsoln function. If not, the column corresponding to the largest entry in a random solution is deleted. This is repeated till the null space of Psi is $\{0\}$. See the explanation of the built-in max function on Appendix C Page 451.

5.3   The Analogy

Here's how to count a finite set V. First choose an element $x_1 \in V$, then choose another element $x_2 \in V$ distinct from x_1, then choose another element x_3 from V distinct from x_1 and x_2, and proceed in this way until a sequence that exhausts $\mathcal{V}$ is obtained. This will construct sequences $\phi_0 = ()$, $\phi_1, \phi_2, \ldots$ of form

$$\phi_k = (x_1, x_2, \ldots, x_k)$$

which satisfy:

(a) Each ϕ_k is constructed from the previous one by adjoining a new element of V on the end.

(b) The elements of each sequence ϕ_k are **distinct**.

(c) The process continues until a sequence $\phi = \phi_\nu$ **exhausts** V.

Condition (b) ensures that every element of V appears *at most once* in the sequence ϕ; condition (c) means that every element of V appears *at least once* in the sequence. Hence, the last sequence ϕ contains each element of V exactly once. Picking the elements in a different order produces a different sequence, say (y, z, x) instead of (x, y, z), but the length ν of the last sequence will be the same no matter what order is used. This length is called the *cardinality* of the finite set V: it is the number of distinct elements in V.

Now let's do the analogous thing in matrix theory. Instead of a finite set V we have a subspace $\mathcal{V} \subset \mathbf{F}^{n \times 1}$. Instead of the sequence ϕ we have a basis $\Phi \in \mathbf{F}^{n \times \nu}$. The counting procedure used above is analogous to the extension argument used on Page 163 to prove the existence of a basis.

Condition (b) in the counting procedure is not right in the context of matrix theory. It is used to insure that no elements x_j is repeated; this is not analogous to the condition that the columns of Φ be distinct. The correct analog of (b) is the condition that the columns of Φ be *independent*. Here's the definition.

Definition 168A The sequence $(X_1, X_2, \ldots, X_k)$ of vectors is called **independent** iff the only solution $z_1, z_2, \ldots, z_k$ of

$$z_1 X_1 + z_2 X_2 + \cdots + z_k X_k = 0$$

is the trivial solution $z_1 = z_2 = \cdots = z_k = 0$. If $X_j = \mathrm{col}_j(\Phi)$ and $z_j = \mathrm{entry}_j(Z)$, then the block multiplication law 53B says that

$$\Phi Z = z_1 X_1 + z_2 X_2 + \cdots + z_k X_k.$$

This means that the condition that the columns of Φ be independent is the same as the condition that the null space of Φ be the zero subspace: $\mathcal{N}(\Phi) = \{0\}$.

Of course, **dependent** means "not independent".[3] What we call *independence* is sometimes called *linear independence* to distinguish it from other kinds of independence; we do not consider other kinds of independence in this book. The definition of independence makes sense for rectangular matrices X_j so long as they are all the same size. However, we usually apply the definition only to column matrices (vectors).

Condition (c) in the counting procedure is also not right in the context of matrix theory. A subspace $\mathcal{V}$ is infinite (except when $\mathcal{V} = \{0\}$) and no finite sequence could possibly exhaust it. The correct analog of (c) is the condition that the columns of Φ *span* $\mathcal{V}$. Here's the definition.

Definition 168B A sequence $(X_1, X_2, \ldots, X_k)$ of vectors is said to **span** the subspace $\mathcal{V}$ iff the elements of $\mathcal{V}$ are precisely the linear combinations of $X_1, X_2, \ldots, X_k$. If (and only if) $(X_1, X_2, \ldots, X_k)$ spans $\mathcal{V}$, then a X is an element of $\mathcal{V} \iff$ it has the form

$$X = z_1 X_1 + z_2 X_2 + \cdots + z_k X_k$$

for some numbers $z_1, z_2, \ldots, z_k$. Again, by the block multiplication law 53B, the condition that the columns of Φ span $\mathcal{V}$ is the same as the condition that the range of Φ be $\mathcal{V}$: $\mathcal{R}(\Phi) = \mathcal{V}$.

---

[3]WARNING: A common mistake is to write *independent* when you mean *dependent*.

Independence and Spanning

Suppose that $V \subset \mathbf{F}^{n \times 1}$ is a subspace and that $(X_1, X_2, \ldots, X_\nu)$ is a sequence of elements of $\mathbf{F}^{n \times 1}$. Form the matrix

$$\Phi = \begin{bmatrix} X_1 & X_2 & \cdots & X_\nu \end{bmatrix}$$

whose columns are the elements of this sequence. Then

$$(X_1, X_2, \ldots, X_\nu) \quad \text{is independent} \quad \Longleftrightarrow \quad \mathcal{N}(\Phi) = \{0\}$$

and

$$(X_1, X_2, \ldots, X_\nu) \quad \text{spans } V \quad \Longleftrightarrow \quad \mathcal{R}(\Phi) = V.$$

Now let's review the proof (Theorem 163A) of the existence of a basis for a subspace V. First choose an element $X_1 \in V$, then choose another element $X_2 \in V$ independent from X_1, then choose another element X_3 from V independent from X_1 and X_2, and proceed in this way until a sequence that spans V is obtained. This constructs sequences $\Phi_0, \Phi_1, \Phi_2, \ldots$ of form

$$\Phi_k = \begin{bmatrix} X_1 & X_2 \cdots & X_k \end{bmatrix}$$

satisfying

(a) Each Φ_k is constructed from the previous one by adjoining a new element of V on the end.

(b) The columns of each matrix Φ_k are **independent**.

(c) The process continues until the columns of some Φ_k **span** V.

Condition (b) ensures that each Φ_k satisfies $\mathcal{N}(\Phi_k) = \{0\}$; condition (c) ensures that the last matrix Φ_ν satisfies $\mathcal{R}(\Phi_\nu) = V$. Hence, the last matrix Φ_ν is a basis for V. There are now infinitely many choices in (i) but every basis has the same number ν of columns. This number ν is the *dimension* of V; it is the largest number of independent vectors in V.

Exercise 169A Show that two vectors X_1, X_2 are dependent if and only if one is a multiple of the other.

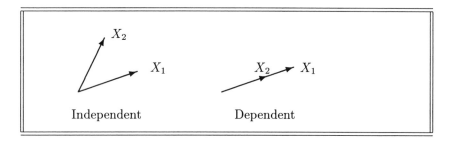

Exercise 170A Let $W_1, W_2, W_3 \in \mathbf{F}^{1 \times n}$ be given by

$$
\begin{aligned}
W_1 &= \begin{bmatrix} 1 & 2 & \cdots & n \end{bmatrix} \\
W_2 &= \begin{bmatrix} n+1 & n+2 & \cdots & 2n \end{bmatrix} \\
W_3 &= \begin{bmatrix} 2n+1 & 2n+2 & \cdots & 3n \end{bmatrix}.
\end{aligned}
$$

Is the sequence (W_1, W_2, W_3) independent?

Exercise 170B Show that if $\mathcal{V} \subset \mathcal{W}$, then $\dim(\mathcal{V}) \leq \dim(\mathcal{W})$. Is the converse true?

Exercise 170C Show that if $\mathcal{V} \subset \mathcal{W}$ and $\dim(\mathcal{V}) = \dim(\mathcal{W})$ then $\mathcal{V} = \mathcal{W}$.

Exercise 170D Let $\#(V)$ denote the cardinality of the finite set V. Explain the formula

$$
\#(V_1 \cup V_2) + \#(V_1 \cap V_2) = \#(V_1) + \#(V_2).
$$

Here $V_1 \cap V_2$ denotes the intersection of V_1 and V_2, that is, the set of elements common to both sets, and $V_1 \cup V_2$ denotes the union of the V_1 and V_2, that is, the set of elements in one or the other (or both).

Exercise 170E Let $\mathcal{V}_1$ and $\mathcal{V}_2$ denote subspaces of $\mathbf{F}^{n \times 1}$. Prove that

$$
\dim(\mathcal{V}_1 + \mathcal{V}_2) + \dim(\mathcal{V}_1 \cap \mathcal{V}_2) = \dim(\mathcal{V}_1) + \dim(\mathcal{V}_2).
$$

5.4   Rank and Nullity

Rank and Nullity

Definition 171A The **rank** of a matrix is the dimension of its range:

$$\text{rank}(A) = \dim \mathcal{R}(A).$$

The **nullity** of a matrix is the dimension of its null space:

$$\text{nullity}(A) = \dim \mathcal{N}(A).$$

Example 171B For the $m \times n$ matrix $D = D_{m,n,r}$ in biequivalence normal form, the rank is r, the number of nonzero entries, and the nullity is $n - r$. (See Theorem 141C.)

Example 171C More generally, for an $m \times n$ matrix R in RREF, the rank is the number r of nonzero rows, and the nullity is the number $n - r$ of free columns. (See Section 4.6 Page 136 and Theorem 147A.)

Invariance of Rank and Nullity

Theorem 171D *Biequivalent matrices have the same rank and nullity.*

Proof: By Corollary 144A.   □

Corollary 171E *The rank of a matrix is the number of nonzero rows in its RREF. The nullity of a matrix is the number of free columns in its RREF.*

Rank Nullity Relation

Corollary 172A *Any matrix A with n columns satisfies*

$$\text{rank}(A) + \text{nullity}(A) = n.$$

Proof: This is true for a matrix in biequivalence normal form (see Example 171B). Hence, by Theorem 171D, it must be true in general, since every matrix is biequivalent to a matrix in biequivalence normal form.   □

The rank of an $m \times n$ matrix A is less than or equal to the number m of rows in A and less than or equal to the number n of columns in A:

$$\text{rank}(A) \leq \min(m, n).$$

The following corollary says when we have equality.

Maximal Rank

Corollary 172B *Suppose $A \in \mathbf{F}^{m \times n}$. Then*

(1) $\text{rank}(A) = n$ *if and only if* $\mathcal{N}(A) = \{0\}$;

(2) $\text{rank}(A) = m$ *if and only if* $\mathcal{R}(A) = \mathbf{F}^{m \times 1}$.

Proof: If $\mathcal{V} \subset \mathbf{F}^{n \times 1}$, then $\mathcal{V} = \{0\}$ if and only if $\dim(\mathcal{V}) = 0$. If $\mathcal{W} \subset \mathbf{F}^{m \times 1}$, then $\mathcal{W} = \mathbf{F}^{m \times 1} \iff \dim(\mathcal{W}) = m$.   □

Transpose and Rank

Corollary 172C *A matrix and its transpose have the same rank.*

Proof: If A and B are biequivalent, so are their transposes A^T and B^T:

$$A^\mathsf{T} = \left(QBP^{-1}\right)^\mathsf{T} = \left(P^{-1}\right)^\mathsf{T} B^\mathsf{T} Q^\mathsf{T}.$$

But the biequivalence normal form $D = D_{m,n,r}$ and its transpose $D^\mathsf{T} = D_{n,m,r}$ have the same rank.   □

▷ **Exercise 173A** Find the rank and nullity of $A = \begin{bmatrix} 1 & 0 & 3 & 4 & 5 \\ 0 & 1 & 1 & 2 & 3 \\ 1 & 1 & 4 & 6 & 8 \end{bmatrix}$.

▷ **Exercise 173B** Prove that Φ is a basis for $\mathcal{N}(A)$ if and only if

(1) $A\Phi = 0$;

(2) $\mathrm{rank}(\Phi) = \mathrm{nullity}(A)$;

(3) $\mathrm{nullity}(\Phi) = 0$.

Using MINIMAT

▷ **Exercise 173C** The rank of a matrix is less than or equal to both the number of its rows and the number of its columns. Confirm that a randomly chosen matrix has maximal rank.

▷ **Exercise 173D** Redo Exercise 165A and explain what happened.

Exercise 173E For each of the matrices in Exercise 160B predict the rank and then confirm your answer by transforming A to RREF and counting the number of nonzero rows.

▷ **Exercise 173F** Confirm Theorem 172C by generating matrices A as in the last exercise and then computing the ranks of A and A'.

5.5   One-Sided Inverses

If A is invertible and $B = A^{-1}$, then both the equations

$$BA = I \qquad\qquad AB = I$$

hold. When A is not a square matrix, it can happen that there is a B for which only one of these equations holds.

Definition 174A Let A be an $m \times n$ matrix. A **left inverse** to A is an $n \times m$ matrix Γ such that

$$\Gamma A = I_n$$

the $n \times n$ identity matrix. A **right inverse** to A is an $n \times m$ matrix Ω such that

$$A\Omega = I_m$$

the $m \times m$ identity matrix. A left inverse or right inverse is sometimes called a **one-sided inverse** to emphasize that it need not be a true inverse.

Example 174B Consider

$$\Gamma = \begin{bmatrix} 1 & 0 & \alpha \\ 0 & 1 & \beta \end{bmatrix}, \qquad A = \begin{bmatrix} 1 & 0 \\ 0 & 1 \\ 0 & 0 \end{bmatrix}.$$

Then $\Gamma A = I$ for any choice of α and β; this shows that A has infinitely many left inverses. Similarly, if

$$A = \begin{bmatrix} 1 & 0 & 0 \\ 0 & 1 & 0 \end{bmatrix}, \qquad \Omega = \begin{bmatrix} 1 & 0 \\ 0 & 1 \\ \alpha & \beta \end{bmatrix},$$

then $A\Omega = I$.

One-sided Inverses and Maximal Rank

Theorem 174C *Suppose* $A \in \mathbf{F}^{m \times n}$. *Then*

(1) *A has a left inverse if and only if* $\operatorname{rank}(A) = n$.

(2) *A has a right inverse if and only if* $\operatorname{rank}(A) = m$.

This is proved in Exercise 177B below. The exercises also show how to find a one-sided inverse when there is one. The strategy is to show that the property of having a one-sided inverse is invariant under biequivalence and thereby reduce the problem to the case where A is in biequivalence normal form.

Independent Columns

Corollary 175A *A matrix Φ has independent columns if and only if it has a left inverse. In particular, a matrix is invertible if and only if it is square and its columns are independent.*

Proof: Suppose that Φ has ν columns. By Definition 168A, Theorem 172B, and Theorem 174C:

$$\text{the columns of } \Phi \text{ are independent} \iff \mathcal{N}(\Phi) = \{0\}$$
$$\iff \text{rank}(\Phi) = \nu$$
$$\iff \Phi \text{ has a left inverse.}$$

According to Theorem 94A and the following Remark 94B, a square matrix is invertible if and only if its null space is zero, and hence, if and only if its columns are independent. $\square$

Remark 175B Left inverses are related to bases for a subspace. Assume that $\mathcal{V} \subset \mathbf{F}^{n \times 1}$ is a subspace and that $\Phi \in \mathbf{F}^{n \times \nu}$ is a basis for $\mathcal{V}$. Since $\mathcal{V} = \mathcal{R}(\Phi)$, the subspace $\mathcal{V}$ is precisely the set of all $X \in \mathbf{F}^{n \times 1}$ for which the inhomogeneous system

$$X = \Phi Z \tag{$*$}$$

has a solution $Z \in \mathbf{F}^{\nu \times 1}$. Since $\mathcal{N}(\Phi) = \{0\}$, the solution Z is unique. We can find Z by solving the system $(*)$, but, if we had to do this for many different values of X, it would be more efficient to find a left inverse Γ for Φ and apply it to both sides of $(*)$:

$$\Gamma X = \Gamma \Phi Z = IZ = Z.$$

This can also be of use in finding the transition matrix between two different bases.

Theorem 175C *Assume that $\mathcal{V} \subset \mathbf{F}^{n \times 1}$ is a subspace and that $\Phi, \Psi \in \mathbf{F}^{n \times \nu}$ are two bases for $\mathcal{V}$. Let Γ be any left inverse to Φ. Then*

$$T = \Gamma \Psi$$

is the transition matrix from Ψ to Φ.

Proof: The transition matrix (see Definition 157A) satisfies $\Psi = \Phi T$. Multiply by Γ.   □

Remark 176A It is quite remarkable that we can take *any* left inverse Γ to find T even though T is uniquely determined by Ψ and Φ. This sometimes happens in mathematics. A problem has a unique answer, but to find it we must make a (somewhat) arbitrary choice.

Exercise 176B Let

$$\Gamma = \begin{bmatrix} -8 & 2 & 5 \\ 4 & -1 & -2 \end{bmatrix}, \quad A = \begin{bmatrix} 1 & 3 \\ 2 & 7 \\ 1 & 2 \end{bmatrix}.$$

Compute ΓA and $A\Gamma$.

Exercise 176C Assume $\Gamma A = I$ and $AX = Y$ where

$$\Gamma = \begin{bmatrix} 3 & 2 & 5 \\ 1 & 1 & 3 \end{bmatrix}, \quad Y = \begin{bmatrix} 1 \\ 7 \\ 3 \end{bmatrix}.$$

Find X.

▷ **Exercise 176D** Suppose that Γ is a left inverse for $A \in \mathbf{F}^{m \times n}$ and that $Y \in \mathbf{F}^{m \times 1}$ and $X \in \mathbf{F}^{n \times 1}$ satisfy $AX = Y$. What is X?

▷ **Exercise 176E** Suppose that Ω is a right inverse for $A \in \mathbf{F}^{m \times n}$ and that $Y \in \mathbf{F}^{m \times 1}$. Find an $X \in \mathbf{F}^{n \times 1}$ such that $AX = Y$.

▷ **Exercise 176F** What's the difference between the last two questions?

▷ **Exercise 176G** Assume that $A \in \mathbf{F}^{m \times n}$. Show that

(1) if A has a left inverse, then $\mathcal{N}(A) = \{0\}$; and

(2) if A has a right inverse, then $\mathcal{R}(A) = \mathbf{F}^{m \times 1}$.

▷ **Exercise 176H** Assume that $A = QDP^{-1}$ where Q and P are invertible. Prove that

(1) if S is a left inverse for D, then PSQ^{-1} is a left inverse for A, and

(2) if S is a right inverse for D, then PSQ^{-1} is a right inverse for A.

▷ **Exercise 177A** Suppose that $D = D_{m,n,r}$ is in biequivalence normal form as Definition 105B. Show that

(1) If $r = n$, then $D^\top$ is a left inverse for D.

(2) If $r = m$, then $D^\top$ is a right inverse for D

▷ **Exercise 177B** Prove Theorem 174C.

▷ **Exercise 177C** Show that if Γ is a left inverse to A and Ω is a right inverse to A, then A is invertible and $\Gamma = \Omega = A^{-1}$. Hint: See Definition 38A.

▷ **Exercise 177D** Show that a square matrix A is invertible if and only if it has a one sided inverse and that A^{-1} is the only left inverse to A and the only right inverse.

▷ **Exercise 177E** For each of the following matrices A either find a left inverse Γ or find an $X \neq 0$ with $X \in \mathcal{N}(A)$.

(1) $\quad A = \begin{bmatrix} 1 & 2 & 3 \\ 2 & 4 & 6 \end{bmatrix}$, $\qquad$ (3) $\quad A = \begin{bmatrix} 2 & 1 & 3 \\ 1 & 1 & 2 \end{bmatrix}$,

(2) $\quad A = \begin{bmatrix} 1 & 2 \\ 2 & 4 \\ 3 & 6 \end{bmatrix}$, $\qquad$ (4) $\quad A = \begin{bmatrix} 2 & 1 \\ 1 & 1 \\ 3 & 2 \end{bmatrix}$.

▷ **Exercise 177F** For each of the matrices A of the last problem either find a right inverse Ω or find a Y such that $Y \notin \mathcal{R}(A)$.

Exercise 177G For each of the four pairs (D, A) of matrices on Pages 111-112 *either*

- find a left inverse for D and use it to find a left inverse for A; *or*

- find a nonzero element of $\mathcal{N}(D)$ and use it to find a nonzero element of $\mathcal{N}(A)$.

Exercise 177H For each of the four pairs (D, A) of matrices on Pages 111-112 *either*

- find a right inverse for D and use it to find a right inverse for A; *or*

- find a vector $\notin \mathcal{R}(D)$ and use it to find a vector $\notin \mathcal{R}(A)$.

Exercise 178A Show that a matrix has a right inverse if and only if its transpose has a left inverse.

▷ **Exercise 178B** Suppose $A \in \mathbf{F}^{m \times n}$ and $B \in \mathbf{F}^{n \times m}$ and that their product $AB \in \mathbf{F}^{m \times m}$ is invertible. Construct a right inverse for A. Construct a left inverse for B.

▷ **Exercise 178C** Let Γ be a left inverse for A. Find a right inverse for Γ.

Exercise 178D Show that if $n \neq m$ and $A \in \mathbf{F}^{m \times n}$ has a one-sided inverse, then A has infinitely many one-sided inverses.

▷ **Exercise 178E** Suppose that $A \in \mathbf{F}^{n \times n}$ is a square matrix. Show that the columns of A are independent if and only if the rows of A are independent. Hint: The transpose of an invertible matrix is invertible.

Exercise 178F Suppose that $B \in \mathbf{F}^{n \times k}$ is obtained from any invertible matrix $A \in \mathbf{F}^{n \times n}$ by discarding some of its columns. Show that B is a basis for $\mathcal{R}(B)$.

Using MINIMAT

Exercise 178G Generate matrices A as in Exercise 160B. For each of these do two things: First, either find a left inverse or a nonzero element of the null space. Second, either find a right inverse or an element not in the range.

Exercise 178H Repeat the previous exercise replacing A by its transpose, that is, generate the matrices A as before but execute the command

```
#>    A = A'
```

before doing the problem.

▷ **Exercise 178I** Assume that $A \in \mathbf{F}^{m \times n}$ and $C \in \mathbf{F}^{n \times m}$. Show that if CA is invertible, then $\Gamma = (CA)^{-1}C$ is a left inverse to A.

▷ **Exercise 178J** Most $m \times n$ matrices A with $m > n$ (more rows than columns) have infinitely many left inverses. Confirm this using MINIMAT.

Exercise 178K Most $m \times n$ matrices A with $m < n$ (more columns than rows) have infinitely many right inverses. Confirm this using MINIMAT.

▷ **Exercise 178L** Write a .M function inv1 so that the command

```
#>   B = inv1(A)
```

assigns to B a left inverse to A when the rank of A is the number of its columns and a right inverse to A when the rank of A is the number of its rows.

Exercise 179A For each of the matrices A_i in Exercise 141A either find a right inverse Ω_i to A_i or a vector $Y_i \notin \mathcal{R}(A_i)$.

Exercise 179B For each of the matrices B_i in Exercise 141B either find a left inverse Γ_i to B_i or a nonzero vector $X_i \in \mathcal{N}(B_i)$.

5.6   Equivalence

Each of the three kinds of equivalence have important invariants: biequivalent matrices have the same rank, left equivalent matrices have the same null space, and right equivalent matrices have the same range. In this section we prove the converses.

Rank and Biequivalence

Theorem 179C *Two matrices of the same size are biequivalent if and only if they have the same rank.*

Proof: 'Only if' is Theorem 171D. For 'if' assume that A and B are matrices of size $m \times n$ and rank r. By Theorem 106D, each is biequivalent to a matrix in biequivalence normal form. Since $D = D_{m,n,r}$ is the only $m \times n$ matrix in biequivalence normal form with rank r, it follows that A and B are both biequivalent to D. Hence, by Theorem 104C they are biequivalent to each other.   □

Corollary 164B says that a basis Φ for the null space of a matrix A can be extended to a basis for the ambient space $\mathbf{F}^{n \times 1}$ by adjoining columns. The next lemma makes explicit what columns may be adjoined.

Lemma 179D *Assume that $A \in \mathbf{F}^{m \times n}$ and that*

$$\Phi = \begin{bmatrix} X_{r+1} & X_{r+2} & \cdots & X_n \end{bmatrix} \in \mathbf{F}^{n \times (n-r)}$$

is a basis for the null space $\mathcal{N}(A)$. Extend Φ to a square matrix

$$P = \begin{bmatrix} \Psi & \Phi \end{bmatrix} = \begin{bmatrix} X_1 & \cdots & X_r, X_{r+1} & \cdots & X_n \end{bmatrix} \in \mathbf{F}^{n \times n}$$

by adjoining the columns $X_1, X_2, \ldots, X_r$ of

$$\Psi = \left[\begin{array}{cccc} X_1 & X_2 & \cdots & X_r \end{array}\right]$$

at the left. Then the matrix P is invertible if and only if the columns of the matrix

$$A\Psi = \left[\begin{array}{cccc} AX_1 & AX_2 & \cdots & AX_r \end{array}\right]$$

are independent.

Proof: Assume that P is invertible. Suppose that

$$z_1 AX_1 + z_2 AX_2 + \cdots + z_r AX_r = 0. \tag{1}$$

Then $z_1 X_1 + z_2 X_2 + \cdots + z_r X_r$ lies in the null space $\mathcal{N}(A)$. As Φ is a basis for this null space there are numbers $z_{r+1}, z_{r+2}, \ldots, z_n$ such that

$$z_1 X_1 + z_2 X_2 + \cdots + z_r X_r = z_{r+1} X_{r+1} + z_{r+2} X_{r+2} + \cdots + z_n X_n. \tag{2}$$

But P is invertible, so its columns are independent. Hence all the z_i vanish. This proves that $AX_1, AX_2, \ldots, AX_r$ are independent.

Conversely, assume that P is not invertible so that (2) has a nontrivial solution $z_1, \ldots, z_n$. Since the vectors $X_{r+1}, \ldots, X_n$ on the right are independent at least one of the coefficients $z_1, \ldots, z_r$ on the left is nonzero. Multiply (2) by A. The right side now vanishes, which proves that the vectors $AX_1, \ldots, AX_r$ are dependent. $\square$

Null Space and Left Equivalence

Theorem 180A *Two matrices of the same size are left equivalent if and only if they have the same null space.*

Proof: For "only if" see Theorem 128A; we prove "if". Assume that the matrices $A, B \in \mathbf{F}^{m \times n}$ have the same null space: $\mathcal{N}(A) = \mathcal{N}(B)$. We must construct an invertible Q with $A = QB$. Let Φ be a basis for the null space $\mathcal{N}(A) = \mathcal{N}(B)$. By Theorem 164A, choose $\Psi \in \mathbf{F}^{n \times r}$ so that $P = \left[\begin{array}{cc} \Psi & \Phi \end{array}\right]$ is invertible. By Lemma 179D, $\mathcal{N}(A\Psi) = \mathcal{N}(B\Psi) = \{0\}$. By Theorem 164A again, there are invertible matrices Q_1, Q_2 of form

$$Q_1 = \left[\begin{array}{cc} A\Psi & \Psi_1 \end{array}\right], \qquad Q_2 = \left[\begin{array}{cc} A\Phi & \Psi_2 \end{array}\right].$$

Let $Q = Q_1 Q_2^{-1}$. Then $A\Psi = QB\Psi$. Since $A\Phi = B\Phi = 0$ we have $AP = QBP$. Now cancel P. $\square$

Range and Right Equivalence

Theorem 181A *Two matrices of the same size are right equivalent if and only if they have the same range.*

Proof: For 'only if' see Theorem 128B; we prove 'if'. Assume that the matrices $A, B \in \mathbf{F}^{m \times n}$ have the same range: $\mathcal{R}(A) = \mathcal{R}(B)$. We must construct an invertible P with $B = AP$. Let $\Psi \in \mathbf{F}^{m \times r}$ be a basis for $\mathcal{R}(A) = \mathcal{R}(B)$. Let Y_j be the jth column of Ψ:

$$\Psi = \begin{bmatrix} Y_1 & Y_2 & \dots & Y_r \end{bmatrix}$$

Choose $X_j \in \mathbf{F}^{n \times 1}$ such that $AX_j = Y_j$. Let

$$\Psi_1 = \begin{bmatrix} X_1 & X_2 & \dots & X_r \end{bmatrix} \in \mathbf{F}^{n \times r}$$

and let

$$\Phi_1 = \begin{bmatrix} X_{r+1} & X_{r+2} & \dots & X_n \end{bmatrix} \in \mathbf{F}^{n \times (n-r)}$$

be basis for the null space $\mathcal{N}(A)$ of A. Form the matrix

$$P_1 = \begin{bmatrix} \Psi_1 & \Phi_1 \end{bmatrix} \in \mathbf{F}^{n \times n}.$$

By Lemma 179D, this matrix is invertible. Clearly, $A\Psi_1 = \Psi$ and $A\Phi_1 = 0$. Similarly, we may find an invertible matrix P_2 of form

$$P_2 = \begin{bmatrix} \Psi_2 & \Phi_2 \end{bmatrix} \in \mathbf{F}^{n \times n}.$$

where $B\Psi_2 = \Psi$ and $B\Phi_2 = 0$. Now choose any $X \in \mathbf{F}^{n \times 1}$. Write it in block form

$$X = \begin{bmatrix} W \\ Z \end{bmatrix}$$

where $W \in \mathbf{F}^{r \times 1}$, $Z \in \mathbf{F}^{(n-r) \times 1}$. Then

$$AP_1 X = A\Psi_1 W + A\Phi_1 Z = \Psi W + 0$$

and

$$BP_2 X = B\Psi_2 W + B\Phi_2 Z = \Psi W + 0$$

so (as this works for any X) $AP_1 = BP_2$ as required. $\square$

▷ **Exercise 182A** For each of the matrices on Pages 111-112, give its rank. There are four sets and each has six matrices $(A, D, P, P^{-1}, Q, Q^{-1})$, so there are 24 questions here.

Exercise 182B There are six pairs of matrices in Exercise 141A. For each pair A_i, A_j either find an invertible matrix $Q = Q_{ij}$ such that $A_i = QA_j$ or find a vector $X = X_{ij}$ that is in the null space of one of the matrices and not the other.

Exercise 182C There are six pairs of matrices in Exercise 141B. For each pair B_i, B_j either find an invertible matrix $P = P_{ij}$ such that $B_i = B_j P$ or find a vector $Y = Y_{ij}$ that is in the range of one of the matrices and not the other.

5.7   Uniqueness of the RREF

The Gauss-Jordan Decomposition theorem was stated on Page 105, but only the existence not the uniqueness of the decomposition was proved there. We give another proof of existence here that has the advantage of making the uniqueness transparent. Before reading this section you should reread the proof of Theorem 147A.

Gauss-Jordan Decomposition

Theorem 182D *Every matrix is left equivalent to a unique matrix in reduced row echelon form.*

Proof: Choose $A \in \mathbf{F}^{m \times n}$. The theorem says that there is a matrix $R \in \mathbf{F}^{m \times n}$ in RREF and an invertible matrix $Q \in \mathbf{F}^{m \times m}$ such that

$$A = QR.$$

It also asserts that the R (but not the Q) is unique; that is, if $A = \widetilde{Q}\widetilde{R}$ (with $\widetilde{Q}$ invertible and $\widetilde{R}$ in RREF), then $\widetilde{R} = R$.

Let $A_j = \mathrm{col}_j(A)$ denote the jth column of A:

$$A = \begin{bmatrix} A_1 & A_2 & \cdots & A_n \end{bmatrix},$$

and let $W_j \subset \mathbf{F}^{m \times 1}$ denote the subspace spanned by the first j columns of A:

$$W_j = \mathcal{R}\left(\begin{bmatrix} A_1 & A_2 & \cdots & A_j \end{bmatrix}\right).$$

Let $j_1 < j_2 < \cdots < j_r$ be those values of j at which the rank jumps:

$$W_{j_i} = W_j \quad \text{for } j_i \leq j < j_{i+1}; \qquad W_{j_i} \neq W_{j_{i+1}}.$$

Define $Q_i \in \mathbf{F}^{m \times 1}$ by

$$Q_i = A_{j_i} \tag{3}$$

and define $\Psi \in \mathbf{F}^{m \times r}$ by

$$\Psi = \begin{bmatrix} Q_1 & Q_2 & \cdots & Q_r \end{bmatrix} \in \mathbf{F}^{m \times r}.$$

By the definition of j_i, we have that

$$\mathcal{R}\left(\begin{bmatrix} Q_1 & Q_2 & \cdots Q_i \end{bmatrix}\right) = W_{j_i} = \mathcal{R}\left(\begin{bmatrix} A_1 & A_2 & \cdots A_j \end{bmatrix}\right)$$

for $j_i \leq j < j_{i+1}$. From this, two things follow. First, Ψ is a basis for the range $\mathcal{R}(A)$. Second, for $j_i < j < j_{i+1}$, there are $unique$ numbers $c_{1j}, c_{2j}, \ldots, c_{ij}$ such that

$$A_j = c_{1j}Q_1 + c_{2j}Q_2 + \cdots + c_{ij}Q_i. \tag{4}$$

Extend Ψ to an invertible matrix

$$Q = \begin{bmatrix} \Psi & Q_{r+1} & Q_{r+2} & \cdots & Q_m \end{bmatrix} \in \mathbf{F}^{m \times m}$$

by adjoining additional columns. Define R by

$$R = Q^{-1}A$$

and let $R_j = \mathrm{col}_j(R)$ denote the jth column of R:

$$R = \begin{bmatrix} R_1 & R_2 & \cdots & R_n \end{bmatrix}.$$

By the block multiplication law 53B,

$$A_j = QR_j$$

for $j = 1, 2, \ldots, n$. Multiply (3) by Q^{-1} to obtain

$$E_i = R_{j_i} \tag{5}$$

for $i = 1, 2, \ldots, r$ where $E_i = \mathrm{col}_i(I_m)$ is ith column of the identity matrix I_m. Multiply (4) by Q^{-1} to obtain

$$R_j = c_{1j}E_1 + c_{2j}E_2 + \cdots + c_{ij}E_i \tag{6}$$

for $j_i < j < j_{i+1}$. Equations (5) and (6) say that R is in RREF.

Conversely, if $A = QR$ where R satisfies (5) and (6), then A satisfies (3) and (4). (Multiply by Q.) Since the entries c_{ij} and the indices j_i are determined uniquely by A, this shows that R is unique.   □

▷ **Exercise 184A** Suppose that $X, Y \in \mathbf{F}^{5 \times 1}$ are independent, that

$$A = \begin{bmatrix} 0 & X & 3X & X - Y & Y & X + Y \end{bmatrix} \in \mathbf{F}^{5 \times 6},$$

and that $MA = R$ where M is invertible and R is in RREF. What is R?

5.8   More Exercises

Characterizations of the Rank

Exercise 184B Prove that the rank of a matrix A is the largest number r such that A has r independent columns.

Exercise 184C Prove that the rank of a matrix A is the largest number r such that A has r independent rows.

Exercise 184D Prove that the rank of a matrix A is the largest number r such that A has an invertible $r \times r$ submatrix. (A $p \times q$ **submatrix** of a matrix A is a matrix $B \in \mathbf{F}^{p \times q}$ formed by discarding $m - p$ of the rows of A and then discarding $n - q$ of the columns.)

Exercise 184E Prove that the rank of a matrix $A \in \mathbf{F}^{m \times n}$ is the smallest number r such that A has a factorization

$$A = \Lambda V, \qquad \Lambda \in \mathbf{F}^{m \times r}, \qquad V \in \mathbf{F}^{r \times n}.$$

A Block Inverse Formula

Exercise 184F Suppose that $B \in \mathbf{F}^{p \times q}$ and $C \in \mathbf{F}^{q \times p}$. Assume that the matrices $I_p - BC$ and $I_q - CB$ are both invertible. Prove that the matrices $(I - BC)^{-1}B$ and $B(I - CB)^{-1}$ commute.

Exercise 184G Let $B \in \mathbf{F}^{p \times q}$ and $C \in \mathbf{F}^{q \times p}$. Assume that $I_p - BC$ and $I_q - CB$ are both invertible. Prove that the matrix

$$A = \begin{bmatrix} I_p & B \\ C & I_q \end{bmatrix}$$

is invertible and find a formula for its inverse.

Exercise 185A In the preceding exercises it is not necessary to assume that both $I_p - BC$ and $I_q - CB$ are invertible, but only that one of them is. Prove that if $I_p - BC$ is invertible, then so is $I_q - CB$.

Geometry and Independence

Exercise 185B Show that $X_1, X_2 \in \mathbf{F}^{2 \times 1}$ are dependent if and only if they lie on the same line through the origin.

Exercise 185C Using rectangular coordinates we may establish a correspondence between pairs (x, y) of real numbers and points in the plane. In these coordinates a line will have an equation of form $ax + by + c = 0$ where either $a \neq 0$ or $b \neq 0$ (or both). Using the notations

$$
L = \begin{bmatrix} a & b & c \end{bmatrix}, \qquad P = \begin{bmatrix} x \\ y \\ 1 \end{bmatrix},
$$

we may write the equation of the line in matrix notation:

$$
LP = 0.
$$

If $d \neq 0$, the rows $L, dL \in \mathbf{R}^{1 \times 3}$ determine the same line. Let L_1, L_2, L_3 represent three distinct lines. Show that the rows L_1, L_2, L_3 are dependent if and only if the corresponding lines are either concurrent (intersect in a point) or parallel.

Exercise 185D A circle in the real (x, y) plane has an equation of form

$$
x^2 + y^2 + ax + by + c = 0.
$$

Find an equation for the circle through the three points $(x_1, y_1) = (1, 0)$, $(x_2, y_2) = (0, 1)$ and $(x_3, y_3) = (0, 0)$. (Hint: a, b, c are the unknowns.)

Exercise 185E What happens when you try to find the circle through the three points $(x_1, y_1) = (2, 0)$, $(x_2, y_2) = (0, 2)$ and $(x_3, y_3) = (1, 1)$?

Exercise 185F An equation of form

$$
d(x^2 + y^2) + ax + by + c = 0.
$$

defines a circle only if $d \neq 0$ and a line if $d = 0$ and either a or b (or both) is nonzero. Repeat the previous exercises with this new form.

Exercise 185G When does the previous equation define a circle of positive radius? A point? The empty set?

Exercise 186A Suppose that C_1, C_2, and C_3 are three circles in the plane and that each pair C_i, C_j of these circles intersect in two points. Let L_{ij} be the line through the two points of intersection of C_i and C_j. Show that the three lines L_{ij} are either concurrent or parallel.

Matrix Representation on a Subspace

In the next two exercises assume that $A \in \mathbf{F}^{m \times n}$, that $\mathcal{V} \subset \mathbf{F}^{n \times 1}$ and $\mathcal{W} \subset \mathbf{F}^{m \times 1}$ are subspaces, and that $A\mathcal{V} \subset \mathcal{W}$.

Exercise 186B Suppose that $\Phi \in \mathbf{F}^{n \times \nu}$ is a basis for $\mathcal{V}$ and $\Psi \in \mathbf{F}^{m \times \mu}$ is a basis for $\mathcal{W}$. Show that there is unique matrix $B \in \mathbf{F}^{\mu \times \nu}$ such that

$$A\Phi = \Psi B.$$

Hint: If $X_k = \mathrm{col}_k(\Phi)$ and $Z_k = \mathrm{col}_k(B)$, then $AX_k = \Psi Z_k$.

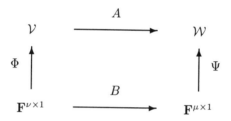

Exercise 186C Suppose that $\Phi_1, \Phi_2 \in \mathbf{F}^{n \times \nu}$ are two bases for $\mathcal{V}$ and that $\Psi_1, \Psi_2 \in \mathbf{F}^{m \times \mu}$ are two bases for $\mathcal{W}$. Show that the corresponding matrices $B_1, B_2 \in \mathbf{F}^{\mu \times \nu}$ of Exercise 186B are biequivalent.

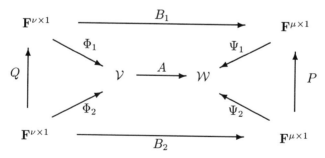

Real Rank vs. Complex Rank

Exercise 187A A real matrix is *a fortiori* a complex matrix but the definition of the set $\mathcal{R}(A)$ depends on whether $\mathbf{F} = \mathbf{R}$ or $\mathbf{F} = \mathbf{C}$. Hence, a real matrix apparently has two ranks, one when $\mathbf{F} = \mathbf{R}$ and another when $\mathbf{F} = \mathbf{C}$. Show that these are the same. Hint: If X and Y are real and of the same size, then $X + iY = 0 \iff X = Y = 0$.

Chapter Summary

For a matrix $A \in \mathbf{F}^{m \times n}$ with $n \leq m$ the following are equivalent:

- *The null space $\mathcal{N}(A)$ of A is the zero subspace $\{0\}$.*

- *The columns of A are independent.*

- *The matrix A has a left inverse.*

- *A has maximal rank: $r = n$.*

For a matrix $A \in \mathbf{F}^{m \times n}$ with $m \leq n$ the following are equivalent:

- *The range $\mathcal{R}(A)$ of A is $\mathbf{F}^{m \times 1}$.*

- *The columns of A span $\mathbf{F}^{m \times 1}$.*

- *The matrix A has a right inverse.*

- *A has maximal rank: $r = m$.*

For a square matrix $A \in \mathbf{F}^{n \times n}$ the above equivalent conditions hold if and only if A is invertible.

Two matrices of the same size are

- *biequivalent iff they have the same rank;*

- *left equivalent iff they have the same null space; and*

- *right equivalent iff they have the same range.*

6

GEOMETRY

In this chapter we study a new operation, the inner product. This operation, together with the operations from Chapter 2, enables us to recast elementary geometry in the language of matrices. This is a double-edged sword: it enables us to use geometrical intuition to guide our matrix calculations and also to prove geometrical theorems using matrix theory.

The inner product is easiest to understand for real matrices, so we consider them first. The complex case is similar, but the definitions involve complex conjugation. For the most part the reader can assume that all numbers are real $(\mathbf{F} = \mathbf{R})$ and simply ignore the overbars. The overbars denote complex conjugation; a real number $z \in \mathbf{R}$ is its own complex conjugate: $z = \bar{z}$.

In freshman calculus there is an operation called the **dot product**. It assigns a scalar (real number) to two vectors $\mathbf{x}$ and $\mathbf{y}$ via the formula

$$\mathbf{x} \cdot \mathbf{y} = \|\mathbf{x}\| \, \|\mathbf{y}\| \cos \theta$$

where $\|\mathbf{x}\|$ is the magnitude of the vector $\mathbf{x}$ and θ is the angle between the vectors $\mathbf{x}$ and $\mathbf{y}$. It can be computed by the formula

$$\mathbf{x} \cdot \mathbf{y} = x_1 y_1 + x_2 y_2 + x_3 y_3$$

where

$$\mathbf{x} = x_1 \mathbf{i} + x_2 \mathbf{j} + x_3 \mathbf{k}, \quad \mathbf{y} = y_1 \mathbf{i} + y_2 \mathbf{j} + y_3 \mathbf{k}.$$

The inner product of two vectors $X, Y \in \mathbf{R}^{3 \times 1}$ is defined by the same formula

$$\langle X, Y \rangle = x_1 y_1 + x_2 y_2 + x_3 y_3$$

where

$$X = \begin{bmatrix} x_1 \\ x_2 \\ x_3 \end{bmatrix}, \qquad Y = \begin{bmatrix} y_1 \\ y_2 \\ y_3 \end{bmatrix}.$$

We won't use the vector notation ($\mathbf{x}$, $\mathbf{i}$, etc.) in this book because we develop a theory that works in $\mathbf{R}^{n \times 1}$ not just in $\mathbf{R}^{3 \times 1}$. Another reason for using the notation $\langle X, Y \rangle$ in place of the notation $\mathbf{x} \cdot \mathbf{y}$ is that the inner product has a further generalization to functions via the formula

$$\langle f, g \rangle = \int_a^b f(x) g(x) \, dx.$$

The dot notation here would be confused with multiplication of functions. (The product of two functions is another function; the inner product of two functions is a number.) The theory of inner products of functions is beyond the scope of this book and is normally studied in courses in partial differential equations.

6.1   Inner Products and Norms

The operations of transpose $X^\top$ and conjugate transpose X^* are used heavily in this chapter. The reader should review the properties listed in Theorems 44A and 45D.

Real Inner Products

Definition 190A The **inner product** of two real matrices $X, Y \in \mathbf{R}^{p \times q}$ is defined by

$$\langle X, Y \rangle = \sum_{i=1}^p \sum_{j=1}^q \operatorname{entry}_{ij}(X) \operatorname{entry}_{ij}(Y).$$

We usually use this definition with $p = n$ and $q = 1$: for $X, Y \in \mathbf{R}^{n \times 1}$ the inner product is defined by

$$\langle X, Y \rangle = x_1 y_1 + x_2 y_2 + \cdots + x_n y_n$$

where

$$X = \begin{bmatrix} x_1 \\ x_2 \\ \vdots \\ x_n \end{bmatrix}, \qquad Y = \begin{bmatrix} y_1 \\ y_2 \\ \vdots \\ y_n \end{bmatrix}.$$

Since

$$Y^{\top} = \begin{bmatrix} y_1 & y_2 & \cdots & y_n \end{bmatrix}$$

we have the formula

$$\langle X, Y \rangle = X^{\top} Y = Y^{\top} X$$

for $X, Y \in \mathbf{R}^{n \times 1}$.

Inner Product Laws (Real)

Theorem 191A *The real inner product satisfies the following:*

$\langle X, X \rangle > 0$ if $X \neq 0$	*(Positivity)*
$\langle X, Y \rangle = \langle Y, X \rangle$	*(Symmetry)*
$\langle aX + bY, Z \rangle = a \langle X, Z \rangle + b \langle Y, Z \rangle$	*(Linearity)*
$\langle AX, Y \rangle = \langle X, A^{\top} Y \rangle$	*(Transpose Law)*

The first three laws hold for $X, Y, Z \in \mathbf{R}^{p \times q}$ and $a, b \in \mathbf{R}$; the last holds for $X \in \mathbf{R}^{n \times 1}$, $Y \in \mathbf{R}^{m \times 1}$, and $A \in \mathbf{R}^{m \times n}$.

Proof: Exercise 194A.

Complex Inner Products

If X is real and nonzero, then $\langle X, X \rangle > 0$ as it is a sum of squares. When $\mathbf{F} = \mathbf{C}$ a sum of squares can be negative or zero or not even real. To avoid this a different formula for the inner product is used when $\mathbf{F} = \mathbf{C}$; it agrees with the formula in the previous section when $\mathbf{F} = \mathbf{R}$.

Definition 191B The **inner product** of two complex matrices $X, Y \in \mathbf{C}^{p \times q}$ is defined by

$$\langle X, Y \rangle = \sum_{i=1}^{p} \sum_{j=1}^{q} \text{entry}_{ij}(X) \overline{\text{entry}_{ij}(Y)}.$$

We usually use this definition with $p = n$ and $q = 1$: the **inner product** of $X \in \mathbf{C}^{n \times 1}$ and $Y \in \mathbf{C}^{n \times 1}$ is

$$\langle X, Y \rangle = \sum_{j=1}^{n} x_j \bar{y}_j$$

where $x_j = \mathrm{entry}_j(X)$ is the jth entry of X and $y_j = \mathrm{entry}_j(Y)$ is the jth entry of Y. This agrees with the real inner product when Y is real, since then $\bar{y}_j = y_j$. This formula can be written

$$\langle X, Y \rangle = Y^* X.$$

The laws for the complex inner product are a little different from the laws for the real inner product.

Inner Product Laws (Complex)

Theorem 192A *The complex inner product satisfies the following:*

$$\langle X, X \rangle > 0 \text{ if } X \neq 0 \qquad \text{(Positivity)}$$
$$\langle X, Y \rangle = \overline{\langle Y, X \rangle} \qquad \text{(Semisymmetry)}$$
$$\langle aX + bY, Z \rangle = a\langle X, Z \rangle + b\langle Y, Z \rangle \quad \text{(Linearity)}$$
$$\langle Z, aX + bY \rangle = \bar{a}\langle Z, X \rangle + \bar{b}\langle Z, Y \rangle \quad \text{(Semilinearity)}$$
$$\langle AX, Y \rangle = \langle X, A^* Y \rangle \qquad \text{(Conjugate Transpose)}$$

The first four laws hold for $X, Y, Z \in \mathbf{C}^{p \times q}$ and $a, b \in \mathbf{C}$; the last holds for $X \in \mathbf{C}^{n \times 1}$, $Y \in \mathbf{C}^{m \times 1}$, and $A \in \mathbf{C}^{m \times n}$.

It is still true that $\langle X, X \rangle \geq 0$ (in particular, it is real) since for any complex number $z = x + iy$ (with x and y real) we have

$$z\bar{z} = (x + iy)(x - iy) = x^2 + y^2.$$

Thus $z\bar{z} = |z|^2$ is real and positive (except when $x = y = 0$). Hence

$$\langle X, X \rangle = \sum_{j=1}^{n} x_j \bar{x}_j = \sum_{j=1}^{n} |x_j|^2 > 0$$

except when $X = 0$. (Here $x_j = \mathrm{entry}_j(X)$.)

Norms

Definition 193A The **norm** of a matrix X is the number

$$\|X\| = \sqrt{\langle X, X \rangle},$$

Some authors call this the **length** or **magnitude** of X.

Norm Laws

Theorem 193B *The norm satisfies the following:*

$\|0\| = 0$	*(Zero)*
$\|X\| > 0$ *if* $X \neq 0$	*(Positivity)*
$\|aX\| = \|a\| \, \|X\|$	*(Homogeneity)*
$\|X + Y\| \leq \|X\| + \|Y\|$	*(Triangle Inequality)*
$|\langle X, Y \rangle| \leq \|X\| \, \|Y\|$	*(Schwarz Inequality)*

Here a is a scalar, and X and Y are matrices of the same size.

Proof: Most of this is developed in the exercises. We prove the real case of the Schwarz Inequality. Form the real valued function

$$f(x) = \|X + xY\|^2 = \langle X + xY, X + xY \rangle.$$

By the positivity law for inner products, we have that $f(x) \geq 0$ for all real numbers x. By the linearity and symmetry laws,

$$f(x) = ax^2 + bx + c$$

where

$$a = \langle Y, Y \rangle = \|Y\|^2, \quad b = 2 \langle X, Y \rangle, \quad c = \langle X, X \rangle = \|X\|^2.$$

Thus $f(x)$ is a quadratic function which does not take negative values: the graph $y = f(x)$ is a parabola that lies above the x=axis. Hence, the equation $f(x) = 0$ does not have two real roots. By the quadratic formula, the roots of $f(x) = 0$ are

$$x = \frac{-b \pm \sqrt{b^2 - 4ac}}{2a}$$

and we must have $b^2 - 4ac \leq 0$ (or else there would be two real roots). This says

$$4 \langle X, Y \rangle^2 \leq 4\|X\|^2 \|Y\|^2.$$

Now divide by 4 and take the square root.   □

▷ **Exercise 194A** Prove the real inner product laws 191A.

Exercise 194B Prove the Schwarz inequality for complex matrices. (Hint: Replace X by cX where the complex number c is chosen so that $\langle cX, Y \rangle$ is real. Then imitate the proof of the real case.)

▷ **Exercise 194C** Prove the formula

$$\|X + Y\|^2 = \|X\|^2 + \|Y\|^2 + 2 \langle X, Y \rangle.$$

when X and Y are real. What is the analog for X and Y complex?

Exercise 194D Prove the triangle inequality. (Hint: Square both sides and use the Schwarz inequality.)

Exercise 194E (Parallelogram Law) Prove the formula

$$\|X + Y\|^2 + \|X - Y\|^2 = 2\|X\|^2 + 2\|Y\|^2$$

Remark 194F Given any nonzero matrix X we can find a multiple cX of X of norm one: since $\|cX\| = |c|\|X\|$, take $c = 1/\|X\|$. Dividing a matrix by its norm is sometimes called **normalizing** it.

6.2   Geometric Interpretation

It is customary to identify $\mathbf{R}^{2 \times 1}$ with the Euclidean plane via the introduction of *rectangular coordinates* as shown in Figure 3. The column vector $X = \begin{bmatrix} x_1 \\ x_2 \end{bmatrix}$ is identified with the directed line segment from the origin $(0,0)$ to the point (x_1, x_2).

To add two vectors

$$X = \begin{bmatrix} x_1 \\ x_2 \end{bmatrix}, \qquad Y = \begin{bmatrix} y_1 \\ y_2 \end{bmatrix}$$

we form the parallelogram having as three of its vertices the points $(0,0)$, (x_1, x_2), and $(y_1, y_2,)$. The fourth vertex is (x_1+x_2, y_1+y_2); its coordinates are the components of $X + Y$. See Figure 3 on Page 196.

To multiply a vector X by a scalar a simply extend the line segment from $(0,0)$ to (x_1, x_2) to one that is a times as long; if $a < 0$, reflect in the origin. The result has components (ax_1, ax_2) representing the product aX and shares a common line with the points $(0,0)$ and (x_1, x_2). See Figure 3.

By the Pythogorean Theorem, the norm of X is the distance from the point with coordinates (x_1, x_2) to the origin:

$$\|X\| = \sqrt{x_1^2 + x_2^2}, \qquad X = \begin{bmatrix} x_1 \\ x_2 \end{bmatrix}.$$

Form the triangle with vertices 0, X, Y. The lengths of the sides of this triangle $\|X\|$, $\|Y\|$, and $\|X - Y\|$. See Figure 4 on Page 197. The **Law of Cosines** from trigonometry asserts that

$$\|X - Y\|^2 = \|X\|^2 + \|Y\|^2 - 2\|X\|\,\|Y\|\cos\theta.$$

By Exercise 194C,

$$\|X - Y\|^2 = \|X\|^2 + \|Y\|^2 - 2\langle X, Y \rangle.$$

Comparing these two we see that the inner product

$$\langle X, Y \rangle = x_1 y_1 + x_2 y_2$$

of two vectors $X = (x_1, x_2)$ and $Y = (y_1, y_2)$ has the interpretation

$$\langle X, Y \rangle = \|X\|\,\|Y\|\cos\theta$$

where θ is the angle between the vectors X and Y. When $\theta = \pi/2$ the Law of Cosines reduces to the Pythogorean Formula. The **triangle inequality**

$$\|X + Y\| \le \|X\| + \|Y\|$$

says that the length of one of the sides is less than or equal the sum of the lengths of the other two. See Figure 4 on Page 197.

These geometric interpretations can guide us in higher dimensions as well. To illustrate we interpret addition of vectors in $\mathbf{R}^{3\times 1}$ geometrically.

Representing Vectors Geometrically

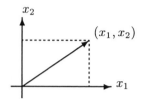

Geometric Vector Addition

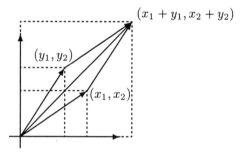

Geometric Scalar Multiplication

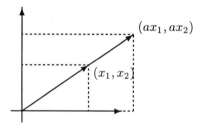

Figure 3

The Inner Product

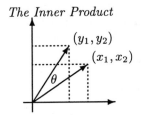

$$\langle X, Y \rangle = \|X\| \, \|Y\| \cos \theta$$

The Law of Cosines

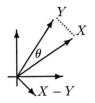

$$\|X - Y\|^2 = \|X\|^2 + \|Y\|^2 - 2 \langle X, Y \rangle.$$

The Triangle Inequality

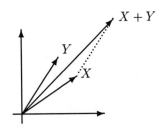

$$\|X + Y\| \leq \|X\| + \|Y\|$$

Figure 4

Points $X \in \mathbf{R}^{3 \times 1}$ can be identified with points in three dimensional Euclidean space. For $X, Y \in \mathbf{R}^{3 \times 1}$ the four points $0, X, Y, X + Y$ form a parallelogram. In particular, since a parallelogram is a plane figure by definition, this means that the four points are coplanar.

▷ **Exercise 198A** Interpret the Schwarz inequality geometrically.

▷ **Exercise 198B** The formula

$$\|X + Y\|^2 + \|X - Y\|^2 = 2\|X\|^2 + 2\|Y\|^2$$

is called the **Parallelogram Law** because it relates the diagonals of a parallelogram to its edges. Explain.

6.3   Unitary Matrices

A lot of computation is usually required to invert a matrix. However, there are certain matrices whose inverse is easy to compute.

Unitary Matrix

Definition 198C A matrix Q is **unitary** iff its conjugate transpose is its inverse:
$$Q^* = Q^{-1}.$$

Remark 198D For a real matrix $Q^\top = Q^*$. Hence, a real matrix Q is unitary if and only if $Q^\top = Q^{-1}$.

Theorem 198E *The unitary matrices satisfy:*

- *The identity matrix is unitary.*

- *The inverse of a unitary matrix is unitary.*

- *A product of unitary matrices is unitary.*

Proof: By Theorems 42A and 45D.   □

Theorem 199A *For a square matrix $Q \in \mathbf{F}^{n \times n}$ the following conditions are equivalent:*

(1) *The matrix Q is unitary.*

(2) *The equation $\langle QX, QY \rangle = \langle X, Y \rangle$ holds for all $X, Y \in \mathbf{F}^{n \times 1}$.*

(3) *The equation $\|QX\| = \|X\|$ holds for all $X \in \mathbf{F}^{n \times 1}$.*

Proof: We shall do the case $\mathbf{F} = \mathbf{R}$. The case $\mathbf{F} = \mathbf{C}$ is left as an exercise. Assume (1). Then $Q^* = Q^{-1}$. Then

$$\langle QX, QY \rangle = \langle Q^*QX, Y \rangle = \langle IX, Y \rangle = \langle X, Y \rangle.$$

Therefore (2).

Assume (2). Then

$$\|QX\|^2 = \langle QX, QX \rangle = \langle X, X \rangle = \|X\|^2.$$

Therefore (3).

Assume (3). Then

$$\begin{aligned}
2\langle QX, QY \rangle &= \|QX + QY\|^2 - \|QX\|^2 - \|QY\|^2 \\
&= \|Q(X + Y)\|^2 - \|QX\|^2 - \|QY\|^2 \\
&= \|X + Y\|^2 - \|X\|^2 - \|Y\|^2 \\
&= 2\langle X, Y \rangle.
\end{aligned}$$

Therefore (2).

Assume (2). Then

$$\langle Q^*QX, Y \rangle = \langle X, Y \rangle$$

for all $X, Y \in \mathbf{R}^{n \times 1}$. Take E_j to be the jth column of the identity matrix:

$$E_j = \mathrm{col}_j(I_n) \quad (j=1,2,\dots n)$$

and $X = E_i$, $Y = E_j$. Then

$$\mathrm{entry}_{ij}(Q^*Q) = \langle Q^*QE_i, E_j \rangle = \langle QE_i, QE_j \rangle = \langle E_i, E_j \rangle = \mathrm{entry}_{ij}(I_n)$$

so $Q^*Q = I$. Therefore (1).   □

▷ **Question 199B** Why was it enough to check just one of the two equations $QQ^* = Q^*Q = I$?

▷ **Exercise 200A** Show that for any real number θ the matrix

$$R(\theta) = \begin{bmatrix} \cos\theta & -\sin\theta \\ \sin\theta & \cos\theta \end{bmatrix}$$

is unitary. Show more generally that $R(\theta + \phi) = R(\theta)R(\phi)$ and $R(0) = I$ so $R(\theta)^{-1} = R(-\theta) = R(\theta)^\top$. Does every real 2×2 unitary matrix have the form $Q = R(\theta)$?

▷ **Exercise 200B** Show that a real matrix $Q = \begin{bmatrix} q_{11} & q_{12} \\ q_{21} & q_{22} \end{bmatrix}$ is unitary iff $q_{11}^2 + q_{21}^2 = q_{12}^2 + q_{22}^2 = 1$ and $q_{11}q_{12} + q_{21}q_{22} = 0$.

Exercise 200C (Hadamard Matrices) Show that the 4×4 matrix

$$Q = \frac{1}{2} \begin{bmatrix} 1 & 1 & 1 & 1 \\ 1 & 1 & -1 & -1 \\ 1 & -1 & 1 & -1 \\ 1 & -1 & -1 & 1 \end{bmatrix}$$

is unitary. Find an 8×8 unitary matrix with no zero entry.

6.4   Orthonormal Bases

Recall from Section 6.2 that the inner product $\langle X, Y \rangle$ of two column vectors X and Y has the interpretation

$$\langle X, Y \rangle = \|X\| \, \|Y\| \cos\theta$$

where θ is the angle between X and Y. Because two vectors are orthogonal iff the angle between them is $\pm\pi/2$, one says that matrices X and Y are *orthogonal* iff their inner product is zero.

Definition 200D Two vectors X and Y are said to be **orthogonal** (written $X \perp Y$) iff their inner product is zero:

$$X \perp Y \iff \langle X, Y \rangle = 0.$$

A sequence of vectors is called **pairwise orthogonal** iff any two distinct elements of the sequence are orthogonal. A vector X is called **normalized** iff its norm $\|X\| = \sqrt{\langle X, X \rangle}$ is one. A sequence of vectors $(U_1, U_2, \ldots, U_k)$ is called **orthonormal** iff

$$\langle U_i, U_j \rangle = \begin{cases} 0 & \text{if } i \neq j, \\ 1 & \text{if } i = j. \end{cases}$$

In other words, an orthonormal sequence is a pairwise orthogonal sequence whose elements are normalized.

For example, for any real number θ, the sequence (U_1, U_2) where

$$U_1 = \begin{bmatrix} \cos\theta \\ \sin\theta \end{bmatrix}, \quad U_2 = \begin{bmatrix} -\sin\theta \\ \cos\theta \end{bmatrix},$$

is orthonormal.

Lemma 201A *Let $U_j = \mathrm{col}_j(\Phi)$ be the jth column of a matrix $\Phi \in \mathbf{F}^{n \times \mu}$, that is,*

$$\Phi = \begin{bmatrix} U_1 & U_2 & \cdots U_k \end{bmatrix}.$$

Then

$$\langle U_j, U_i \rangle = \mathrm{entry}_{ij}(\Phi^*\Phi).$$

Proof:

$$\begin{aligned} \mathrm{entry}_{ij}(\Phi^*\Phi) &= \mathrm{row}_i(\Phi^*)\,\mathrm{col}_j(\Phi) \\ &= \mathrm{col}_i(\Phi)^*\,\mathrm{col}_j(\Phi) \\ &= \langle \mathrm{col}_j(\Phi), \mathrm{col}_i(\Phi) \rangle \\ &= \langle U_j, U_i \rangle \qquad \square \end{aligned}$$

Orthonormal Columns

Theorem 201B *A matrix Φ has orthonormal columns iff its conjugate transpose Φ^* is a left inverse for Φ:*

$$\Phi^*\Phi = I.$$

In particular, a matrix is unitary if and only if it is square and its columns are orthonormal.

Proof: By Lemma 201A, $\Phi^*\Phi = I$ if and only if

$$\langle U_j, U_i \rangle = \begin{cases} 0 & \text{if } i \neq j, \\ 1 & \text{if } i = j. \end{cases}$$

(Theorem 201B should be compared with Corollary 175A.)   □

▷ **Question 202A** Do the rows of a unitary matrix form an orthonormal sequence?

Definition 202B Suppose $\mathcal{V} \subset \mathbf{F}^{n \times 1}$ is a subspace. An **orthonormal basis** for $\mathcal{V}$ is a basis for $\mathcal{V}$ whose columns are orthonormal.

Example 202C Any $n \times n$ unitary matrix is an orthonormal basis for $\mathbf{F}^{n \times 1}$. Any matrix obtained from a unitary matrix by deleting some of its columns is an orthonormal basis for its range. (Theorem 210A below says that every orthonormal basis is obtained this way.)

Remark 202D Suppose that $\Phi \in \mathbf{F}^{n \times \nu}$ is a basis for $\mathcal{V}$. Then any $X \in \mathcal{V}$ can be uniquely expressed as a linear combination of the columns of Φ; this means that we can find a unique $Z \in \mathbf{F}^{\nu \times 1}$ such that

$$X = \Phi Z. \qquad (*)$$

Normally, we must solve a system of inhomogeneous equations to find the entries of Z. However, when the basis Φ is orthonormal, there is a much easier way to find the coefficients. The equation $\Phi^*\Phi = I$ says that Φ^* is a left inverse for Φ. To solve the equation $(*)$ for Z, we need only multiply by Φ^*:

$$X = \Phi Z \implies Z = \Phi^* X.$$

Compare this with Remark 175B. Here is another way of looking at this:

Theorem 202E (Orthonormal Expansion)   *Assume that*

$$\Phi = \begin{bmatrix} U_1 & U_2 & \cdots U_\nu \end{bmatrix}$$

is an orthonormal basis for $\mathcal{V}$. *Then for* $X \in \mathcal{V}$ *the coefficients in the expansion*

$$X = \sum_{k=1}^{\nu} z_k U_k$$

are given by

$$z_j = \langle X, U_j \rangle.$$

Proof: We choose $j = 1, 2, \ldots, \nu$ and compute the inner product of X with U_j using the expansion:

$$\langle X, U_j \rangle = \left\langle \sum_{k=1}^{\nu} z_k U_k, U_j \right\rangle.$$

By the properties of the inner product,

$$\langle X, U_j \rangle = \sum_{k=1}^{\nu} z_k \langle U_k, U_j \rangle.$$

Since the U_j are pairwise orthogonal, $\langle U_k, U_j \rangle = 0$ for $k \neq j$, so only one of the terms on the right survives:

$$\langle X, U_j \rangle = z_j \langle U_j, U_j \rangle.$$

As $\langle U_j, U_j \rangle = 1$, this is the desired result.   □

Corollary 203A *An orthonormal sequence is independent.*

Proof: If $\sum_k z_k U_k = 0$, then all $z_k = 0$ by Theorem 202E. For an alternate proof compare Corollary 175A with Theorem 201B.   □

▷ **Question 203B** Is a sequence whose elements are pairwise orthogonal an independent sequence?

▷ **Exercise 203C (Parseval's Formula)** Assume that

$$\Phi = \begin{bmatrix} U_1 & U_2 & \cdots U_\nu \end{bmatrix}$$

is an orthonormal basis for $\mathcal{V}$ and suppose X and Y are two elements of $\mathcal{V}$. Prove that

$$\langle X, Y \rangle = \sum_{k=1}^{\nu} \langle X, U_k \rangle \langle U_k, Y \rangle.$$

In particular,

$$\|X\| = \sqrt{\sum_{k=1}^{\nu} |\langle X, U_k \rangle|^2}.$$

▷ **Exercise 203D** Let A be a rectangular matrix. Show that the rows of A form an orthonormal sequence if and only if $AA^* = I$.

Exercise 204A Prove the following generalization of Theorem 199A to nonsquare matrices: For a rectangular matrix $\Phi \in \mathbf{F}^{n \times \mu}$ the following conditions are equivalent:

(0) The columns of Φ are orthonormal.

(1) The conjugate transpose Φ^* is a left inverse to Φ.

(2) The equation $\langle \Phi Z_1, \Phi Z_2 \rangle = \langle Z_1, Z_2 \rangle$ holds for all $Z_1, Z_2 \in \mathbf{F}^{\mu \times 1}$.

(3) The equation $\|\Phi Z\| = \|Z\|$ holds for all $Z \in \mathbf{F}^{\mu \times 1}$.

▷ **Exercise 204B** Calculate $\langle X_i, X_j \rangle$ where

$$
X_1 = \begin{bmatrix} a \\ 0 \\ 0 \end{bmatrix}, \qquad X_2 = \begin{bmatrix} b \\ d \\ 0 \end{bmatrix}, \qquad X_3 = \begin{bmatrix} c \\ e \\ f \end{bmatrix}.
$$

What are the triangular unitary matrices? What are the diagonal unitary matrices?

6.5   The Gram-Schmidt Decomposition

In this section we'll learn an algorithm which converts *any* basis Ψ for a subspace $\mathcal{V}$ into an orthonormal basis Φ for the same subspace. The algorithm is called the **Gram-Schmidt Process**. The Gram-Schmidt Process is like the algorithm of Gauss-Jordan Elimination described in Theorem 84C, in that it uses the elementary operations of Chapter 3. There are three crucial differences:

- Gauss-Jordan Elimination acts on the rows of a matrix via elementary row operations, whereas the Gram-Schmidt Process acts on the columns;

- not all of the elementary column operations are allowed in the Gram-Schmidt Process; and

- Gauss-Jordan Elimination can be applied to any matrix, whereas the Gram-Schmidt process fails if it is applied to a matrix with dependent columns.

Positive Triangular Matrices

The first step in explaining the Gram-Schmidt Process is to specify the allowable column operations. Recall Definition 48E that a square matrix T is called **triangular** iff all its entries below the diagonal vanish, that is, iff $\text{entry}_{ij}(T) = 0$ for $j < i$. When one says a complex number is positive, one means that it is *real* and positive.

Positive Triangular

Definition 205A A square matrix T is called **positive triangular** iff it is triangular and all its diagonal entries are positive, $\text{entry}_{ii}(T) > 0$.

Theorem 205B (Factorization Theorem) *A matrix T is positive triangular if and only if it is a product of "upper shears'*

$$E = \text{Shear}(I, p, q, c), \qquad q > p,$$

and "positive scales"

$$E = \text{Scale}(I, p, c), \qquad c > 0.$$

Proof: Mimic the proof of the Factorization Theorem 93B. Note which operations are required to transform a positive triangular matrix to the identity.   □

Theorem 205C *The positive triangular matrices satisfy the following:*

- *the identity matrix is positive triangular;*

- *the inverse of an positive triangular matrix is positive triangular;*

- *a product of positive triangular matrices is positive triangular.*

Proof: The identity matrix is obviously positive triangular. By Theorem 205B, a matrix is positive triangular if and only if factors as

$$T = E_1 E_2 \cdots E_k$$

where $E_1, E_2, \ldots, E_k$ are either upper shears or positive scales. Two such products

$$(E_1 E_2 \cdots E_k)(F_1 F_2 \cdots F_h) = E_1 E_2 \cdots E_k F_1 F_2 \cdots F_h$$

is another product of this form, so the product of two positive triangular matrices is positive triangular. The inverse

$$(E_1 E_2 \cdots E_k)^{-1} = E_k^{-1} \cdots E_2^{-1} E_1^{-1}$$

of such a product is another such, so the inverse of a positive triangular matrix is positive triangular.   □

Positive Triangular Equivalence

Definition 206A Let Ψ and Φ be two matrices of the same size. We say that Ψ is **positive triangular right equivalent** to Φ iff there is a positive triangular matrix T with

$$\Psi = \Phi T.$$

Theorem 206B *This relation satisfies the following:*

(Reflexive Law) $A \equiv A$;

(Symmetric Law *if $A \equiv B$, then $B \equiv A$.*

(Transitive Law) *if $A \equiv B$ and $B \equiv C$, then $A \equiv C$.*

Here $A \equiv B$ means "A is positive triangular right equivalent to B" .

Proof: As in Theorem 104C.   □

The Gram-Schmidt Process

Gram-Schmidt Decomposition

Theorem 206C *Any matrix with independent columns is positive triangular right equivalent to a unique matrix with orthonormal columns.*

Proof: In other words, if $\Psi \in \mathbf{F}^{n \times \nu}$ has independent columns, then Ψ may be written uniquely in the form

$$\Psi = \Phi T \tag{1}$$

where $\Phi \in \mathbf{F}^{n \times \nu}$ has orthonormal columns and $T \in \mathbf{F}^{\nu \times \nu}$ is positive triangular. The proof gives an algorithm, called the **Gram-Schmidt Process**, for finding the decomposition. The algorithm tells how to choose the elementary column operations to transform a given matrix Ψ to a matrix Φ with orthonormal columns.

Let V_j denote the jth column of given matrix Ψ; let U_j and Z_j denote the jth columns of the matrices Φ and T to be constructed:

$$\begin{aligned}
\Psi &= \begin{bmatrix} V_1 & V_2 & \cdots & V_\nu \end{bmatrix}, \\
\Phi &= \begin{bmatrix} U_1 & U_2 & \cdots & U_\nu \end{bmatrix}, \\
T &= \begin{bmatrix} Z_1 & Z_2 & \cdots & Z_\nu \end{bmatrix}.
\end{aligned}$$

Denote the entries of T by

$$z_{jk} = \text{entry}_{jk}(T) = \text{entry}_j(Z_k).$$

Then equation (1) can be rewritten as

$$V_k = \sum_{j=1}^{k} z_{jk} U_j. \tag{1.k}$$

First we solve (1.1) for z_{11} and U_1; then we solve (1.2) for z_{12}, z_{22} and U_2; then we solve (1.3) for z_{13}, z_{23}, z_{33} and U_3; and so on. The process is inductive, meaning that to solve (1.k), we use the fact that $U_1, U_2, \ldots, U_{k-1}$ have already been found. Since the columns of Φ are to be orthonormal, we require

$$\langle U_i, U_j \rangle = 0, \qquad i \neq j, \tag{2}$$

and

$$\|U_j\| = 1. \tag{3}$$

As the hypothesis of induction we assume (2) and (3) for $i, j < k$. Take the inner product of both sides of (1.k) with U_j:

$$z_{kj} = \langle V_k, U_j \rangle.$$

For $j < k$ this defines z_{jk}. Now define

$$W_k = V_k - \sum_{j<k} z_{jk} U_j. \tag{4}$$

By induction, the terms on the right in (4) are known. By $(1.k)$,

$$W_k = z_{kk} U_k.$$

Since T is to be *positive* triangular, we require $z_{kk} > 0$. Thus z_{kk} must be the norm of W_k, and U_k must be the normalization of W_k:

$$z_{kk} = \|W_k\|, \qquad U_k = \|W_k\|^{-1} W_k.$$

(If $W_k = 0$, then the columns of Ψ are dependent and the algorithm fails.) We have solved $(1.k)$. $\square$

Remark 208A The formula (4) for W_k may be written as

$$W_k = V_k - \Phi_{k-1}^* Z_k. \tag{4$'$}$$

(This form is used in Listing 11 on Page 212.)

Remark 208B Let

$$\begin{aligned} \Psi_k &= \begin{bmatrix} V_1 & V_2 & \cdots & V_k \end{bmatrix} \\ \Phi_k &= \begin{bmatrix} U_1 & U_2 & \cdots & U_k \end{bmatrix}. \end{aligned}$$

The construction of U_k from V_k and Ψ_k is uniquely determined by the following conditions:

(a) $\Phi_k^* \Phi_k = I_k$;

(b) $\mathcal{R}(\Phi_k) = \mathcal{R}(\Psi_k)$;

(c) $\langle V_k, U_k \rangle > 0$.

Condition (a) says that Φ has orthonormal columns, condition (b) says that T is triangular, and condition (c) says that T has positive diagonal entries.

Geometric Interpretation

Example 208C The case $\nu = 3$ has a nice geometric interpretation. The matrix T will have the form

$$T = \begin{bmatrix} z_{11} & z_{12} & z_{13} \\ 0 & z_{22} & z_{23} \\ 0 & 0 & z_{33} \end{bmatrix}$$

with positive entries on the diagonal. Let V_j and U_j be respectively the jth column of the given basis Ψ and the jth column of the basis Φ to be found:

$$\Psi = \begin{bmatrix} V_1 & V_2 & V_3 \end{bmatrix}$$
$$\Phi = \begin{bmatrix} U_1 & U_2 & U_3 \end{bmatrix}.$$

The equation $\Psi = \Phi T$ may be written as

$$V_1 = z_{11}U_1$$

$$V_2 = z_{12}U_1 + z_{22}U_2$$

$$V_3 = z_{13}U_1 + z_{23}U_2 + z_{33}U_3.$$

The vector V_1 determines a line L, and the vectors V_1 and V_2 determine a plane M. The vector U_1 is the unit normal in the line L in the direction V_1. The vector U_2 lies in the plane M and is the unit normal to the line L on the same side of L as V_2. The vector U_3 is the unit normal to the plane M on the same side as V_3.

▷ **Question 209A** What does "on the same side as" mean?

The Gram-Schmidt process tells us to "project and normalize". For example, vector $W_2 = V_2 - \langle V_2, U_2 \rangle U_1$ is difference of V_2 with the projection of V_2 along U_1, and U_2 is obtained by normalizing W_2. Let's go through the steps in Example 208C:

(1) To find U_1 we normalize V_1:

$$U_1 \qquad\qquad V_1 = z_{11}U_1$$

(2) To find W_2 we subtract from V_2 its projection on the line along U_1. Then we find U_2 by normalizing W_2.

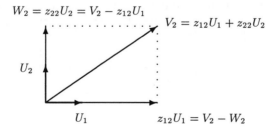

(3) To find W_3 we subtract from V_3 its projection on the plane containing U_1 and U_2. Then we find U_3 by normalizing W_3.

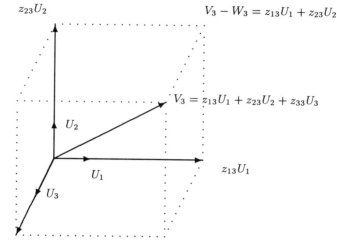

$$W_3 = z_{33}U_3 = V_3 - z_{13}U_1 - z_{23}U_2$$

The following corollary to the Gram-Schmidt Process says that any matrix with orthonormal columns may be obtained from some unitary matrix by discarding some of its columns.

Orthonormal Extension

Corollary 210A *A matrix with orthonormal columns may be extended to a unitary matrix by adjoining additional columns.*

Proof: By Corollary 164B, extend the given matrix to an invertible matrix by adding additional columns. Apply the Gram-Schmidt process to this invertible matrix. This will not affect the original columns because they are already orthonormal.   □

▷ **Exercise 211A** Write each of the following matrices as ΦT where Φ has orthonormal columns and T is positive triangular.

$$
(1) \begin{bmatrix} 1 & 7 \\ 1 & 7 \\ 1 & 3 \\ 1 & 3 \end{bmatrix}. \quad
(2) \begin{bmatrix} 1 & 7 & 17 \\ 1 & 7 & 11 \\ 1 & 3 & 5 \\ 1 & 3 & -1 \end{bmatrix}. \quad
(3) \begin{bmatrix} 1 & 7 & 17 & 20 \\ 1 & 7 & 11 & -2 \\ 1 & 3 & 5 & -6 \\ 1 & 3 & -1 & -12 \end{bmatrix}.
$$

6.6   Using MINIMAT

The command

```
#>   c = Y'*X
```

will assign to the variable c the inner product

$$c = \langle X, Y \rangle$$

of two column vectors X and Y. If the vectors X and Y are real, the command c=X'*Y also works, but in the complex case, this command gives the complex conjugate of the inner product. The command

```
#>   s = sqrt(X'*X)
```

uses MINIMAT's built-in square root function to compute the norm

$$s = \|X\| = \sqrt{\langle X, X \rangle}$$

of the column vector X.

▷ **Exercise 211B** Generate a random 5×3 matrix and compute the result of applying the Gram-Schmidt Process to its columns. Do not use the .M function gs. Follow Example 208C.

▷ **Exercise 211C** Using MINIMAT construct a random 2×5 matrix and find an orthonormal basis matrix for its null space. (You may use MINIMAT's Nbasis1 .M function, but do not use the .M function gs.

Function 211D The .M function gs, shown in Listing 11, implements the Gram-Schmidt process. The call

```
#> [Phi T] = gs(Psi)
```

produces a matrix Phi with orthonormal columns and an invertible triangular matrix T such that Psi=Phi*T (provided that Psi has independent columns).

▷ **Exercise 211E** Test gs on a random 5×3 matrix.

The Gram-Schmidt Process using MINIMAT

```
function [Phi, T] = gs(Psi)

    [n, nu] = size(Psi);
    Phi=zeros(n,nu); T=zeros(nu,nu);
    for k=1:nu
        V = Psi(:,k);
        Z = Phi'*V;
        W = V -  Phi*Z;
        c = sqrt(W'*W);
        if c<1.0E-9
            disp('dependent columns')
            return
        else
            T(:, k) = Z;
            T(k,k) = c;
            Phi(:,k) =  W/c;
        end
    end
```

Listing 11

6.7   Projection (General)

Definition 212A Let $\mathcal{V}$ be a subspace of $\mathbf{F}^{n \times 1}$. A **complement** to $\mathcal{V}$ is a subspace $\mathcal{W}$ of $\mathbf{F}^{n \times 1}$ such that

$$\mathcal{V} + \mathcal{W} = \mathbf{F}^{n \times 1} \tag{1}$$

and

$$\mathcal{V} \cap \mathcal{W} = \{0\} \tag{2}$$

The definition is symmetric in $\mathcal{V}$ and $\mathcal{W}$, so $\mathcal{W}$ is a complement to $\mathcal{V}$ if and only if $\mathcal{V}$ is a complement to $\mathcal{W}$. Subspaces satisfying (1) and (2) are called **complementary**.

Theorem 213A *Assume that V and W are complementary. Then every $Z \in \mathbf{F}^{n \times 1}$ can be written uniquely in the form*

$$Z = X + Y, \qquad\qquad X \in V, \quad Y \in W. \qquad (3)$$

The element $X \in V$ defined by equation (3) is called the **projection** *of Z on V along W. See Figure 5 on Page 218.*

Proof: By definition, the subspace $V + W$ consists of all elements $X + Y$ where $X \in V$ and $Y \in W$. Hence, equation (1) says that every $Z \in \mathbf{F}^{n \times 1}$ can be written as $Z = X + Y$ where $X \in V$ and $Y \in W$. Equation (2) says that this representation is unique: if $Z = X_1 + Y_1 = X_2 + Y_2$, then

$$X_1 - X_2 = Y_2 - Y_1 \in V \cap W = \{0\},$$

so $X_1 - X_2 = Y_2 - Y_1 = 0$, so $X_1 = X_2$ and $Y_1 = Y_2$. $\qquad\qquad$ $\square$

Remark 213B If X is the projection of Z on V along W, then $Z - X$ is the projection of Z on W along V.

Theorem 213C *Let $V, W \subset \mathbf{F}^{n \times 1}$ be subspaces, Φ be a basis for V, and Ψ be a basis for W. Then W is a complement to V if and only if the matrix*

$$Q = \begin{bmatrix} \Phi & \Psi \end{bmatrix} \qquad (4)$$

is invertible.

Proof: We prove the following sharper result:

$$\mathcal{R}(Q) = \mathbf{F}^{n \times 1} \iff V + W = \mathbf{F}^{n \times 1} \qquad (5)$$

and

$$\mathcal{N}(Q) = \{0\} \iff V \cap W = \{0\}. \qquad (6)$$

Let $\nu = \dim(V)$ and $\mu = \dim(W)$, so

$$\Phi \in \mathbf{F}^{n \times \nu}, \qquad \Psi \in \mathbf{F}^{n \times \mu}.$$

In block matrix notation we write

$$\tilde{Z} = \begin{bmatrix} \tilde{X} \\ \tilde{Y} \end{bmatrix}, \qquad\qquad Q\tilde{Z} = \Phi\tilde{X} + \Psi\tilde{Y}, \qquad (7)$$

where $\widetilde{X} \in \mathbf{F}^{\nu \times 1}$, $\widetilde{Y} \in \mathbf{F}^{\mu \times 1}$, $\widetilde{Z} \in \mathbf{F}^{n \times 1}$. This proves that

$$\mathcal{R}(Q) = \mathcal{R}(\Phi) + \mathcal{R}(\Psi),$$

which proves (5). Also,

$$\widetilde{Z} \in \mathcal{N}(Q) \iff \Phi \widetilde{X} = -\Psi \widetilde{Y},$$

which proves (6).   □

Corollary 214A *Complementary subspaces* $\mathcal{V}$ *and* $\mathcal{W}$ *have complementary dimensions:*

$$\dim(\mathcal{V}) + \dim(\mathcal{W}) = n.$$

Theorem 214B *Let* $\mathcal{V}$ *and* $\mathcal{W}$ *be complementary subspaces of* $\mathbf{F}^{n \times 1}$ *as above. Then there is a unique matrix* $\Pi \in \mathbf{F}^{n \times n}$ *such that for every* $Z \in \mathbf{F}^{n \times 1}$ *the projection of* Z *on* $\mathcal{V}$ *along* $\mathcal{W}$ *is* ΠZ. *The matrix* Π *is called the* **projection matrix** *on* $\mathcal{V}$ *along* $\mathcal{W}$.

Proof: First we show uniqueness. The matrix Π must satisfy

$$\Pi X = X \text{ for } X \in \mathcal{V}, \qquad\qquad \Pi Y = 0 \text{ for } Y \in \mathcal{W}.$$

Choose bases

$$\Phi = \begin{bmatrix} X_1 & X_2 & \cdots & X_\mu \end{bmatrix}, \qquad \Psi = \begin{bmatrix} Y_1 & Y_2 & \cdots & Y_\nu \end{bmatrix},$$

for $\mathcal{V}$ and $\mathcal{W}$ and form $Q = \begin{bmatrix} \Phi & \Psi \end{bmatrix}$ as in (4). Since $\Pi X_i = X_i$ and $\Pi Y_j = 0$ block multiplication gives

$$\Pi Q = \begin{bmatrix} X_1 & X_2 & \cdots & X_\mu & 0 & \cdots & 0 \end{bmatrix}.$$

Multiply on the right by Q^{-1}:

$$\Pi = \begin{bmatrix} \Phi & 0 \end{bmatrix} Q^{-1}. \tag{8}$$

This shows uniqueness. For existence, define Π by (8) and argue backwards.
□

Remark 214C To construct Π we chose bases Φ for $\mathcal{V}$ and Ψ for $\mathcal{W}$. However, the answer Π is independent of the choice of Φ and Ψ, but depends only on the subspaces $\mathcal{V}$ and $\mathcal{W}$.

Exercise 215A The subspaces $V, W \subset \mathbf{F}^{3 \times 1}$ have bases

$$\Phi = \begin{bmatrix} -4 & 1 \\ 3 & -2 \\ 1 & -3 \end{bmatrix}, \qquad \Psi = \begin{bmatrix} -1 \\ 1 \\ 1 \end{bmatrix},$$

respectively. The inverse of the matrix matrix $Q = [\ \Phi \quad \Psi\ ]$ is

$$Q^{-1} = \begin{bmatrix} 1 & 2 & -1 \\ -2 & -3 & 1 \\ -7 & -11 & 5 \end{bmatrix}.$$

(1) Find the projection matrix Π on V along W.

(2) Write each of the following vectors Z in the form $Z = X + Y$ where $X \in V$ and $Y \in W$.

$$\begin{bmatrix} -6 \\ 4 \\ 1 \end{bmatrix}, \qquad \begin{bmatrix} 0 \\ 0 \\ 1 \end{bmatrix}, \qquad \begin{bmatrix} -2 \\ 1 \\ 3 \end{bmatrix}, \qquad \begin{bmatrix} -2 \\ 2 \\ 10 \end{bmatrix}.$$

(3) Calculate $\Pi\Phi$, $\Pi\Psi$, and Π^2.

Exercise 215B Repeat the previous exercise with

$$V = \left\{ \begin{bmatrix} x_1 \\ x_2 \\ 0 \end{bmatrix} : x_1, x_2 \in \mathbf{F} \right\}, \qquad W = \left\{ \begin{bmatrix} 0 \\ 0 \\ y \end{bmatrix} : y \in \mathbf{F} \right\}.$$

Choose Φ and Ψ so that $Q = I$.

Exercise 215C Define four subspaces of $\mathbf{F}^{2 \times 1}$ by

$$V_1 = \mathcal{N}(A), \qquad V_2 = \mathcal{N}(B), \qquad V_3 = \mathcal{R}(U), \qquad V_4 = \mathcal{R}(V),$$

where

$$A = \begin{bmatrix} 1 & 2 \end{bmatrix}, \qquad B = \begin{bmatrix} 2 & 1 \end{bmatrix}, \qquad U = \begin{bmatrix} 1 \\ 1 \end{bmatrix}, \qquad V = \begin{bmatrix} 1 \\ -1 \end{bmatrix}.$$

Let $Z = \begin{bmatrix} 3 \\ 4 \end{bmatrix}$. For each $j < k$ write Z in the form $Z = X + Y$ where $X \in V_j$ and $Y \in V_k$. (There are six problems here.) For each of the six problems draw a crude graph showing the lines V_j and V_k and the vectors Z, X, Y.

Exercise 216A Continue the notation of Exercise 215C. For each $j < k$ find the projection matrix on $\mathcal{V}_j$ along $\mathcal{V}_k$ and the projection matrix on $\mathcal{V}_k$ along $\mathcal{V}_j$.

Exercise 216B For each of the projections Π in Exercise 216A check that $\Pi^2 = \Pi$. Which satisfy $\Pi = \Pi^*$?

The following exercises establish a correspondence between pairs of complementary subspaces and matrices that satisfy $\Pi^2 = \Pi$. (Such matrices are called *idempotents*.)

Exercise 216C Show that the projection matrix Π on $\mathcal{V}$ along $\mathcal{W}$ is the unique matrix satisfying

$$\mathcal{R}(\Pi) = \mathcal{V}; \qquad \mathcal{N}(\Pi) = \mathcal{W}; \qquad \Pi^2 = \Pi.$$

Exercise 216D Suppose that $\Pi^2 = \Pi$ and let $\mathcal{V} = \mathcal{R}(\Pi)$ and $\mathcal{W} = \mathcal{N}(\Pi)$. Show that the subspaces $\mathcal{V}$ and $\mathcal{W}$ are complementary.

Exercise 216E Show that if $\Pi^2 = \Pi$, then $(I - \Pi)^2 = I - \Pi$.

Exercise 216F Suppose that Π is the projection matrix on $\mathcal{V}$ along $\mathcal{W}$. Show that the projection matrix on $\mathcal{W}$ along $\mathcal{V}$ is $I - \Pi$.

The next two exercises explain what happens when one sided inverses are multiplied in the wrong order.

Exercise 216G Suppose that Γ is a left inverse to A. Show that $A\Gamma$ is a projection matrix on the range $\mathcal{R}(A)$.

Exercise 216H Suppose that Ω is a right inverse to A. Show that $I - \Omega A$ is a projection matrix on the null space $\mathcal{N}(A)$.

6.8   Projection (Orthogonal)

Just as there are many one-sided inverses, so are there many complements. However, one of them is the simplest.

Definition 216I Let $\mathcal{V} \subset \mathbf{F}^{n \times 1}$ be a subspace. The **orthogonal complement** of $\mathcal{V}$ is the subspace

$$\mathcal{V}^{\perp} = \{Y \in \mathbf{F}^{n \times 1} : \langle Y, X \rangle = 0 \quad \forall X \in \mathcal{V}\}.$$

Here the symbol $\forall$ is read "for all". Geometrically, $\mathcal{V}^{\perp}$ consists of all those vectors perpendicular to every vector in $\mathcal{V}$. For example, if $\mathcal{V}$ is a plane through the origin in $\mathbf{F}^{3 \times 1}$, then $\mathcal{V}$ is the line through the origin perpendicular to this plane.

▷ **Question 217A** Why is $\mathcal{V}^\perp$ a subspace?

Lemma 217B *For any matrix $A \in \mathbf{F}^{m \times n}$ we have*

$$\mathcal{R}(A)^\perp = \mathcal{N}(A^*).$$

Proof: Assume $Y \in \mathcal{R}(A)^\perp$. Then $\langle Y, AX \rangle = 0$ for all X. But

$$\langle Y, AX \rangle = \langle A^*Y, X \rangle,$$

so $A^*Y = 0$, i.e. $Y \in \mathcal{N}(A^*)$. Conversely, assume $Y \in \mathcal{N}(A^*)$. Then $A^*Y = 0$, so $\langle Y, AX \rangle = 0$ for all X, so $Y \in \mathcal{R}(A)^\perp$. □

Lemma 217C *The orthogonal complement $\mathcal{V}^\perp$ is a complement to $\mathcal{V}$. In fact, if*

$$Q = \begin{bmatrix} U_1 & U_2 & \cdots & U_n \end{bmatrix}$$

is a unitary matrix whose first ν columns

$$\Phi = \begin{bmatrix} U_1 & U_2 & \cdots & U_\nu \end{bmatrix}$$

form an orthonormal basis for $\mathcal{V}$, then its last $n - \nu$ columns

$$\Psi = \begin{bmatrix} U_{\nu+1} & U_{\nu+2} & \cdots & U_n \end{bmatrix}$$

form an orthonormal basis for $\mathcal{V}^\perp$.

Proof: Any $Y \in \mathbf{F}^{n \times 1}$ has the form

$$Y = \sum_{j=1}^n y_j U_j, \qquad y_j = \langle Y, U_j \rangle.$$

Then $Y \in \mathcal{V}^\perp$ iff $\langle Y, U_j \rangle = 0$ for $j = 1, 2, \ldots, \nu$, because these first ν columns form a basis for $\mathcal{V}$. In other words, $Y \in \mathcal{V}^\perp \iff Y \in \mathcal{R}(\Psi)$. □

Remark 217D Any $Z \in \mathbf{F}^{n \times 1}$ may be written as $Z = X + Y$ where $X \in \mathcal{V}$ and $Y \in \mathcal{V}^\perp$ (see Figure 5 on Page 218) as follows:

$$Z = \sum_{j=1}^n z_j U_j, \qquad X = \sum_{j=1}^\nu z_j U_j, \qquad Y = \sum_{j=\nu+1}^n z_j U_j,$$

where $z_j = \langle Z, U_j \rangle$ as in Theorem 202E. In matrix notation

$$X = \begin{bmatrix} \Phi & 0 \end{bmatrix} Q^* Z, \qquad Y = \begin{bmatrix} 0 & \Psi \end{bmatrix} Q^* Z.$$

Since $Q^* = Q^{-1}$, this agrees with equation (8) on Page 214.

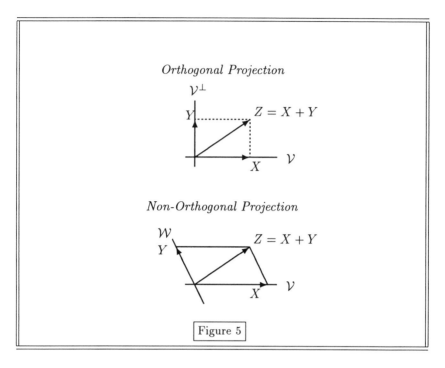

Orthogonal Projection

Non-Orthogonal Projection

Figure 5

Definition 218A The projection of Z on $\mathcal{V}$ along $\mathcal{V}^\perp$ is called the **orthogonal projection** of Z on $\mathcal{V}$.

Theorem 218B *Let Π be the projection matrix on $\mathcal{V}$ along $\mathcal{W}$. Then*

$$\mathcal{W} = \mathcal{V}^\perp \iff \Pi = \Pi^*.$$

Proof: If $\Pi = \Pi^*$, then, by Lemma 217B,

$$\mathcal{W} = \mathcal{N}(\Pi) = \mathcal{N}(\Pi^*) = \mathcal{R}(\Pi)^\perp = \mathcal{V}^\perp.$$

The converse is left as an exercise.   □

Theorem 218C (Left Inverse Theorem)   *A matrix $A \in \mathbf{F}^{m \times n}$ has a left inverse if and only if $A^*A \in \mathbf{F}^{n \times n}$ is invertible. When defined, the matrix*

$$\Gamma = (A^*A)^{-1} A^*$$

is a left inverse for A. The matrix

$$\Pi = A\Gamma = A (A^*A)^{-1} A^*$$

is the orthogonal projection matrix on $\mathcal{R}(A)$.

Proof: Since $\langle AX, AX \rangle = \langle A^*AX, X \rangle$, we see that $\mathcal{N}(A) = \mathcal{N}(A^*A)$. But A^*A is a square matrix. Thus A^*A is invertible $\iff \mathcal{N}(A) = 0 \iff A$ has a left inverse. The formula $\Gamma A = I$ is immediate. $\qquad \square$

▷ **Question 219A** Assume that a rectangular matrix $A \in \mathbf{F}^{m \times n}$ has orthonormal columns. What is A^*A? AA^*?

Theorem 219B (Right Inverse Theorem) *A matrix $A \in \mathbf{F}^{m \times n}$ has a right inverse if and only if $AA^* \in \mathbf{F}^{m \times m}$ is invertible. When defined, the matrix*

$$\Omega = A^* \left(AA^* \right)^{-1}$$

is a right inverse for A. The matrix

$$\Pi = \Omega A = A^* \left(AA^* \right)^{-1} A$$

is the orthogonal projection matrix on the orthogonal complement $\mathcal{N}(A)^\perp$ of the null space.

Proof: Exercise 220A.

▷ **Exercise 219C** The subspace $\mathcal{V} \subset \mathbf{F}^{3 \times 1}$ has basis $\Phi = \begin{bmatrix} 3 & 2 \\ 2 & 0 \\ 7 & 4 \end{bmatrix}$.

(1) Find the orthogonal projection matrix Π on $\mathcal{V}$.

(2) Write each of the following matrices Z in the form $Z = X + Y$ where $X \in \mathcal{V}$ and $Y \in \mathcal{V}^\perp$.

$$\begin{bmatrix} -6 \\ 4 \\ 1 \end{bmatrix}, \quad \begin{bmatrix} 0 \\ 0 \\ 1 \end{bmatrix}, \quad \begin{bmatrix} -2 \\ 1 \\ 3 \end{bmatrix}, \quad \begin{bmatrix} -2 \\ 2 \\ 10 \end{bmatrix}.$$

Exercise 219D For each of the following find the orthogonal projection matrix on $\mathcal{R}(A)$ and on $\mathcal{R}(A)^\perp$.

(1) $A = \begin{bmatrix} a \\ b \end{bmatrix}$, **(2)** $A = \begin{bmatrix} a \\ b \\ c \end{bmatrix}$, **(3)** $A = \begin{bmatrix} a_1 & a_2 \\ b_1 & b_2 \\ c_1 & c_2 \end{bmatrix}$.

Exercise 219E For each of the matrices A of Exercise 219D, find the orthogonal projection matrix on $\mathcal{N}(A^*)$ and on $\mathcal{N}(A^*)^\perp$.

▷ **Exercise 220A** Prove Theorem 219B.

Exercise 220B Show that if Π is the projection matrix on $\mathcal{V}$ along $\mathcal{V}^\perp$, then $I - \Pi$ is the projection on $\mathcal{V}^\perp$ along $\mathcal{V}$

Exercise 220C Show that the orthogonal complement satisfies the following laws:

(1) $\mathcal{V} \subset \mathcal{W} \implies \mathcal{W}^\perp \subset \mathcal{V}^\perp$.

(2) $\left(\mathcal{V}^\perp\right)^\perp = \mathcal{V}$.

(3) $(\mathcal{V} + \mathcal{W})^\perp = \mathcal{V}^\perp \cap \mathcal{W}^\perp$.

(4) $(\mathcal{V} \cap \mathcal{W})^\perp = \mathcal{V}^\perp + \mathcal{W}^\perp$.

▷ **Exercise 220D** Each of the following is a consequence of Lemma 217B. Why?

$$\mathcal{R}(A^*)^\perp = \mathcal{N}(A), \qquad \mathcal{R}(A^*) = \mathcal{N}(A)^\perp, \qquad \mathcal{R}(A) = \mathcal{N}(A^*)^\perp.$$

Exercise 220E Recall Definition 150A: a **co-basis** for a subspace $\mathcal{V} \subset \mathbf{F}^{n \times 1}$ is a matrix $\Lambda \in \mathbf{F}^{k \times n}$ such that

$$\mathcal{N}(\Lambda) = \mathcal{V}, \qquad \mathcal{R}(\Lambda) = \mathbf{F}^{k \times 1}.$$

Show that Λ is a co-basis for $\mathcal{V}$ if and only if Λ^* is a basis for $\mathcal{V}^\perp$.

Using MINIMAT

▷ **Exercise 220F** Generate a random 5×3 matrix Φ and a random 5×2 matrix Ψ. Then find the projection matrix Π on $\mathcal{R}(\Phi)$ along $\mathcal{R}(\Psi)$. Check your answer.

▷ **Exercise 220G** Generate a random 5×3 matrix Φ and compute the orthogonal projection matrix Π on $\mathcal{R}(\Phi)$. Check your answer.

6.9   Least Squares

The title of this section refers to two related problems. The first problem concerns finding the best possible approximate solution X to an inhomogeneous system $AX = Y$ when there is no exact solution; the second problem is a constrained minimization problem.

The Best Approximate Solution

Assume that $A \in \mathbf{F}^{m \times n}$ has maximal rank $n < m$. Then, given Y, the inhomogeneous system $AX = Y$ is **overdetermined**, meaning that there are more equations than unknowns. We expect that there will be no solution X because $\mathcal{R}(A) \neq \mathbf{F}^{m \times 1}$. We seek the best possible approximate solution X, that is, the solution which makes $AX - Y$ as small as possible. We measure the size of $AX - Y$ using the norm $\|AX - Y\|$. In other words, we want to solve the following

---

Problem of the Best Approximate Solution.

Given $A \in \mathbf{F}^{m \times n}$ and $Y \in \mathbf{F}^{m \times 1}$:

minimize $\|AX - Y\|$ for $X \in \mathbf{F}^{n \times 1}$.

The solution to this problem is given by the following

Theorem 221A *Assume $A \in \mathbf{F}^{m \times n}$ has rank n so that*

$$\Gamma = (A^*A)^{-1} A^*$$

is a left inverse for A. Then for each $Y \in \mathbf{F}^{m \times 1}$ the value of X where $\|AX - Y\|$ is smallest is $X_0 = \Gamma Y$. See Figure 6.

Proof: In other words, the function $F(X) = \|AX - Y\|^2$ assumes its minimum at $X_0 = \Gamma Y$. To prove this we choose $X \neq X_0$ and show that $F(X_0) < F(X)$. Let $\hat{X} = X - X_0$. The function

$$g(t) = F(X_0 + t\hat{X}) = \|AX_0 - Y\|^2 + 2t \left\langle AX_0 - Y, \hat{X} \right\rangle + t^2 \|\hat{X}\|^2$$

is quadratic in t. Its derivative at 0 is

$$g'(0) = 2 \left\langle AX_0 - Y, \hat{X} \right\rangle = 2 \left\langle A\Gamma Y - Y, \hat{X} \right\rangle = 2 \left\langle Y - Y, \hat{X} \right\rangle = 0.$$

The second derivative is $g''(0) = 2\|A\hat{X}\|^2$ which is positive, so

$$F(X_0) = g(0) < g(1) = F(X)$$

as required. $\qquad\qquad\qquad\qquad\qquad\qquad\qquad\qquad\qquad\qquad$ $\square$

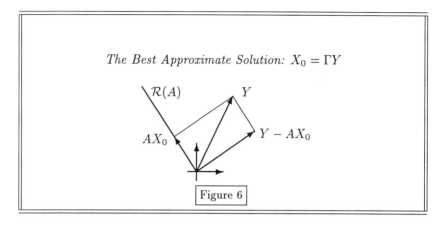

The Best Approximate Solution: $X_0 = \Gamma Y$

Figure 6

Remark 222A As X varies in $\mathbf{F}^{n \times 1}$ the point AX varies in $\mathcal{R}(A)$. Since $\|AX - Y\|$ assumes its minimum at $X = X_0 = \Gamma Y$, the point AX_0 is the closest point of $\mathcal{R}(A)$ to the point Y. In other words, $\|AX_0 - Y\|$ is the distance from Y to $\mathcal{R}(A)$. By Theorem 218C, the orthogonal projection of Y on $\mathcal{R}(A)$ is $AX_0 = A\Gamma Y$. See Figure 6.

The Closest Point

Assume that $A \in \mathbf{F}^{m \times n}$ has maximal rank $m < n$. The system $AX = Y$ is **underdetermined**, meaning that there are not enough equations to determine X uniquely; we will find the solution X that is closest to the origin. We must solve the

Problem of the Closest Point.

Minimize $\|X\|$ subject to the constraint $AX = Y$.

The solution to this problem is given by the following

Theorem 222B *Suppose that $A \in \mathbf{F}^{m \times n}$ has rank m so that*

$$\Omega = A^* \left(AA^*\right)^{-1}$$

is a right inverse for A. Then the point $X \in A^{-1}\{Y\}$ for which $\|X\|$ is smallest is $X_0 = \Omega Y$. See Figure 7.

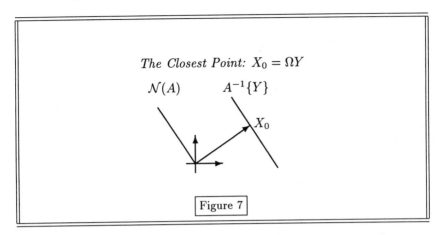

The Closest Point: $X_0 = \Omega Y$

$\mathcal{N}(A)$ $A^{-1}\{Y\}$

X_0

Figure 7

Proof: Let $X_0 = \Omega Y$. Then $AX_0 = A\Omega Y = Y$ so $X_0 \in A^{-1}\{Y\}$. Any other point $X \in A^{-1}(Y)$ has form $X = X_0 + \hat{X}$ where $\hat{X} \in \mathcal{N}(A)$. Define

$$g(t) = \|X_0 + t\hat{X}\|^2 = \|X_0\| + 2 + 2t\left\langle X_0, \hat{X} \right\rangle + t^2\|\hat{X}\|^2$$

for $t \in \mathbf{R}$. Then

$$g'(0) = 2\left\langle X_0, \hat{X} \right\rangle = 2\left\langle \Omega Y, \hat{X} \right\rangle = 0$$

since

$$\left\langle \Omega Y, \hat{X} \right\rangle = \left\langle A^* \left(AA^*\right)^{-1} Y, \hat{X} \right\rangle = \left\langle \left(AA^*\right)^{-1} Y, A\hat{X} \right\rangle$$

and $\hat{X} \in \mathcal{N}(A)$. Moreover, the second derivative is $g''(0) = 2\|\hat{X}\|^2 > 0$. Hence,

$$\|X_0\|^2 = g(0) < g(1) = \|X_0 + \hat{X}\|^2 = \|X\|^2$$

as required. $\square$

Remark 223A According to Theorem 219B, the orthogonal projection of $X \in A^{-1}\{Y\}$ on $\mathcal{N}(A)^{\perp}$ is $X_0 = \Omega AX = \Omega Y$. The reader should show that this orthogonal projection is the intersection of $A^{-1}\{Y\}$ with $\mathcal{N}(A)^{\perp}$:

$$\{\Omega Y\} = A^{-1}\{Y\} \cap \mathcal{N}(A)^{\perp}.$$

See Figure 7. In other words, $X_0 = \Omega Y$ is the unique point in $A^{-1}\{Y\}$ that is orthogonal to all the elements of the null space $\mathcal{N}(A)$. This is the value of $X \in A^{-1}(Y)$ for which $\|X\|$ is smallest, i.e. such that X is closest to the origin.

Exercise 224A Minimize the function $F(x) = \|Ax - Y\|$ where $x \in \mathbf{F}$ and $A, Y \in \mathbf{F}^{2 \times 1}$ are given by

$$A = \begin{bmatrix} 3 \\ 4 \end{bmatrix}, \qquad Y = \begin{bmatrix} 5 \\ 5 \end{bmatrix}.$$

Exercise 224B Minimize the function $F(X) = \|AX - Y\|$ where $X \in \mathbf{F}^{2 \times 1}$ and $A, Y \in \mathbf{R}^{3 \times 2}$ are given by

$$A = \begin{bmatrix} 1 & 0 \\ 0 & 1 \\ 3 & 4 \end{bmatrix}, \qquad Y = \begin{bmatrix} 1 \\ 2 \\ 3 \end{bmatrix}.$$

Exercise 224C Find the point on the line $3x_1 + 4x_2 = 5$ which is closest to the origin. Find the distance from the origin to this line.

Exercise 224D Find the point on the plane $2x_1 + 3x_2 + 4x_3 = 5$ which is closest to the origin. Find the distance from the origin to this plane.

Exercise 224E Find the point on the line $x_1 + 2x_2 + 3x_3 = 6$, $x_1 - x_2 = 3$ which is closest to the origin. Find the distance from the origin to this line.

Exercise 224F Minimize the function $F(X) = \|AX - Y\|$ where

$$A = \begin{bmatrix} B \\ 0_{(m-n) \times n} \end{bmatrix}, \qquad Y = \begin{bmatrix} Y_1 \\ Y_2 \end{bmatrix},$$

with $B \in \mathbf{F}^{n \times n}$ invertible, $Y_1 \in \mathbf{F}(n \times 1$, $Y_2 \in \mathbf{F}^{(m-n) \times 1}$.

Exercise 224G Minimize $\|X\|$ subject to the constraint $AX = Y$ if

$$A = \begin{bmatrix} B & 0_{(n-m) \times m} \end{bmatrix} \qquad X = \begin{bmatrix} X_1 \\ X_2 \end{bmatrix},$$

where $B \in \mathbf{F}^{m \times m}$ is invertible, $X_1 \in \mathbf{F}^{m \times 1}$, $X_2 \in \mathbf{F}^{(n-m) \times 1}$.

Exercise 224H A surveyor makes more measurements than are strictly necessary. For example, to find the altitudes of four points x_1, x_2, x_3, x_4 a surveyor made four direct measurements and three relative measurements to obtain

$$x_1 = 1.01, \qquad x_2 = 2.98, \qquad x_3 = 5.02, \qquad x_4 = 6.03,$$

$$x_2 - x_1 = 2.01, \qquad x_3 - x_1 = 3.96, \qquad x_4 - x_2 = 3.01.$$

The observations contain error. There is no solution to these seven equations. Express the equations in the form $AX = Y$ and find X which minimizes $\|AX - Y\|$. (This is what the surveyor decides is the most accurate answer.)

Exercise 225A A scientist makes three measurements to measure two quantities x_1 and x_2. The theory says that each measurement produces a known linear expression in the two quantities. Least square approximation is used to compute an answer. What will be the answer if

$$2x_1 + x_2 = 7.1, \qquad x_1 + x_2 = 4.9, \qquad 2x_1 - x_2 = 1.2?$$

Will the computed answer change if the first equation is multiplied by 9?

Exercise 225B Prove that a matrix $A \in \mathbf{F}^{m \times n}$ has a left inverse if and only if its conjugate transpose $A^* \in \mathbf{F}^{n \times m}$ has a right inverse. In fact, $B \in \mathbf{F}^{n \times m}$ is a left inverse for A if and only if B^* is a right inverse for A^*. Is this true for the ordinary transpose?

Exercise 225C Show that for any matrix A, the orthogonal complement of its null space $\mathcal{N}(A)$ is the range of its conjugate transpose:

$$\mathcal{N}(A)^\perp = \mathcal{R}(A^*).$$

Using MINIMAT

Exercise 225D Generate a random 5×3 matrix A and a random 5×1 matrix Y. Find X_0 which minimizes $\|AX - Y\|$. Check your answer by comparing $\|AX_0 - Y\|$ with $\|AX - Y\|$ for four randomly chosen values of $X \in \mathbf{F}^{3 \times 1}$.

Exercise 225E Generate a random 3×5 matrix A and a random 3×1 matrix Y. Find $X = X_0$ which satisfies $AX = Y$ and for which $\|X\|$ is smallest. Check your answer by comparing $\|X_0\|$ with $\|X\|$ for four randomly chosen values of $X \in A^{-1}(Y)$. (Use the call X=randsoln(A,Y) to generate a random element of $A^{-1}(Y)$.)

6.10   More Exercises

Exercise 225F Assume that $A \in \mathbf{F}^{n \times 2}$ so that $A^*A \in \mathbf{F}^{2 \times 2}$. Show that $\det(A^*A) \geq 0$. Hint: Let $X = \mathrm{col}_1(A)$ and $Y = \mathrm{col}_2(A)$.

Submultiplicative Inequality

▷ **Exercise 225G (Submultiplicative Inequality)**   Assume $A \in \mathbf{F}^{m \times n}$ and $B \in \mathbf{F}^{n \times p}$. Show that

$$\|AB\| \leq \|A\| \, \|B\|.$$

Exercise 226A Show that if A is a square matrix and k is a positive integer, then $\|A^k\| \leq \|A\|^k$.

Exercise 226B Show that the inequality in the previous exercise fails for negative powers.

Norms

A **norm** is a function which assigns to each matrix $X \in \mathbf{F}^{m \times n}$ a nonnegative number $\ell(X)$ satisfying

(Positivity) $\ell(X) > 0$ if $X \neq 0$,

(Homogeneity) $\ell(bX) = |b|\ell(X)$,

(Triangle Inequality) $\ell(X + Y) \leq \ell(X) + \ell(Y)$,

for $b \in \mathbf{F}$ $X, Y \in \mathbf{F}^{m \times n}$. According to Theorem 193B, the function

$$\ell(X) = \|X\| = \sqrt{\langle X, X \rangle}$$

is a norm.

Exercise 226C Suppose that $g(t)$ is a twice continuously differentiable real valued function of a real variable such that $g''(t) \geq 0$ for $0 \leq t \leq 1$. show that

$$g(t) \leq (1 - t)g(0) + tg(1)$$

for $0 \leq t \leq 1$.

Exercise 226D Define $\|X\|_p$ by

$$\|X\|_p = \left(\sum_{ij} |x_{ij}|^p \right)^{1/p}$$

where $x_{ij} = \text{entry}_{ij}(X)$ and the sum is over all the entries of X. In particular, $\|X\|_2 = \|X\|$. Show that if X and Y are of the same size and have only positive entries, then

$$\|(1 - t)X + tY\|_p \leq (1 - t)\|X\|_p + t\|Y\|_p$$

for $p \geq 1$ and $0 \leq t \leq 1$.

Exercise 226E Show that $\ell(X) = \|X\|_p$ is a norm. Hint: Show that it suffices to prove the triangle inequality when X and Y have only positive entries.

Exercise 227A Define

$$\|X\|_\infty = \max_{ij} |x_{ij}|.$$

where $x_{ij} = \text{entry}_{ij}(X)$ and the maximum is over all the entries of X. Show that $\ell(X) = \|X\|_\infty$ is a norm and that

$$\lim_{p \to \infty} \|X\|_p = \|X\|_\infty.$$

Exercise 227B Show that $\|X\|_p$ is not a norm if $p < 1$. Hint: Take $X = \text{col}_1(I_2)$ and $Y = \text{col}_2(I_2)$.

Exercise 227C Graph, on the same axes, the equations

$$\|X\|_p = 1, \qquad X \in \mathbf{R}^{2 \times 1}$$

for $p = 0.5,\ 1,\ 1.5,\ 2,\ 4,\ \infty$.

Exercise 227D According to the Principle of **Lagrange Multipliers**, to maximize $f(X)$ subject to the constraint $g(X) = 0$, we should form the function

$$F(X, \lambda) = f(X) + \lambda g(X)$$

and find its critical points. Using this principle, show that the maximum of $|HX|$ subject to the constraint $\|X\|_p = 1$ is $\|H\|_q$. Here $H \in \mathbf{F}^{1 \times n}$, X ranges over $\mathbf{F}^{n \times 1}$, $1 \le p \le \infty$, and q is defined by

$$\frac{1}{p} + \frac{1}{q} = 1.$$

As a consequence, prove that

$$|HX| \le \|H\|_q \|X\|_p$$

for $H \in \mathbf{F}^{1 \times n}$ and $X \in \mathbf{F}^{n \times 1}$. This is called **Holder's Inequality**.

Pauli Matrices and Quaternions

Exercise 227E The **Pauli matrices** are

$$J_0 = \begin{bmatrix} 1 & 0 \\ 0 & 1 \end{bmatrix}, \ J_1 = \begin{bmatrix} 0 & 1 \\ -1 & 0 \end{bmatrix}, \ J_2 = \begin{bmatrix} i & 0 \\ 0 & -i \end{bmatrix}, \ J_3 = \begin{bmatrix} 0 & -i \\ -i & 0 \end{bmatrix}.$$

For $X = \begin{bmatrix} x_0 & x_1 & x_2 & x_3 \end{bmatrix} \in \mathbf{R}^{1 \times 4}$, define $\mathbf{J}(X) \in \mathbf{C}^{2 \times 2}$ by

$$\mathbf{J}(X) = x_0 J_0 + x_1 J_1 + x_2 J_2 + x_3 J_3.$$

The matrices $\mathbf{J}(X)$ are called the **quaternions**. Prove that

(1) $J_0^* = J_0$ and $J_i^* = -J_i$ for $i = 1, 2, 3$.

(2) $J_0^2 = J_0$, $J_i^2 = -J_0$ for $i = 1, 2, 3$.

(3) $J_i J_j = -J_j J_i$ for $1 \leq i < j \leq 3$.

(4) $J_1 J_2 = J_3$, $J_2 J_3 = J_1$, $J_3 J_1 = J_2$.

(5) $\mathbf{J}(X)\mathbf{J}(X)^* = \mathbf{J}(X)^*\mathbf{J}(X) = \|X\|^2 J_0 = \det\big(\mathbf{J}(X)\big) J_0$.

Conclude that $\mathbf{J}(X)$ is unitary iff $\|X\|^2 = 1$.

Exercise 228A Show that a matrix $U \in \mathbf{C}^{2 \times 2}$ is unitary if it has the form

$$U = e^{i\theta} \mathbf{J}(X)$$

where θ is real and $X \in \mathbf{R}^{1 \times 4}$ satisfies $\|X\| = 1$.

Exercise 228B Prove the converse of the last exercise. Hint: Let

$$U = \begin{bmatrix} a & b \\ c & d \end{bmatrix}, \quad U^* = \begin{bmatrix} \bar{a} & \bar{c} \\ \bar{b} & \bar{d} \end{bmatrix}.$$

Use the equation $U^*U = J_0$ to show that $|a| = |d|$. Then find θ so $e^{-i\theta}U$ has form

$$e^{-i\theta}U = \begin{bmatrix} \alpha & \beta \\ \gamma & \bar{\alpha} \end{bmatrix}.$$

Chapter Summary

Unitary Matrices

Each of the following conditions on a square matrix $Q \in \mathbf{F}^{n \times n}$ is equivalent to its being unitary:

- $Q^* = Q^{-1}$;

- Q *has orthonormal columns;*

- $\langle QX, QY \rangle = \langle X, Y \rangle$ *for all* $X, Y \in \mathbf{F}^{n \times 1}$;

- $\|QX\| = \|X\|$ *for all* $X \in \mathbf{F}^{n \times 1}$.

Independent and Orthonormal

These are analogous:

- *A square matrix is invertible iff its columns are independent.*

- *A square matrix is unitary iff its columns are orthonormal.*

Gram-Schmidt Decomposition

A matrix Φ with independent columns can be written in the form

$$\Phi = \Psi T$$

where Ψ has orthonormal columns and T is positive triangular.

Complements and Projections

For any subspace $\mathcal{V}$ there is a one-one correspondence between complements $\mathcal{W}$ to $\mathcal{V}$ and projection matrices Π on $\mathcal{V}$:

$$\mathcal{R}(\Pi) = \mathcal{V}, \qquad \mathcal{N}(\Pi) = \mathcal{W}.$$

The orthogonal complement $\mathcal{W} = \mathcal{V}^\perp$ corresponds to orthogonal projection $\Pi^* = \Pi$.

7

DETERMINANTS-I

In this chapter we study a function called the **determinant** with many interesting properties. The most important property of the determinant is that a square matrix $A \in \mathbf{F}^{n \times n}$ is invertible if and only if its determinant $\det(A)$ is not zero. Determinants are hard to compute directly from the definition, and for most purposes it is not necessary to compute them. However, there are clever ways for computing determinants: one of these computes the determinant of a matrix at the same time that it transforms the matrix to row echelon form via Gauss-Jordan elimination.

7.1   Permutations

A **permutation** of length n is a finite sequence of length n

$$\sigma = (\sigma(1), \sigma(2), \ldots, \sigma(n))$$

where each integer from 1 to n appears exactly once on the right. The permutation which lists the numbers in their usual order is the **identity permutation** and is denoted by ε.

$$\varepsilon = (1, 2, \ldots, n).$$

The **composition** of two permutations

$$\sigma = (\sigma(1), \sigma(2), \ldots, \sigma(n)), \qquad \tau = (\tau(1), \tau(2), \ldots, \tau(n))$$

of the same length is the permutation $\sigma \circ \tau$ defined by

$$\sigma \circ \tau(j) = \sigma(\tau(j))$$

for $j = 1, 2, \ldots, n$. The **inverse** of a permutation σ is the permutation σ^{-1} of the same length defined by

$$j = \sigma(i) \iff i = \sigma^{-1}(j).$$

Examples 232A There are $n!$ permutations of length n. When $n = 3$ they are

$$(1, 2, 3), \quad (2, 3, 1), \quad (3, 1, 2), \quad (1, 3, 2), \quad (3, 2, 1), \quad (2, 1, 3).$$

The composition $\sigma \circ \tau$ of

$$\sigma = (3, 1, 2), \qquad \tau = (3, 2, 1),$$

is

$$\sigma \circ \tau = (2, 1, 3).$$

The composition operation is not commutative:

$$\tau \circ \sigma = (1, 3, 2).$$

The inverses of these permutations are

$$\sigma^{-1} = (2, 3, 1), \quad \tau^{-1} = \tau, \quad (\sigma \circ \tau)^{-1} = \sigma \circ \tau, \quad (\tau \circ \sigma)^{-1} = \tau \circ \sigma.$$

Permutation Laws

Theorem 232B *Permutations satisfy the following:*

$\sigma^{-1} \circ \sigma = \sigma \circ \sigma^{-1} = \varepsilon$	*(Definition of Inverse)*
$(\sigma \circ \tau) \circ \rho = \sigma \circ (\tau \circ \rho)$	*(Associative Law)*
$\varepsilon \circ \sigma = \sigma$	*(Left Multiplicative Identity)*
$\sigma \circ \varepsilon = \sigma$	*(Right Multiplicative Identity)*
$\varepsilon^{-1} = \varepsilon$	*(Inverse of the Identity)*
$(\sigma^{-1})^{-1} = \sigma$	*(Double Inverse Law)*
$(\sigma \circ \tau)^{-1} = \tau^{-1} \circ \sigma^{-1}$	*(Inverse Antimultiplicative Law)*

Sign of a Permutation

Definition 233A The **sign of the permutation** σ is the number $\mathrm{sgn}(\sigma) = (-1)^k$ where k is the number of crossings in σ: A **crossing** of σ is a pair (i, j) with $i < j$ but $\sigma(i) > \sigma(j)$. (Here $i, j = 1, 2, \ldots, n$ where n is the length of the permutation.)

▷ **Question 233B** What is the sign of the identity permutation?

For example, we compute the sign of the permutation

$$\sigma = (4, 2, 1, 3)$$

which is of length 4. We make a table of the six pairs (i, j) with $i < j$:

(i, j)	$(\sigma(i), \sigma(j))$	$\sigma(i) > \sigma(j)$?
$(1, 2)$	$(4, 2)$	Y
$(1, 3)$	$(4, 1)$	Y
$(1, 4)$	$(4, 3)$	Y
$(2, 3)$	$(2, 1)$	Y
$(2, 4)$	$(2, 3)$	N
$(3, 4)$	$(1, 3)$	N

Because there are four Yes's in the column on the right, the number of crossings is $k = 4$, so the sign of σ is

$$\mathrm{sgn}(\sigma) = (-1)^k = (-1)^4 = +1.$$

A quicker way to find the sign of a permutation is with a **crossing diagram**. We place the numbers $1, 2, \ldots, n$ in two adjacent identical columns, and for each $i = 1, 2, \ldots, n$ we draw an arrow from i on the left to $j = \sigma(i)$ on the right. The number of crossings of σ is the number of crossings in the diagram. (To count correctly, always jiggle the lines so that no three of them go through the same point.) For example, we draw the crossing diagram for σ:

$$
\begin{array}{l}
1 = \sigma(3) \\
2 = \sigma(2) \\
3 = \sigma(4) \\
4 = \sigma(1)
\end{array}
\tag{1}
$$

Theorem 233C (Crossing Diagram Principle)   *In constructing a crossing diagram for the purpose of computing the sign of a permutation*

σ *the line we draw from* i *to* $\sigma(i)$ *need not be straight: any system of curves connecting each* i *on the left to the corresponding* $\sigma(i)$ *on the right suffices to compute the sign of* σ *so long as no two of them are tangent anywhere and no three of them have a common point.*

Proof: To see this, deform the straight-line crossing diagram into some other diagram with the same beginning and ending points for the arrows. This will probably introduce some new crossings as we push one of the curves across another. Each time we do this we introduce *two* new crossings, so we do not change the number $(-1)^k$, since $(-1)^{k+2} = (-1)^k$. For example, diagram (1) has four crossings and

$$
\begin{array}{ll}
1 = \sigma(3) \\
2 = \sigma(2) \\
3 = \sigma(4) \\
4 = \sigma(1)
\end{array}
\tag{2}
$$

is a crossing diagram for the same permutation $\sigma = (3, 2, 4, 1)$ with six crossings. Both (1) and (2) have an even number of crossings. Indeed, *any* crossing diagram for this permutation σ will have an *even number* of crossings.   □

Multiplication Laws for the Sign

Theorem 234A *The sign function satisfies:*

- *The sign of the identity is one:*

$$\mathrm{sgn}(\varepsilon) = 1.$$

- *the sign of the inverse is the inverse of the sign:*

$$\mathrm{sgn}(\sigma^{-1}) = \mathrm{sgn}(\sigma)^{-1} = \mathrm{sgn}(\sigma).$$

- *The sign of the composition is the product of the signs:*

$$\mathrm{sgn}(\sigma \circ \tau) = \mathrm{sgn}(\sigma)\,\mathrm{sgn}(\tau).$$

Proof: That $\sigma(\varepsilon) = 1$ is clear since there are no crossings in the identity permutation ε and $(-1)^0 = 1$. To prove that σ and σ^{-1} have the same sign, note that we can convert the crossing diagram of σ into a crossing diagram for σ^{-1} by reversing the directions of the arrows.

It remains to prove that the sign of the composition is the product of the signs. In other words, if k is the number of crossings for σ and if ℓ is the number of crossings for τ, then

$$\operatorname{sgn}(\sigma \circ \tau) = (-1)^{k+l} = (-1)^k(-1)^l = \operatorname{sgn}(\sigma)\operatorname{sgn}(\tau).$$

We must be careful here because $k + \ell$ is *not* necessarily the number of crossings in $\sigma \circ \tau$. We shall use Theorem 233C.

If we draw a crossing diagram for τ to the left of a crossing diagram for σ and connect the arrows coming in to the middle column with those leaving it, we obtain a crossing diagram for $\sigma \circ \tau$. For example, with $\sigma = (4, 2, 1, 3)$ and $\tau = (1, 3, 4, 2)$ we get:

$$
\begin{array}{ll}
1 \longrightarrow 1 = \tau(1) & 1 \quad\quad 1 = \sigma(3) \\
2 \quad\quad 2 = \tau(4) & 2 \quad\quad 2 = \sigma(2) \\
3 \quad\quad 3 = \tau(2) & 3 \quad\quad 3 = \sigma(4) \\
4 \quad\quad 4 = \tau(3) & 4 \quad\quad 4 = \sigma(1)
\end{array}
\tag{3}
$$

We connect the arrows in diagram (3) to get a crossing diagram for $\sigma \circ \tau$:

$$
\begin{array}{l}
1 \longrightarrow 1 = \sigma(\tau(2)) \\
2 \quad\quad\quad 2 = \sigma(\tau(4)) \\
3 \quad\quad\quad 3 = \sigma(\tau(3)) \\
4 \quad\quad\quad 4 = \sigma(\tau(1))
\end{array}
\tag{4}
$$

Here is another crossing diagram for $\sigma \circ \tau$:

$$
\begin{array}{l}
1 \quad\quad 1 = \sigma(\tau(2)) \\
2 \quad\quad 2 = \sigma(\tau(4)) \\
3 \quad\quad 3 = \sigma(\tau(3)) \\
4 \quad\quad 4 = \sigma(\tau(1))
\end{array}
\tag{5}
$$

Diagram (4) has six crossings whereas diagram (5) has only four crossings. By Theorem 233C, either of the diagrams (4) or (5) show that $\operatorname{sgn}(\sigma \circ \tau) = +1$.) $\quad\square$

Transpositions

The simplest permutations of all (after the identity permutation) are the transpositions. A **transposition** is a permutation that leaves all but two of the numbers $i = 1, 2, \ldots, n$ fixed and interchanges those two. Thus a permutation

$$\tau : \{1, 2, \ldots, n\} \to \{1, 2, \ldots, n\}$$

of length n is a transposition iff there are distinct numbers p and q such that

$$\tau(p) = q,$$
$$\tau(q) = p,$$
$$\tau(i) = i \quad \text{for } i \neq p, q.$$

A transposition τ is its own inverse:

$$\tau \circ \tau = \varepsilon, \quad \text{so} \quad \tau^{-1} = \tau.$$

Theorem 236A *The sign of a transposition is* -1.

Proof: Suppose the transposition τ interchanges p and q and that $p < q$. The pairs (i, j) with $i < j$ but $\sigma(i) > \sigma(j)$ are

$$(i, j) = (p, p+1), \ (p, p+2), \ldots, (p, q-1), \ (p, q),$$
$$(p+1, q), \ (p+2, q), \ldots, (q-1, q).$$

There are an odd number $(2(q - p) - 1)$ of such pairs so $\mathrm{sgn}(\tau) = -1$. For example, the transposition $\tau = (4, 2, 3, 1)$ interchanges 1 and 4.

$$\begin{array}{ll}
1 & 1 = \tau(4) \\
2 & 2 = \tau(2) \\
3 & 3 = \tau(3) \\
4 & 4 = \tau(1)
\end{array} \qquad (6)$$

Its crossing diagram (6) has five crossings so $\mathrm{sgn}(\tau) = -1$ as claimed. $\square$

Theorem 236B (Factorization Theorem)   *Every permutation*

$$\sigma : \{1, 2, \ldots, n\} \to \{1, 2, \ldots, n\}$$

may be decomposed as a finite composition

$$\sigma = \tau_1 \circ \tau_2 \circ \cdots \circ \tau_k$$

of transpositions.

Proof: If $\sigma(1) \neq 1$, then define $\tau_1 : \{1, 2, \ldots, n\} \to \{1, 2, \ldots, n\}$ to be the transposition that interchanges 1 and $\sigma(1)$; if $\sigma(1) = 1$, then let τ_1 be the identity permutation. In either case, the composition $\sigma_1 = \tau_1 \circ \sigma$ has the property that $\sigma_1(1) = 1$.

Now $\sigma_1(2) \neq 1$, since $\sigma_1(1) = 1$. Repeat the process with 2 in place of 1 and σ_1 in place of σ. This produces τ_2 which is either a transposition or the identity permutation, so that $\sigma_2 = \tau_2 \circ \tau_1 \circ \sigma$ satisfies $\sigma_2(i) = i$ for $i = 1, 2$.

Repeat $n - 1$ times to achieve

$$\tau_{n-1} \circ \tau_{n-2} \circ \cdots \circ \tau_2 \circ \tau_1 \circ \sigma(i) = i$$

for $i = 1, 2, \ldots, n - 1$. This must also be true for $i = n$, so the permutation on the left is the identity. Now as $\tau^{-1} = \tau$ if τ is either a transposition or the identity we may compose on the left with τ_{n-1}, $\tau_{n-2}, \ldots, \tau_2$, τ_1 successively to achieve

$$\sigma = \tau_1 \circ \tau_2 \circ \cdots \circ \tau_{n-1}.$$

Now delete the factors τ_j which equal the identity and renumber. □

Using MINIMAT

We can represent permutations in MINIMAT using integer row vectors as in

```
#>  s = [3 5 2 4 1], t = [4 2 3 1 5], e = 1:5
```

MINIMAT's submatrix notation can be used to compute the composition:

```
#> st = s(t), ts = t(s), et = e(t), se = s(e)
   st = [4 5 2 3 1]
   ts = [3 5 2 1 4]
   et = [4 2 3 1 5]
   se = [3 5 2 4 1]
```

Function 237A The .M function next, shown in Listing 12, computes the next permutation (in alphabetical order) after its input. For example, the command

```
#> c=[1 2 3 4 5]; c = next(c)
   c = [1 2 3 5 4]
```

replaces c by the next permutation in alphabetical order.[1]

Function 237B The .M function signum, shown in Listing 12 computes the sign of its input. The command

---

[1] This algorithm is taken from E. W. Dijkstra, *A Discipline of Programming*, Prentice Hall (1976).

The Next Permutation in MINIMAT

```
function c = next(c)

 [m n] = size(c); % should have m=1
 i = n-1; while c(i) >= c(i+1), i = i-1; end;
 j = n;   while c(j) <= c(i),   j = j-1; end;
 x = c(i);  c(i) = c(j);  c(j) = x;
 i = i+1; j = n;
 while i < j,
        x = c(i);  c(i) = c(j);  c(j) = x;
        i = i+1;   j = j-1;
 end
```

The Sign of a Permutation in MINIMAT

```
function s = signum(c)

 [m n] = size(c);
 s = 1;
 for i = 1:n
        for j = i+1:n
                if c(i) > c(j), s = -s; end
        end
 end
```

> Listing 12

```
#>   e=signum(s)
```

assigns to e the sign of of the permutation s.

▷ **Exercise 238A** Write a .M function perminv so that the command

```
#>   t=perminv(s)
```

assigns to t the inverse of the permutation s.

Exercise 239A Write a `.M` function `randperm` so that the call

```
#> s = randperm(n)
```

produces a random[2] permutation of $1, 2, \ldots, n$.

7.2 Determinant Defined

Let A be a square matrix, say $A \in \mathbf{F}^{n \times n}$, and let

$$a_{ij} = \mathrm{entry}_{ij}(A)$$

be the entry in the ith row and jth column of A.

Definition 239B The **determinant** $\det(A)$ is the sum

$$\det(A) = \sum_{\sigma} \mathrm{sgn}(\sigma) \prod_{i=1}^{n} a_{i,\sigma(i)}$$

of the signed elementary products of A.

The formula in Definition 239B may be explained as follows. Each permutation σ of length n determines an **elementary product** of A. It is the product

$$\prod_{i=1}^{n} a_{i,\sigma(i)} = a_{1,\sigma(1)} \cdot a_{2,\sigma(2)} \cdots a_{n,\sigma(n)}$$

consisting of one factor from each row of A, namely, the one in the $\sigma(i)$th column. The **signed elementary product** of A for σ is the product of the (unsigned) elementary product with the sign $\mathrm{sgn}(\sigma)$ of the permutation σ. Since there are $n!$ permutations of length n, the determinant $\det(A)$ is a sum of $n!$ terms, one signed elementary product for each permutation σ.

Example 239C There are two permutations of length 2, namely $(1, 2)$ and $(2, 1)$. The former is the identity permutation and has sign $+1$; the latter has sign -1. Hence, the determinant of a 2×2 matrix A is given by a sum of two terms

$$\det(A) = a_{11}a_{22} - a_{12}a_{21}.$$

---

[2]If you want all $n!$ permutations to be equally likely, this is tricky. See Nijenhuis and Wilf, *Combinatorial Algorithms*, Academic Press (1975).

Example 240A There are six permutations of length 3. Three of these have sign $+1$: $(1,2,3)$, $(2,3,1)$, $(3,1,2)$, and the other three have sign -1: $(1,3,2)$, $(3,2,1)$, $(2,1,3)$. Hence, the determinant of a 3×3 matrix A is given by a sum of six terms

$$\det(A) = a_{11}a_{22}a_{33} + a_{12}a_{23}a_{31} + a_{13}a_{21}a_{32}$$
$$-a_{11}a_{23}a_{32} - a_{13}a_{22}a_{31} - a_{12}a_{21}a_{33}$$

Exercise 240B Make a table of the $24 = 4!$ permutations of length 4 and compute the sign of each one. Then write a formula for the determinant of a 4×4 matrix. You may want to use MINIMAT to help with the calculation. See Function 237A.

Easy Properties

A matrix A is called **triangular** iff all its entries below the diagonal are zero, that is, if $\text{entry}_{ij}(A) = 0$ for $i > j$. The entries $\text{entry}_{ii}(A)$ are the called the **diagonal entries** of A.

Theorem 240C *The determinant of a triangular matrix is the product of its diagonal entries.*

Proof: Suppose the square matrix A is triangular. Then $\text{entry}_{ij}(A) = 0$ for $j < i$. Now if a permutation σ is not the identity permutation, there must be some i with $\sigma(i) < i$. Then the corresponding elementary product $\prod_i \text{entry}_{i,\sigma(i)}(A)$ vanishes as one of its factors is zero. Thus the only elementary product that does not vanish is the one corresponding to the identity permutation. Thus

$$\det(A) = \prod_{i=1}^{n} \text{entry}_{ii}(A)$$

as required.   □

Example 240D $\det \left(\begin{bmatrix} a & b & c \\ 0 & d & e \\ 0 & 0 & f \end{bmatrix} \right) = adf.$

Corollary 241A *The determinant of the identity matrix $I = I_n$ is one:*

$$\det(I) = 1.$$

Proof: The identity matrix is triangular.   □

Theorem 241B *(Additivity)* *If A, B, and C are square matrices of the same size (say $n \times n$), identical in all rows except one (say the kth row):*

$$\text{row}_i(C) = \text{row}_i(B) = \text{row}_i(A) \quad \text{for } i \neq k,$$

and if the kth row of C is the sum of the kth rows of A and B:

$$\text{row}_k(C) = \text{row}_k(A) + \text{row}_k(B),$$

then the determinant of C is sum of the determinants of A and B:

$$\det(C) = \det(A) + \det(B).$$

Proof: For each elementary product we have

$$\prod_{i=1}^{n} \text{entry}_{i,\sigma(i)}(C) = \prod_{i=1}^{n} \text{entry}_{i,\sigma(i)}(A) + \prod_{i=1}^{n} \text{entry}_{i,\sigma(i)}(B),$$

since

$$\text{entry}_{i,\sigma(i)}(C) = \text{entry}_{i,\sigma(i)}(A) = \text{entry}_{i,\sigma(i)}(B)$$

for $i \neq k$, and

$$\text{entry}_{k,\sigma(k)}(C) = \text{entry}_{k,\sigma(k)}(A) + \text{entry}_{k,\sigma(k)}(B).$$

Now multiply by $\text{sgn}(\sigma)$ and sum over σ.   □

Example 241C

$$\det\left(\begin{bmatrix} a_1+b_1 & a_2+b_2 & a_3+b_3 \\ c_{21} & c_{22} & c_{23} \\ c_{31} & c_{32} & c_{33} \end{bmatrix}\right) =$$

$$= \det\left(\begin{bmatrix} a_1 & a_2 & a_3 \\ c_{21} & c_{22} & c_{23} \\ c_{31} & c_{32} & c_{33} \end{bmatrix}\right) + \det\left(\begin{bmatrix} b_1 & b_2 & b_3 \\ c_{21} & c_{22} & c_{23} \\ c_{31} & c_{32} & c_{33} \end{bmatrix}\right).$$

Theorem 241D *(Scaling)* *If B results from A by multiplying a row of A by the number c, then*

$$\det(B) = c\det(A).$$

Proof: In other words, if

$$\text{row}_i(B) = \text{row}_i(A) \text{ for } i \neq k,$$
$$\text{row}_k(B) = c\,\text{row}_k(A),$$

then $\det(B) = \det(C)$. This is because every elementary product in B is c times the corresponding elementary product of A.   □

Example 242A

$$\det\left(\begin{bmatrix} ca_{11} & ca_{12} & ca_{13} \\ a_{21} & a_{22} & a_{23} \\ a_{31} & a_{32} & a_{33} \end{bmatrix}\right) = c\,\det\left(\begin{bmatrix} a_{11} & a_{12} & a_{13} \\ a_{21} & a_{22} & a_{23} \\ a_{31} & a_{32} & a_{33} \end{bmatrix}\right).$$

Corollary 242B *Any matrix having a row that vanishes identically has determinant zero.*

Theorem 242C *(Row Interchange)   The determinant* $\det(A)$ *of a matrix* A *is an* **alternating** *function the rows* A. *This means the following: Suppose that the matrix* B *results from the matrix* A *by interchanging two distinct rows, say the pth row and the qth row where* $p \neq q$. *That is,*

$$\text{row}_p(B) = row_q(A),$$
$$\text{row}_q(B) = row_p(A),$$
$$\text{row}_i(B) = row_i(A)$$

for all $i \neq p, q$. *Then*

$$\det(B) = -\det(A).$$

Proof: Let τ be the transposition that interchanges p and q. Then

$$\text{row}_i(B) = \text{row}_{\tau(i)}(A)$$

for each i. Hence, for each i and each permutation σ,

$$\text{entry}_{i,\sigma(i)}(B) = \text{entry}_{\tau(i),\sigma(i)}(A).$$

Thus

$$\prod_{i=1}^{n} \text{entry}_{i,\sigma(i)}(B) = \prod_{i=1}^{n} \text{entry}_{\tau(i),\sigma(i)}(A)$$
$$= \prod_{j=1}^{n} \text{entry}_{j,\sigma(\tau(j))}(A)$$

where we have made a change in the dummy variable $j = \tau(i)$ (and so $i = \tau(j)$ as τ is a transposition.) In other words, the elementary product of B corresponding to the permutation σ equals the elementary product of A corresponding to the permutation $\sigma \circ \tau$. Now

$$\operatorname{sgn}(\sigma \circ \tau) = \operatorname{sgn}(\sigma)\operatorname{sgn}(\tau) = -\operatorname{sgn}(\sigma)$$

by Theorem 234A and Theorem 236A. Hence, the signed elementary product of B corresponding to the permutation σ is the negative of the signed elementary product of A corresponding to the permutation $\sigma \circ \tau$. Now sum over σ. (As σ runs over the set of all permutations, so does $\sigma \circ \tau$.)   □

Example 243A

$$\det\left(\begin{bmatrix} d & e & f \\ a & b & c \\ g & h & i \end{bmatrix}\right) = -\det\left(\begin{bmatrix} a & b & c \\ d & e & f \\ g & h & i \end{bmatrix}\right).$$

Corollary 243B *Any matrix having two equal rows has determinant zero.*

Proof: Swapping these rows leaves the matrix unchanged but reverses the sign of the determinant.   □

Determinant of the Transpose

Theorem 243C *A matrix A and its transpose $A^\top$ have the same determinant:*

$$\det(A^\top) = \det(A)$$

Proof: For each permutation σ and each i,

$$\operatorname{entry}_{i,\sigma(i)}(A^\top) = \operatorname{entry}_{\sigma(i),i}(A) = \operatorname{entry}_{j,\sigma^{-1}(j)}(A)$$

where $j = \sigma(i)$. Hence,

$$\prod_{i=1}^{n} \operatorname{entry}_{i,\sigma(i)}(A^\top) = \prod_{j=1}^{n} \operatorname{entry}_{j,\sigma^{-1}(j)}(A),$$

in other words the elementary product of $A^\top$ corresponding to the permutation σ equals the elementary product of A corresponding to the permutation σ^{-1}. Since a permutation and its inverse have the same sign, the signed elementary products are also equal. Now sum over σ.   □

Example 244A $\det\left(\begin{bmatrix} a & b & c \\ d & e & f \\ g & h & i \end{bmatrix}\right) = \det\left(\begin{bmatrix} a & d & g \\ b & e & h \\ c & f & i \end{bmatrix}\right).$

▷ **Question 244B** If B results from A by interchanging two columns, what is the relation between the determinants?

Computing Determinants

The formula for the determinant of a matrix is a sum of $n!$ n-fold products. To compute the determinant from the definition is quite inefficient. For example, to compute the determinant of a 5×5 matrix, we would have to add up 120 numbers, each of which is a product of 5 different entries of the matrix. Fortunately, there are better ways to compute the determinant.

Recall the three kinds of elementary row operations (scale, swap, shear) defined in Section 3.1, the corresponding elementary matrices (Section 3.2), and the Fundamental Theorem 76B.

Lemma 244C *Suppose that $A \in \mathbf{F}^{n \times n}$ and that EA results from A by a single elementary row operation. Then*

(1) $EA = \mathrm{Scale}(A, p, c) \implies \det(EA) = c \cdot \det(A)$;

(2) $EA = \mathrm{Swap}(A, p, q) \implies \det(EA) = -\det(A)$;

(3) $EA = \mathrm{Shear}(A, p, q, c) \implies \det(EA) = \det(A)$.

Proof: Item (1) is Theorem 241D, and item (2) is Theorem 242C. If $EA = \mathrm{Shear}(A, p, q, c)$, Theorem 241B gives that $\det(EA) = \det(A) + c \det(C)$ where C has two identical rows. But $\det(C) = 0$ by Corollary 243B. This proves item (3).   □ .

Corollary 244D *The determinant of the elementary matrix E is given by*

(1) $\det(E) = c$, *if $E = \mathrm{Scale}(I, p, c)$*;

(2) $\det(E) = -1$, *if $E = \mathrm{Swap}(I, p, q)$*;

(3) $\det(E) = 1$, *if $E = \mathrm{Shear}(I, p, q, c)$.*

Proof: Take $A = I$ in Lemma 244C.   □

Corollary 244E *Let $E, A \in \mathbf{F}^{n \times n}$ with E elementary. Then*

$$\det(EA) = \det(E)\det(A).$$

Proof: By Lemma 244C and Corollary 244D.   □

Example 245A Suppose

$$A = \begin{bmatrix} a & b & c \\ d & e & f \\ g & h & i \end{bmatrix}, \qquad EA = \begin{bmatrix} a & b & c \\ d+ug & e+uh & f+ui \\ g & h & i \end{bmatrix}.$$

Then $EA = Shear(A, 2, 3, u)$ and $\det(EA) = \det(A) + u\det(C)$ where

$$C = \begin{bmatrix} a & b & c \\ g & h & i \\ g & h & i \end{bmatrix}$$

But $\det(C) = 0$, since C has two identical rows. Hence, $\det(EA) = \det(A)$. Since $\det(E) = 1$, this is the result predicted by Corollary 244E.

Remark 245B Corollary 244E gives us an algorithm for computing the determinant of a matrix that is far more efficient than adding up the $n!$ signed elementary products which occur in Definition 239B. The algorithm is the same as the one used by MINIMAT when it calculates its built-in det function. The algorithm is as follows. We generate a sequence $A_0, A_1, A_2, \ldots$ of matrices with $A_0 = A$ and each A_{k+1} obtained from A_k by an elementary row operation. For each k we will find the number d_k such that

$$\det(A) = d_k \det(A_k).$$

Then $d_0 = 1$ (since $A_0 = A$) and

(1) $d_{k+1} = c^{-1}d_k$ if $A_{k+1} = \text{Scale}(A_k, p, c)$;

(2) $d_{k+1} = -d_k$ if $A_{k+1} = \text{Swap}(A_k, p, q)$;

(3) $d_{k+1} = d_k$ if $A_{k+1} = \text{Shear}(A_k, p, q, c)$.

After a certain stage it will be easy to evaluate $\det(A_k)$, say, because it is triangular or has a row of zeroes. At this point we use the formula $\det(A) = d_k \det(A_k)$ to evaluate $\det(A)$.

The fact that $d_{k+1} = c^{-1}d_k$ (rather than cd_k) is a bit confusing, but if we perform the correct shears, we needn't do any scales at all. Also, we usually don't have to do swaps. Since shears don't change the determinant, the algorithm is quite easy in practice.

Example 246A We transform the matrix

$$A_0 = \begin{bmatrix} 0 & 5 & 1 \\ 4 & 6 & 8 \\ 2 & 3 & 7 \end{bmatrix}$$

to a triangular matrix in three steps:

$$A_1 = \begin{bmatrix} 4 & 6 & 8 \\ 0 & 5 & 1 \\ 2 & 3 & 7 \end{bmatrix} \qquad \text{(swap)}$$

$$A_2 = \begin{bmatrix} 2 & 3 & 4 \\ 0 & 5 & 1 \\ 2 & 3 & 7 \end{bmatrix} \qquad \text{(scale)}$$

$$A_3 = \begin{bmatrix} 2 & 3 & 4 \\ 0 & 5 & 1 \\ 0 & 0 & 3 \end{bmatrix} \qquad \text{(shear)}$$

Since A_3 is triangular, $\det(A_3) = 30$. Hence,

$$\det(A_0) = -\det(A_1) = -2\det(A_2) = -2\det(A_3) = -60. \qquad \square$$

Determinants and Invertibility

Corollary 246B *A square matrix is invertible if and only if its determinant is not zero.*

Proof: We may write A in the form

$$A = E_1 E_2 \cdots E_l R$$

where the matrices E_k are elementary and the matrix R is in reduced row echelon form. Hence, $\det(A) = m \det(R)$ where the multiplier m is the product of the determinants $\det(E_k)$:

$$m = \det(E_1)\det(E_2) \cdots \det(E_l).$$

The multiplier m is nonzero since each factor $\det(E_k)$ is. Hence, $\det(A) \neq 0$ if and only if $\det(R) \neq 0$. If A is invertible, then R is the identity matrix so $\det(R) = 1 \neq 0$. If A is not invertible, then R has a row of zeros so $\det(R) = 0$. $\qquad \square$

Theorem 247A *The determinant of a product of square matrices is the product of the determinants:*

$$\det(AB) = \det(A)\det(B)$$

for any square matrices A and B of the same size.

Proof: Write A in the form

$$A = E_1 E_2 \cdots E_l R$$

as in the previous proof. There are two cases.

- If A is invertible, then R is the identity so

$$AB = E_1 E_2 \cdots E_l B$$

so

$$\det(AB) = \det(E_1)\det(E_2)\cdots\det(E_l)B = \det(A)\det(B).$$

- If A is not invertible, then $\det(A) = 0$ and R has a row of zeros and hence RB has a row of zeros so RB is not invertible. But then there must be a nonzero X with $RBX = 0$. Hence,

$$ABX = E_1 E_2 \cdots E_l RBX = 0.$$

This says that AB is not invertible so that $\det(AB) = 0$. Thus $\det(AB) = \det(A)\det(B)$ as both sides are zero.   □

Multiplication Laws for the Determinant

Corollary 247B *The determinant function satisfies the following:*

- $\det(I) = 1$;
- $\det(A^{-1}) = \det(A)^{-1}$;
- $\det(AB) = \det(A)\det(B)$.

Proof: Only the law $\det(A^{-1}) = \det(A)^{-1}$ for the inverse is not an earlier theorem. However, this law follows from the other two laws as

$$\det(A)\det(A^{-1}) = \det(AA^{-1}) = \det(I) = 1. \qquad \Box$$

▷ **Exercise 248A** Use the algorithm of Remark 245B to find the determinant of the matrix

$$A = \begin{bmatrix} 1 & 2 & 3 & 4 \\ 3 & 1 & 4 & 2 \\ 4 & 1 & -1 & 1 \\ 0 & 1 & 0 & 2 \end{bmatrix}.$$

▷ **Exercise 248B** Prove the Triangular Invertibility Theorem 49B.

Using MINIMAT

Function 248C The .M function `detm`, shown in Listing 13, computes the determinant of its input. After the MINIMAT command

```
#>  d = detm(A)
```

the variable d holds the determinant of A. This function produces the same result as MINIMAT's built-in `det` function.   It uses the algorithm employed in Remark 245B; it maintains the invariant relation $d \cdot \det(A) = 1$. Note how `detm` is patterned after the .M function `gj` (Listing 2).

▷ **Exercise 248D** Use MINIMAT to confirm the properties of the determinant function described in this chapter.

Exercise 248E Write a .M function `detslow` that computes the determinant of its input using the definition. You may use the .M functions `next` and `signum` defined in Section 7.1. Make sure to test your function by confirming that it gets the same answer as MINIMAT's built-in function `det` or the .M function `detm`. Warning: This function will be very slow.

Chapter 15 contains more material on determinants. It may be studied now, but the material covered there is not required in the rest of the book.

Determinant in MINIMAT

```
function d = detm(A)
% invariant relation: d*det(A) = constant
[m n] = size(A);   d=1;
for k=1:n
    [y,h] = max(abs(A(k:m, k)));  h=k-1+h;
    if  y < 1.0E-9  % (i.e if y == 0)
        d=0; return
    else
        if (k~=h)
            A([k h],:) = A([h k],:);  % swap
            d=-d;
        end
        c = A(k,k);
        A(k,:) = A(k,:)/c;            % scale
        d=c*d;
        for i =  k+1:m                % shear
            A(i,:) = A(i,:) - A(i,k)*A(k,:);
        end
    end % if
end  % for
```

> Listing 13

7.3   More Exercises

Exercise 249A Suppose $A \in \mathbf{F}^{3 \times 2}$ and $B \in \mathbf{F}^{2 \times 3}$. Prove that $\det(AB) = 0$. What about $\det(BA)$?

Exercise 249B Prove that if a matrix A has the block triangular form

$$A = \begin{bmatrix} A_{11} & A_{12} \\ 0 & A_{22} \end{bmatrix}$$

where A_{11} and A_{22} are square, then $\det(A) = \det(A_{11}) \det(A_{22})$.

Wedge Product

Definition 250A The **wedge product** $X \wedge Y \in \mathbf{F}^{3 \times 1}$ of $X, Y \in \mathbf{F}^{3 \times 1}$ is defined by

$$X \wedge Y = \begin{bmatrix} x_2 y_3 - x_3 y_2 \\ x_3 y_1 - x_1 y_3 \\ x_1 y_2 - x_2 y_1 \end{bmatrix}$$

where $x_j = \mathrm{entry}_j(X)$, $y_j = \mathrm{entry}_j(Y)$.

▷ **Exercise 250B** When $\mathbf{F} = \mathbf{R}$, the wedge product is essentially what is called the **cross product** in vector analysis. The cross product is usually defined by the following rules:

(1) The cross product of two vectors is perpendicular to each of them.

(2) The length of the cross product is the area of the parallelogram determined by the factors.

(3) The triple formed by two vectors followed by their cross product forms a right-hand triple.

Justify this by proving the following:

(1) $\langle X \wedge Y, X \rangle = \langle X \wedge Y, Y \rangle = 0$.

(2) $\|X \wedge Y\|^2 + \langle X, Y \rangle^2 = \|X\|^2 \|Y\|^2$ so that $\|X \wedge Y\| = \|X\| \|Y\| \sin \theta$ where θ is the angle between X and Y.

(3) $\det(M) > 0$ where $M \in \mathbf{R}^{3 \times 3}$ is defined by $\mathrm{col}_1(M) = X$, $\mathrm{col}_2(M) = Y$, $\mathrm{col}_3(M) = X \wedge Y$.

Exercise 250C Let $Q \in \mathbf{R}^{3 \times 3}$ be real unitary. Show that $\det(Q) = \pm 1$ and that

$$(QX) \wedge (QY) = \pm Q(X \wedge Y)$$

for $X, Y \in \mathbf{R}^{3 \times 1}$. Hint: First prove $\det(QM) = \det(Q) \det(M)$.

▷ **Exercise 250D** Show that the sequence (X, Y) is independent if and only if $X \wedge Y \neq 0$.

Exercise 250E Assume $X \wedge Y \neq 0$. Show that

$$\Phi = \begin{bmatrix} X & Y \end{bmatrix} \in \mathbf{F}^{3 \times 2}$$

is a basis for null space $\mathcal{N}(A)$ of the matrix

$$A = (X \wedge Y)^\top \in \mathbf{F}^{1 \times 3}.$$

Exercise 250F Assume $\mathbf{F} = \mathbf{R}$ and $X \wedge Y \neq 0$. Show that the sequence $(X, Y, X \wedge Y)$ is independent.

▷ **Exercise 250G** Show that the previous result is false when $\mathbf{F} = \mathbf{C}$.

Real Equivalence

Recall the definitions of left equivalence, right equivalence, and biequiv-alence from Definition 104A. They define A and B to be *equivalent* iff there exist invertible P and Q satisfying a certain matrix equation. We say that A and B are **equivalent over F** iff P and Q may be chosen from **F**. For example, A and B are *right equivalent over* **F** iff there is an invertible $P \in \mathbf{F}^{n \times n}$ with $AP = B$. A priori, it seems that it could happen that two real matrices $A, B \in \mathbf{R}^{m \times n}$ are equivalent over **C** but not over **R**. The following exercises show that this does not happen.

Exercise 251A Suppose that

$$P = P_0 + iP_1$$

where $P_0, P_1 \in \mathbf{R}^{n \times n}$ are real. Show that if P is invertible, then there are real numbers ξ for which the real matrix $P_0 + \xi P_1$ is invertible. Hint: Is the polynomial $f(\xi) = \det(P_0 + \xi P_1)$ identically zero?

Exercise 251B Show that *If two real matrices $A, B \in \mathbf{R}^{m \times n}$ are equivalent over* **C***, then they are equivalent over* **R**. This works for each of the three kinds of equivalence: left equivalence, right equivalence, and biequiv-alence. Hints: Assume $AP = B$. Equate real and imaginary parts. Consider $P(\xi) = P_0 + \xi P_1$ where $\xi \in \mathbf{R}$. For biequivalence two approaches work: we can equate real and imaginary parts in $AP = QB$ or we can use Exercise 187A.

▷ **Question 251C** Why is there no theory of when two complex matrices A and B are equivalent (or similar) over **R**?

Chapter Summary

Properties of Determinants

- $\det(I) = 1$.
- $\det(AB) = \det(A)\det(B)$
- $E = \text{Scale}(I, p, c) \implies \det(E) = c$
- $E = \text{Swap}(I, p, q) \implies \det(E) = -1$
- $E = \text{Shear}(I, c, p, q) \implies \det(E) = 1$
- $\det(A^\top) = \det(A)$.

Consequences of these Properties

- $\det(A^{-1}) = \det(A)^{-1}$.

- *A matrix is invertible iff its determinant is nonzero.*

- *The determinant of a triangular matrix is the product of its diagonal entries. In particular, the determinant of a diagonal matrix is the product of its diagonal entries.*

8

DIAGONALIZATION

Many problems are easy to solve for diagonal matrices, so the strategy for solving them in the general case is to find an equivalent problem involving a diagonal matrix. In this chapter we learn how to transform a matrix into a diagonal matrix in such a way as to preserve the essential structure of such problems.

In Chapter 9 we shall solve certain problems by diagonalization. These include

- the linear system

$$\dot{X} = AX$$

  of differential equations, and

- the problem of finding a pth root C of a matrix A:

$$C^p = A.$$

For diagonal matrices A these problems are easy. The techniques of this chapter often enable us to reduce these problems to the diagonal case.

253

8.1   Similarity

Similarity

Definition 254A Let $A \in \mathbf{F}^{n \times n}$ and $B \in \mathbf{F}^{n \times n}$ be square matrices of the same size. We say A is **similar** to B iff there is an invertible matrix $P \in \mathbf{F}^{n \times n}$ satisfying

$$A = PBP^{-1}.$$

▷ **Question 254B** What's the difference between biequivalence and similarity?

▷ **Question 254C** Which matrices are similar to the identity matrix? Which matrices are biequivalent to the identity matrix?

Diagonalizable

Definition 254D A matrix A is called **diagonalizable** iff it is similar to a diagonal matrix; that is, iff there are matrices P and D satisfying

$$A = PDP^{-1}$$

where P is invertible and D is diagonal.

▷ **Question 254E** Is a diagonal matrix diagonalizable?

Remark 254F For the real matrix $J \in \mathbf{R}^{2 \times 2}$ in Example 262C below, there is a complex invertible matrix P such that $P^{-1}JP$ is diagonal, but no such real matrix P. To make this distinction we say that a matrix $A \in \mathbf{F}^{n \times n}$ is **diagonalizable over** $\mathbf{F}$ iff there is an invertible matrix $P \in \mathbf{F}^{n \times n}$ and a diagonal matrix $D \in \mathbf{F}^{n \times n}$ such that

$$A = PDP^{-1}.$$

With this terminology the real matrix J is diagonalizable over $\mathbf{C}$ but not diagonalizable over $\mathbf{R}$. *Most, but not all, matrices are diagonalizable over* $\mathbf{C}$. (Example 263A below gives an example of a nondiagonalizable matrix.)

Remark 255A It is easy to make examples of diagonalizable matrices by choosing P and D first. For example, take

$$P = \begin{bmatrix} 2 & 1 \\ 1 & 1 \end{bmatrix}, \qquad P^{-1} = \begin{bmatrix} 1 & -1 \\ -1 & 2 \end{bmatrix}, \qquad D = \begin{bmatrix} 7 & 0 \\ 0 & 4 \end{bmatrix}.$$

The matrix

$$A = PDP^{-1} = \begin{bmatrix} 10 & 6 \\ 3 & 1 \end{bmatrix}$$

is diagonalizable.

Theorem 255B *Similarity satisfies the following laws:*

(Reflexive Law) $A \equiv A$;

(Symmetric Law) *if* $A \equiv B$, *then* $B \equiv A$;

(Transitive Law) *if* $A \equiv B$ *and* $B \equiv C$, *then* $A \equiv C$.

Here $A \equiv B$ *abbreviates "A is similar to B".*

Proof: Exercise. (See Theorem 104C.)

▷ **Exercise 255C** Assume that $A_2 = P_2 A_1 P_2^{-1}$, $A_3 = P_3 A_1 P_3^{-1}$, $A_4 = P_4 A_2 P_4^{-1}$. Find P such that $A_4 = P A_3 P^{-1}$.

8.2 Eigenvalues and Eigenvectors

Our ultimate aim is to develop a theory which tells us how to **diagonalize** a given matrix A, that is, how to find matrices P and D with D diagonal, P invertible, and $A = PDP^{-1}$. As you may have guessed, this isn't always possible (otherwise why would someone have invented the word *diagonalizable?*), so our theory should also tell us when it is. The theory of similarity gives invariants of a matrix under similarity. The simplest of these invariants are the *eigenvalues* of the matrix explained in this section.

Let $A \in \mathbf{F}^{n \times n}$ be a square matrix. If a number $\lambda \in \mathbf{F}$ and nonzero column vector $X \in \mathbf{F}^{n \times 1}$ satisfy the equation

$$AX = \lambda X,$$

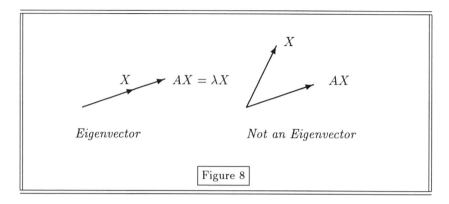

Figure 8

we say that λ is an *eigenvalue* of A and that X is an *eigenvector* of A for the eigenvalue λ. See Figure 8. This equation can be rewritten in the form

$$(\lambda I - A)X = 0 \qquad (EE)$$

where $I = I_n$ is the $n \times n$ identity matrix. This equation is called the **eigenequation**. A square matrix is not invertible iff its null space is not the zero subspace. Hence, we may reformulate the definition as follows:

Eigen Terminology

Definition 256A A number $\lambda \in \mathbf{F}$ is called an **eigenvalue** of the square matrix A iff the matrix $\lambda I - A$ is not invertible. The subspace

$$\mathcal{E}_\lambda(A) = \mathcal{N}(\lambda I - A)$$

is called the **eigenspace** of λ. An element of $\mathcal{E}_\lambda(A)$ is called an **eigenvector** of A corresponding to the eigenvalue λ.

▷ **Question 256B** What are the eigenvalues and eigenvectors of the identity matrix? the zero matrix?

Theorem 256C (Diagonal Eigenvalues)   *The eigenvalues of a diagonal matrix are its diagonal entries; the corresponding columns of the identity matrix are corresponding eigenvectors.*

Proof: Exercise 260C.

Example 257A With $n = 3$ and $j = 2$ we have

$$\begin{bmatrix} \lambda_1 & 0 & 0 \\ 0 & \lambda_2 & 0 \\ 0 & 0 & \lambda_3 \end{bmatrix} \begin{bmatrix} 0 \\ 1 \\ 0 \end{bmatrix} = \lambda_2 \begin{bmatrix} 0 \\ 1 \\ 0 \end{bmatrix}.$$

This shows that the second column E_2 of the 3×3 identity matrix is an eigenvector of the above diagonal matrix.

Invariance of Eigenvalues

Theorem 257B *Similar matrices have the same eigenvalues.*

Proof: Suppose A and B are similar, say

$$A = PBP^{-1}$$

with P invertible. Then $\lambda I - A$ and $\lambda I - B$ are similar as well:

$$\lambda I - A = P(\lambda I - B)P^{-1}.$$

Similar matrices have the same rank (they are biequivalent), so $\lambda I - A$ is invertible if and only if $\lambda I - B$ is. □

Remark 257C By Theorem 142B, $P\mathcal{N}(\lambda I - B) = \mathcal{N}(\lambda I - A)$, that is,

$$P\mathcal{E}_\lambda(B) = \mathcal{E}_\lambda(A).$$

This says that P transforms eigenvectors of B to eigenvectors of A. The following diagram helps avoid confusion:

$$
\begin{array}{ccc}
\mathcal{E}_\lambda(A) \subset \mathbf{F}^{n \times 1} & \xrightarrow{\ A\ } & \mathcal{E}_\lambda(A) \subset \mathbf{F}^{n \times 1} \\[2mm]
P \Big\uparrow & & \Big\uparrow P \\[2mm]
\mathcal{E}_\lambda(B) \subset \mathbf{F}^{n \times 1} & \xrightarrow{\ B\ } & \mathcal{E}_\lambda(B) \subset \mathbf{F}^{n \times 1}
\end{array}
$$

If $\widetilde{X}$ is an eigenvector for B, then $P\widetilde{X}$ is an eigenvector for A. Compare with Remark 143A.

Diagonalization Theorem

Theorem 258A *A square matrix A is diagonalizable if and only if there is an invertible matrix P of the same size whose columns are eigenvectors of A. For such a P, the matrix $P^{-1}AP$ is diagonal.*

Remark 258B In view of the fact that the columns of a square matrix are independent iff that matrix is invertible (Corollary 175A), we may reformulate this as follows: *A matrix $A \in \mathbf{F}^{n \times n}$ is diagonalizable if and only if it has n independent eigenvectors.*

Proof of 258A: Assume that $X_1, X_2, \ldots, X_n$ are the columns of an invertible matrix

$$P = \begin{bmatrix} X_1 & X_2 & \ldots & X_n \end{bmatrix},$$

and that each X_j is an eigenvector of A with eigenvalue λ_j:

$$AX_j = \lambda_j X_j. \qquad\qquad (EE.j)$$

Form the diagonal matrix D with entries λ_j on the diagonal:

$$D = \operatorname{diag}(\lambda_1, \lambda_2, \ldots, \lambda_n).$$

Then

$$
\begin{aligned}
AP &= \begin{bmatrix} AX_1 & AX_2 & \ldots & AX_n \end{bmatrix} && \text{by 54B} \\[2mm]
&= \begin{bmatrix} \lambda_1 X_1 & \lambda_2 X_2 & \ldots & \lambda_n X_n \end{bmatrix} && \text{by } (EE.j) \\[2mm]
&= PD && \text{by 47D}
\end{aligned}
$$

so $A = PDP^{-1}$ as P is assumed invertible. Conversely, assume that A is diagonalizable, say $A = PDP^{-1}$ with P invertible and D as above. Then as before

$$
\begin{aligned}
AP &= \begin{bmatrix} AX_1 & AX_2 & \ldots & AX_n \end{bmatrix} && \text{by 54B} \\[2mm]
PD &= \begin{bmatrix} \lambda_1 X_1 & \lambda_2 X_2 & \ldots & \lambda_n X_n \end{bmatrix} && \text{by 47D}
\end{aligned}
$$

and $AP = PD$ by hypothesis, so $(EE.j)$ holds. This shows that the columns $X_j = \operatorname{col}_j(P)$ of P are eigenvectors.   □

▷ **Exercise 259A** Let A, D, P, be given by

$$P = \begin{bmatrix} 2 & 1 \\ 1 & 1 \end{bmatrix}, \qquad D = \begin{bmatrix} 3 & 0 \\ 0 & 7 \end{bmatrix}, \qquad A = \begin{bmatrix} -1 & 8 \\ -4 & 11 \end{bmatrix},$$

so that $A = PDP^{-1}$. What are the eigenvalues of D? What are the eigenvalues of A? Find an invertible matrix whose columns are eigenvectors of D. Find an invertible matrix whose columns are eigenvectors of A.

▷ **Exercise 259B** The matrix equation

$$\begin{bmatrix} \lambda_1 & 0 & 0 \\ 0 & \lambda_2 & 0 \\ 0 & 0 & \lambda_3 \end{bmatrix} \begin{bmatrix} x_1 \\ x_2 \\ x_3 \end{bmatrix} = \lambda_2 \begin{bmatrix} x_1 \\ x_2 \\ x_3 \end{bmatrix}.$$

states that X is an eigenvector of the diagonal matrix D for the eigenvalue λ_2. Show that if the diagonal entries λ_1, λ_2, and λ_3 are all different, then this equation only holds when $x_1 = x_3 = 0$. What can you conclude about the entries x_1, x_2, x_3 of X when $\lambda_1 = \lambda_2 \neq \lambda_3$? When $\lambda_1 \neq \lambda_2 = \lambda_3$? When $\lambda_1 = \lambda_3 \neq \lambda_2$?

Exercise 259C Let A, D, P, be given by

$$P = \begin{bmatrix} 2 & 1 & 2 \\ 3 & 3 & 4 \\ 1 & 3 & 3 \end{bmatrix}, \qquad P^{-1} = \begin{bmatrix} -3 & 3 & -2 \\ -5 & 4 & -2 \\ 6 & -5 & 3 \end{bmatrix},$$

$$D = \begin{bmatrix} -1 & 0 & 0 \\ 0 & 1 & 0 \\ 0 & 0 & 2 \end{bmatrix}, \qquad A = \begin{bmatrix} 25 & -22 & 14 \\ 42 & -37 & 24 \\ 24 & -21 & 14 \end{bmatrix},$$

so that $A = PDP^{-1}$. What are the eigenvalues of D? What are the eigenvalues of A? Find an invertible matrix whose columns are eigenvectors of D. Find an invertible matrix whose columns are eigenvectors of A.

▷ **Exercise 259D** Each of X, Y, Z is an eigenvector of A. Find the corresponding eigenvalue.

$$A = \begin{bmatrix} 3 & 1 & -2 \\ -6 & -6 & 18 \\ -2 & -3 & 9 \end{bmatrix}, \qquad X = \begin{bmatrix} -1 \\ 6 \\ 2 \end{bmatrix}, \qquad Y = \begin{bmatrix} -1 \\ 3 \\ 1 \end{bmatrix}, \qquad Z = \begin{bmatrix} 0 \\ 2 \\ 1 \end{bmatrix}.$$

▷ **Exercise 259E** Which of X, Y and Z are eigenvectors of A?

$$A = \begin{bmatrix} -12 & 7 & 32 \\ 32 & -14 & -71 \\ -12 & 6 & 29 \end{bmatrix}, \qquad X = \begin{bmatrix} 1 \\ 2 \\ 0 \end{bmatrix}, \qquad Y = \begin{bmatrix} 2 \\ -5 \\ 2 \end{bmatrix}, \qquad Z = \begin{bmatrix} 1 \\ -3 \\ 1 \end{bmatrix}.$$

▷ **Exercise 260A** The eigenvalues of the matrix $A = \begin{bmatrix} 13 & 6 & 30 \\ 1 & 2 & 3 \\ -4 & -2 & -9 \end{bmatrix}$
are $\lambda_1 = 1$, $\lambda_2 = 2$, $\lambda_3 = 3$. Diagonalize A; that is, find an invertible
matrix P so that $P^{-1}AP$ is diagonal. Check your answer by comparing
AP and PD where $D = \mathrm{diag}(\lambda_1, \lambda_2, \lambda_3)$

Exercise 260B In each of the following you are given a matrix A and
its eigenvalues. You are to diagonalize A; that is, you must find an invertible matrix P so that $P^{-1}AP$ is diagonal. Can the matrix P have
integer entries? Check your answer by comparing AP and PD where
$D = \mathrm{diag}(\lambda_1, \lambda_2, \lambda_3)$. (You may use MINIMAT's `randsoln` function, but
not the `eig` function.)

(1) $A = \begin{bmatrix} 13 & 6 & 30 \\ 1 & 2 & 3 \\ -4 & -2 & -9 \end{bmatrix}$, $\lambda_1 = 1$, $\lambda_2 = 2$, $\lambda_3 = 3$.

(2) $A = \begin{bmatrix} -11 & -6 & -30 \\ -1 & 0 & -3 \\ 4 & 2 & 11 \end{bmatrix}$, $\lambda_1 = 1$, $\lambda_2 = 0$, $\lambda_3 = -1$.

(3) $A = \begin{bmatrix} -8 & -6 & -24 \\ 0 & 1 & 0 \\ 3 & 2 & 9 \end{bmatrix}$, $\lambda_1 = \lambda_2 = 1$, $\lambda_3 = 0$.

(4) $A = \begin{bmatrix} -2 & 0 & 2 \\ 0 & 1 & 0 \\ -4 & 0 & 4 \end{bmatrix}$, $\lambda_1 = 0$, $\lambda_2 = 1$, $\lambda_3 = 2$.

(5) $A = \begin{bmatrix} 4 & 0 & -1 \\ 0 & 1 & 0 \\ 2 & 0 & 1 \end{bmatrix}$, $\lambda_1 = 3$, $\lambda_2 = 1$, $\lambda_3 = 2$.

▷ **Exercise 260C** Prove Theorem 256C.

8.3   Computing Eigenvalues

In Chapter 14 we shall study the problem of diagonalization more carefully.
Here we show how to diagonalize a 2×2 matrix. This illustrates the main
ideas. Larger problems, even 3×3 matrices, involve so much computation
that it is generally not feasible to do them by hand.

A number λ is an eigenvalue of a square matrix A iff the eigenequation

$$(\lambda I - A)X = 0 \qquad (EE)$$

has a nonzero solution X. The key property of the determinant function is that

$$\mathcal{N}(B) \neq \{0\} \iff \det(B) = 0.$$

Applying this property with $B = \lambda I - A$, we see that the values of λ for which the eigenequation has a nonzero solution X are the roots of the equation

$$\det(\lambda I - A) = 0. \qquad (CE)$$

This equation is called the **characteristic equation** of the matrix A; the left hand side is a polynomial of degree n called the **characteristic polynomial** of the matrix A. Generally, it is very hard to compute the characteristic polynomial, and even harder to find its roots. However, when $n = 2$, this isn't so hard. We'll study the general case later in Section 8.4.

When $n = 2$, the matrix A is given by

$$A = \begin{bmatrix} a & b \\ c & d \end{bmatrix}.$$

The system (EE) above has matrix of coefficients

$$(\lambda I - A) = \begin{bmatrix} \lambda - a & -b \\ -c & \lambda - d \end{bmatrix}.$$

To obtain the characteristic equation (CE) we equate the determinant to zero:

$$(\lambda - a)(\lambda - d) - bc = 0. \qquad (CE.2)$$

The following algorithm will diagonalize a 2×2 matrix A (when A is diagonalizable).

(1) Find the roots λ_1 and λ_2 of the quadratic equation (CE.2) either by factoring or the Quadratic Formula.

(2) If these two eigenvalues are equal, then either A is already diagonal (of form λI) or else A is not diagonalizable.

(3) If the eigenvalues λ_i ($i = 1, 2$) are distinct, plug each of them in for λ in the system (EE) and solve the resulting linear homogeneous system. Choose any (it doesn't matter which) nonzero solution X_i of this system:

$$(\lambda_1 I - A)X_1 = 0, \qquad (\lambda_2 I - A)X_2 = 0. \qquad (EE.2)$$

(4) This gives
$$A = PDP^{-1}$$
where
$$D = \begin{bmatrix} \lambda_1 & 0 \\ 0 & \lambda_2 \end{bmatrix}, \qquad P = \begin{bmatrix} X_1 & X_2 \end{bmatrix}$$

Example 262A Suppose A is already diagonal. Then $b = c = 0$ and the characteristic equation (CE.2) takes the form
$$(\lambda - a)(\lambda - d) = 0$$
which has the two solutions $\lambda_1 = a$ and $\lambda_2 = d$. Any diagonal matrix with nonzero entries on the diagonal will serve as P; in particular, we can take $P = I$ the identity matrix.

Example 262B Let
$$A = \begin{bmatrix} -4 & 6 \\ -3 & 5 \end{bmatrix}$$
so the characteristic equation (CE) is
$$0 = (-4 - \lambda)(5 - \lambda) + 18 = \lambda^2 - \lambda - 2.$$

The roots are $\lambda_1 = -1$ and $\lambda_2 = 2$ so the equations $(A - \lambda_i I)X = 0$ resulting from the eigenequations (EE.2) have coefficient matrices
$$A - \lambda_1 I = \begin{bmatrix} -3 & 6 \\ -3 & 6 \end{bmatrix}, \qquad A - \lambda_2 I = \begin{bmatrix} -6 & 6 \\ -3 & 3 \end{bmatrix}$$
with nonzero solutions
$$X_1 = \begin{bmatrix} 2 \\ 1 \end{bmatrix}, \qquad X_2 = \begin{bmatrix} 1 \\ 1 \end{bmatrix}.$$

Hence, $A = PDP^{-1}$ with
$$P = \begin{bmatrix} 2 & 1 \\ 1 & 1 \end{bmatrix}, \qquad D = \begin{bmatrix} -1 & 0 \\ 0 & 2 \end{bmatrix}.$$

Example 262C Let
$$J = \begin{bmatrix} 0 & 1 \\ -1 & 0 \end{bmatrix}.$$

The characteristic polynomial of J is
$$\det(\lambda I - J) = \lambda^2 + 1$$

so its eigenvalues are $\pm i$. There is no real matrix P such that $P^{-1}JP$ is diagonal; if there were, the eigenvalues of J would be real. There is a complex P:

$$P^{-1}JP = \begin{bmatrix} i & 0 \\ 0 & -i \end{bmatrix} \text{ for } P = \begin{bmatrix} 1 & 1 \\ i & -i \end{bmatrix}.$$

The real matrix J is diagonalizable over $\mathbf{C}$, but not diagonalizable over $\mathbf{R}$.

Example 263A (A Nondiagonalizable Matrix)   The matrix

$$A = \begin{bmatrix} d & 1 \\ 0 & d \end{bmatrix}$$

is not diagonalizable. The characteristic polynomial is

$$\det(\lambda I - A) = (\lambda - d)^2.$$

The only eigenvalue is d, so if $A = PDP^{-1}$ with D diagonal, we must have $D = dI$. But then $PDP^{-1} = PdIP^{-1} = dPP^{-1} = dI = D \neq A$ which is a contradiction.

▷ **Exercise 263B** Diagonalize each of the following 2×2 matrices.

(1) $A = \begin{bmatrix} 1 & 2 \\ -1 & 4 \end{bmatrix}$ (2) $A = \begin{bmatrix} 2 & 1 \\ 1 & 1 \end{bmatrix}$ (3) $A = \begin{bmatrix} 0 & 1 \\ 1 & 0 \end{bmatrix}$

(4) $A = \begin{bmatrix} 3 & 4 \\ -4 & 3 \end{bmatrix}$ (5) $A = \begin{bmatrix} 1 & 2 \\ 0 & 3 \end{bmatrix}$ (6) $A = \begin{bmatrix} 4 & 0 \\ 5 & 6 \end{bmatrix}$

Exercise 263C Use the formula for the determinant of a 3×3 matrix, which appears on Page 240, to compute the characteristic polynomial $\det(A - \lambda I)$ of the 3×3 matrix A from Exercise 259C. Also find $\det(D - \lambda I)$ and factor these polynomials. Hint: What are the roots of

$$f(\lambda) = (\lambda - \lambda_1)(\lambda - \lambda_2)(\lambda - \lambda_3)?$$

Exercise 263D Find the characteristic polynomials of each of the five matrices A in Exercise 260B.

Exercise 263E Find the characteristic polynomial and the eigenvalues of each of the following matrices A:

(1) $A = \begin{bmatrix} 2 & 3 \\ 3 & 1 \end{bmatrix}$ (3) $A = \begin{bmatrix} 1 & 2 & 3 \\ 0 & 4 & 5 \\ 0 & 0 & 6 \end{bmatrix}$

(2) $A = \begin{bmatrix} 2 & 3 \\ -3 & 1 \end{bmatrix}$ (4) $A = \begin{bmatrix} 7 & 0 & 0 \\ 0 & 2 & 3 \\ 0 & 3 & 1 \end{bmatrix}$

Remark 264A (Block by Block Diagonalization)   From the block multiplication law 51A we obtain

$$\begin{bmatrix} A & 0 \\ 0 & B \end{bmatrix} \begin{bmatrix} X \\ Y \end{bmatrix} = \begin{bmatrix} AX \\ BY \end{bmatrix}$$

if $A \in \mathbf{F}^{n \times n}$, $B \in \mathbf{F}^{m \times m}$, $X \in \mathbf{F}^{n \times 1}$, $Y \in \mathbf{F}^{m \times 1}$. If $AX = \lambda X$, then

$$\begin{bmatrix} A & 0 \\ 0 & B \end{bmatrix} \begin{bmatrix} X \\ 0 \end{bmatrix} = \lambda \begin{bmatrix} X \\ 0 \end{bmatrix}.$$

Similarly, if $BY = \mu Y$, then

$$\begin{bmatrix} A & 0 \\ 0 & B \end{bmatrix} \begin{bmatrix} 0 \\ Y \end{bmatrix} = \mu \begin{bmatrix} 0 \\ Y \end{bmatrix}.$$

This means that to diagonalize a block diagonal matrix, we need only diagonalize each of its blocks.

▷ **Exercise 264B** Use the principle 264A of block by block diagonalization to diagonalize each of the following matrices:

(1) $\quad A = \begin{bmatrix} 1 & 2 & 0 & 0 \\ -1 & 4 & 0 & 0 \\ 0 & 0 & -5 & 9 \\ 0 & 0 & -6 & 10 \end{bmatrix}$ (2) $\quad A = \begin{bmatrix} -5 & 9 & 0 & 0 \\ -6 & 10 & 0 & 0 \\ 0 & 0 & 1 & 2 \\ 0 & 0 & -1 & 4 \end{bmatrix}$

(3) $\quad A = \begin{bmatrix} 1 & 2 & 0 & 0 \\ -1 & 4 & 0 & 0 \\ 0 & 0 & 1 & 2 \\ 0 & 0 & -1 & 4 \end{bmatrix}$ (4) $\quad A = \begin{bmatrix} 1 & 0 & 0 & 0 \\ 0 & 4 & 0 & 0 \\ 0 & 0 & 1 & 2 \\ 0 & 0 & -1 & 4 \end{bmatrix}$

(5) $\quad A = \begin{bmatrix} 4 & 0 & 0 \\ 0 & 1 & 2 \\ 0 & -1 & 4 \end{bmatrix}$ (6) $\quad A = \begin{bmatrix} 1 & 2 & 0 \\ -1 & 4 & 0 \\ 0 & 0 & 4 \end{bmatrix}$

Using MINIMAT

The built-in function `eig` tries to compute the eigenvalues and eigenvectors of its input. The command

```
#> [P D] = eig(A)
```

causes MINIMAT to diagonalize A if possible. The first output P holds
the eigenvectors of A. The second output D is a diagonal matrix whose
entries are the eigenvalues of A (in the appropriate order). The subsequent
command

```
#> P*D*P^(-1),   A
```

will tell you if eig was successful; if it was, you should get two copies of
the same matrix. In general, P and D will be complex matrices, even when
A is real. We shall not explain how MINIMAT computes these matrices:[1]
just accept that MINIMAT does something here that usually works (but
not always) and which you can always check. The one-output command

```
#> L = eig(A)
```

produces a row matrix whose entries are the eigenvalues of A. The subsequent command

```
#> D, diag(L)
```

should produce two copies of the same matrix.

Exercise 265A Try out MINIMAT's eig function:

```
#> A = [1 2 3; 4 5 6; 7 8 9]
#> [P D] = eig(A)
#> D, P^(-1)*A*P
#> P*D*P^(-1),  A
```

What should happen on the last line?

Exercise 265B Find the value of the scalar d so that the following computation in MINIMAT gives A*X=d*X.

```
#> I=eye(4), E=I(:,3), P=rand(4,4)
#> A=P*diag([1 3 5 8])*inv(P), X=P*E
```

Exercise 265C Diagonalize the matrices in Exercise 263B using MINI-
MAT's built-in eig function and compare the results with the results you
got by hand. (Since the solution X_i of the system $(A - \lambda_i I)X_i = 0$ is
only defined up to a constant and since you can use the eigenvalues in any
order, these may not be the same. You can make them agree by rescaling
the columns of P and possibly interchanging them.)

▷ **Exercise 265D** Generate matrices in MINIMAT with the commands

---

[1]MINIMAT does *not* use the method described in Section 8.3.

```
#> D=diag([1,3,7]), P=rand(3,3), A=P*D*inv(P)
```

Confirm that the columns of P are eigenvectors of A.

▷ **Exercise 266A** Generate a diagonal matrix in MINIMAT with the command

```
#> D=diag([3,5,0])
```

and generate a random invertible 3×3 matrix P and a matrix A with the commands

```
#> P=rand(3,3), A=P*D*inv(P)
```

Confirm that $d = 3, 5, 0$ are eigenvalues of A (without using MINIMAT's eig function).

Exercise 266B Write a .M function veig such that the command

```
#>  [P D] = veig(A)
```

produces an invertible matrix P and a diagonal matrix D which satisfy $A = P*D*P^{(-1)}$. You may use the command

```
#>  L = eig(A)
```

but not the command

```
#>  [P D] = eig(A)
```

You may, however, use the .M function **randsoln** which produces a random element of the null space of its input.

8.4   The Characteristic Polynomial

The *characteristic polynomial* can be used to compute the eigenvalues of a matrix. We repeat the definition.

Characteristic Polynomial

Definition 267A Suppose that $A \in \mathbf{F}^{n \times n}$ is a square matrix. The **characteristic polynomial** of A is the polynomial $p(\xi)$ in the indeterminate ξ defined by

$$p(\xi) = \det(\xi I - A).$$

Here $I = I_n$ is the $n \times n$ identity matrix. The characteristic polynomial is of degree n in the indeterminate ξ.

Theorem 267B *The roots of the characteristic polynomial of a matrix are the eigenvalues of that matrix.*

Proof: The number λ is an eigenvalue $\iff \mathcal{E}_\lambda(A) = \mathcal{N}(\lambda I - A) \neq 0$ $\iff \lambda I - A$ is not invertible $\iff \det(\lambda I - A) = 0$.   □

Corollary 267C *Every square matrix has at least one (possibly complex) eigenvalue.*

This is because a nonconstant polynomial always has at least one (complex) root. This fact is known as the *Fundamental Theorem of Algebra*. Its proof is outside the scope of this book.

Invariance of the Characteristic Polynomial

Theorem 267D *Similar matrices have the same characteristic polynomial.*

Proof: If $A = PBP^{-1}$, then

$$\begin{aligned}
\xi I - A &= \xi I - PBP^{-1} \\
&= \xi P P^{-1} - PBP^{-1} \\
&= P(\xi I - B)P^{-1}
\end{aligned}$$

so

$$\det(\xi I - A) = \det(P) \det(\xi I - B) \det(P^{-1}) = \det(\xi I - B).$$

$\square$

Definition 268A The **trace** of a square matrix $A \in \mathbf{F}^{n \times n}$ is the sum of its diagonal entries:

$$\text{trace}(A) = \sum_{k=1}^{n} \text{entry}_{kk}(A).$$

The trace is linear:

$$\text{trace}(aA + bB) = a\,\text{trace}(A) + b\,\text{trace}(B)$$

for $A, B \in \mathbf{F}^{n \times n}$ and $a, b \in \mathbf{F}$.

Theorem 268B *The determinant of a matrix is the product of its eigenvalues. The trace of a matrix is the sum of its eigenvalues.*

Proof: The characteristic polynomial of an $n \times n$ matrix is a monic polynomial of degree n. This means that it has the form

$$\det(\xi I - A) = \xi^n + p_1 \xi^{n-1} + \cdots + p_{n-1}\xi + p_n. \tag{1}$$

(The word **monic** means that the leading coefficient is one.) Set $\xi = 0$ to obtain

$$p_n = (-1)^n \det(A). \tag{2}$$

From the definition of the determinant (239B) we obtain that

$$\det(\xi I - A) = \prod_{j=1}^{n}(\xi - a_{jj}) + \cdots$$

where the dots represent terms where ξ appears to order less than $n-1$. This is because any permutation other than the identity permutation omits at least two diagonal terms from the corresponding elementary product. From the last equation we see that

$$\text{trace}(A) = -p_1 \tag{3}$$

gives the coefficient of ξ^{n-1} in $\det(\xi I - A)$. The characteristic polynomial is a monic polynomial whose roots are the eigenvalues $\lambda_1, \lambda_2, \ldots, \lambda_n$. But there is only one such polynomial, so

$$\det(\xi I - A) = \prod_{j=1}^{n} (\xi - \lambda_j).$$

Multiply out the right hand side to get

$$p_1 = -\sum_{j=1}^{n} \lambda_j, \qquad p_n = (-1)^n \prod_{j=1}^{n} \lambda_j. \tag{4}$$

The theorem follows from (2), (3), and (4).   □

Example 269A The characteristic polynomial has form (1) where the coefficients p_k are certain complicated expressions in the entries $a_{ij} = \text{entry}_{ij}(A)$ of A. For example, when $n = 2$ and

$$A = \begin{bmatrix} a & b \\ c & d \end{bmatrix},$$

we obtain

$$\det \begin{bmatrix} \xi - a & -b \\ -c & \xi - d \end{bmatrix} = (\xi - a)(\xi - d) - bc,$$

so in this case,

$$\begin{aligned} \det(\xi I - A) &= \xi^2 - (a + d)\xi + (ad - bc) \\ &= \xi^2 - \text{trace}(A)\xi + \det(A). \end{aligned}$$

Since

$$\begin{aligned} \det(\xi I - A) &= (\xi - \lambda_1)(\xi - \lambda_2) \\ &= \xi^2 - (\lambda_1 + \lambda_2)\xi + \lambda_1\lambda_2, \end{aligned}$$

we get $-p_1 = \lambda_1 + \lambda_2 = \text{trace}(A)$ and $p_2 = \lambda_1\lambda_2 = \det(A)$ as expected.

Exercise 270A For the 3×3 matrix

$$A = \begin{bmatrix} a_{11} & a_{12} & a_{13} \\ a_{21} & a_{22} & a_{23} \\ a_{31} & a_{32} & a_{33} \end{bmatrix}$$

find the coefficients p_1, p_2, and p_3 so that

$$\det(\xi I - A) = \xi^3 + p_1\xi^2 + p_2\xi + p_3.$$

Express the answer both in terms of the entries a_{ij} and the eigenvalues λ_j.

Exercise 270B Show that the coefficient $p_k = p_k(A)$ of ξ^{n-k} in the characteristic polynomial is of degree k in A. This means that $p_k(tA) = t^k p_k(A)$. Hint: Compare the coefficients of $\det(\xi I - tA)$ and $t^n \det(\xi t^{-1} I - A)$.

▷ **Exercise 270C** Show that if A is real and λ is an eigenvalue of A, then so is $\bar{\lambda}$.

Using MINIMAT

▷ **Exercise 270D** Confirm that the trace of a square matrix is the sum of its eigenvalues.

▷ **Exercise 270E** Confirm that the determinant of a square matrix is the product of the eigenvalues.

8.5   Multiplicity

Now we can say when a matrix is diagonalizable. Let A be an $n \times n$ matrix and factor its characteristic polynomial:

$$\det(\xi I - A) = \prod_{j=1}^{n} (\xi - \lambda_j). \tag{4}$$

Of course, $\lambda_1, \lambda_2, \ldots, \lambda_n$ are the eigenvalues of A, but there will be repetitions if there is a multiple root. Let $\alpha(\lambda, A)$ be the number of times λ appears in the list; this is called the **algebraic multiplicity** of the eigenvalue λ for the matrix A. Notice that we always have

$$n = \sum_{\lambda} \alpha(\lambda, A)$$

where each eigenvalue λ is taken once in the sum. This is because $\alpha(\lambda, A)$ is the number of times the factor $\xi - \lambda$ appears in equation (4) and the characteristic polynomial has degree n.

The number λ is an eigenvalue of A iff the eigenspace $\mathcal{E}_\lambda(A) = \mathcal{N}(\lambda I - A)$ is nonzero. The dimension

$$\nu(\lambda, A) = \dim \mathcal{E}_\lambda(A)$$

is called the **geometric multiplicity** of the eigenvalue λ for the matrix A. The next theorem is a refinement of Theorem 257B.

▷ **Question 271A** What is the geometric multiplicity of an eigenvalue of a diagonal matrix?

Invariance of Multiplicities

Theorem 271B *Similar matrices have the same eigenvalues each with the same algebraic multiplicity and the same geometric multiplicity.*

Proof: Equation (4) may be written

$$\det(\xi I - A) = \prod_{i=1}^{m}(\xi - \lambda_i)^{\alpha(\lambda_i, A)}$$

where the list $\lambda_1, \lambda_2, \ldots, \lambda_m$ contains each eigenvalue of A exactly once. If A and B are similar, then $\det(\xi I - A) = \det(\xi I - B)$, so $\alpha(\lambda_i, A) = \alpha(\lambda_i, B)$. For the geometric multiplicities, note that if A and B are similar, then so are $\lambda I - A$ and $\lambda I - B$ for any $\lambda \in \mathbf{F}$. Similar matrices are, in particular, biequivalent. Hence, $\lambda I - A$ and $\lambda I - B$ have the same nullity; in other words, $\nu(\lambda, A) = \nu(\lambda, B)$.   □

▷ **Question 271C** The matrices $A = \operatorname{diag}(4, 4, 5)$ and $B = \operatorname{diag}(4, 5, 5)$ have the same eigenvalues. Are they similar?

The nullity of a diagonal matrix is the number of zero diagonal entries and the determinant of a diagonal matrix is the product of its diagonal

entries. It follows that for a diagonal matrix D, the geometric and algebraic multiplicities agree:
$$\nu(\lambda, D) = \alpha(\lambda, D).$$
Hence, this is true for a diagonalizable matrix as well. The following theorem says that the converse is true as well.

Diagonalizability Theorem

Theorem 272A *For any matrix the geometric multiplicity of an eigenvalue is at most its algebraic multiplicity. A matrix is diagonalizable if and only if the geometric multiplicity of each of its eigenvalues is the same as the algebraic multiplicity of that eigenvalue.*

In other words, $\nu(\lambda, A) \leq \alpha(\lambda, A)$, and the matrix A is diagonalizable $\iff \nu(\lambda, A) = \alpha(\lambda, A)$ for all λ. Theorem 272A will be proved in Chapter 13. (See Exercise 392A.)

Remark 272B Once we know all the eigenvalues of A, here is how to decide if A is diagonalizable and how to diagonalize A if so.

(1) Arrange the eigenvalues of A in a list $(\lambda_1, \lambda_2, \ldots, \lambda_m)$ without repetitions. This means

$$\det(\lambda I - A) = 0 \iff \lambda = \lambda_i \quad \text{for some } i$$

and $\lambda_i \neq \lambda_j$ for $i \neq j$.

(2) For each i construct a basis Φ_i for the eigenspace $\mathcal{E}_{\lambda_i}(A)$. Each basis satisfies

$$A\Phi_i = \lambda_i \Phi_i.$$

(3) Form the matrix
$$P = \begin{bmatrix} \Phi_1 & \Phi_2 & \cdots & \Phi_m \end{bmatrix}.$$

The matrix P has independent columns. Hence, by Corollary 175A, it is invertible if and only if it is square.

(4) The matrix A is diagonalizable if and only if P is invertible. When defined, $P^{-1}AP$ is diagonal. See Theorem 258A.

Item (3) – that P has independent columns – is proved in Chapter 13. (See Exercise 392B.) This is not completely obvious. Of course, the columns of each Φ_i are independent, but that does not prove that the matrix P formed by catenating them has independent columns.

8.6 More Exercises

▷ **Exercise 273A** True or False?

(1) [T] [F] All diagonal matrices are diagonalizable.

(2) [T] [F] All diagonalizable matrices are diagonal.

(3) [T] [F] All invertible matrices are diagonalizable.

(4) [T] [F] All diagonalizable matrices are invertible.

(5) [T] [F] Similar matrices must have the same eigenvalues.

(6) [T] [F] Similar matrices must have the same eigenvectors.

(7) [T] [F] Similar matrices must be biequivalent.

(8) [T] [F] Similar matrices must have the same nullity.

(9) [T] [F] Square biequivalent matrices must be similar.

Exercise 273B What are the eigenvalues of

$$W = \begin{bmatrix} x & y \\ -y & x \end{bmatrix}?$$

Suppose that $A \in \mathbf{R}^{2\times 2}$ is a real matrix with eigenvalues $x \pm yi$ (where $x, y \in \mathbf{R}$). Show that A is similar (over $\mathbf{R}$) to W, that is, that there is a real invertible matrix $P \in \mathbf{R}^{2\times 2}$ with $A = PWP^{-1}$.

▷ **Exercise 273C** Find P so that $A = PDP^{-1}$ where

$$
\begin{aligned}
A &= \operatorname{diag}(\lambda_3, \lambda_1, \lambda_4, \lambda_2) \\
D &= \operatorname{diag}(\lambda_1, \lambda_2, \lambda_3, \lambda_4)
\end{aligned}
$$

are similar. Hint: You must find P with $A = PDP^{-1}$. Mimic the proof of the Diagonalization Theorem 258A.

▷ **Exercise 274A** Suppose that

$$A = \text{diag}(\lambda_{\sigma(1)}, \lambda_{\sigma(2)}, \ldots, \lambda_{\sigma(n)})$$
$$D = \text{diag}(\lambda_1, \lambda_2, \ldots, \lambda_n)$$

where $((\sigma(1), \sigma(2), \ldots, \sigma(n))$ is some permutation of $(1, 2, \ldots, n)$. Find P with $A = PDP^{-1}$.

Exercise 274B Prove that if P is a permutation matrix and D is a diagonal matrix, then PDP^{-1} is diagonal.

▷ **Exercise 274C** Prove that the geometric multiplicities form a complete system of invariants for similarity of diagonalizable matrices. This means that two diagonalizable matrices are similar if and only if they have the same eigenvalues each with the same geometric multiplicity.

▷ **Exercise 274D** Show that a complex 2×2 matrix is either diagonalizable or else it is similar to a matrix B of form

$$B = \begin{bmatrix} \lambda & 1 \\ 0 & \lambda \end{bmatrix}.$$

▷ **Exercise 274E (Triangular Eigenvalues)** Show that the eigenvalues of a triangular matrix are its diagonal entries.

▷ **Exercise 274F** Suppose that $A \in \mathbf{F}^{m \times n}$ and $B \in \mathbf{F}^{n \times m}$. Show that AB and BA have the same nonzero eigenvalues.

▷ **Exercise 274G** Find the eigenvalues of AB and BA where

$$A = \begin{bmatrix} 1 & 2 & 0 \\ 2 & 1 & 2 \end{bmatrix}, \qquad B = \begin{bmatrix} 1 & -4 \\ 0 & 1 \\ 1 & 7 \end{bmatrix}.$$

▷ **Exercise 274H** Using MINIMAT verify that AB and BA have the same nonzero eigenvalues by computing the eigenvalues of AB and BA where A is a randomly chosen 3×5 matrix and B is a randomly chosen 5×3 matrix.

Real Similarity

Exercise 274I Say that square matrices $A, B \in \mathbf{F}^{n \times n}$ are **similar over F** iff there is an invertible matrix $P \in \mathbf{F}^{n \times n}$ with

$$A = PBP^{-1}.$$

Show that if two real square matrices $A, B \in \mathbf{R}^{n \times n}$ are similar over $\mathbf{C}$, then they are similar over $\mathbf{R}$. (Compare this with Exercise 251B.)

Chapter Summary

Similarity Invariants

Similar matrices have the same characteristic polynomial, trace, and determinant, and the same eigenvalues with the same (geometric and algebraic) multiplicities. The trace of a matrix is the sum of its diagonal entries and also the sum of its eigenvalues. The determinant of a matrix is the product of its eigenvalues.

Multiplicity

The geometric multiplicity of an eigenvalue is less than or equal to its algebraic multiplicity. A matrix is diagonalizable if and only if the geometric multiplicity of each eigenvalue is equal to its algebraic multiplicity.

Diagonalization

Let $A \in \mathbf{F}^{n \times n}$ and $P \in \mathbf{F}^{n \times \nu}$ be given by

$$P = \begin{bmatrix} \Phi_1 & \Phi_2 & \cdots & \Phi_m \end{bmatrix}.$$

Here Φ_i is a basis for the eigenspace $\mathcal{E}_{\lambda_i}(A)$ and the list $\lambda_1, \lambda_2, \ldots, \lambda_m$ contains each eigenvalue of A exactly once. The columns of P are independent so that P is invertible if and only if it is square. When P is square, $P^{-1}AP$ is diagonal.

9

DIFFERENTIAL EQUATIONS

Just as an inhomogeneous system of linear equations can be written in matrix form as

$$Y = AX,$$

so can a homogeneous linear system of differential equations be written in matrix form as

$$\dot{U} = AU.$$

There is an important difference, however. In the case of the linear system $Y = AX$, the natural equivalence relation is biequivalence. If $A = QBP^{-1}$, then the systems $Y = AX$ and $\tilde{Y} = B\tilde{X}$ are related by the change of variables $Y = Q\tilde{Y}$ and $X = P\tilde{X}$. In the case of the system $\dot{U} = AU$ of differential equations, the natural equivalence relation is similarity. If $A = PBP^{-1}$, then the equations $\dot{U} = AU$ and $\dot{V} = BV$ are related via the change of variables $U = PV$.

9.1   Derivatives

Let $U = U(t) \in \mathbf{F}^{p \times q}$ be a curve of matrices, that is, a function that assigns to each real number $t \in \mathbf{R}$ a matrix $U(t) \in \mathbf{F}^{p \times q}$. The derivative $\dot{U} = \dot{U}(t) \in \mathbf{F}^{p \times q}$ can be defined using a matrix analog of the definition

from calculus:
$$\dot{U}(t) = \lim_{h \to 0} \frac{1}{h}\left(U(t+h) - U(t)\right).$$

The derivative $\dot{U}$ can be computed by differentiating the entries of U. For example,

$$\text{if } U(t) = \begin{bmatrix} 1 + t^2 \\ t^3 \\ 5 \\ t^2 + t^4 \end{bmatrix}, \text{ then } \dot{U}(t) = \begin{bmatrix} 2t \\ 3t^2 \\ 0 \\ 2t + 4t^3 \end{bmatrix}.$$

This provides a handy notation for writing linear systems of **ordinary differential equations**. Suppose we are given a system of n differential equations in n unknown functions

$$u_1(t), u_2(t), \ldots, u_n(t)$$

in the following form

$$\begin{aligned}
\dot{u}_1(t) &= a_{11}u_1(t) + a_{12}u_2(t) + \cdots + a_{1n}u_n(t) \\
\dot{u}_2(t) &= a_{21}u_1(t) + a_{22}u_2(t) + \cdots + a_{2n}u_n(t) \\
&\vdots \\
\dot{u}_n(t) &= a_{n1}u_1(t) + a_{n2}u_2(t) + \cdots + a_{nn}u_n(t)
\end{aligned}$$

with initial conditions at $t = 0$ given by

$$u_1(0) = u_{01}, \qquad u_2(0) = u_{02}, \qquad \ldots, \qquad u_n(0) = u_{0n},$$

where the initial values $u_{01}, u_{02}, \ldots, u_{0n}$ are given constants. We define the **matrix of coefficients** $A \in \mathbf{F}^{n \times n}$ by

$$A = \begin{bmatrix} a_{11} & a_{12} & \cdots & a_{1n} \\ a_{21} & a_{22} & \cdots & a_{2n} \\ & & \ddots & \\ a_{n1} & a_{n2} & \cdots & a_{nn} \end{bmatrix},$$

and the **vector of unknowns** $U(t) \in \mathbf{F}^{n \times 1}$ and the **vector of initial conditions** $X_0 \in \mathbf{F}^{n \times 1}$ by

$$U(t) = \begin{bmatrix} u_1(t) \\ u_2(t) \\ \vdots \\ u_n(t) \end{bmatrix}, \qquad X_0 = \begin{bmatrix} u_{01} \\ u_{02} \\ \vdots \\ u_{0n} \end{bmatrix}.$$

Then our system of differential equations and initial conditions can be written more compactly in matrix form as follows:

$$\dot{U}(t) = AU(t), \qquad U(0) = X_0.$$

For example, the following system of two differential equations in two unknowns

$$\begin{aligned} \dot{u}_1(t) &= 3u_1(t) + 4u_2(t) \\ \dot{u}_2(t) &= 5u_1(t) + 6u_2(t) \end{aligned}$$

with initial conditions $u_1(0) = 7$, $u_2(0) = 1$, can be written in matrix form with

$$A = \begin{bmatrix} 3 & 4 \\ 5 & 6 \end{bmatrix}, \qquad U(t) = \begin{bmatrix} u_1(t) \\ u_2(t) \end{bmatrix}, \qquad X_0 = \begin{bmatrix} 7 \\ 1 \end{bmatrix}.$$

9.2 Similarity and Differential Equations

Similarity provides a way of changing variables. If $A = PBP^{-1}$, $X = PY$, and $\hat{X} = P\hat{Y}$, then

$$\hat{X} = AX \iff \hat{Y} = BY.$$

In this section we use this observation to solve differential equations.

Similarity of Differential Equations

Theorem 279A *Let $A, B \in \mathbf{F}^{n \times n}$ be similar matrices, say*

$$A = PBP^{-1}$$

where $P \in \mathbf{F}^{n \times n}$ is invertible. Then the two differential equations

$$\dot{U}(t) = AU(t), \qquad \dot{V}(t) = BV(t),$$

are related as follows. If $V = V(t)$ is a solution of the system $\dot{V} = BV$, then $U = U(t)$ defined by

$$U(t) = PV(t)$$

is a solution of the system $\dot{U} = AU$.

Proof: $\dot{U}(t) = P\dot{V}(t) = PBV(t) = (PBP^{-1})PV(t) = AU(t)$ □

Remark 280A Suppose the solution $V(t)$ of $\dot{V} = BV$ satisfies the initial condition $V(0) = Y_0$. Then the corresponding solution $U(t) = PV(t)$ of $\dot{U} = AU$ satisfies the initial condition

$$U(0) = PV(0) = PY_0 = X_0.$$

This observation can be used to solve an initial value problem.

If A is diagonalizable, it is easy to solve the differential equation $\dot{U} = AU$. This is because it is easy to solve the differential equation $\dot{V} = DV$ when D is diagonal.

Theorem 280B *If the matrix D is diagonal, say*

$$D = \begin{bmatrix} \lambda_1 & 0 & \cdots & 0 \\ 0 & \lambda_2 & \cdots & 0 \\ & & \ddots & \\ 0 & 0 & \cdots & \lambda_n \end{bmatrix},$$

then the matrix differential equation

$$\dot{V}(t) = DV(t)$$

separates into n equations in one unknown,

$$\dot{v}_i(t) = \lambda_i v_i(t) \qquad (i = 1, 2, \ldots, n),$$

for the entries $v_i(t) = \text{entry}_i(V(t))$. The solution is

$$V(t) = \begin{bmatrix} v_1(t) \\ v_2(t) \\ \vdots \\ v_n(t) \end{bmatrix}$$

where

$$v_i(t) = v_i(0)e^{\lambda_i t}.$$

Now here's a method for solving the system $\dot{U}(t) = AU(t)$ with initial condition $U(0) = 0$:

(1) Diagonalize A using (say) MINIMAT or the technique explained in Section 8.3. (If this step fails because A is not diagonalizable, a more complicated method is required.)

(2) Step (1) gave P and D with $A = PDP^{-1}$ and D diagonal. Solve the related system $\dot{V}(t) = DV(t)$ with initial condition $V(0) = P^{-1}X_0$. This is easy by Theorem 280B.

(3) By Theorem 279A the solution is given by $U(t) = PV(t)$.

Remark 281A It is easy to make a mistake and write $V = PU$ instead of $U = PV$. The following diagram is useful to keep things straight:

$$
\begin{array}{ccc}
U = PV \in \mathbf{F}^{n \times 1} & \xrightarrow{\quad A \quad} & \dot{U} = AU \in \mathbf{F}^{n \times 1} \\[1em]
P \Big\uparrow & & \Big\uparrow P \\[1em]
V \in \mathbf{F}^{n \times 1} & \xrightarrow{\quad D \quad} & \dot{V} = DV \in \mathbf{F}^{n \times 1}
\end{array}
$$

The diagram represents the equation $A = PDP^{-1}$ and helps us remember that $U = PV$. We used such a diagram (for similar reasons) in Remark 143A.

Example 281B We solve the linear system of differential equations

$$
\begin{aligned}
\dot{u}_1(t) &= -4u_1(t) + 6u_2(t) \\
\dot{u}_2(t) &= -3u_1(t) + 5u_2(t)
\end{aligned}
$$

with initial conditions

$$
u_1(0) = 9, \qquad u_2(0) = 7.
$$

This system has form $\dot{U}(t) = AU(t)$ where

$$
A = \begin{bmatrix} -4 & 6 \\ -3 & 5 \end{bmatrix}, \qquad U(t) = \begin{bmatrix} u_1(t) \\ u_2(t) \end{bmatrix}.
$$

This matrix A was diagonalized in Example 262B with the result that $A = PDP^{-1}$ where

$$
D = \begin{bmatrix} -1 & 0 \\ 0 & 2 \end{bmatrix}, \qquad P = \begin{bmatrix} 2 & 1 \\ 1 & 1 \end{bmatrix}, \qquad P^{-1} = \begin{bmatrix} 1 & -1 \\ -1 & 2 \end{bmatrix}.
$$

The system $\dot{V}(t) = DV(t)$ consists of two equations

$$
\dot{v}_1(t) = -v_1(t), \qquad \dot{v}_2(t) = 2v_2(t),
$$

each involving only one of the components $v_i(t)$ of $V(t)$. We find the appropriate initial condition for the transformed problem:

$$V(0) = P^{-1}U(0) = \begin{bmatrix} 1 & -1 \\ -1 & 2 \end{bmatrix} \begin{bmatrix} 9 \\ 7 \end{bmatrix} = \begin{bmatrix} 2 \\ 5 \end{bmatrix},$$

and solve it:

$$v_1(t) = 2e^{-t}, \qquad\qquad v_2(t) = 5e^{2t}.$$

Now we transform back to find the solution:

$$U(t) = PV(t) = \begin{bmatrix} 2 & 1 \\ 1 & 1 \end{bmatrix} \begin{bmatrix} 2e^{-t} \\ 5e^{2t} \end{bmatrix} = \begin{bmatrix} 4e^{-t} + 5e^{2t} \\ 2e^{-t} + 5e^{2t} \end{bmatrix}.$$

Expressed in terms of the original unknowns the solution is

$$u_1(t) = 4e^{-t} + 5e^{2t}, \qquad\qquad u_2(t) = 2e^{-t} + 5e^{2t}.$$

▷ **Exercise 282A** Following Example 281B, solve the linear system of differential equations

$$\begin{aligned} \dot{u}_1(t) &= 11u_1(t) + 4u_2(t) \\ \dot{u}_2(t) &= -8u_1(t) - u_2(t) \end{aligned}$$

subject to the initial condition

$$u_1(0) = 0, \qquad u_2(0) = 1.$$

Exercise 282B For each of the matrices A in Exercise 263B, solve the matrix differential equation $\dot{U}(t) = AU(t)$ with initial condition $U(0) = \begin{bmatrix} 3 \\ 5 \end{bmatrix}$.

Exercise 282C For each of the matrices A in Exercise 260B, solve the matrix differential equation $\dot{U}(t) = AU(t)$ with initial condition $U(0) = \begin{bmatrix} 3 \\ 5 \\ 1 \end{bmatrix}$.

Exercise 282D One approach to solving a differential equation

$$\dot{U} = AU$$

is to look for special solutions in the form

$$U(t) = e^{\lambda t}X.$$

Show that U is a solution if and only if X is an eigenvector of A corresponding to the eigenvalue λ:

$$AX = \lambda X.$$

A solution to $\dot{U} = AU$ of form $U(t) = e^{\lambda t}X$ where X is an eigenvector of A is called an **eigensolution**.

Example 283A In this terminology we can rephrase the algorithm given in the previous section for solving the problem

$$\dot{U}(t) = AU(t), \qquad U(0) = X_0.$$

(1) Find the eigensolutions $U_j(t)$.

(2) Find coefficients c_j so that $X_0 = \sum_j c_j U_j(0)$.

(3) The solution is $U(t) = \sum_j c_j U_j(t)$.

To illustrate, we solve the system $\dot{U}(t) = AU(t)$ where

$$A = \begin{bmatrix} -4 & 6 \\ -3 & 5 \end{bmatrix}, \qquad U(0) = \begin{bmatrix} 9 \\ 7 \end{bmatrix}.$$

This matrix A was diagonalized in Example 262B. The eigensolutions are

$$U_1(t) = e^{-t}\begin{bmatrix} 2 \\ 1 \end{bmatrix}, \qquad U_2(t) = e^{2t}\begin{bmatrix} 1 \\ 1 \end{bmatrix}.$$

The solution is

$$U(t) = 2U_1(t) + 5U_2(t).$$

The coefficients $2, 5$ were found by solving the inhomogeneous system

$$U(0) = c_1 U_1(0) + c_2 U_2(0)$$

for c_1, c_2.

Exercise 283B For each of the matrices A in Exercise 260B, find the eigensolutions. Then solve the differential equation $\dot{U}(t) = AU(t)$ with initial condition $U(0) = \begin{bmatrix} 3 \\ 5 \\ 1 \end{bmatrix}$.

Exercise 283C A function $X(t)$ is said to be **periodic** of period $\tau > 0$ iff $X(t + \tau) = X(t)$ for all t. Consider the system $\dot{U} = DU$ where D is the diagonal matrix $D = \mathrm{diag}(2i, 3i)$.

(1) Find all the eigensolutions.

(2) Show that every solution has period 2π.

(3) Which solutions have period π? $2\pi/3$?

9.3   Similarity and Powers

Recall that for any nonnegative integer p and any square matrix B the pth power B^p of B is the product of p copies of B:

$$B^p = \underbrace{B \cdot B \cdots B}_{p}.$$

This can be defined inductively by

$$B^0 = I, \qquad B^{p+1} = B^p B.$$

Similarity of Powers

Theorem 284A *Let A, B and P be square matrices of the same size and assume that P is invertible. Then*

$$P(AB)P^{-1} = (PAP^{-1})(PBP^{-1}).$$

Hence, for any nonnegative integer p,

$$(P^{-1}BP)^p = P^{-1}B^p P.$$

Proof: The first equation results from the associative law. The second equation results from repeated application of the first:

$$(P^{-1}BP)^p = \cdots (P^{-1}BP)(P^{-1}BP) \cdots$$

and the adjacent P and P^{-1} cancel: $PP^{-1} = I$. For example,

$$
\begin{aligned}
(P^{-1}BP)^3 &= (P^{-1}BP)(P^{-1}BP)(P^{-1}BP) \\
&= P^{-1}B(PP^{-1})B(PP^{-1})BP \\
&= P^{-1}BIBIBP \\
&= P^{-1}BBBP \\
&= P^{-1}B^3 P \qquad \square
\end{aligned}
$$

Corollary 284B *If A and B are similar, then so are A^p and B^p.*

Proof: If A and B are similar, then $A = PBP^{-1}$ for some invertible matrix P, so the theorem asserts that $A^p = PB^pP^{-1}$ as required. $\square$

Theorem 285A (Powers of Diagonals)   *If D is a diagonal matrix, then so is D^p. The diagonal entries of D^p are obtained from the corresponding entries of D by taking the pth power.*

Proof: In other words, if

$$D = \begin{bmatrix} \lambda_1 & 0 & \cdots & 0 \\ 0 & \lambda_2 & \cdots & 0 \\ & & \ddots & \\ 0 & 0 & \cdots & \lambda_n \end{bmatrix},$$

then

$$D^p = \begin{bmatrix} \lambda_1^p & 0 & \cdots & 0 \\ 0 & \lambda_2^p & \cdots & 0 \\ & & \ddots & \\ 0 & 0 & \cdots & \lambda_n^p \end{bmatrix}.$$

This follows by repeated application of the principle that to multiply diagonal matrices you multiply the diagonal entries. (This is Theorem 46B.) $\square$

Corollary 285B *Let p be a positive integer. Then any diagonalizable matrix $A \in \mathbf{C}^{n \times n}$ has a pth root; that is, there is a matrix $B \in \mathbf{C}^{n \times n}$ satisfying $B^p = A$.*

Proof: We are asked to solve the following problem:

Given a diagonalizable matrix A and a positive integer p, find a matrix B such that
$$B^p = A.$$

We use a strategy like the one we used to solve a system of linear differential equations.

(1) We find a diagonal matrix D similar to A:

$$A = PDP^{-1}$$

where

$$D = \mathrm{diag}(d_1, d_2, \ldots, d_n).$$

We can do this since we assume that A is diagonalizable.

(2) We solve the analogous problem for D in place of A:

$$C^p = D$$

where

$$C = \text{diag}(c_1, c_2, \ldots, c_n)$$

and $c_i^p = d_i$ for $i = 1, 2, \ldots, n$. The equation $D = C^p$ follows from Theorem 46B.

(3) We transform the solution of the diagonal problem to a solution of the original problem:

$$B = PCP^{-1}$$

The fact that $A = B^p$ follows from Theorem 284A:

$$B^p = (PCP^{-1})^p = PC^pP^{-1} = PDP^{-1} = A. \qquad \square$$

▷ **Question 286A** How many pth roots can we find by this method?

Exercise 286B For each of the matrices A in Exercise 263B, find a matrix B, possibly complex, with $B^2 = A$.

Exercise 286C For each of the matrices A in Exercise 260B, find a matrix B, possibly complex, with $B^2 = A$.

▷ **Exercise 286D** Let $P \in \mathbf{F}^{3\times 3}$ be invertible and $D \in \mathbf{F}^{3\times 3}$ be defined by

$$D = \begin{bmatrix} 1 & 0 & 0 \\ 0 & 1 & 0 \\ 0 & 0 & 0 \end{bmatrix}.$$

What is D^2? What is A^2 where $A = PDP^{-1}$?

Exercise 286E Show that there are infinitely many 2×2 matrices J each having exactly two nonzero entries such that $J^2 = -I$.

▷ **Exercise 286F** Show that if λ is an eigenvalue of A, then λ^p is an eigenvalue of A^p.

Using MINIMAT

▷ **Exercise 287A** Confirm Theorem 284A in MINIMAT.

▷ **Exercise 287B** What is the result of the following?

```
#> D=eye(3), D(3,3)=0, D^2
#> P=rand(3,3), A=P^(-1)*D*P, A^2
```

▷ **Exercise 287C** Find a matrix B such that

$$B^3 = A$$

where A is the matrix

$$A = \begin{bmatrix} -27 & -328 & -3438 & 888 \\ -70 & -819 & -8595 & 2220 \\ -210 & -2538 & -26350 & 6808 \\ -840 & -10152 & -105656 & 27296 \end{bmatrix}.$$

▷ **Exercise 287D** How did I figure out such a complicated matrix A with integer entries and such a simple diagonalization?

9.4 Matrix Polynomials

Suppose $f(\xi)$ is a **polynomial**, say

$$f(\xi) = c_0 + c_1\xi + c_2\xi^2 + \cdots + c_k\xi^k,$$

where the coefficients $c_0, c_1, c_2, \ldots, c_k$ are numbers in $\mathbf{F}$. We can write this polynomial using the *sigma* notation as

$$f(\xi) = \sum_{j=0}^{k} c_j\xi^j.$$

Since the jth power A^j of a square matrix A is meaningful, it makes sense to evaluate a polynomial at $\xi = A$. We simply substitute the matrix A for the indeterminate ξ to get a matrix $f(A)$:

$$f(A) = c_0 I + c_1 A + c_2 A^2 + \cdots + c_k A^k$$

or in sigma notation

$$f(A) = \sum_{j=0}^{k} c_j A^j.$$

Similarity of Functions

Theorem 288A *Let A and P be square matrices of the same size and assume that P is invertible. Then*

$$f(P^{-1}AP) = P^{-1}f(A)P.$$

Proof: Theorem 284A says that

$$(P^{-1}AP)^j = P^{-1}A^jP.$$

Multiply this equation by c_j and sum over $j = 0, 1, 2, \ldots, k$.   □

Corollary 288B *If A and B are similar, then so are $f(A)$ and $f(B)$.*

Proof: If A and B are similar, then $B = P^{-1}AP$ for some invertible P, so the theorem asserts that $f(B) = P^{-1}f(A)P$, as required.   □

Theorem 288C *For any polynomials f and g we have*

$$(f + g)(A) = f(A) + g(A) \qquad and \qquad (fg)(A) = f(A)g(A).$$

Proof: In other words,

$$\sum_j (c_j + b_j)A^j = \sum_j c_j A^j + \sum_j b_j A^j$$

and

$$\sum_j \left(\sum_{p+q=j} c_p b_q \right) A^j = \left(\sum_p c_q A^p \right) \left(\sum_q b_q A^q \right).$$

The polynomials in A obey the same algebraic laws as polynomials in ξ. This is because the powers of A commute with one another.   □

Functions of Diagonals

Theorem 288D *If D is a diagonal matrix, then so is $f(D)$; the diagonal entries of $f(D)$ are obtained from the corresponding entries of D by applying f.*

Proof: In other words, if

$$D = \begin{bmatrix} \lambda_1 & 0 & \cdots & 0 \\ 0 & \lambda_2 & \cdots & 0 \\ & & \ddots & \\ 0 & 0 & \cdots & \lambda_n \end{bmatrix},$$

then

$$f(D) = \begin{bmatrix} f(\lambda_1) & 0 & \cdots & 0 \\ 0 & f(\lambda_2) & \cdots & 0 \\ & & \ddots & \\ 0 & 0 & \cdots & f(\lambda_n) \end{bmatrix}.$$

The case where $f(\xi) = \xi^p$ is Theorem 285A. The general case follows from the fact that the operations of matrix addition and scalar multiplication are performed entrywise. □

Exercise 289A Show that if λ is an eigenvalue of A, then $f(\lambda)$ is an eigenvalue of $f(A)$.

9.5   Matrix Power Series

A **power series** is an "infinite" polynomial

$$f(\xi) = \sum_{j=0}^{\infty} c_j \xi^j.$$

The theory of power series is studied in calculus. Here are the most important examples:

$$\frac{1}{1-\xi} = \sum_{j=0}^{\infty} \xi^j \qquad\qquad \text{(geometric)}$$

$$e^\xi = \sum_{j=0}^{\infty} \frac{\xi^j}{j!} \qquad\qquad \text{(exponential)}$$

$$\sin(\xi) = \sum_{k=0}^{\infty} \frac{(-)^k \xi^{2k+1}}{(2k+1)!} \qquad\qquad \text{(sine)}$$

$$\cos(\xi) = \sum_{k=0}^{\infty} \frac{(-)^k \xi^{2k}}{(2k)!} \qquad\qquad \text{(cosine)}$$

▷ **Question 290A** In each of these examples find the coefficient c_j of ξ^j.

We can substitute the matrix A into a power series just as we did for polynomials:

$$f(A) = \sum_{j=0}^{\infty} c_j A^j.$$

Power series work much like polynomials, but we must be careful about convergence. The last three series above converge for any value of ξ, but the geometric series only converges when $|\xi| < 1$. As long as we are careful about convergence (a topic which is beyond the scope of this book) the theorems about matrix polynomials that were enumerated in the previous chapter generalize to matrix power series.

▷ **Exercise 290B** Prove that $e^{i\theta} = \cos\theta + i\sin\theta$. Hint:

$$e^{\xi} = \sum_{k=0}^{\infty} \frac{\xi^{2k}}{(2k)!} + \sum_{k=0}^{\infty} \frac{\xi^{2k+1}}{(2k+1)!}.$$

▷ **Exercise 290C** Let $z = x + iy$. How should $w = \rho + i\theta$ be chosen so that $e^w = z$? Here x, y, ρ, θ are all real.

▷ **Exercise 290D** Find the power series expansions for the hyperbolic functions sinh and cosh. Hint: Compare the formulas

$$\cosh(\xi) = \frac{e^{\xi} + e^{-\xi}}{2}, \qquad \sinh(\xi) = \frac{e^{\xi} - e^{-\xi}}{2},$$

with the formulas

$$\cos(\xi) = \frac{e^{i\xi} + e^{-i\xi}}{2}, \qquad \sin(\xi) = \frac{e^{i\xi} - e^{-i\xi}}{2i}.$$

9.6   The Matrix Exponential

Perhaps the most important matrix power series is the exponential

$$e^A = \sum_{j=0}^{\infty} \frac{A^j}{j!}.$$

It is especially important when tA (where t is a real number) is substituted for A:

$$e^{tA} = \sum_{j=0}^{\infty} \frac{t^j A^j}{j!}.$$

Here's why it's so important:

Theorem 291A *For* $A \in \mathbf{F}^{n \times n}$ *the solution* $\Omega = \Omega(t) \in \mathbf{F}^{n \times n}$ *of the matrix differential equation*

$$\dot{\Omega}(t) = A\Omega(t)$$

with initial condition

$$\Omega(0) = I$$

(the identity matrix) is given by

$$\Omega(t) = e^{tA}.$$

Proof: In other words, the function e^{tA} satisfies the following conditions:

$$e^{0A} = e^0 = I$$

the identity matrix, and

$$\frac{d}{dt}e^{tA} = Ae^{tA}.$$

These properties generalize facts from calculus (which is the special case $n = 1$). Written without the sigma notation, the formula for e^{tA} is

$$e^{tA} = I + tA + \frac{t^2}{2}A^2 + \frac{t^3}{6}A^3 + \frac{t^4}{24}A^4 + \cdots.$$

This makes it clear that $e^{tA} = I$ when $t = 0$. Differentiate the series termwise and factor out A:

$$
\begin{aligned}
\frac{d}{dt}e^{tA} &= A + tA^2 + \frac{t^2}{2}A^3 + \frac{t^2}{6}A^4 + \cdots \\
&= A\left(I + tA + \frac{t^2}{2}A^2 + \frac{t^3}{6}A^3 + \cdots\right) \\
&= Ae^{tA}.
\end{aligned}
$$

(The termwise differentiation can be rigorously justified.)   □

Corollary 291B *For a square matrix* $A \in \mathbf{F}^{n \times n}$ *the general solution of the differential equation*

$$\dot{U}(t) = AU(t)$$

is given by

$$U(t) = e^{tA}U(0).$$

Corollary 292A *For any square matrix the identities*

$$e^{0A} = I$$
$$e^{-tA} = \left(e^{tA}\right)^{-1}$$
$$e^{(t+s)A} = e^{tA}e^{sA}$$

hold for all numbers t and s.

Proof: The fact that $e^{tA} = I$ when $t = 0$ follows by plugging into the power series. Hence, the second equation follows from the third by substituting $s = -t$. The last equation can be proved by noting that both sides $\Omega(t) = e^{(t+s)A}$ and $\Omega(t) = e^{tA}e^{sA}$ satisfy the equation $\dot{\Omega} = A\Omega$ with initial condition $\Omega(0) = e^{sA}$. Equality follows from the fact[1] that this initial value problem has a unique solution. It also follows by reading tA for A and sA for B in Exercise 294G.   □

Remark 292B By Theorem 288D, we have

$$e^{tD} = \begin{bmatrix} e^{t\lambda_1} & 0 & \cdots & 0 \\ 0 & e^{t\lambda_2} & \cdots & 0 \\ & & \ddots & \\ 0 & 0 & \cdots & e^{t\lambda_n} \end{bmatrix},$$

where

$$D = \begin{bmatrix} \lambda_1 & 0 & \cdots & 0 \\ 0 & \lambda_2 & \cdots & 0 \\ & & \ddots & \\ 0 & 0 & \cdots & \lambda_n \end{bmatrix}.$$

By Theorem 288A, we have

$$e^{tA} = Pe^{tD}P^{-1}$$

where

$$A = PDP^{-1}.$$

These formulas allow us to compute e^{tA} quite explicitly when A is diagonalizable.

---

[1]This is called the *Existence and Uniqueness Theorem for Ordinary Differential Equations*. It is proved in books on differential equations.

Example 293A We compute e^{tA} for the matrix

$$A = \begin{bmatrix} -4 & 6 \\ -3 & 5 \end{bmatrix}.$$

The matrix A was diagonalized in Example 262B with the result that $A = PDP^{-1}$ where

$$D = \begin{bmatrix} -1 & 0 \\ 0 & 2 \end{bmatrix}, \qquad P = \begin{bmatrix} 2 & 1 \\ 1 & 1 \end{bmatrix}, \qquad P^{-1} = \begin{bmatrix} 1 & -1 \\ -1 & 2 \end{bmatrix}.$$

We have

$$e^{tD} = \begin{bmatrix} e^{-t} & 0 \\ 0 & e^{2t} \end{bmatrix}$$

so

$$e^{tA} = Pe^{tD}P^{-1} = \begin{bmatrix} 2 & 1 \\ 1 & 1 \end{bmatrix} \begin{bmatrix} e^{-t} & 0 \\ 0 & e^{2t} \end{bmatrix} \begin{bmatrix} 1 & -1 \\ -1 & 2 \end{bmatrix}$$

or

$$e^{tA} = \begin{bmatrix} 2e^{-t} - e^{2t} & -2e^{-t} + 2e^{2t} \\ e^{-t} - e^{2t} & -e^{-t} + 2e^{2t} \end{bmatrix}.$$

▷ **Exercise 293B** Solve the differential equation $\dot{U}(t) = AU(t)$ of Example 293A with initial condition $U(0) = \begin{bmatrix} 9 \\ 7 \end{bmatrix}$.

▷ **Exercise 293C** Find e^{tA} if $A = \begin{bmatrix} 1 & 2 \\ -1 & 4 \end{bmatrix}$.

Exercise 293D For each of the matrices A of Exercise 263B find e^{tA}. Check your answer by evaluating at $t = 0$ (you should get the identity matrix) and differentiating (you should get Ae^{tA}).

Exercise 293E For each of the matrices A of Exercise 260B find e^{tA}.

▷ **Exercise 293F** The eigenvalues of a diagonalizable matrix A are 1, 4, and 9. What are the eigenvalues of e^{tA}?

▷ **Exercise 293G** Let $N = \begin{bmatrix} 0 & 1 \\ 0 & 0 \end{bmatrix}$. Prove that

$$e^{tN} = \begin{bmatrix} 1 & t \\ 0 & 1 \end{bmatrix}$$

▷ **Exercise 294A** Let $J = \begin{bmatrix} 0 & 1 \\ -1 & 0 \end{bmatrix}$. Prove that

$$e^{tJ} = \begin{bmatrix} \cos t & \sin t \\ -\sin t & \cos t \end{bmatrix}$$

Hint: See Exercise 290B.

▷ **Exercise 294B** Let $K = \begin{bmatrix} 0 & 1 \\ 1 & 0 \end{bmatrix}$. Prove that

$$e^{tK} = \begin{bmatrix} \cosh t & \sinh t \\ \sinh t & \cosh t \end{bmatrix}$$

▷ **Exercise 294C** Let $N \in \mathbf{F}^{n \times n}$ be the matrix with ones on the first superdiagonal and zeros elsewhere:

$$\text{entry}_{ij}(N) = \begin{cases} 1 & \text{if } j = i + 1, \\ 0 & \text{otherwise.} \end{cases}$$

Compute e^{tN}.

▷ **Exercise 294D** Compute $(A+B)^2$. What happens if A and B commute?

▷ **Exercise 294E** Compute $(A+B)^3$. What happens if A and B commute?

▷ **Exercise 294F** Prove the **Binomial Theorem**: If $AB = BA$, then

$$(A + B)^n = \sum_{p+q=n} \frac{n!}{p!q!} A^p B^q.$$

▷ **Exercise 294G** Prove that if $AB = BA$, then

$$e^{A+B} = e^A e^B.$$

Show that this is not true when A and B do not commute by calculating e^{A+B} and $e^A e^B$ when

$$A = \begin{bmatrix} 0 & 1 \\ 0 & 0 \end{bmatrix}, \qquad B = \begin{bmatrix} 0 & 0 \\ 1 & 0 \end{bmatrix}.$$

Exercise 294H Show that e^A is invertible.

Exercise 294I Show that if B is invertible and diagonalizable, then there is a complex matrix A such that e^A. Hint: See Exercise 290C. (The hypothesis that B is diagonalizable may be dropped. See Exercise 395A.)

Exercise 295A Find a real matrix A such that $e^A = \begin{bmatrix} -2 & 0 \\ 0 & -2 \end{bmatrix}$.

Exercise 295B Find a complex matrix A such that $e^A = \begin{bmatrix} -2 & 0 \\ 0 & -3 \end{bmatrix}$
but show that there is no real matrix A satisfying this equation. Hint: See Exercise 270C.

Exercise 295C Prove that if $AB = BA$, then

$$\sin(A + B) = \sin(A)\cos(B) + \cos(A)\sin(B).$$

Exercise 295D The formula

$$e^A = \lim_{p \to \infty} \left(I + \frac{1}{p}A \right)^p$$

works for square matrices as well as for real numbers. Prove this formula under the hypothesis that A is diagonalizable.

▷ **Exercise 295E** Assume that $C(t) = A(t)B(t)$. Compute $\dot{C}(t)$.

▷ **Exercise 295F** Assume that $C(t) = (A + tB)^2$. Compute $\dot{C}(0)$.

▷ **Exercise 295G** Assume that $C(t) = (A + tB)^{-1}$. Compute $\dot{C}(0)$. Hint: $C(t)(A + tB) = I$ is independent of t.

▷ **Exercise 295H** Assume that $C(t) = (A + tB)^p$ and $AB = BA$. Compute $\dot{C}(0)$.

Exercise 295I Prove that $\det(e^A) = e^{\operatorname{trace}(A)}$.

▷ **Exercise 295J** Find all diagonal 2×2 matrices such that $A^2 = -A$. Find a nondiagonal matrix A such that $A^2 = -A$.

▷ **Exercise 295K** Suppose that $A \in \mathbf{F}^{n \times n}$ satisfies $A^2 = -A$. Find e^{tA}.

Exercise 295L Prove that

$$\frac{d}{dt} e^{tA} e^{tB} e^{-tA} e^{-tB} \bigg|_{t=0} = 0$$

and that

$$\frac{d}{ds} e^{\sqrt{s}A} e^{\sqrt{s}B} e^{-\sqrt{s}A} e^{-\sqrt{s}B} \bigg|_{s=0} = AB - BA.$$

Hint: Both formulas will follow from a formula like

$$e^{tA} e^{tB} e^{-tA} e^{-tB} = I + t^2(AB - BA) + \cdots$$

where $+\cdots$ represents a power series in t containing only powers t^k where $k \geq 3$. (For $k \geq 3$ the derivatives of t^k and $(\sqrt{s})^k$ vanish at 0.) The definition of the exponential gives

$$e^{tA} = I + tA + \frac{t^2}{2}A^2 + \cdots$$

with similar formulas for e^{tB}, e^{-tA}, e^{-tB}.

In the following four exercises, the submultiplicative inequality of Exercise 225G will be useful. Recall (Page 13) that the absolute value of a complex number $z = x + iy$ is $|z| = \sqrt{z\bar{z}}$. Also, make sure you understand the complex exponential as explained in the exercises of Section 9.5. Do each exercise first in the 1×1 case, then in the diagonal case, and finally in the diagonalizable case. In Exercise 395C we will learn how to drop the hypothesis that A is diagonalizable.

Exercise 296A Show that a square matrix A satisfies $\lim_{t\to\infty} \|e^{tA}\| = 0$ if and only if all the eigenvalues of A have negative real part. Hint: If λ has negative real part, then $\lim_{t\to\infty} |e^{t\lambda}| = 0$.

Exercise 296B Show that a square matrix A satisfies $\lim_{k\to\infty} \|A^k\| = 0$ if and only if all the eigenvalues of A have absolute value strictly less than one.

Exercise 296C Let A be an invertible square matrix. We say that A has **bounded iterates** iff there is a constant $c > 0$ such that $\|A^k\| \leq c$ for all integers k. (Here k ranges over *all* integers, positive or negative.) Show that A has bounded iterates if and only if A is diagonalizable and all its eigenvalues have absolute value one.

Exercise 296D Show that there is a constant $C > 0$ such that $\|e^{tA}\| \leq C$ for all $t \in \mathbf{R}$ if and only if A is diagonalizable and all its eigenvalues are pure imaginary.

Using MINIMAT

MINIMAT's built-in function `expm` computes the matrix exponential defined above. The command

```
#> Q=expm(A)
```

causes the variable Q to receive the value e^A. In mathematics the notations $\exp(A)$ and e^A have the same meaning, but not in MINIMAT. The effect of the command

```
#> F=exp(A)
```

is completely different; it computes the elementwise exponential

$$\text{entry}_{ij}(F) = \exp(\text{entry}_{ij}(A))$$

which is generally quite different from $\text{entry}_{ij}(\exp(A))$. The elementwise exponential is not very useful in mathematics, but it is handy in MINIMAT. If L is a row matrix, then the commands

```
#> diag(exp(L)),  expm(diag(L))
```

return the same value.

Thus MINIMAT does not use the notation $f(A)$ in the way mathematicians do. Rather, MINIMAT applies most built-in functions (like exp) elementwise. For example, the Matlab command B=cos(A) will assign to B a matrix whose (i, j)-entry is the *cosine* of the (i, j)-entry of A. The result is quite different from the result of substituting the matrix A into the power series for the cosine, *even when A is diagonal*. For example, if

$$A = \begin{bmatrix} \alpha & 0 \\ 0 & \beta \end{bmatrix},$$

then

$$\cos(A) = \begin{bmatrix} \cos(\alpha) & 0 \\ 0 & \cos(\beta) \end{bmatrix} \neq \begin{bmatrix} \cos(\alpha) & \cos(0) \\ \cos(0) & \cos(\beta) \end{bmatrix}$$

MINIMAT does not have a built-in function cosm to compute the power series $\cos(A)$, but one could write a .M function to do the job. (See Exercise 299D.)

Using MINIMAT's eig function we can compute the matrix exponential another way. The command

```
#> [P D]= eig(A)
```

finds matrices P and D satisfying $A = PDP^{-1}$. The command

```
#> L=diag(D)
```

picks off the diagonal entries. The statement

```
#> W=diag(exp(L))
```

will produce a diagonal matrix whose entries are the exponentials of the entries of L. This is the exponential of D. Then the command

```
#> V= P*W*P^(-1)
```

produces the exponential of $V = e^A$ of A.

▷ **Exercise 298A** What is the result of the following?

```
#> L=rand(1,3), D=diag(L), W=diag(exp(L)), expm(D)
```

▷ **Exercise 298B** What is the result of the following?

```
#> A=zeros(3,3), W=expm(A), F=exp(A)
```

▷ **Exercise 298C** Use MINIMAT's elementwise exponential `exp` to compute the matrix exponential of a random diagonal matrix and compare the result to the output of MINIMAT's matrix exponential `expm`.

▷ **Exercise 298D** Let A be a random 4×4 matrix. Compute e^A in two ways (one using `eig` and the other using `expm`) and compare the results.

▷ **Exercise 298E** For small A the sum of the first few terms of the series for e^A ought to be a good approximation for e^A. Confirm this.

▷ **Exercise 298F** Use MINIMAT to confirm that it is *not* in general true that the exponential of a sum is the product of the exponentials.

Exercise 298G Write a .M function to compute the exponential of a matrix using the definition:

$$e^A = \sum_{j=0}^{\infty} \frac{A^j}{j!}$$

The series will converge more rapidly when A is small, so it is a good idea to use the formula

$$e^A = \left(e^{A/m}\right)^m$$

where m is large enough to make the entries of the exponent smaller than $1/n$. Since computers can only do finite (not infinite) sums, the trick is to figure out how many terms you should take in the sum. Do not use the built-in `expm` function.

Exercise 298H Write a .M function to compute the exponential of a matrix by diagonalizing it. Do not use the built-in `expm` function.

▷ **Exercise 298I** Use MINIMAT to confirm the formula for the derivative of an exponential.

▷ **Exercise 298J** Confirm the equation

$$e^{tJ} = \begin{bmatrix} \cos t & \sin t \\ -\sin t & \cos t \end{bmatrix}, \qquad J = \begin{bmatrix} 0 & 1 \\ -1 & 0 \end{bmatrix}$$

by computing both sides for a randomly chosen value of t.

▷ **Exercise 299A** Use MINIMAT to construct nondiagonal matrices A and B which commute. Then confirm that the exponential of the sum is the product of the exponentials when the matrices commute.

▷ **Exercise 299B** The formula

$$e^A = \lim_{p \to \infty} \left(I + \frac{1}{p} A \right)^p$$

works for square matrices as well as for real numbers. (See Exercise 295D.) Confirm this using MINIMAT.

Exercise 299C Write a .M function to compute the exponential of a matrix using the formula in the previous exercise. Since computers cannot take limits, the trick is to figure out how big p should be.

Exercise 299D Write .M functions sinm and cosm to compute the (matrix) power series $\sin(A)$ and $\cos(A)$ and confirm that $\sin^2(A) + \cos^2(A) = I$.

9.7 The Companion Matrix

Given a monic polynomial p there are many matrices for which p is the characteristic polynomial. One of them is the *companion matrix* of p defined in this section. There is a relation between the companion matrix and higher order differential equations. Consider an nth order differential equation

$$x^{(n)} + p_1 x^{(n-1)} + \cdots + p_{n-1} x^{(1)} + p_n x^{(0)} = 0 \tag{1}$$

where $x^{(k)}$ denotes the kth derivative of the unknown function $x = x^{(0)}$. We may write this as a first order system by defining new variables u_k $(k = 1, 2, \ldots, n)$ by:

$$u_k = x^{(k-1)}.$$

The system takes the form

$$\begin{aligned}
\dot{u}_1 &= u_2 \\
\dot{u}_2 &= u_3 \\
&\ \ \vdots \\
\dot{u}_{n-1} &= u_n \\
\dot{u}_n &= -p_n u_1 - p_{n-1} u_2 - \cdots - p_1 u_n
\end{aligned}$$

If we write this in matrix form, we get

$$\dot{U} = CU \tag{2}$$

where $\text{entry}_j(U) = u_j$ and C is the *companion matrix* of the polynomial $p(\xi) = \xi^n + p_1\xi^{n-1} + \cdots + p_{n-1}\xi + p_n$. Here's the definition:

Definition 300A Let

$$p(\xi) = \xi^n + p_1\xi^{n-1} + p_2\xi^{n-2} \cdots + p_{n-1}\xi + p_n$$

be a monic polynomial of degree n. The **companion matrix** of this polynomial is the matrix $C \in \mathbf{F}^{n \times n}$ defined by

$$C = \begin{bmatrix} 0 & 1 & \cdots & 0 & 0 \\ 0 & 0 & \cdots & 0 & 0 \\ & & \ddots & & \\ 0 & 0 & \cdots & 0 & 1 \\ -p_n & -p_{n-1} & \cdots & -p_2 & -p_1 \end{bmatrix}.$$

The companion matrix has all its entries zero except for the ones just above the diagonal and the last row. The last row consists of the negatives of the coefficients of the polynomial in reverse order. The polynomial $p(\xi)$ is **monic**, which means that the leading coefficient is one. This leading coefficient is not used in the companion matrix.

Example 300B If $p(\xi) = \xi^2 + b\xi + c$, then

$$C = \begin{bmatrix} 0 & 1 \\ -c & -b \end{bmatrix}.$$

▷ **Question 300C** What is the characteristic polynomial of C?

Example 300D If $p(\xi) = \xi^3 + a\xi^2 + b\xi + c$, then

$$C = \begin{bmatrix} 0 & 1 & 0 \\ 0 & 0 & 1 \\ -c & -b & -a \end{bmatrix}.$$

▷ **Question 300E** What is the characteristic polynomial of C?

Remark 300F It is convenient to use the notation

$$p\left(\frac{d}{dt}\right) = \frac{d^n}{dt^n} + p_1\frac{d^{n-1}}{dt^{n-1}} + p_2\frac{d^{n-2}}{dt^{n-2}} + \cdots + p_{n-1}\frac{d}{dt} + p_n$$

so that equation (1) takes the form

$$p\left(\frac{d}{dt}\right)x = 0.$$

The proof of the next theorem shows that there is a relation between the solutions of the differential equation $p(d/dt)x = 0$ and the algebraic equation $p(\xi) = 0$.

Theorem 301A *The characteristic polynomial of the companion matrix C of the monic polynomial $p(\xi)$ is $p(\xi)$:*

$$\det(\xi I - C) = p(\xi).$$

Proof: We will prove this here under the additional hypothesis that the roots of p are distinct. This actually implies the general case, since both sides of the equation to be proved depend continuously on the coefficients of p, and, by perturbing p slightly, we can arrange that it has distinct roots. Another proof, using expansion by cofactors, avoids the hypothesis that the roots are distinct. See Theorem 424A.

We begin by solving equation (1). We look for special solutions of the form $x(t) = e^{\lambda t}$. For such an x, the kth derivative is given by $x^{(k)}(t) = \lambda^k x(t)$, so equation (1) may be rewritten as

$$p(\lambda)e^{\lambda t} = 0.$$

Thus there is a solution for every root of $p(\xi)$. Let $\lambda_1, \lambda_2, \ldots, \lambda_n$ be these roots. Since there is exactly one monic polynomial of degree n having prescribed roots, we obtain

$$p(\xi) = (\xi - \lambda_1)(\xi - \lambda_2)\cdots(\xi - \lambda_n) \qquad (3)$$

Let U_j be the solution of equation (2) corresponding to the solution $x_j(t) = e^{\lambda_j t}$ of equation (1).

$$U_j = e^{\lambda_j t}X_j, \qquad X_j = \begin{bmatrix} 1 \\ \lambda_j \\ \lambda_j^2 \\ \vdots \\ \lambda_j^{n-1} \end{bmatrix}.$$

Then

$$\lambda_j e^{\lambda_j t}X_j = \lambda_j U_j = \dot{U}_j = CU_j = e^{\lambda_j t}CX_j$$

by equation (1) and the chain rule from calculus. Canceling $e^{\lambda_j t}$ gives

$$\lambda_j X_j = C X_j.$$

In other words, $\lambda_1, \lambda_2, \ldots, \lambda_n$ are the eigenvalues of C. Since the eigenvalues of C are the precisely the roots of its characteristic polynomial (see Theorem 267B) we have

$$\det(\xi I - A) = (\xi - \lambda_1)(\xi - \lambda_2) \cdots (\xi - \lambda_n) \tag{4}$$

Now use equations (3) and (4). $\square$

Corollary 302A *The eigenvalues of the companion matrix of a polynomial are the roots of that polynomial.*

Example 302B The solutions of the equation $\ddot{x} - b^2 x = 0$ are linear combinations of $x_1(t) = e^{bt}$ and $x_2(t) = e^{-bt}$. Here $p(\xi) = \xi^2 - b^2$ and the companion matrix and its eigenvectors are

$$C = \begin{bmatrix} 0 & 1 \\ b^2 & 0 \end{bmatrix}, \qquad X_1 = \begin{bmatrix} 1 \\ b \end{bmatrix}, \qquad X_2 = \begin{bmatrix} 1 \\ -b \end{bmatrix}.$$

Example 302C The solutions of the equation $\ddot{x} + \omega^2 x = 0$ are linear combinations of $x_1(t) = \cos(\omega t)$ and $x_2(t) = \sin(\omega t)/\omega$. This equation fits the form of the previous example, if we take $b = i\omega$. This is consistent with the formula

$$e^{\pm i\omega t} = \cos(\omega t) \pm i \sin(\omega t).$$

Exercise 302D Diagonalize the companion matrix of $p(\xi) = \xi^4 - 1$.

Exercise 302E Diagonalize the companion matrix of $p(\xi) = \xi^4 + 1$.

Exercise 302F Diagonalize the companion matrix of

$$p(\xi) = (\xi - \alpha)(\xi - \beta)(\xi - \gamma).$$

Exercise 302G A mass m_1 is hung from a spring (attached to the ceiling) with spring constant k_1. A second mass m_2 is hung from the first mass from a spring with spring constant k_2. If y_i denotes the displacement from equilibrium of m_i, then Hooke's law (ignoring gravity) gives the equations of motion

$$\begin{aligned} m_1 \ddot{y}_1 &= -k_1 y_1 + k_2(y_2 - y_1) \\ m_2 \ddot{y}_2 &= -k_2(y_2 - y_1) \end{aligned}$$

(1) Let $u_1 = y_1$, $u_2 = \dot{y}_1$ $u_3 = y_2$, $u_4 = \dot{y}_2$ and write the system in the form $\dot{U} = AU$.

(2) Find the eigensolutions (see Exercise 282D) if $k_1 = 21$, $k_2 = 15$, $m_1 = 7$, $m_2 = 3$.

(3) Find the solution satisfying the initial conditions $y_1(0) = 1$, $y_2(0) = -1$, $\dot{y}_1(0) = \dot{y}_2(0) = 0$.

Using MINIMAT

Function 303A The .M function compan, shown in Listing 14, computes the companion matrix of its input. The command

        #> C=compan(p)

assigns to C the companion matrix of the polynomial

$$f(t) = t^n + p(1) * t^{n-1} + p(2) * t^{n-2} + \cdots + p(n-1) * t + p(n)$$

whose coefficients appear (as indicated) in the $1 \times n$ row matrix p.

▷ **Exercise 303B** Use the .M function compan and MINIMAT'S eig function to find the roots of the polynomial represented by p.

The Companion Matrix in MINIMAT

```
function C = compan(p)

    p = p(:).';          % make sure p is a row
    n = max(size(p));

    if  n > 1
        C = zeros(n,n);
        for i=  1:n-1, C(i,i+1)=1; end
    else
        C = 1;
    end
    C(n,:) = -p(n:-1:1);
```

Listing 14

Chapter Summary

Functions of Matrices

A polynomial $f(\xi)$ may be evaluated at a square matrix A to produce another square matrix $f(A)$:

$$f(\xi) = \sum_j c_j \xi^j \implies f(A) = \sum_j c_j A^j$$

Evaluation at a matrix commutes with the usual algebraic operations

$$(f + g)(A) = f(A) + g(A), \qquad (fg)(A) = f(A)g(A).$$

One way to evaluate $f(A)$ is to diagonalize A and use the rules

$$f(PBP^{-1}) = Pf(B)P^{-1}$$

$$D = \text{diag}(\lambda_1, \ldots, \lambda_n) \implies f(D) = \text{diag}(f(\lambda_1), \ldots, f(\lambda_n)).$$

Linear Differential Equations

$A = PBP^{-1}$, $U = PV$, $\dot{V} = BV \implies \dot{U} = AU$.

$AX = \lambda X$, $U(t) = e^{\lambda t} X \implies \dot{U} = AU$.

$U(t) = e^{tA} X_0 \implies \dot{U} = AU, \quad U(0) = X_0$.

$\Omega(t) = e^{tA} \implies \dot{\Omega} = A\Omega, \quad \Omega(0) = I$.

The Companion Matrix

The nth order differential equation $p(d/dt)x = 0$ in a single unknown x may be written as a system $\dot{U} = CU$ where C is the companion matrix of $p(\xi)$. If $p(\lambda) = 0$, then

- $p(d/dt)x(t) = 0$ *where* $x(t) = e^{\lambda t}$;

- $CX = \lambda X$ *where* $\text{entry}_j(X) = \lambda^{j-1}$;

- $\dot{U} = CU$ *where* $U = e^{\lambda t} X$.

Hence, the eigenvalues of C are the roots of p, and p is the characteristic polynomial of C.

10

HERMITIAN MATRICES

Matrices A and B are said to be *unitarily similar* iff there is a unitary matrix P with $A = PBP^{-1}$. The theory of unitary similarity is easier than the theory of general similarity provided we restrict attention to *Hermitian matrices*, that is, those matrices A that satisfy $A^* = A$.

10.1 Hermitian Matrices Defined

Hermitian

Definition 305A A **Hermitian matrix** is a square matrix A that is equal to its conjugate transpose:

$$A = A^*.$$

A **symmetric matrix** is a square matrix A that is equal to its ordinary transpose:

$$A = A^\top.$$

For a real matrix $A^\top = A^*$. Hence, a real matrix is Hermitian if and only if it is symmetric: when $\mathbf{F} = \mathbf{R}$, the two concepts agree. When $\mathbf{F} = \mathbf{C}$ the concept of *Hermitian matrix* is more important.

Example 306A A matrix $A \in \mathbf{C}^{2\times2}$ is Hermitian iff it has the form

$$A = \begin{bmatrix} a & b+ic \\ b-ic & d \end{bmatrix}$$

where a, b, c, d are real.

▷ **Question 306B** Give an example of a Hermitian matrix A that is not symmetric and an example of a symmetric matrix B that is not Hermitian.

Example 306C A matrix $A \in \mathbf{F}^{3\times3}$ is symmetric if it has the form

$$A = \begin{bmatrix} a & b & c \\ b & d & e \\ c & e & f \end{bmatrix}.$$

Theorem 306D *A matrix $A \in \mathbf{F}^{n\times n}$ is symmetric if and only if*

$$\mathrm{entry}_{ij}(A) = \mathrm{entry}_{ji}(A)$$

for $i, j = 1, 2, \ldots, n$. A matrix $A \in \mathbf{F}^{n\times n}$ is Hermitian if and only if

$$\mathrm{entry}_{ij}(A) = \overline{\mathrm{entry}_{ji}(A)}$$

for $i, j = 1, 2, \ldots, n$.

Proof: $\mathrm{entry}_{ij}(A) = \mathrm{entry}_{ji}(A^\top) = \overline{\mathrm{entry}_{ji}(A^*)}$ □

Corollary 306E *The diagonal entries of a Hermitian matrix are real.*   □

Theorem 306F *A matrix $A \in \mathbf{F}^{n\times n}$ is Hermitian if and only if it satisfies the equation*

$$\langle AX, Y \rangle = \langle X, AY \rangle \tag{*}$$

for all column vectors $X, Y \in \mathbf{F}^{n\times1}$.

Proof: By Theorem 192A,

$$\langle AX, Y \rangle = \langle X, A^*Y \rangle.$$

Hence, $A = A^* \implies (*)$. Conversely, assume that $(*)$ holds for all X and Y. Let

$$E_j = \text{col}_j(I_n) \quad (j=1,2,\ldots n)$$

denote the jth column of the identity matrix. Take $X = E_i$ and $Y = E_j$ in $(*)$.. Then

$$
\begin{aligned}
\text{entry}_{ij}(A) &= \langle AE_i, E_j \rangle \\
&= \langle E_i, AE_j \rangle \\
&= \overline{\text{entry}_{ji}(A)} \\
&= \text{entry}_{ij}(A^*)
\end{aligned}
$$

for $i, j = 1, 2, \ldots n$, so $A = A^*$. $\qquad\square$

▷ **Exercise 307A** Suppose

$$
A = \begin{bmatrix} a_{11} & a_{12} & a_{13} \\ a_{21} & a_{22} & a_{33} \\ a_{31} & a_{32} & a_{33} \end{bmatrix}, \quad E_1 = \begin{bmatrix} 1 \\ 0 \\ 0 \end{bmatrix}, \quad E_2 = \begin{bmatrix} 0 \\ 1 \\ 0 \end{bmatrix}.
$$

Compute AE_1, AE_2, $\langle AE_1, E_2 \rangle$, $\langle E_1, AE_2 \rangle$.

▷ **Exercise 307B** Show that $A = B + B^*$ is Hermitian.

Exercise 307C Show that for any matrix $B \in \mathbf{F}^{m \times n}$ both of the matrices $B^*B \in \mathbf{F}^{n \times n}$ and $BB^* \in \mathbf{F}^{m \times m}$ are Hermitian. (This works even if B is not square.)

Exercise 307D Show that a diagonal matrix is Hermitian if and only if it is real.

Exercise 307E Give an example of two matrices $A, B \in \mathbf{F}^{2 \times 2}$ such that A and B are Hermitian but the product AB is not.

Exercise 307F Let $A, B \in \mathbf{F}^{n \times n}$ be Hermitian, $I = I_n$ be the $n \times n$ identity matrix, and assume that $I - AB$ is invertible. Show that the matrix $B(I - AB)^{-1}$ is Hermitian.

10.2 Unitary Diagonalization

The theory for diagonalizing a Hermitian matrix is much easier than the general theory of diagonalization. In fact, a Hermitian matrix can always be "unitarily" diagonalized. This is the content of the Spectral Theorem 317C. Here we prove this under an additional hypothesis as a warmup.

Unitary Similarity

Definition 308A The matrix $A \in \mathbf{F}^{n \times n}$ is **unitarily similar** to the matrix $B \in \mathbf{F}^{n \times n}$ iff there is an unitary matrix $P \in \mathbf{F}^{n \times n}$ such that

$$A = PBP^{-1}.$$

A square matrix A is **unitarily diagonalizable** iff it is unitarily similar to a diagonal matrix.

If the word *unitary* is replaced by the word *invertible* we recover the definitions of *similar* and *diagonalizable*.

Theorem 308B *Unitary similarity satisfies the following:*

(Reflexive Law) $A \equiv A$;

(Symmetric Law) *if* $A \equiv B$*, then* $B \equiv A$;

(Transitive Law) *if* $A \equiv B$ *and* $B \equiv C$*, then* $A \equiv C$.

Here $A \equiv B$ *is an abbreviation for "A is unitarily similar to B".*

The Diagonalization Theorem 258A asserts that a matrix A is diagonalizable if and only if there is an invertible matrix P whose columns are eigenvectors of A. In the present situation we have the following variant. The proof is word for word the same.

Unitary Diagonalization Theorem

Theorem 308C *A square matrix A is unitarily diagonalizable if and only if there is a unitary matrix P of the same size whose columns are eigenvectors of A.*

Remark 308D In view of the fact that the columns of a square matrix are orthonormal iff that matrix is unitary (Theorem 201B), we may reformulate this as follows: *A matrix $A \in \mathbf{F}^{n \times n}$ is unitarily diagonalizable if and only it has n orthonormal eigenvectors.*

Theorem 309A (Hermitian Eigenvalues) *The eigenvalues of a Hermitian matrix are real. In particular, the eigenvalues of a real symmetric matrix are real.*

Proof: Suppose $A = A^*$ and $AX = \lambda X$ where $X \neq 0$. We must show that $\lambda = \bar{\lambda}$. Here is the argument:

$$
\begin{aligned}
\lambda \langle X, X \rangle &= \langle \lambda X, X \rangle \\
&= \langle AX, X \rangle \\
&= \langle X, A^* X \rangle \\
&= \langle X, AX \rangle \\
&= \langle X, \lambda X \rangle \\
&= \bar{\lambda} \langle X, X \rangle
\end{aligned}
$$

Now divide by $\langle X, X \rangle$.   □

Theorem 309B (Hermitian Eigen Orthogonality)   *If two eigenvectors of a Hermitian matrix correspond to distinct eigenvalues, they are orthogonal.*

Proof: Suppose $A = A^* \in \mathbf{F}^{n \times 1}$, and that $X, Y \in \mathbf{F}^{n \times 1}$ satisfy

$$
AX = \lambda X, \qquad AY = \mu Y, \qquad \lambda \neq \mu.
$$

We must show that $\langle X, Y \rangle = 0$. Here is the argument:

$$
\begin{aligned}
\lambda \langle X, Y \rangle &= \langle \lambda X, Y \rangle \\
&= \langle AX, Y \rangle \\
&= \langle X, A^* Y \rangle \\
&= \langle X, AY \rangle \\
&= \langle X, \mu Y \rangle \\
&= \mu \langle X, Y \rangle,
\end{aligned}
$$

so $\lambda \langle X, Y \rangle = \mu \langle X, Y \rangle$. Thus $\langle X, Y \rangle = 0$, since λ and μ are assumed to be distinct.   □

Corollary 309C (Generic Diagonalization)   *If a Hermitian matrix $A \in \mathbf{F}^{n \times n}$ has n distinct eigenvalues, it is unitarily diagonalizable.*

Proof: By Theorem 309B, the eigenvectors corresponding to these distinct eigenvalues are pairwise orthogonal. By normalizing them we obtain an orthonormal sequence. Hence, the matrix A is unitarily diagonalizable by the Theorem 201B and the Theorem 308C.   □

Remark 310A The corollary is true even without the hypothesis that the eigenvalues are distinct. The Spectral Theorem 317C says that *any* Hermitian matrix is unitarily diagonalizable.

▷ **Question 310B** Is the converse of Corollary 309C true?

▷ **Exercise 310C** For each of the following matrices A, find a unitary matrix so that $D = P^{-1}AP$ is diagonal.

$$(1)\ A = \begin{bmatrix} 4 & 1 \\ 1 & 4 \end{bmatrix}. \quad (2)\ A = \begin{bmatrix} 8.2 & 2.4 \\ 2.4 & 6.8 \end{bmatrix}. \quad (3)\ A = \begin{bmatrix} 8.2 & 2.4 & 0 \\ 2.4 & 6.8 & 0 \\ 0 & 0 & 7 \end{bmatrix}.$$

Exercise 310D Can $\xi^2 + \xi + 6$ be the characteristic polynomial of a Hermitian matrix?

▷ **Exercise 310E** What's wrong with the following proof of the Spectral Theorem? Assume that $A \in \mathbf{F}^{n \times n}$ is Hermitian. Let $\lambda_1, \ldots, \lambda_m$ be the distinct eigenvalues and (by Gram-Schmidt) let Φ_i be an orthonormal basis for the ith eigenspace $\mathcal{E}_{\lambda_i}(A)$. By Theorem 309B, the matrix

$$P = \begin{bmatrix} \Phi_1 & \Phi_2 & \cdots & \Phi_m \end{bmatrix}$$

has orthonormal columns, so by Theorem 201B, it is unitary. Since the columns of P are eigenvectors of A, $P^{-1}AP$ is diagonal by Theorem 308C.

Exercise 310F Show that the eigenvalues of a unitary matrix have absolute value one.

Using MINIMAT

Exercise 310G The command[1]

```
#> P=qr(rand(n,n)), D=diag(rand(1,n))
```

assigns to the variable P a random $n \times n$ real unitary matrix and to D a random $n \times n$ diagonal matrix. Execute these commands and confirm that A=P*D*P^(-1) is Hermitian and that P*P', P'*P, and eye(n) are the same.

▷ **Exercise 310H** How can we generate a random 5×5 Hermitian matrix?

▷ **Exercise 310I** What is the result of the following?

---

[1] The built-in function qr is explained on Page 364.

```
#> D=diag(rand(1,5)), Q=qr(rand(5,5)), A=Q*D*Q'
#> [P D1]= eig(A)
#> Q'*Q, Q*Q', P'*P, P*P'
```

▷ **Exercise 311A** Generate a random real 3×3 Hermitian matrix A, and then find a real unitary matrix P such that $P^{-1}AP$ is diagonal. You may use MINIMAT's built-in `eig` function. Hint: $B + B^\mathsf{T}$ is Hermitian. *Warning: The matrix P produced by the command*

```
#> [P D]=eig(A)
```

may not be unitary.

Exercise 311B Write a `.M` function `heig` such that when A is Hermitian with distinct eigenvalues, the command

```
#>  [P D] = heig(A)
```

produces unitary matrix P and a diagonal matrix D which satisfy A = P*D*P^(-1). You may use the built-in function `eig`.

Exercise 311C Repeat Exercise 311B under the following more stringent conditions. You may use the command

```
#> L = eig(A)
```

but not the command

```
#>  [P D] = eig(A)
```

You may, however, use the `.M` function `randsoln` which produces a random element of the null space of its input.

Remark 311D The `.M` function `schur` described in 315F implements the Schur Triangularization Theorem 312A discussed below. This function, in particular, solves Exercise 311C. The Exercises 311B and 311C don't require you to implement Theorem 312A but only to use Theorem 309B.

10.3   Schur's Theorem

Recall that matrices $A, R \in \mathbf{F}^{n \times n}$ are called *unitarily similar* iff there is a unitary matrix $P \in \mathbf{F}^{n \times n}$ such that $A = PRP^{-1}$. Recall also that a matrix R is called **triangular** iff all its entries below the diagonal vanish, that is, iff

$$\mathrm{entry}_{ij}(R) = 0 \qquad \text{for } i > j.$$

Schur Triangularization

Theorem 312A *Assume that* $\mathbf{F} = \mathbf{C}$. *Then any square matrix is unitarily similar to a triangular matrix.*

Proof: In other words, if $A \in \mathbf{F}^{n \times n}$, then there is a unitary matrix P and a triangular matrix R such that

$$A = PRP^{-1}.$$

The diagonal entries of R are the eigenvalues of R (and of A).

Lemma 312B *Suppose that* $P_1 \in \mathbf{C}^{n \times n}$ *is a invertible matrix whose first column is an eigenvector of* A. *Then the matrix*

$$A_1 = P_1^{-1} A P_1$$

has form

$$A_1 = \begin{bmatrix} \lambda_1 & H \\ 0 & B \end{bmatrix}$$

where $H \in \mathbf{C}^{1 \times (n-1)}$, $0 = 0_{(n-1) \times 1} \in \mathbf{F}^{(n-1) \times 1}$, *and* $B \in \mathbf{C}^{(n-1) \times (n-1)}$.

Proof: Let $E_1 \in \mathbf{F}^{n \times 1}$ be the first column of the identity matrix, and let X_1 be the first column of P_1:

$$E_1 = \mathrm{col}_1(I_n), \qquad X_1 = \mathrm{col}_1(P_1) = P_1 E_1.$$

The hypothesis is that X_1 is an eigenvector of A:

$$A X_1 = \lambda_1 X_1.$$

Then

$$
\begin{aligned}
\mathrm{col}_1(A_1) &= (P_1^{-1} A P_1) E_1 \\
&= P_1^{-1} A X_1 \\
&= P_1^{-1} \lambda_1 X_1 \\
&= \lambda_1 P_1^{-1} X_1 \\
&= \lambda_1 E_1
\end{aligned}
$$

so A_1 has the desired form.

Lemma 313A *There is a unitary matrix P_1 as in Lemma 312B.*

Proof: Let λ_1 be an eigenvalue of A and X_1 be a corresponding eigenvector. Normalize if necessary so that $\|X_1\| = 1$. By Theorem 210A, there is a unitary matrix P_1 whose first column is X_1:

$$P_1 E_1 = X_1.$$

Remark 313B A more efficient way of constructing P_1 is via Theorem 357A.

Now we complete the proof of Theorem 312A by induction on the number n of columns (and rows) of the matrix A. Assume that A is $n \times n$. Apply Lemmas 312B and 313A. The matrix B of Lemma 312B is $(n-1) \times (n-1)$. By the induction hypothesis B is unitarily equivalent to a triangular matrix S:

$$B = QSQ^{-1}$$

where $Q \in \mathbf{F}^{(n-1)\times(n-1)}$ is unitary. Let

$$W = \begin{bmatrix} 1 & 0 \\ 0 & Q \end{bmatrix}, \qquad R = \begin{bmatrix} \lambda_1 & HQ \\ 0 & S \end{bmatrix},$$

where we have again used block matrix notation. Then W is unitary, R is triangular, and

$$
\begin{aligned}
WRW^{-1} &= \begin{bmatrix} 1 & 0 \\ 0 & Q \end{bmatrix} \begin{bmatrix} \lambda_1 & HQ \\ 0 & S \end{bmatrix} \begin{bmatrix} 1 & 0 \\ 0 & Q^* \end{bmatrix} \\
&= \begin{bmatrix} \lambda_1 & H \\ 0 & QSQ^* \end{bmatrix} \\
&= \begin{bmatrix} \lambda_1 & H \\ 0 & B \end{bmatrix} \\
&= A_1 \\
&= P_1 A P_1^{-1}.
\end{aligned}
$$

Thus $A = PRP^{-1}$ where $P = P_1^{-1}W$ as required.   □

Remark 313C With A_1, B, and λ as in Lemma 312B we have

$$\det(\xi I - A) = \det(\xi I - A_1) = (\xi - \lambda_1)\det(\xi I - B).$$

This means that the eigenvalues of A are those of B together with λ_1. Thus the diagonal entries of the matrix R are the eigenvalues of A and can appear in any order we like. (The proof is word for word the same as the proof of Theorem 312A.) In the proof of the Block Diagonalization Theorem (Theorem 384C below), it will be convenient to require that repeated eigenvalues are adjacent on the diagonal.

Theorem 314A (Real Triangularization) *Suppose that* $\mathbf{F} = \mathbf{R}$. *Then any square matrix having only real eigenvalues is unitarily similar to a triangular matrix.*

Proof: The proof of Theorem 312A goes through almost word for word but requires one small modification. To apply the induction hypothesis to B, we need to show that its eigenvalues are real. The proof (in the complex case) shows that the eigenvalues of B are $\lambda_2, \ldots, \lambda_n$, that is, the eigenvalues of A are those of B together with λ_1. Hence, the hypothesis that A has only real eigenvalues implies that B has only real eigenvalues. □

▷ **Question 314B** Do we have to worry about real eigenvectors as well?

Example 314C The matrix $A = \begin{bmatrix} 1 & 2 \\ -1 & 4 \end{bmatrix}$ has eigenvectors $X = \begin{bmatrix} 2 \\ 1 \end{bmatrix}$ and $Y = \begin{bmatrix} 1 \\ 1 \end{bmatrix}$. An appropriate multiple of either one can be chosen as the first column of a unitary matrix that triangularizes A:

$$P^{-1}AP = \begin{bmatrix} 2 & 3 \\ 0 & 3 \end{bmatrix}, \qquad P = 1/\sqrt{5} \begin{bmatrix} 2 & -1 \\ 1 & 2 \end{bmatrix},$$

$$Q^{-1}AQ = \begin{bmatrix} 3 & -3 \\ 0 & 2 \end{bmatrix}, \qquad Q = 1/\sqrt{2} \begin{bmatrix} 1 & 1 \\ 1 & -1 \end{bmatrix}.$$

Example 314D Suppose that

$$A = \begin{bmatrix} 3 & -1 & 4 \\ 1 & 1 & 0 \\ 2 & -2 & 2 \end{bmatrix}, \qquad P_1 = 1/\sqrt{2} \begin{bmatrix} 1 & 1 & 0 \\ 1 & -1 & 0 \\ 0 & 0 & \sqrt{2} \end{bmatrix}.$$

Then P_1 is unitary and has $X = \mathrm{col}_1(P)$ is an eigenvector of A. Calculate

$$A_1 = P_1^{-1}AP_1 = \begin{bmatrix} 1 & \sqrt{2} & 1 \\ 0 & 1 & \sqrt{2} \\ 0 & \sqrt{2} & 1 \end{bmatrix}.$$

The eigenvalues of the lower right 2×2 block B are $1 \pm \sqrt{2}$. It is easy to diagonalize:

$$B = \begin{bmatrix} 1 & \sqrt{2} \\ \sqrt{2} & 1 \end{bmatrix}, \quad Q = \frac{1}{\sqrt{2}} \begin{bmatrix} 1 & -1 \\ 1 & 1 \end{bmatrix}, \quad Q^{-1}BQ = \begin{bmatrix} 1 + \sqrt{2} & 0 \\ 0 & 1 - \sqrt{2} \end{bmatrix}.$$

Hence, $P^{-1}AP$ is diagonal where $P = P_1^{-1}W$, and

$$W = \begin{bmatrix} 1 & 0 \\ 0 & Q \end{bmatrix} = \frac{1}{\sqrt{2}} \begin{bmatrix} \sqrt{2} & 0 & 0 \\ 0 & 1 & -1 \\ 0 & 1 & 1 \end{bmatrix},$$

In the following exercises you may *not* use the .M function schurtri.m described in 315F.

▷ **Exercise 315A** The eigenvalues of the matrix

$$A = \begin{bmatrix} 9.0 & -3.0 & -2.5 & -0.5 \\ 5.5 & 0.5 & 0.0 & -3.0 \\ 3.0 & 0.0 & -0.5 & 0.5 \\ 1.5 & 1.5 & 0.0 & 0.0 \end{bmatrix}$$

are $3, 3, 1, 2$. Find a unitary matrix P and a triangular matrix R such that $A = PRP^{-1}$.

Exercise 315B For each of the matrices A in Exercise 263B, find a unitary matrix P, such that $P^{-1}AP$ is triangular.

Exercise 315C For each of the matrices A in Exercise 260B, find a unitary matrix P such that $P^{-1}AP$ is triangular.

▷ **Exercise 315D** Suppose that $B = P^{-1}AP$ is triangular. Find a unitary matrix Q such that $Q^{-1}AQ$ is triangular. Hint: Use Gram-Schmidt.

Exercise 315E Suppose that the real 2×2 matrix A has distinct real eigenvalues. Show that there are exactly eight distinct real unitary matrices P such that $P^{-1}AP$ is (upper) triangular.

Using MINIMAT

Function 315F The .M function schur, shown in Listing 15, implements the Schur Triangularization Theorem 312A. The command

```
#> [P, R]=schur(A)
```

produces a unitary matrix P and a triangular matrix R with $A = PRP^{-1}$.

You may use the built-in function

```
L = eig(A)
```

Schur Triangularization in MINIMAT

```
function [P, R] = schurtri(A)

[n, n]=size(A);
P=eye(n); R=A;        % A=P*R*R^(-1)
for k=0:n-2
      B=R(k+1:n,k+1:n);
      lambda = max(eig(B));
      X = randsoln(lambda*eye(n-k)-B);
      c = sqrt(X'*X);
      E = zeros(n-k,1);  E(1) = c;
      U=R(1:k,k+1); Y = [U; X];  Z = [U; E];
      W = flip(Y,Z);
      P = P*W';   R = W*R*W';
end
```

Listing 15

---

to compute the eigenvalues of A and then invoke schur. Be warned that schur may fail because of roundoff error. If randsoln mistakenly thinks that its input is invertible, it will return zero as an eigenvector.

The function schur invokes the .M function flip explained in 360C below. This function has the following properties:

(1) The call W=flip(Y,Z) assigns a unitary matrix to W.

(2) If the inputs Y and Z have the same norm, then W*Y equals Z.

(3) If the inputs Y and Z agree in the first k entries, then W has the block form

$$W = \begin{bmatrix} I_k & 0 \\ 0 & Q \end{bmatrix}$$

where I_k is the $k \times k$ identity matrix and $Q \in \mathbf{F}^{(n-k) \times (n-k)}$ is unitary.

10.4   Spectral Theorem

Now we prove that a Hermitian matrix is unitarily diagonalizable. We already proved a special case (Theorem 309C).

Lemma 317A *A matrix that is unitarily similar to a Hermitian matrix is itself Hermitian.*

Proof: Assume that $B^* = B$ and $A = PBP^{-1}$ where $P^{-1} = P^*$. Then

$$
\begin{aligned}
A^* \; &= \; \left(PBP^{-1}\right)^* \\
&= \; \left(PBP^*\right)^* \\
&= \; P^{**}B^*P^* \\
&= \; PBP^{-1} \\
&= \; A.
\end{aligned}
$$

This shows that A is Hermitian.   □

Recall that a matrix D is *triangular* iff $\text{entry}_{ij}(D) = 0$ for $i > j$.

Lemma 317B *A triangular Hermitian matrix D is real diagonal.*

Proof: By the definition of D^* we have

$$
\text{entry}_{ij}(D) = \text{entry}_{ij}(D^*) = \overline{\text{entry}_{ji}(D)}.
$$

Then $\text{entry}_{ij}(D) = 0$ if $i \neq j$; for then either $i > j$, so $\text{entry}_{ij}(D) = 0$ or $i < j$, so $\text{entry}_{ji}(D) = 0$. Also $\text{entry}_{ii}(D)$ is real, as a complex number d is real $\iff d = \bar{d}$. For example, if

$$
D = \begin{bmatrix} a & b & c \\ 0 & d & e \\ 0 & 0 & f \end{bmatrix}, \qquad
D^* = \begin{bmatrix} \bar{a} & 0 & 0 \\ \bar{b} & \bar{d} & 0 \\ \bar{c} & \bar{e} & \bar{f} \end{bmatrix},
$$

so that if $D = D^*$, then $b = c = e = 0$ and a, d, f are real.   □

Spectral Theorem

Theorem 317C *A matrix is Hermitian if and only if it is unitarily similar to a real diagonal matrix.*

Proof: A real diagonal matrix is Hermitian, so 'if' follows immediately from the Lemma 317A. For the converse assume that A is Hermitian. By the Schur Triangularization Theorem 312A, we may find a triangular matrix D and a unitary matrix P with $A = PDP^{-1}$. We show that D is diagonal. Since $A = A^*$, $D = P^{-1}AP$, and $P^{-1} = P^*$, Lemma 317A yields that $D^* = D$. But D is triangular, so by Lemma 317B, the matrix D is diagonal and real.   □

Remark 318A The proof works when $\mathbf{F} = \mathbf{R}$. If $A \in \mathbf{R}^{n \times n}$ is symmetric, then there is a unitary $P \in \mathbf{R}^{n \times n}$ such that PAP^{-1} is diagonal.

Example 318B The matrix A is Hermitian and X is an eigenvector:

$$A = \begin{bmatrix} 4 & -1 & 1 \\ -1 & 4 & -1 \\ 1 & -1 & 4 \end{bmatrix}, \qquad X = 1/\sqrt{2} \begin{bmatrix} -1 \\ 0 \\ 1 \end{bmatrix}.$$

Then $AX = \lambda X$ where $\lambda = 3$. The matrix P_1 given by

$$P_1 = \frac{1}{\sqrt{2}} \begin{bmatrix} -1 & 1 & 0 \\ 0 & 0 & \sqrt{2} \\ 1 & 1 & 0 \end{bmatrix}$$

is unitary and has X as its first column. The matrix

$$A_1 = P_1^{-1}AP_1 = \begin{bmatrix} 3 & 0 & 0 \\ 0 & 5 & -\sqrt{2} \\ 0 & -\sqrt{2} & 5 \end{bmatrix}$$

is again Hermitian. The problem of diagonalizing A has been reduced to the problem of diagonalizing the 2×2 matrix B in the lower right corner. For this we have

$$B = \begin{bmatrix} 5 & -\sqrt{2} \\ -\sqrt{2} & 5 \end{bmatrix}, \quad Q = 1/\sqrt{2} \begin{bmatrix} 1 & 1 \\ -1 & 1 \end{bmatrix}, \quad Q^{-1}BQ = \begin{bmatrix} 3 & 0 \\ 0 & 6 \end{bmatrix}.$$

Hence, $P^{-1}AP = \mathrm{diag}(3, 3, 6)$ where $P = P_1 P_2$, and

$$P_2 = \frac{1}{\sqrt{2}} \begin{bmatrix} \sqrt{2} & 0 & 0 \\ 0 & 1 & 1 \\ 0 & -1 & 1 \end{bmatrix}.$$

10.5 Normal Spectral Theorem

Normal Matrix

Definition 319A A square matrix is called **normal** iff it commutes with its conjugate transpose:

$$A^*A = AA^*.$$

Remark 319B A Hermitian matrix $A \in \mathbf{F}^{n \times n}$ is certainly normal: it is equal to its conjugate transpose.

Example 319C The conjugate transpose of a diagonal matrix is diagonal. Any two diagonal matrices commute. Hence, a diagonal matrix is normal. A diagonal matrix is Hermitian if and only if it is real.

It is easy to check if a matrix A is normal: we need only compute AA^* and A^*A and see if they are equal. The Normal Spectral Theorem, proved in this section, says that this easily verified condition is necessary and sufficient for the matrix A to be unitarily diagonalizable.

Lemma 319D *A matrix that is unitarily similar to a normal matrix is itself normal.*

Proof: Assume that $BB^* = B^*B$ and $A = PBP^{-1}$ where $P^{-1} = P^*$. Then

$$
\begin{aligned}
A^* &= \left(PBP^{-1}\right)^* \\
&= (PBP^*)^* \\
&= P^{**}B^*P^* \\
&= PB^*P^{-1}
\end{aligned}
$$

so

$$
\begin{aligned}
AA^* &= (PBP^{-1})(PB^*P^{-1}) \\
&= PBB^*P^{-1} \\
&= PB^*BP^{-1} \\
&= (PB^*P^{-1})(PBP^{-1}) \\
&= A^*A.
\end{aligned}
$$

This shows that A is normal.   □

Lemma 320A *Suppose that D is triangular and normal. Then D is diagonal.*

Proof: The matrix $A \in \mathbf{F}^{n \times n}$ has form

$$D = \begin{bmatrix} d & H \\ 0 & B \end{bmatrix}, \qquad D^* = \begin{bmatrix} \bar{d} & 0 \\ H^* & B^* \end{bmatrix},$$

where $d \in \mathbf{F}$, $H \in \mathbf{F}^{1 \times (n-1)}$, and $B \in \mathbf{F}^{(n-1) \times (n-1)}$. By block multiplication

$$\text{entry}_{11}(DD^*) = d\bar{d}, \qquad \text{entry}_{11}(D^*D) = d\bar{d} + \|H\|^2.$$

From $DD^* = D^*D$ follows $H = 0$ whence

$$D = \begin{bmatrix} d & 0 \\ 0 & B \end{bmatrix}.$$

Now B is triangular and normal (since D is), so we may repeat the argument with D. Hence, by induction, B (and hence D) is diagonal as required.   □

Normal Spectral Theorem

Theorem 320B *Assume that $\mathbf{F} = \mathbf{C}$. A matrix $A \in \mathbf{F}^{n \times n}$ is normal if and only if it is unitarily diagonalizable, that is, unitarily similar to a diagonal matrix $D \in \mathbf{F}^{n \times n}$.*

Proof: As in the Hermitian case 317C.   □

Remark 320C Theorem 320B fails for real matrices. If a, b are real, then the matrix

$$A = \begin{bmatrix} a & b \\ -b & a \end{bmatrix}$$

is normal:

$$AA^* = \begin{bmatrix} a^2 + b^2 & 0 \\ 0 & a^2 + b^2 \end{bmatrix} = A^*A.$$

However, its eigenvalues are $a \pm ib$. If $b \neq 0$, these are not real, so there is no real matrix P with $P^{-1}AP$ diagonal.

▷ **Question 320D** Is a unitary matrix unitarily diagonalizable?

10.6   Invariants

In Section 8.5 we introduced the notions of geometric multiplicity of an eigenvalue. According to Theorem 271B, similar matrices have the same eigenvalues, each with the same geometric multiplicity. In this section we prove that for normal matrices the converse holds. We repeat the definition of geometric multiplicity:

Definition 321A The **geometric multiplicity** $\nu_\lambda(A)$ of an eigenvalue λ in a matrix A is the dimension of the corresponding space of eigenvectors:

$$\nu_\lambda(A) = \text{nullity}(\lambda I - A) = \dim \mathcal{E}_\lambda(A)$$

where $\mathcal{E}_\lambda(A) = \mathcal{N}(\lambda I - A)$ is the eigenspace of λ.

Example 321B For a diagonal matrix the geometric multiplicity of an eigenvalue is the number of times it is repeated on the diagonal.

Remark 321C The geometric multiplicity of the eigenvalue zero is the nullity of the matrix.

The next theorem says that these multiplicities form a complete system of invariants for normal matrices up to unitary similarity.

Normal Similarity Test

Theorem 321D *Two normal matrices are unitarily similar if and only if they have the same eigenvalues each with the same geometric multiplicity.*

Proof: 'Only if' is Lemma 271B. 'If' follows from Lemma 271B, the Normal Spectral Theorem 320B and the observation in Example 321B. This argument works whenever the Spectral Theorem applies, namely when $\mathbf{F} = \mathbf{C}$, or when $\mathbf{F} = \mathbf{R}$ and the matrices are Hermitian and not just normal. When $\mathbf{F} = \mathbf{R}$, an additional argument is required, since A and B cannot be unitarily diagonalizable over $\mathbf{R}$ if they have nonreal eigenvalues. We'll develop the argument in the exercises of Section 10.7.

10.7   More Exercises

Exercise 322A Suppose that A is Hermitian and that the eigenvalues of A have been arranged in a list $(\lambda_1, \lambda_2, \ldots, \lambda_m)$ without repetitions. Suppose that, for each i, Φ_i is an orthonormal basis for the eigenspace $\mathcal{E}_{\lambda_i}(A)$. Form the matrix

$$P = \begin{bmatrix} \Phi_1 & \Phi_2 & \cdots & \Phi_m \end{bmatrix}.$$

Show that P is unitary and that $P^{-1}AP$ is diagonal. (This should be compared with the algorithm following Theorem 272A.) Hint: Why is P square?

Real Normal Matrices

Exercise 322B Suppose that $A \in \mathbf{R}^{n \times n}$ is real and that $\lambda \in \mathbf{C}$ is an eigenvalue of A of geometric multiplicity k. Show that $\bar{\lambda}$ is also an eigenvalue of A of geometric multiplicity k.

Exercise 322C Suppose that $A \in \mathbf{C}^{n \times n}$ is normal and that $\lambda, \nu \in \mathbf{C}$ are distinct eigenvalues of A with corresponding eigenvectors $X, Y \in \mathbf{C}^{n \times 1}$:

$$AX = \lambda X, \qquad AY = \mu Y, \qquad \lambda \neq \mu.$$

Show that X and Y are orthogonal:

$$\langle X, Y \rangle = 0.$$

Exercise 322D Suppose that $A \in \mathbf{R}^{n \times n}$ is real and normal. that $\lambda = \alpha + i\beta$ is a nonreal eigenvalue for A, and that $Z = X + iY$ is a corresponding eigenvector. (Here α, β, X, Y are real.) Show that

$$\|X\| = \|Y\|, \quad \langle X, Y \rangle = 0, \quad AX = \alpha X - \beta Y, \quad AY = \beta X + \alpha Y.$$

(The hypothesis is that $AZ = \lambda Z$ and $\beta \neq 0$.)

Exercise 322E A matrix $D \in \mathbf{R}^{n \times n}$ is said to be in **normal block diagonal form** iff it is a block diagonal

$$D = \operatorname{diag}(\Lambda_1, \Lambda_2, \ldots, \Lambda_m)$$

where each Λ_k is either 1×1 or else is 2×2 of form

$$\Lambda_k = \begin{bmatrix} \alpha_k & \beta_k \\ -\beta_k & \alpha_k \end{bmatrix}$$

where $\beta_k \neq 0$. Show that: (1) such a matrix D is normal; (2) its real eigenvalues are the 1×1 blocks Λ_k in D, and its nonreal eigenvalues are $\alpha_k \pm i\beta_k$ where Λ_k is a 2×2 block in D; and (3) the number of repetitions of a 2×2 block Λ_k is the geometric multiplicity of the eigenvalue $\alpha_k + i\beta_k$.

Exercise 323A Suppose that $A \in \mathbf{R}^{n \times n}$ is real and normal. Show that there is a matrix D in normal block diagonal form and a real unitary matrix P with $A = PDP^{-1}$.

We say that two matrices A and B are **unitarily similar over F** iff there is a unitary matrix $P \in \mathbf{F}^{n \times n}$ with $A = PBP^{-1}$. (Compare with Section 7.3.)

Corollary 323B (Real Unitary Similarity)[2]   *Assume $A, B \in \mathbf{R}^{n \times n}$ are real and normal. Then they are unitarily similar over* $\mathbf{C}$ *if and only if they are unitarily similar over* $\mathbf{R}$. *Hint: Use Theorem 321D and Exercise 322E.*

Positive Semidefinite Matrices

A matrix B is **positive semidefinite** iff it is Hermitian and $\langle BX, X \rangle \geq 0$ for all X. A matrix B is called **positive definite** iff it is Hermitian and $\langle BX, X \rangle > 0$ for all nonzero X.

Exercise 323C Show that for any matrix A the matrix $B = A^*A$ is positive semidefinite.

Exercise 323D Show that a positive semidefinite matrix B has a nonnegative determinant. Is the converse true? (Hint: The determinant is the product of the eigenvalues.)

Exercise 323E Show that $\det(A^*A) \geq 0$ for $A \in \mathbf{F}^{m \times n}$. (When $m = 2$ this is the **Schwartz inequality**.)

Exercise 323F Show that a Hermitian matrix B is positive semidefinite if and only if all its eigenvalues are nonnegative and is positive definite if and only if all is eigenvalues are positive. Show that a positive semidefinite matrix is positive definite if and only if it is invertible.

Exercise 323G Show that a matrix B is positive semidefinite if and only if $B = R^2$ where R is Hermitian.

Exercise 323H Show that a matrix B is positive definite if and only if $B = e^A$ where A is Hermitian.

---

[2]This corollary is true even without the hypothesis that A and B are normal, but the proof is much harder. See Helene Shapiro: *A survey of canonical forms and invariants for unitary similarity*, Linear Algebra and its Applications (North Holland) **144** (1991) 101-167.

Skew-Hermitian matrices

Definition 324A A matrix A is called **skew-Hermitian** iff it satisfies $A^* = -A$.

Exercise 324B Show that the eigenvalues of a skew-Hermitian matrix are pure imaginary.

Exercise 324C Find all real 2×2 skew-Hermitian matrices.

Exercise 324D Show that if f is a real polynomial, and A is a square matrix, then $f(A^*) = f(A)^*$. What happens if the coefficients of f are complex?

Exercise 324E Show that if A and B are skew-Hermitian, then $AB - BA$ is skew-Hermitian.

Exercise 324F Show that A is skew-Hermitian if and only if $U(t) = e^{tA}$ is unitary for all real t. Hint: For half of the proof differentiate $U(t)U(t)^* = I$ and put $t = 0$.

Exercise 324G (Axis of Rotation) There is a one-one correspondence between real skew-Hermitian matrices $A \in \mathbf{R}^{3\times3}$ and column vectors $W \in \mathbf{R}^{3\times1}$:

$$A = \begin{bmatrix} 0 & -c & b \\ c & 0 & -a \\ -b & a & 0 \end{bmatrix}, \qquad W = \begin{bmatrix} a \\ b \\ c \end{bmatrix}.$$

Show that for $X \in \mathbf{R}^{3\times1}$ we have

$$AX = W \wedge X$$

where $A \wedge X$ is the wedge (cross) product defined in 250A. Conclude that $AW = 0$. Show that there is an orthonormal sequence $U, X, Y \in \mathbf{R}^{3\times1}$ and a real number ω such that $W = \omega U$ and

$$\begin{aligned} e^{tA}U &= U \\ e^{tA}X &= \cos(\omega t)X + \sin(\omega t)Y \\ e^{tA}Y &= -\sin(\omega t)X + \cos(\omega t)X \end{aligned}$$

Invariant subspaces

Definition 324H Let $A \in \mathbf{F}^{n\times n}$ and $\mathcal{V} \subset \mathbf{F}^{n\times1}$ be a subspace. The subspace $\mathcal{V}$ is called an **invariant subspace** of A iff

$$A\mathcal{V} \subset \mathcal{V},$$

that is, iff $AX \in \mathcal{V}$ whenever $X \in \mathcal{V}$. The phrase "$\mathcal{V}$ is an invariant subspace of A" is often abbreviated to "$\mathcal{V}$ is A-invariant".

Exercise 325A Let $A = \begin{bmatrix} 2 & 0 \\ 0 & 3 \end{bmatrix}$ and $\mathcal{V} = \mathcal{N}(B)$. For each of the following values of B draw a graph showing $\mathcal{V}$ and $A\mathcal{V}$:

(1) $B = \begin{bmatrix} 1 & 1 \end{bmatrix}$,

(2) $B = \begin{bmatrix} 0 & 1 \end{bmatrix}$,

(3) $B = \begin{bmatrix} 1 & 0 \end{bmatrix}$,

(4) $B = \begin{bmatrix} 2 & 3 \end{bmatrix}$.

Which subspaces are A-invariant?

Exercise 325B Repeat the previous exercise with $A = \begin{bmatrix} 2 & 1 \\ 0 & 2 \end{bmatrix}$.

Exercise 325C Find eight invariant subspaces of the diagonal matrix $D = \operatorname{diag}(1, 2, 3)$.

Exercise 325D Find all of the invariant subspaces for the matrix A of Exercise 325B.

Exercise 325E Show that if the subspace $\mathcal{V}$ is A-invariant and P is an invertible matrix, then the subspace $P^{-1}\mathcal{V}$ is $P^{-1}AP$-invariant.

▷ **Exercise 325F** Suppose that $A \in \mathbf{F}^{n \times n}$ has block form

$$A = \begin{bmatrix} A_{11} & A_{12} \\ A_{21} & A_{22} \end{bmatrix}$$

where $A_{11} \in \mathbf{F}^{k \times k}$, $A_{12} \in \mathbf{F}^{(n-k) \times k}$, $A_{21} \in \mathbf{F}^{k \times (n-k)}$, $A_{22} \in \mathbf{F}^{(n-k) \times (n-k)}$. Show that the subspace

$$\mathcal{V}_k = \mathcal{R}(\Phi_k), \qquad \Phi_k = \begin{bmatrix} I_k \\ 0_{(n-k) \times k} \end{bmatrix}$$

is A-invariant iff $A_{21} = 0$.

▷ **Exercise 325G** If $\mathcal{V}_k$ is A-invariant, find B such that $A\Phi_k = \Phi_k B$.

Exercise 325H Let $A \in \mathbf{F}^{n \times n}$, $\mathcal{V} \subset \mathbf{F}^{n \times 1}$ be a subspace, and $\Phi \in \mathbf{F}^{n \times \nu}$ be a basis for $\mathcal{V}$. Show that $\mathcal{V}$ is A-invariant if and only if there is a matrix $B \in \mathbf{F}^{\nu \times \nu}$ such that $A\Phi = \Phi B$. (See Exercise 186B.)

Exercise 325I Let $A \in \mathbf{F}^{n \times n}$, $\mathcal{V} \subset \mathbf{F}^{n \times 1}$ be an A-invariant subspace, $\Phi_1, \Phi_2 \in \mathbf{F}^{n \times \nu}$ be bases for $\mathcal{V}$, and $A\Phi_i = \Phi_i B_i$ as in Exercise 325H. Show that B_1 and B_2 are similar. (See Exercise 186C.)

Recall from Definition 216I that the orthogonal complement of a subspace $V \subset \mathbf{F}^{n \times 1}$ is the subspace

$$V^{\perp} = \{Y \in \mathbf{F}^{n \times 1} : \langle Y, X \rangle = 0 \quad \forall\, X \in V\}.$$

Exercise 326A Show that $(PV)^{\perp} = P(V^{\perp})$ if P is unitary. Give an example which shows that this is false (in general) if P is not unitary.

▷ **Exercise 326B** Suppose that A is Hermitian and that V is an A-invariant subspace. Show that $V^{\perp}$ is an A-invariant subspace.

Exercise 326C Show that for any square matrix, normal or not, if a subspace $V \subset \mathbf{F}^{n \times 1}$ is A-invariant, then $V^{\perp}$ is A^*-invariant.

Exercise 326D (Lagrange Interpolation) Assume that $a_1, a_2, \ldots, a_m$ are distinct numbers and that $f(\xi)$ is a polynomial of form

$$f(\xi) = \sum_k b_k \prod_{j \neq k} \frac{(\xi - a_j)}{(a_k - a_j)}.$$

What is $f(a_k)$?.

Exercise 326E Show that if A is normal, then there is a polynomial f such that $A^* = f(A)$. Hint: When $\mathbf{F} = \mathbf{C}$, use the Normal Spectral Theorem. When $\mathbf{F} = \mathbf{R}$, find a real polynomial f such that $f(\lambda) = \bar{\lambda}$ for each eigenvalue of A.

Exercise 326F Show that for a normal matrix A, if a subspace $V \subset \mathbf{F}^{n \times 1}$ is A-invariant, then it is A^*-invariant.

Exercise 326G Redo Exercise 326B, reading "normal" for "Hermitian".

Conic sections

Exercise 326H Verify the formulas

$$\langle AX, Y \rangle = a x_1 y_1 + b(x_1 y_2 + x_2 y_1) + c x_2 y_2$$

and

$$\langle AX, X \rangle = a x_1^2 + 2b x_1 x_2 + c x_2^2$$

for real matrices $A \in \mathbf{R}^{2 \times 2}$ and $X, Y \in \mathbf{R}^{2 \times 1}$ given by

$$A = \begin{bmatrix} a & b \\ b & c \end{bmatrix}, \quad X = \begin{bmatrix} x_1 \\ x_2 \end{bmatrix}, \quad Y = \begin{bmatrix} y_1 \\ y_2 \end{bmatrix}.$$

Exercise 327A As θ varies between 0 and 2π, the column vector

$$X(\theta) = \begin{bmatrix} \cos\theta \\ \sin\theta \end{bmatrix}$$

traverses the circle $\|X\|^2 = 1$. Choose a real matrix $A \in \mathbf{R}^{2\times2}$ and form the function

$$f(\theta) = \langle AX(\theta), X(\theta) \rangle .$$

The function f is π-periodic (this means that $f(\theta + \pi) = f(\theta)$), since $X(\theta + \pi) = -X(\theta)$. Assume that A is symmetric:

$$A = \begin{bmatrix} a & b \\ b & c \end{bmatrix} .$$

Calculate the maximum value attained by the function $f(\theta)$ and the values of θ for which this maximum is attained. Similarly for the minimum. Show that for these values of θ, $X(\theta)$ is an eigenvector of A.

Exercise 327B Continue the notation of the preceding exercise. Let

$$R(\theta) = \begin{bmatrix} \cos\theta & -\sin\theta \\ \sin\theta & \cos\theta \end{bmatrix} .$$

For what values of θ is $R(-\theta)AR(\theta)$ diagonal?

Exercise 327C Continue the notation of the preceding exercise. The set of $X \in \mathbf{R}^{2\times1}$ satisfying the equation $\langle AX, X \rangle = 1$ is either (1) an ellipse, (2) a hyperbola, (3) two parallel lines, or (4) empty. How can you tell which of these four possibilities occurs from the eigenvalues of A? Show that $ac - b^2$ is the product of the eigenvalues and that $a + c$ is their sum. How can we tell which of the four possibilities occurs from these numbers?

Exercise 327D Let $A \in \mathbf{R}^{3\times3}$ be symmetric. Show that the set of all $X \in \mathbf{R}^{3\times3}$ satisfying $\langle AX, X \rangle = 1$ is (1) an ellipsoid, if A has three positive eigenvalues; (2) a hyperboloid of one sheet, if A has one negative and two positive eigenvalues; (3) a hyperboloid of two sheets, if A has one positive and two negative eigenvalues; (4) an elliptic cylinder, if A has one zero and two positive eigenvalues; (5) a hyperbolic cylinder ,if A has one zero, one positive, and one negative eigenvalue; (6) two parallel planes, if A has one positive and two zero eigenvalues; and (7) empty, if A has no positive eigenvalue.

Exercise 327E Assume that the real symmetric matrix has eigenvalues of opposite sign, so that the equation $\langle AX, X \rangle = 0$ defines an elliptic cone through the origin. If

$$A = \begin{bmatrix} a & b & d \\ b & c & e \\ d & e & f \end{bmatrix}, \qquad X = \begin{bmatrix} x \\ y \\ 1 \end{bmatrix},$$

then equation $\langle AX, X \rangle = 0$ has form

$$ax^2 + 2bxy + cy^2 + 2dx + 2ey + f = 0.$$

This quadratic equation in two variables defines a **conic section**, the intersection of the cone $\langle AX, X \rangle = 0$ with the plane $z = 1$. Show that there is a unitary matrix Q, so that these equation have the form

$$\tilde{a}\tilde{x}^2 + \tilde{c}\tilde{y}^2 + \tilde{f}\tilde{z}^2 = 0$$

$$\alpha\tilde{x} + \beta\tilde{y} + \gamma\tilde{z} = 1$$

where

$$QX = \begin{bmatrix} \tilde{x} \\ \tilde{y} \\ \tilde{z} \end{bmatrix}$$

and all the coefficients are real. Say when, in terms of the new coefficients, the result is an ellipse, hyperbola, or parabola.

Chapter Summary

Schur Triangularization

Every square matrix is unitarily similar to a triangular matrix.

Spectral Theorem

(Normal) *A matrix A is unitarily similar to a diagonal matrix if and only if it is normal: $AA^* = A^*A$.*

(Hermitian) *A matrix A is unitarily similar to a real diagonal matrix if and only if it is Hermitian: $A = A^*$.*

Complete System of Invariants

Two normal matrices and are unitarily similar iff they have the same eigenvalues each with the same geometric multiplicity. The geometric multiplicity of an eigenvalue λ in the matrix A is the dimension

$$\nu_\lambda(A) = \dim \mathcal{E}_\lambda(A)$$

of the corresponding eigenspace.

11

TRIANGULAR MATRICES

This chapter is analogous to Chapter 3. The main difference is that we are going to find normal forms using fewer operations. For example, in Chapter 3, we learned how to transform a matrix to *reduced row echelon form* using elementary row operations; in this chapter, we will see how to transform a matrix to a normal form called *leading entry normal form* without using swaps. For many purposes, the leading entry normal form is as useful as the reduced row echelon form, and it can be computed more quickly. This is why computer programmers are interested in the material in this chapter. They abhor moving data around in the computer without acting on it. They don't like to swap rows.

Two theorems proved in this chapter are from very diverse parts of mathematics. These are the LU *(LUP)* Decomposition from numerical linear algebra and the Bruhat *(LPU)* Decomposition from the theory of Lie groups. Despite a superficial similarity they are really quite different, as we will see.

331

11.1   Definitions

Triangular Matrices

Definition 332A A square matrix is **upper triangular** iff all the entries below the diagonal vanish and **lower triangular** iff all the entries above the diagonal vanish. In other words, a matrix

- U is upper triangular iff $\text{entry}_{ij}(U) = 0$ for $i > j$, and

- L is lower triangular iff $\text{entry}_{ij}(L) = 0$ for $i < j$.

Example 332B The matrix L is lower triangular and the matrix U is upper triangular:

$$L = \begin{bmatrix} * & 0 & 0 \\ * & * & 0 \\ * & * & * \end{bmatrix}, \qquad U = \begin{bmatrix} * & * & * \\ 0 & * & * \\ 0 & 0 & * \end{bmatrix}.$$

Here the asterisks represent any numbers, zero or not.

Unitriangular Matrices

Definition 332C A square matrix N is **strictly triangular** iff all its entries on or below the diagonal are zero. A matrix U **unitriangular** iff $U = I + N$ where N is strictly triangular and I is the identity matrix, that is, iff all the entries below the diagonal are zero and all the entries on the diagonal are one.

Example 332D The matrix N is strictly triangular; the matrix U is unitriangular.

$$N = \begin{bmatrix} 0 & * & * \\ 0 & 0 & * \\ 0 & 0 & 0 \end{bmatrix}, \qquad U = \begin{bmatrix} 1 & * & * \\ 0 & 1 & * \\ 0 & 0 & 1 \end{bmatrix}.$$

11.2   Factorization

Recall from Chapter 3 that there are three kinds of **elementary matrices.**

Scale The elementary matrix Scale(I, p, c) results from the identity matrix by multiplying the pth row by c

Swap The elementary matrix Swap(I, p, q) results from the identity matrix by interchanging rows p and q.

Shear The elementary matrix Shear(I, p, q, c) results from the identity matrix by adding c times the qth row to the pth row.

The Fundamental Theorem 76B on Row Operations says that the matrix which results by multiplying a matrix A on the left by an elementary matrix is the same as the matrix which results by applying the corresponding elementary row operation to A. In other words,

- $E = \text{Scale}(I, p, c) \implies EA$ results from A by multiplying the pth row by c;

- $E = \text{Swap}(I, p, q) \implies EA$ results from A by interchanging rows p and q;

- $E = \text{Shear}(I, p, q, c) \implies EA$ results from A by adding c times the qth row to the pth row.

The Fundamental Theorem 99A on Column Operations says that multiplying on the right performs the corresponding column operation:

- $E = \text{Scale}(I, p, c) \implies AE$ results from A by multiplying the pth column by c;

- $E = \text{Swap}(I, p, q) \implies AE$ results from A by interchanging columns p and q;

- $E = \text{Shear}(I, p, q, c) \implies AE$ results from A by adding c times the pth column to the qth column.

Definition 333A An **upper shear** is an upper triangular shear matrix. A **lower shear** is a lower triangular shear matrix.

Remark 333B According to the Fundamental Theorem, if E is an upper shear, then EA results from A by adding a multiple of some row to a row above it. Upper row shears correspond to right column shears. If E is an upper shear, then AE results from A by adding a multiple of some column to a column on its right.

Unitriangular Factorization

Theorem 334A *A matrix is unitriangular if and only if it is a product of upper shears.*

Triangular Factorization

Theorem 334B *A matrix is invertible lower triangular if and only if it is a product of lower shear and scale matrices.*

Proof: Mimic the proof of the Factorization Theorem 93B. Note which operations are required to transform a triangular matrix to the identity.
□

Definition 334C A **matrix group** is a set

$$\mathcal{G} \subset \mathbf{F}^{n \times n}$$

of invertible matrices such that

- $\mathcal{G}$ contains the identity matrix: $I_n \in \mathcal{G}$.

- $\mathcal{G}$ is closed under taking inverses: $A \in \mathcal{G} \implies A^{-1} \in \mathcal{G}$.

- $\mathcal{G}$ is closed under multiplication: $A, B \in \mathcal{G} \implies AB \in \mathcal{G}$.

Example 334D The set of *all* invertible matrices of size $n \times n$ is a group. See Theorem 40D.

Example 334E The set of *all* unitary matrices of size $n \times n$ is a group. See Theorem 198E.

Example 334F The set of *all* positive triangular matrices of size $n \times n$ is a group. See Theorem 205C.

Theorem 334G *The set of all $n \times n$ unitriangular matrices is a group. The set of all $n \times n$ invertible lower triangular matrices is also a group.*

Proof: A direct argument is sketched in Exercises 336C and 336D. Here we show how this follows from the Factorization Theorem. Let $\mathcal{E}$ denote a set of invertible matrices such that $E^{-1} \in \mathcal{E}$ whenever $E \in \mathcal{E}$. Then the set $\mathcal{G}$ of all products

$$A = E_1 E_2 \cdots E_k$$

where $E_1, E_2, \ldots, E_k \in \mathcal{E}$ is a group:

- $I_n \in \mathcal{G}$ (take $k = 0$);

- $(E_1 E_2 \cdots E_k)^{-1} = E_k^{-1} \cdots E_2^{-1} E_1^{-1}$; and

- $(E_1 E_2 \cdots E_k)(F_1 F_2 \cdots F_h) = E_1 E_2 \cdots E_k F_1 F_2 \cdots F_h$.

Take $\mathcal{E}$ to be the set of all upper shears to prove that the unitriangular matrices form a group. Take $\mathcal{E}$ to consist of all lower shears and scales to prove that the lower triangular matrices form a group. $\quad\square$

Exercise 335A Write $\begin{bmatrix} 1 & x & y \\ 0 & 1 & z \\ 0 & 0 & 1 \end{bmatrix}$ as a product of elementary matrices and find its inverse.

Exercise 335B Assume that a, b, c are nonzero. Write $\begin{bmatrix} a & 0 & 0 \\ x & b & 0 \\ y & z & c \end{bmatrix}$ as a product of elementary matrices and find its inverse.

▷ **Exercise 335C** Do the Hermitian matrices form a group?

▷ **Exercise 335D** Show that the set of all invertible matrices $Q \in \mathbf{F}^{n \times n}$ such that $Q^{-1} = Q^{\top}$ is a group.

Exercise 335E Show that if $c = \cos\theta$ and $s = \sin\theta$, then the matrix

$$Q = \begin{bmatrix} c & s \\ -s & c \end{bmatrix}$$

satisfies $Q^{\top} = Q^{-1}$ as in the previous exercise.

▷ **Exercise 335F** Let D be any matrix. Show that the set of all invertible matrices P such that $PDP^{-1} = D$ is a group. What is this group when D is a diagonal matrix with distinct diagonal entries? Hint: $PD = DP$.

Exercise 336A The **affine group** is the set of matrices T of form

$$T = \begin{bmatrix} L & X_0 \\ 0_{1 \times n} & 1 \end{bmatrix} \in \mathbf{F}^{(n+1) \times (n+1)}$$

where $L \in \mathbf{F}^{n \times n}$ is invertible and $X_0 \in \mathbf{F}^{n \times 1}$. Prove that the affine group is a matrix group.

Exercise 336B The **Euclidean group** is the set of all matrices T of form

$$T = \begin{bmatrix} L & X_0 \\ 0_{1 \times n} & 1 \end{bmatrix} \in \mathbf{R}^{(n+1) \times (n+1)}$$

where $L \in \mathbf{R}^{n \times n}$ is invertible, $L^{\top} = L^{-1}$, and $X_0 \in \mathbf{R}^{n \times 1}$. Prove that the Euclidean group is a matrix group.

▷ **Exercise 336C** Show that the $n \times n$ unitriangular matrices form a group. (You must show that the product of unitriangular matrices is unitriangular, and that the inverse of a unitriangular matrix is unitriangular. If you can't do the general case, do the case $n = 3$.)

Exercise 336D Let B be an invertible upper triangular matrix. Show that there are unique diagonal matrices D_1 and D_2 and unique unitriangular matrices U_1 and U_2 with $B = D_1 U_1 = U_2 D_2$. Show that $D_1 = D_2$, but that usually $U_1 \neq U_2$. Using the previous exercise, conclude that the invertible upper triangular matrices form a group.

11.3   Equivalence

The equivalence relations studied in Chapter 3 generalize easily to other groups: one replaces the requirement that the matrices P and Q in the definition be invertible (see Definition 104A) with the requirement that they belong to some matrix group. This is done in Chapter 12 where the group of invertible matrices is replaced by the group of unitary matrices. In this chapter we study another variant.

Triangular Equivalence

Definition 337A Let $A, B \in \mathbf{F}^{m \times n}$ be two matrices of the same size. We say that

- A is **lower equivalent** to B iff there is an invertible lower triangular matrix L with

$$A = LB.$$

- A is **lower/upper equivalent** to B iff there are invertible matrices L and U with L lower triangular, U unitriangular, and

$$A = LBU^{-1}.$$

Theorem 337B *Both these relations satisfy the following laws:*

(Reflexive Law) $A \equiv A;$

(Symmetric Law) *if $A \equiv B$, then $B \equiv A;$*

(Transitive Law) *if $A \equiv B$ and $B \equiv C$, then $A \equiv C$.*

Here $A \equiv B$ means either "A is lower equivalent to B" or "A is lower/upper equivalent to B",

Proof: The proof of Theorem 104C uses only the fact that the invertible matrices form a group. Hence, it works for other groups. □

Next, we define two normal forms: *leading entry normal form* (LENF) and *rook normal form* (RNF). The former is roughly analogous to reduced row echelon form (RREF) and the latter to biequivalence normal form (BENF).

Rook Normal Form (RNF)

Definition 337C A matrix is in **rook normal form**, abbreviated **RNF**, iff all its entries are either 0 or 1 and it has at most one nonzero entry in every row and at most one nonzero entry in every column.

The terminology is suggested by the game of chess. The rook is the chess piece that moves horizontally and vertically. If we imagine that a matrix in RNF represents a chessboard and the nonzero entries represent rooks, then the rooks do not attack one another. A matrix is in RNF iff it can be obtained from a matrix in BENF (Definition 105B) by permuting the rows and columns.

Example 338A The following matrix is in RNF:

$$D = \begin{bmatrix} 0 & 0 & 1 & 0 & 0 \\ 0 & 0 & 0 & 0 & 0 \\ 0 & 0 & 0 & 1 & 0 \\ 1 & 0 & 0 & 0 & 0 \end{bmatrix}.$$

Recall that the **leading entry** in a row of a matrix is the first (left-most) nonzero entry in that row. A matrix is in leading entry normal form iff every leading entry is 1, and any directly below a leading entry is 0. (It follows that there is at most one leading entry in any column.) Let's say it more precisely.

Leading Entry Normal Form (LENF)

Definition 338B A matrix $R \in \mathbf{F}^{m \times n}$ is said to be in in **leading entry normal form**, abbreviated **LENF**, iff there is a matrix $D \in \mathbf{F}^{m \times n}$, in RNF, such that for each pair (p, q) of indices for which $\mathrm{entry}_{pq}(D) \neq 0$ we have

$$\begin{aligned} \mathrm{entry}_{p,q}(R) &= 1, \\ \mathrm{entry}_{p,j}(R) &= 0 \quad \text{for } j < q, \\ \mathrm{entry}_{i,q}(R) &= 0 \quad \text{for } p < i. \end{aligned}$$

Examples 338C The 4×5 matrix

$$R = \begin{bmatrix} 0 & 0 & 1 & * & * \\ 0 & 0 & 0 & 0 & 0 \\ 0 & 0 & 0 & 1 & * \\ 1 & * & 0 & 0 & * \end{bmatrix},$$

is in LENF; the matrix D of Example 338A is the corresponding RNF. Both of the following are in LENF:

$$
\begin{bmatrix}
1 & * & * & * & * \\
0 & 1 & * & * & * \\
0 & 0 & 1 & * & *
\end{bmatrix}, \qquad
\begin{bmatrix}
1 & * & * \\
0 & 1 & * \\
0 & 0 & 1 \\
0 & 0 & 0
\end{bmatrix}.
$$

For the last two, the corresponding RNF is in BENF.

Leading Entry Decomposition

Theorem 339A *Every matrix is lower equivalent to a unique matrix in LENF (leading entry normal form).*

Proof: In other words any matrix A may be written in the form

$$A = LR$$

where L is invertible and lower triangular and R is in LENF. Moreover, the R in this decomposition is unique: if $LR = \tilde{L}\tilde{R}$, then $R = \tilde{R}$. The uniqueness is a consequence of the uniqueness of the Rook Decomposition and is developed in the Exercise 11.7 below. The proof of existence is much like the proof of the Gauss-Jordan Theorem 84C. By the Fundamental Theorem 76B on Elementary Row Operations, it is enough to show that any matrix can be transformed to LENF by performing lower shears and scales. It is fairly clear that scales and lower shears suffice to transform an arbitrary matrix to LENF. The algorithm which does this is called **Gaussian Elimination**. Compare it with the Gauss-Jordan algorithm (84C). The former is obviously more efficient since it employs no swaps and half the number of shears.

The Gaussian elimination algorithm processes the columns of A consecutively. The following steps process the qth column assume that the first $q - 1$ columns have been processed. Processing the qth column will not change the first $q - 1$ columns.

- First decide if the qth column will contain a leading entry. Starting at the top look for a non zero entry for which all the entries to its left are zero. If no such entry exists, there will be no leading entry in the qth column, and we proceed to the next column. Otherwise, find the highest such leading entry, that is, the one whose row index p is as small as possible.

- Scale so that the (p, q) entry becomes 1.

- Subtract appropriate multiples of the pth row from the lower rows so that all other entries in the qth column below the pth row are zero. Since the entries in the pth row to the left of the qth column are zero, this will not change the first $q - 1$ columns.

These steps are to be repeated until all the columns have been processed.
□

Rook Decomposition

Theorem 340A *Every matrix is lower/upper equivalent to a unique matrix in RNF (rook normal form).*

Proof: In other words, any matrix A may be written in the form

$$A = LDU^{-1}$$

where L is invertible and lower triangular, U is invertible and upper u-nitriangular, and D is in RNF. Moreover, the D in this decomposition is unique: if $LDU^{-1} = \widetilde{L}\widetilde{D}\widetilde{U}^{-1}$, then $D = \widetilde{D}$. The uniqueness is proved in Section 11.5 below; it follows from Lemmas 342C and 343B. The existence of the decomposition $A = LDU^{-1}$ is easy. By the Leading Entry Decomposition 339A, write $A = LR$ where R is in LENF. We need to find a unitriangular U so that RU is in RNF. To do this, go through the rows of R successively (starting at the top) and do right column shears (always subtracting multiples of a column from the columns to its right.) By the time we get to the pth row, the leading entry in that row (if any) is the only nonzero entry in its column: each column operation only changes one entry. When we are done, $D = RU$ is in RNF, so we obtain the desired decomposition $A = LR = LDU^{-1}$.   □

Bruhat Decomposition

Corollary 341A *Every invertible matrix is lower/upper equivalent to a unique permutation matrix.*

Proof: A permutation matrix and an invertible matrix in RNF are the same thing. □

▷ **Exercise 341B** Let

$$A = \begin{bmatrix} 0 & 1 & -4 & 13 & -68 \\ 1 & 3 & -17 & 54 & -276 \\ 2 & -7 & 18 & -61 & 332 \\ 1 & -3 & 7 & -23 & 125 \end{bmatrix}.$$

Find L, D, and U such that L invertible lower triangular, U unitriangular, D is in RNF, and $A = LDU^{-1}$.

Exercise 341C For each of the matrices A of Exercise 88C find L, D, and U as in Theorem 340A. Check your answer by comparing A and LDU^{-1}.

▷ **Exercise 341D** If I know the Rook Decomposition $A = LDU^{-1}$, how do I find the Leading Entry Decomposition $A = LR$?

11.4 The LU Decomposition

The Bruhat Decomposition 341A asserts that any invertible matrix A may be written in the form

$$A = LPU$$

where L is invertible lower triangular, U is upper unitriangular, and P is a permutation matrix. In this section we prove that an invertible matrix can also be written the form $A = LUP$. For most (but not all) matrices the permutation matrix P can be taken to be the identity matrix. For this reason, the latter decomposition is called the **LU Decomposition**.

LU Decomposition

Theorem 342A *Every invertible matrix A may be written in the form*

$$A = LUP$$

where L is invertible lower triangular, U is upper unitriangular, and P is a permutation matrix.

Proof: According to the Fundamental Theorem 76B and the Column Permutation Theorem 101B, the LU Decomposition asserts that any invertible matrix A can be transformed into an upper unitriangular matrix U by performing scales and lower shears (producing L) and swapping columns (producing P). This is fairly obvious: simply do Gaussian elimination (on the rows) as usual, but avoid swaps and upper shears. Usually, this will transform the matrix to an upper unitriangular matrix. If at some point a diagonal entry is zero, swap columns so as to bring a nonzero entry into that diagonal position.   □

11.5   Uniqueness

Corner Rank

Definition 342B For any matrix A the rank $\delta_{pq}(A)$ of the $p \times q$ submatrix in the upper left hand upper of A is called the (p, q)th **corner rank** of A.

Lemma 342C *Lower/upper equivalent matrices have the same corner ranks.*

Proof: Choose $A, B \in \mathbf{F}^{m \times n}$ and integers $p = 1, 2, 3, \ldots, m$ and $q = 1, 2, \ldots, n$. Write A and B in block form

$$A = \left[\begin{array}{cc} A_1 & A_2 \\ A_3 & A_4 \end{array} \right], \qquad B = \left[\begin{array}{cc} B_1 & B_2 \\ B_3 & B_4 \end{array} \right],$$

where $A_1, B_1 \in \mathbf{F}^{p \times q}$. Then the (p,q)th corner ranks are given by

$$\delta_{pq}(A) = \mathrm{rank}(A_1), \qquad \delta_{pq}(B) = \mathrm{rank}(B_1).$$

Now assume that $A = LBU^{-1}$ with L invertible lower triangular and U invertible upper triangular. Decompose L and U in block form:

$$L = \begin{bmatrix} L_1 & 0 \\ L_2 & L_3 \end{bmatrix}, \qquad U = \begin{bmatrix} U_1 & U_2 \\ 0 & U_3 \end{bmatrix},$$

where $L_1 \in \mathbf{F}^{p \times p}$ and $U_1 \in \mathbf{F}^{q \times q}$. The matrices L_1, L_3, U_1 and U_3 are triangular and invertible. Now

$$AU = \begin{bmatrix} U_1 A_1 & * \\ * & * \end{bmatrix}, \qquad LB = \begin{bmatrix} L_1 B1 & * \\ * & * \end{bmatrix}.$$

From $AU = LB$ follows $A_1 U_1 = L_1 B_1$, and hence, $A_1 = L_1 B_1 U_1^{-1}$. In other words, A_1 and B_1 are biequivalent. Hence, by Theorem 179C,

$$\delta_{pq}(A) = \mathrm{rank}(A_1) = \mathrm{rank}(B_1) = \delta_{pq}(B)$$

as required. $\qquad\qquad\qquad\qquad\qquad\qquad\qquad\qquad\qquad\qquad\qquad\qquad$ □

Lemma 343A *For a matrix D in RNF the corner ranks are given by*

$$\delta_{pq}(D) = \sum_{i=1}^{p} \sum_{j=1}^{q} \mathrm{entry}_{ij}(D).$$

Proof: Since the nonzero entries of a matrix in RNF are 1, the formula simply counts the number of nonzero entries in in the upper lefthand $p \times q$ corner. Permuting the rows and columns leaves both the sum of the entries and the rank unchanged. Since every column has at most one nonzero entry, and every row has at most one nonzero entry, the matrix may be transformed to biequivalence form by row and column permutations. This shows that the number of nonzero entries is the rank. $\qquad\qquad\qquad$ □

Corollary 343B *Two matrices in RNF are equal if and only if they have the same invariants δ_{pq}.*

Proof: For a matrix D in RNF, we have that $\mathrm{entry}_{pq}(D) = 1$ if and only if $\delta_{p-1,q}(D) = \delta_{p,q-1}(D) = \delta_{p,q}(D) - 1$. $\qquad\qquad\qquad\qquad\qquad$ □

Lemma 342C and Corollary 343B imply the uniqueness of D in the Rook Decomposition. We have also proved the following corollary which says that the corner ranks δ_{pq} form a complete systems of invariants for lower/upper equivalence.

Corner Ranks and Lower/Upper Equivalence

Corollary 344A *Matrices* $A, B \in \mathbf{F}^{m \times n}$ *are lower/upper equivalent if and only if they have the same corner ranks:*

$$\delta_{pq}(A) = \delta_{pq}(B)$$

for $p = 1, 2, \ldots, m$ and $q = 1, 2, \ldots, n$.

Next, we prove an analog of Corollary 344A for lower equivalence. Let $I_{m,k} \in \mathbf{F}^{m \times k}$ denote the matrix formed from the first k columns of the $m \times m$ identity matrix, and let $\mathcal{W}_k$ denote its range:

$$\mathcal{W}_k = \mathcal{R}\left(I_{m,k}\right), \qquad I_{m,k} = \left[\begin{array}{c} I_k \\ 0_{(m-k) \times k} \end{array} \right].$$

The next lemma gives a geometric interpretation of the group of invertible lower triangular in terms of these subspaces.

Lemma 344B *Suppose that $L \in \mathbf{F}^{m \times m}$ is invertible. Then L is lower triangular if and only if*

$$LW_k = \mathcal{W}_k$$

for $k = 1, 2, \ldots, m$.

Proof: We do the case $m = 2$; the general case is similar. Suppose that $L = \left[\begin{array}{cc} a & b \\ c & d \end{array} \right]$ and $X = I_{2,1} = \left[\begin{array}{c} 1 \\ 0 \end{array} \right]$. Then $\mathcal{W}_1$ consists of all the multiples of X, i.e. all columns whose second entry is zero. But $LX = \left[\begin{array}{c} a \\ c \end{array} \right]$, the first column of L. Hence,

$$LW_1 = \mathcal{W}_1 \iff LX \in \mathcal{W}_1 \iff c = 0,$$

and $c = 0 \iff L$ is lower triangular. $\qquad \qquad \square$

Theorem 180A says that the null space is a complete invariant for left equivalence. This means that two matrices are left equivalent if and only if they have the same null space. The next theorem is an analog for lower equivalence. It says the preimages of the spaces $\mathcal{W}_k$ form a complete system of invariants for lower equivalence. The proof is similar to the proof of Theorem 180A and will be left as Exercise 345E.

Subspaces and Lower Equivalence

Theorem 345A *Suppose that $A, B \in \mathbf{F}^{m \times n}$. Then A and B are lower equivalent if and only if*

$$A^{-1}\mathcal{W}_k = B^{-1}\mathcal{W}_k$$

for $k = 0, 1, 2, \ldots, m$.

▷ **Exercise 345B** Find all the $\delta_{pq}(D)$ if $D = \begin{bmatrix} 0 & 0 & 1 & 0 & 0 \\ 0 & 0 & 0 & 0 & 0 \\ 0 & 0 & 0 & 1 & 0 \\ 1 & 0 & 0 & 0 & 0 \end{bmatrix}$.

▷ **Exercise 345C** A 2×3 matrix A satisfies

$$\delta_{11}(A) = 0, \quad \delta_{12}(A) = \delta_{13}(A) = 1, \quad \delta_{21}(A) = 1, \quad \delta_{22}(A) = \delta_{23}(A) = 2,$$

and is lower/upper equivalent to a matrix D in RNF. Find D.

▷ **Exercise 345D** A matrix $R \in \mathbf{F}^{2 \times 4}$ is in LENF. The space $R^{-1}\mathcal{W}_1$ has a basis

$$\Phi = \begin{bmatrix} c & b & 0 \\ 0 & 1 & 0 \\ a & 0 & 1 \\ 1 & 0 & 0 \end{bmatrix},$$

and deleting the third column gives a basis for the null space $\mathcal{N}(R) = R^{-1}\{0\} = R^{-1}\mathcal{W}_0$. Of course, $R^{-1}\mathcal{W}_2 = R^{-1}\mathbf{F}^{2 \times 1} = \mathbf{F}^{4 \times 1}$. Find R.

Exercise 345E Prove Theorem 345A. Hint: Let $\mathcal{V}_k = A^{-1}\mathcal{W}_k = B^{-1}\mathcal{W}_k$. Then

$$\mathcal{N}(A) = \mathcal{N}(B) = \mathcal{V}_0 \subset \mathcal{V}_1 \subset \cdots \subset \mathcal{V}_m = \mathbf{F}^{n \times 1}.$$

Imitate the proof of Theorem 180A. Find matrices $\Phi_0, \Phi_1, \ldots, \Phi_m$ such that

- Φ_k is a basis for $\mathcal{V}_k$;

- Φ_k is obtained from Φ_{k-1} by adjoining a new column X_k;

- $\begin{bmatrix} AX_1, AX_2, \ldots AX_k \end{bmatrix}$ is a basis for $\mathcal{W}_k$; and

- $\begin{bmatrix} BX_1, BX_2, \ldots BX_k \end{bmatrix}$ is also a basis for $\mathcal{W}_k$.

11.6   Using MINIMAT

Function 346A The .M function rook, shown in in Listing 16, calculates
the Rook Decomposition. The command

```
#> [L, D, U] = rook(A)
```

produces an invertible lower triangular matrix L, a unitriangular U, and a
matrix D in RNF so that the subsequent command

```
#> L*D*U^(-1),   A
```

produces two copies of the same matrix. (Try it!) The two output call

```
#> [L, R] = rook(A)
```

yields the Leading Entry Decomposition.

Exercise 346B Write two .M functions so that the commands

```
#> L=randlow(m), U=randuni(n)
```

return an integer invertible lower triangular matrix L, and an integer unitriangular matrix U.

Exercise 346C Write a .M function randrook so that the command

```
#> R = randrook(m,n,r)
```

produces a random $m \times n$ matrix in RNF of rank r. Remember that such
a matrix has at most one nonzero entry in each row column. Hint: Start
with a matrix in BENF and randomly permute its rows and columns.

▷ **Exercise 346D** Test the rook function using randlow, randrook, and
randuni functions.

Exercise 346E Rewrite the .M function rook (see Listing 16) so that the
RNF D is computed in the loop labeled "loop on columns". Then use the
formula
$$\text{row}_q(R) = \text{row}_q(DU^{-1}) = \text{row}_k(U^{-1})$$
to replace the elementary operations in the loop labeled "loop on rows" by
assignment statements.

Exercise 346F Write a .M function lup so that the command

```
#> [L U P] = lup(A)
```

(with A invertible) produces invertible matrices L and U and a permutation
matrix P with L lower triangular, U unitriangular, and $A = LUP$. Imitate
the proof of the LU Decomposition 342A and use the .M function gjm as
a model.

Rook Decomposition in MINIMAT

```
function [L, D, U] = rook(A)

[m n] = size(A);
R=A; L=eye(m);                    % A = L*R
tol=1.0E-9;   tol2=tol*tol;

for q = 1:n %    loop on columns
    for p=1:m
        c =R(p,q); H=R(p, 1:q-1);
        leading = (H*H'<tol2 & abs(c)>tol);
        if (leading) break; end
    end
    if (leading)
        R(p,:) = R(p,:)/c;        % scale
        L(:,p)=L(:,p)*c;
        for i=p+1:m               % shear
            c = R(i,q);
            R(i,:) = R(i,:) - c*R(p,:);
            L(:,p) = L(:,p) + c*L(:,i);
        end
    end
end

D = R; U=eye(n);                  % R=D*U^(-1)
if nargin == 2, return; end
for p=1:m %  loop on rows
    q=1;
    while (D(p,q) < tol & q<n) q=q+1; end
    for j=q+1:n                   % shear
        c=D(p,j);
        D(:,j) = D(:,j) - c*D(:,q);
        U(:,j) = U(:,j) - c*U(:,q);
    end
end
```

Listing 16

11.7   More Exercises

Back Substitution

▷ **Exercise 348A** Suppose that $Y = AX$ where

$$A = \begin{bmatrix} 5 & a_{12} & a_{13} & a_{14} \\ 0 & 4 & a_{23} & a_{24} \\ 0 & 0 & 3 & a_{34} \\ 0 & 0 & 0 & 2 \end{bmatrix}, \qquad Y = \begin{bmatrix} y_1 \\ y_2 \\ y_3 \\ y_4 \end{bmatrix}.$$

Find X. Hint: Find x_4, then x_3, etc.

▷ **Exercise 348B (Back-Substitution)**   Suppose that $A \in \mathbf{F}^{n \times n}$ is triangular and that its diagonal entries are nonzero. Solve the system $AX = Y$ for X in terms of Y.

▷ **Exercise 348C** Suppose that

$$A = \begin{bmatrix} 5 & a_{12} & a_{13} & a_{14} \\ 0 & 4 & a_{23} & a_{24} \\ 0 & 0 & 0 & a_{34} \\ 0 & 0 & 0 & 2 \end{bmatrix}.$$

Find a nonzero X with $AX = 0$.

▷ **Exercise 348D** Generate a 4×4 matrix with the commands

```
#> A=eye(4)+randsutr(4),
#> A(3,3)=0
```

Find a nonzero 4×1 matrix X with A*X=0.

▷ **Exercise 348E** Assume $A \in \mathbf{F}^{n \times n}$ is triangular but that some diagonal entry, $\text{entry}_{pp}(A) = 0$. Find $X \in \mathbf{F}^{n \times 1}$ with $AX = 0$ but $X \neq 0$.

▷ **Exercise 348F** Use Exercises 348B and 348E to give a proof of the Triangular Invertibility Theorem 49B.

Factorization Theorems

Definition 348G A set $\mathcal{E}$ of invertible matrices is said to **generate** a group $\mathcal{G}$ of invertible matrices iff (1) $\mathcal{E} \subset \mathcal{G}$, and (2) every element of $\mathcal{G}$ is the product of a finite number of elements of $\mathcal{E}$.

Example 348H The Factorization Theorem 93B asserts that the set of elementary matrices (of a given size) generates the group of all invertible matrices of that size.

Example 349A Theorem 334B asserts that the upper shears generate the group of all unitriangular matrices and that the upper shears together with the scales generate the group of invertible triangular matrices.

Example 349B Theorem 365C asserts that the reflection matrices generate the real unitary matrices, and that the reflection matrices and the unitary scales generate the group of complex unitary matrices.

Exercise 349C Show that the swaps generate the group of permutation matrices.

Exercise 349D Show that the swaps and scales generate the group of all matrices having exactly one nonzero entry in each row and exactly one nonzero entry in each column.

Exercise 349E Show that the shears and scales generate the group of invertible matrices.

Exercise 349F Show that the shears generate the group of invertible matrices of determinant one.

Exercise 349G Show that the shears and swaps generate the group of invertible matrices of determinant ± 1.

2×2 **LU and Bruhat**

It is tempting to look for a relation between the LU Decomposition and the Bruhat Decomposition, perhaps a trick to prove one from the other. This is misleading. The permutation matrix P in the Bruhat Decomposition is unique (although L and U might not be), whereas in the LU decomposition none of the matrices is unique. In fact, most square matrices can be written in the form $A = LU$ (P is the identity matrix). This means that if P is prescribed, most matrices A can also be written in the form $A = LUP$ (with a different L and U): we simply write $AP^{-1} = LU$. The following two exercises should help clarify the situation.

Exercise 349H (2×2 **Bruhat Decomposition**) Denote by

$$I = \begin{bmatrix} 1 & 0 \\ 0 & 1 \end{bmatrix}, \quad W = \begin{bmatrix} 0 & 1 \\ 1 & 0 \end{bmatrix},$$

the two 2×2 permutation matrices, and consider matrices

$$A = \begin{bmatrix} a & b \\ c & d \end{bmatrix}, \quad L = \begin{bmatrix} x & 0 \\ y & z \end{bmatrix}, \quad U = \begin{bmatrix} 1 & u \\ 0 & 1 \end{bmatrix}.$$

Assume A is invertible $(ad - bc \neq 0)$ and consider the two nonlinear systems $A = LPU$ with $P = I, W$. Each of these two systems is to be considered as a system of four equations (one for each entry of A) in four unknowns (namely, x, y, z, u). Show that

(1) $A = LIU$ is solvable iff $a \neq 0$, and

(2) $A = LWU$ is solvable iff $a = 0$,

Conclude that exactly one of the two systems $A = LIU$ and $A = LWU$ has a solution. Show that the solution of $A = LIU$ is unique (when it exists), whereas the solution of $A = LWU$ is not.

Exercise 350A $(2 \times 2$ **LU Decomposition**) Continue the notation of Exercise 349H. This time we consider the two systems $A = LUP$ with $P = I, W$. Show that

(3) $A = LUI$ is solvable iff $a \neq 0$, and

(4) $A = LUW$ is solvable iff $b \neq 0$.

Conclude that if both a and b are nonzero (as will be the case for a randomly chosen matrix), then the matrix A may be written in both forms $A = LUI$ and $A = LUW$. (Of course, the L and the U will be different in the two decompositions.)

Related Decompositions

Exercise 350B Define the $m \times m$ **reversal matrix** $W \in \mathbf{F}^{m \times m}$ by

$$\text{entry}_{ij}(W) \quad = 1 \quad \text{if } i + j = m + 1,$$
$$= 0 \quad \text{otherwise.}$$

For example, when $m = 3$,

$$W = \begin{bmatrix} 0 & 0 & 1 \\ 0 & 1 & 0 \\ 1 & 0 & 0 \end{bmatrix}.$$

Show that

(1) W is a permutation matrix,

(2) $W = W^{-1}$, and

(3) WLW is upper triangular, if L is lower triangular.

▷ **Exercise 351A (Alternate Bruhat Decomposition)** An invertible matrix A can be written in the form

$$A = VQU^{-1}$$

where V is invertible upper triangular, U is upper unitriangular, and Q is a permutation matrix.

Exercise 351B (Alternate LU Decomposition) An invertible matrix A may be written in the form

$$A = PLU$$

where L^{T} is unitriangular, U^{T} is invertible lower triangular, and P is a permutation matrix.

Exercise 351C (Another LU Decomposition) An invertible matrix A may be written in the form
$$A = PLU$$

where L is invertible lower triangular, U is upper unitriangular, and P is a permutation matrix. (Hint: Write a triangular matrix as the product of a unitriangular matrix and a diagonal matrix.)

▷ **Exercise 351D (Alternate Rook Decomposition)** Any matrix A can be written in the form

$$A = VDU^{-1}$$

where V is invertible upper triangular, U is upper unitriangular, and D is in RNF.

Exercise 351E For each of the decompositions of this Section 11.7 write a .M function which implements it.

Uniqueness of the LENF

The following five exercises prove the uniqueness of the LENF R in the Leading Entry Decomposition 339A. Assume that

$$LR = \widetilde{L}\widetilde{R}$$

where L and $\widetilde{L}$ are invertible lower triangular, and R and $\widetilde{R}$ are in LENF. The objective is to prove that $R = \widetilde{R}$.

▷ **Exercise 352A** Show that the leading entries of R occur in the same positions as those of $\widetilde{R}$, and that the positions of the rows of R which vanish identically are the same as those of $\widetilde{R}$.

▷ **Exercise 352B** Assume that the kth row of R (and hence of $\widetilde{R}$) vanishes. Form M and $\widetilde{M}$ from L and $\widetilde{L}$ be deleting the kth row and the kth column. Let S and $\widetilde{S}$ result from R and $\widetilde{R}$ by deleting the kth row. Show that M and $\widetilde{M}$ are also triangular and $MS = \widetilde{M}\widetilde{S}$.

▷ **Exercise 352C** Assume that $LU = \widetilde{L}\widetilde{U}$ where L and $\widetilde{L}$ are invertible lower triangular and U and $\widetilde{U}$ are unitriangular. Show that $L = \widetilde{L}$ and $U = \widetilde{U}$.

▷ **Exercise 352D** Show that $R = \widetilde{R}$.

Exercise 352E The proof outlined in Exercises 352A through 352D is computational. Find a conceptual proof that $R = \widetilde{R}$ along the lines of the proof of the uniqueness of the RREF given in Theorem 182D. Use the spaces $A^{-1}\mathcal{W}_k$ introduced in Exercise 345E.

Gershgorin's Theorem

A matrix A is called **diagonally dominant** iff the absolute value of each diagonal entry is greater than the sum of the absolute values of the other entries in its row; i.e.

$$|a_{kk}| > \sum_{j \neq k} |a_{kj}|$$

where $a_{kj} = \operatorname{entry}_{kj}(A)$.

Exercise 352F Assume that

$$1 > |a| + |b|, \qquad 1 > |c| + |d|.$$

Show that

$$|a - ca| > |d - cb|.$$

Hint: The absolute value satisfies $|x - y| \geq |x| - |y|$ and $|x| + |y| \geq |x - y|$.

Exercise 352G Show that if the rows of a diagonally dominant matrix are rescaled by a positive number, then the result is diagonally dominant.

Exercise 352H Show that if a diagonally dominant matrix A is transformed to LENF as in the proof of Theorem 339A, then each intermediate result is diagonally dominant. Hence, $A = LU$ where L is invertible lower triangular and U is unitriangular. In particular, A is invertible.

Remark 353A Here's an easier proof that a diagonally dominant matrix is invertible. If A is not invertible, then $\Pi AX = 0$ for some $X \neq 0$. Suppose that $x_i = \text{entry}_i(X)$ and that the kth entry of X is the largest in absolute value. Rescale so that it is one:

$$1 = x_k = \max_i |x_i|.$$

From $\text{entry}_k(AX) = 0$ conclude that

$$|a_{kk}| = \left|\sum_{i \neq k} a_{ki}x_i\right| \leq \sum_{i \neq k} |a_{ki}|\,|x_i| \leq \sum_{i \neq k} |a_{ki}|$$

contradicting diagonal dominance.

Exercise 353B (Gershgorin's Theorem) The kth **Gershgorin circle** of a square matrix $A \in \mathbf{F}^{n \times n}$ is the set of all $\lambda \in \mathbf{C}$ which satisfy the inequality

$$|\lambda - a_{kk}| \leq \sum_{j \neq k} |a_{kj}|.$$

Show that every eigenvalue of A lies in some Gershgorin circle. Hint: Prove the contrapositive.

Exercise 353C Let $A = D + W$ where D is diagonal and all the diagonal terms of W vanish. Let $A(t) = D + tW$. Show that if A is diagonally dominant, then so is $A(t)$ for $0 \leq t \leq 1$.

Remark 353D It can be shown that the eigenvalues of $A(t)$ vary continuously with t. As t decreases from 1 to 0, the Gershgorin circles of $A(t)$ shrink. It follows that each union of m Gershgorin circles contains m eigenvalues. This is called **Gershgorin's Second Theorem**.

Real Triangular Equivalence

Exercise 353E Recall the two Definitions 337A of triangular equivalence. As in Section 7.3 these appear to depend on $\mathbf{F}$; strictly speaking, we should define **triangular equivalence over F**. Show that *if two real matrices $A, B \in \mathbf{R}^{m \times n}$ are triangularly equivalent over $\mathbf{C}$, then they are triangularly equivalent over $\mathbf{R}$.* This works for both lower equivalence and lower/upper equivalence. Hint: See Exercise 251B.

Chapter Summary

Factorization Theorem

A unitriangular matrix U can be written as a product of upper shears. An invertible lower triangular matrix L can be written as a product of lower shears and scales.

Leading Entry Decomposition

Every matrix A is lower equivalent to a unique matrix R in LENF (leading entry normal form):

$$A = LR.$$

Rook Decomposition

Every matrix A is lower/upper equivalent to a unique matrix D in RNF (rook normal form):

$$A = LDU^{-1}$$

Bruhat Decomposition

Every invertible matrix A is lower/upper equivalent to a unique permutation matrix P:

$$A = LPU^{-1}$$

LU Decomposition

Every invertible matrix A may be factored as $A = LUP$ where P is a permutation matrix. (Here P is not unique.)

Characterization

Two matrices of the same size are lower/upper equivalent if and only if they have the same corner ranks.

12

UNITARY MATRICES

The goal in this chapter is to redo the theory of equivalence, replacing the the word *invertible* by the word *unitary*. We will find that the theorems in Chapter 3 have analogs here. For example, any matrix is left equivalent to a matrix in reduced row echelon form, and any matrix is unitarily left equivalent to a matrix in positive row echelon form.

Recall that in the theory of Gauss-Jordan elimination an important role was played by the elementary matrices. These elementary matrices were of three types – scale, swap, and shear – and are manifestly invertible. Moreover, they are the "building blocks" of invertible matrices in the sense that every invertible matrix is a product of elementary matrices. In this chapter, the analogous role is played by the reflection matrices defined below. They are the building blocks of unitary matrices in the sense that every unitary matrix is a product of reflections. (At least when $\mathbf{F} = \mathbf{R}$; for $\mathbf{F} = \mathbf{C}$ one must throw in unitary scales.)

Throughout this chapter, except where the contrary is indicated, $\mathbf{F}$ may be either the real numbers $\mathbf{R}$ or the complex numbers $\mathbf{C}$.

12.1   Reflections

The reflection matrices play a role analogous to the role played by the elementary matrices in Chapter 3. They are the simplest of all unitary matrices (except for the identity matrix) and are the building blocks from which unitary matrices are constructed. Here's the definition:

Definition 356A Let $N \in \mathbf{F}^{n \times 1}$ be a column with norm one:

$$\|N\|^2 = 1.$$

The **reflection matrix** determined by N is the matrix $Q_N \in \mathbf{F}^{n \times n}$ defined by

$$Q_N = I - 2NN^*.$$

Example 356B If $N \in \mathbf{R}^{2 \times 1}$ is given by

$$N = \begin{bmatrix} 0 \\ 1 \end{bmatrix},$$

then the reflection matrix Q_N determined by N is given by

$$Q_N = \begin{bmatrix} 1 & 0 \\ 0 & -1 \end{bmatrix}.$$

Represent $X \in \mathbf{R}^{2 \times 1}$ graphically with the first coordinate of X plotted on the horizontal axis and the second coordinate plotted on the vertical axis. Then $Q_N X$ is the reflection of X in the horizontal axis. This is shown in Figure 9. The horizontal axis is labeled $N^\perp$ as it represents the vectors that are perpendicular to the vector N.

Theorem 356C lists some of the properties of reflections. Formula (1) says the left multiplication by Q_N is the reflection with axis N. This means that it reverses N and fixes any X that is perpendicular to N as expressed by formulas (2) and (3). The remaining formulas say that the matrix Q_N is (4) Hermitian, (5) a square root of the identity matrix, and (6) unitary.

Theorem 356C (Reflection Laws)   *Let $N \in \mathbf{F}^{n \times 1}$ have unit norm and $Q_N = I - 2NN^* \in \mathbf{F}^{n \times n}$ be the reflection determined by N. Then*

(1) $Q_N X = X - 2 \langle X, N \rangle N$ *for* $X \in \mathbf{F}^{n \times 1}$.

(2) $Q_N N = -N$.

(3) $Q_N X = X$ *if* $\langle N, X \rangle = 0$.

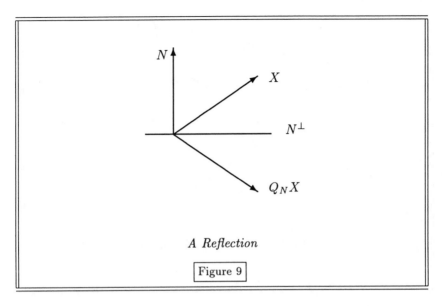

A Reflection

Figure 9

(4) $Q_N^* = Q_N$.

(5) $Q_N^2 = I$.

(6) $Q_N^* = Q_N^{-1}$.

(7) *If the vector N is real, then so is the matrix Q_N.*

Proof: Exercise 359D.

Theorem 357A (Reflection Theorem)   *Assume that $Y, Z \in \mathbf{F}^{n \times 1}$ are distinct and have equal norm:*

$$\|Y\| = \|Z\|$$

and let N be normalized and parallel to $Y - Z$:

$$N = \|Y - Z\|^{-1}(Y - Z).$$

In case $\mathbf{F} = \mathbf{C}$, make the additional hypothesis that $\langle Y, Z \rangle$ is real. Then the reflection Q_N determined by N satisfies

$$Q_N Y = Z.$$

Proof: (See Figure 10.) First, we assume that $\|Y - Z\| = 1$ so that

$$N = Y - Z.$$

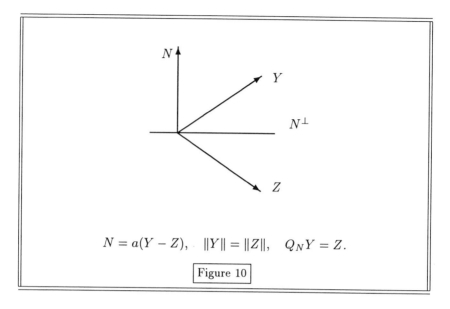

$$N = a(Y - Z), \quad \|Y\| = \|Z\|, \quad Q_N Y = Z.$$

Figure 10

As N has unit norm we obtain

$$1 = \|N\|^2 = \|Y\|^2 - 2\langle Y, Z \rangle + \|Z\|^2,$$

so as $\|Y\| = \|Z\|$ we get

$$2\|Y\|^2 - \langle Z, Y \rangle = 1.$$

Then

$$
\begin{aligned}
QY &= Y - 2\langle N, Y \rangle N \\
&= Y - 2\langle Y - Z, Y \rangle (Y - Z) \\
&= Y - 2(\|Y\|^2 - \langle Z, Y \rangle)(Y - Z) \\
&= Y - (Y - Z) \\
&= Z
\end{aligned}
$$

as required.

Now we eliminate the assumption that $\|Y - Z\| = 1$. Let $a = \|Y - Z\|^{-1}$ so that $N = a(Y - Z)$. Then $N = Y_0 - Z_0$ where $Y_0 = aY$ and $Z_0 = aZ$. By the first part of the proof, $QY_0 = Z_0$. Multiply this equation by a to obtain $QY = Z$.   □

Exercise 359A Find Q as in Theorem 357A for each of the following pairs Y, Z. In each case verify that Q satisfies properties (1)-(7) of Theorem 356C.

(1) $Y = \begin{bmatrix} 1 \\ 1 \end{bmatrix}$, $Z = \begin{bmatrix} 1 \\ -1 \end{bmatrix}$. 　(2) $Y = \begin{bmatrix} 1 \\ 2 \end{bmatrix}$, $Z = \begin{bmatrix} 2 \\ 1 \end{bmatrix}$.

(3) $Y = \begin{bmatrix} 1 \\ 2 \\ 3 \end{bmatrix}$, $Z = \begin{bmatrix} 2 \\ 1 \\ 3 \end{bmatrix}$. 　(4) $Y = \begin{bmatrix} 1 \\ 2 \\ 3 \end{bmatrix}$, $Z = \begin{bmatrix} 3 \\ 1 \\ 2 \end{bmatrix}$.

Exercise 359B Show that a swap is a reflection.

Exercise 359C Show that if $\|Y\| = \|Z\|$, then there is a unitary matrix Q with $QY = Z$.

▷ **Exercise 359D** Prove Theorem 356C.

▷ **Exercise 359E** Find a formula for the reflection Q_N determined by the vector $N \in \mathbf{R}^{2\times1}$ defined by

$$N = \begin{bmatrix} a \\ b \end{bmatrix}, \qquad a^2 + b^2 = 1.$$

Exercise 359F Find a formula for the reflection Q_N determined by the vector $N \in \mathbf{R}^{3\times1}$ defined by

$$N = \begin{bmatrix} a \\ b \\ c \end{bmatrix}, \qquad a^2 + b^2 + c^2 = 1.$$

Exercise 359G Suppose that $\|Y\| = 1$ and that Z is the first column of the identity matrix. Suppose also that $\langle Y, Z \rangle > 0$. Show that the reflection matrix which interchanges Y and Z has from

$$Q_N = \begin{bmatrix} y_1 & Y_1^* \\ Y_1 & I - cY_1Y_1^* \end{bmatrix}, \qquad Y = \begin{bmatrix} y_1 \\ Y_1 \end{bmatrix}$$

where $c = (1 + x_1)^{-1}$.

Householder Reflection in MINIMAT

```
function Q = flip(Y,Z);

    [n, toss] =  size(Y);
    N = Y-Z; c = sqrt(N'*N);
    if c > 1.0E-9
          N = N/c;  % normalize
          Q = eye(n) - 2*N*N';
    else
          Q = eye(n);
    end
```

Listing 17

Using MINIMAT

▷ **Exercise 360A** Generate a random column of unit norm and then construct the reflection matrix $Q \in \mathbf{R}^{4\times4}$ determined by N. Confirm that $QN = -N$ and that $QX = X$ for a random solution X of the single equation $NX = 0$. Confirm that $Q^\mathsf{T} = Q^{-1} = Q$.

▷ **Exercise 360B** Generate two random column matrices Y and Z of the same norm. Following the proof of Theorem 357A, construct a 3×3 reflection matrix such that $QY = Z$. Confirm that $Q^\mathsf{T}Q = QQ^\mathsf{T} = Q^2 = I$ and that $QY = Z$ and $QZ = Y$.

Function 360C The .M function flip, shown in Listing 17, calculates a reflection that interchanges its inputs *provided they have the same norm.* The command

```
#> Q = flip(Y,Z)
```

produces a reflection Q so that multiplication by Q transforms Z to Y and Y to Z.

▷ **Exercise 360D** Test flip.

12.2   Unitary Equivalence

Now we define the key equivalence relations studied in this chapter. If the word *unitarily* is dropped and the word *unitary* is replaced by the word *invertible*, we recover the relation studied in Section 3.7.

Unitary Equivalence

Definition 361A Let A and B be two matrices of the same size, say $A, B \in \mathbf{F}^{m \times n}$. We say that

- A is **unitarily left equivalent** to B iff there is a unitary matrix Q with
$$A = QB.$$

- A is **unitarily right equivalent** to B iff there is a unitary matrix P with
$$A = BP^{-1}.$$

- A is **unitarily biequivalent** to B iff there are unitary matrices Q and P with

$$A = QBP^{-1}.$$

The proof of Theorem 104C (see also Theorem 337B) goes through word for word to prove the following

Theorem 361B *All three of these relations satisfy the following laws:*

(Reflexive Law) $A \equiv A;$

(Symmetric Law) *if $A \equiv B$, then $B \equiv A;$*

(Transitive Law) *if $A \equiv B$ and $B \equiv C$, then $A \equiv C$.*

Here $A \equiv B$ means either "A is unitarily left equivalent to B", "A is unitarily right equivalent to B", or "A is unitarily biequivalent to B".

12.3   Householder Decomposition

The Gauss-Jordan Decomposition 105A says that every matrix is left e-quivalent to a unique matrix in reduced row echelon form. Our aim is to find an analog of this theorem for unitary left equivalence.

Positive Row Echelon Form (PREF)

Definition 362A An $m \times n$ matrix R is in **positive row echelon form**, abbreviated **PREF**, iff

- all the rows that vanish identically (if any) appear below the other (nonzero) rows;

- the leading entry in any row appears to the left of the leading entry of any nonzero row below;

- all the leading entries are positive.

The positive row echelon form differs from reduced row echelon form (Definition 82A) in that it is not required that the entries above a leading entry vanish. It is also not required that the leading entries be 1, but only that they be positive. The following Householder Decomposition Theorem[1] should be compared with the Gauss-Jordan Decomposition 105A.

Householder Decomposition

Theorem 362B *Every matrix is unitarily left equivalent to a unique matrix in PREF (positive row echelon form).*

Proof: In other words, any matrix $A \in \mathbf{F}^{m \times n}$ can be written in the form

$$A = QR$$

---

[1] Also called the **QR-decomposition**

where $Q \in \mathbf{F}^{m \times m}$ is unitary and $R \in \mathbf{F}^{m \times n}$ is in PREF. Moreover, in this decomposition the matrix R is unique: if $\widetilde{Q}\widetilde{R} = QR$, then $\widetilde{R} = R$. The construction is reminiscent of the transformation of a general matrix to RREF by elementary row operations: reflection matrices (and unitary scales when $\mathbf{F} = \mathbf{C}$) play the role of elementary matrices.

Our proof is a description of the algorithm that describes how to choose the reflections to transform A to PREF. Before describing the algorithm let us reformulate Theorem 357A in the form in which we will use it. Suppose that

$$
X = \begin{bmatrix} x_1 \\ \vdots \\ x_{r-1} \\ 0 \\ 0 \\ \vdots \\ 0 \end{bmatrix}, \quad Y = \begin{bmatrix} y_1 \\ \vdots \\ y_{r-1} \\ y_r \\ y_{r+1} \\ \vdots \\ y_m \end{bmatrix}, \quad Z = \begin{bmatrix} y_1 \\ \vdots \\ y_{r-1} \\ z_r \\ 0 \\ \vdots \\ 0 \end{bmatrix}, \quad (*)
$$

where $y_r, z_r > 0$ are positive and $\|Y\| = \|Z\|$. Theorem 357A constructs a reflection matrix Q_N such that $Q_N Y = Z$ and $Q_N Z = Y$. It has the further property that $Q_N X = X$ for every X as in $(*)$, because such an X is orthogonal to $N = \|Y - Z\|^{-1}(Y - Z)$.

The algorithm for transforming A to PREF processes the columns one at a time, in order from left to right. When the kth column is processed, A is replaced by a new, unitarily left equivalent, matrix. This new matrix agrees with the old matrix in the first $k - 1$ columns. The algorithm maintains a count r of the number of leading columns found so far. When the algorithm begins, no leading columns have been found ($r = 0$). The following steps process the kth column.

- First, decide if the kth column is free or leading. If all the entries in the positions at or below the $(r + 1, k)$ entry are zero, then the kth column is free, so we proceed to the next column. In the contrary case, the kth column is leading so we increase r.

- At this point the (k, r) entry u is nonzero. Multiply the rth row by $|u|^{-1}\bar{u}$, so that the (k, r) entry becomes positive. (This replaces A by EA where E is a unitary scale.)

- Denote the kth column by Y as in equation $(*)$ above. Construct a column Z, as in $(*)$, by choosing $z_r > 0$ so that $\|Z\| = \|Y\|$.

- Replace A by $Q_N A$. This leaves the first $k-1$ columns of A unchanged and replaces the kth column Y by Z.

These steps are to be repeated until the number r of leading columns is the number m of rows or until all the columns have been processed.

The proof of uniqueness of the Householder decomposition is very much like the proof of the uniqueness of the Gauss-Jordan decomposition (182D). We omit the details. Let $A \in \mathbf{F}^{m \times n}$ be given, and let $A_j = \mathrm{col}_j(A)$ denote the jth column of A:

$$A = \begin{bmatrix} A_1 & A_2 & \cdots & A_n \end{bmatrix}.$$

Let $1 \le j_1 < j_2 < \cdots < j_r \le n$ be the leading column indices where the rank jumps. Apply the Gram-Schmidt process to the basis for $\mathcal{R}(A)$ whose columns are $A_{j_i} = \mathrm{col}_{j_i}(A)$ to obtain an orthonormal basis Ψ for $\mathcal{R}(A)$. If $R_j = \mathrm{col}(R)$ is the jth column of the PREF, then $A_j = \Psi R_j$. This argument uses the uniqueness of the Gram-Schmidt decomposition as explained in Theorem 206C. If Ψ is extended to a unitary matrix Q by adjoining columns, then $Q^{-1}A$ is the PREF, so the uniqueness proof reproves the existence of the Householder decomposition.   □

Exercise 364A Expand the sketch of the uniqueness proof given in the text to a detailed argument.

▷ **Exercise 364B** A matrix is said to be in **row echelon form** iff it is in positive row echelon form and all the leading entries are 1. Define REF equivalence so that the following becomes a theorem: *Every matrix is REF equivalent to a unique matrix in row echelon form.*

Using MINIMAT

▷ **Exercise 364C** Use MINIMAT to calculate the Householder decomposition of a random 3×5 matrix by imitating the proof in the text.

▷ **Exercise 364D** Repeat the last exercise with a random 4×5 matrix.

▷ **Exercise 364E** Repeat the last exercise with a random 5×3 matrix.

The Householder Decomposition 362B is implemented as a built-in function qr in MINIMAT. The command

```
#>   [Q R]=qr(A)
```

will assign to the variable Q and R the factors in the Householder decomposition. The one-output form

```
#>   Q=qr(A)
```

assigns to Q the unitary part. Hence, the command

```
#>  Q=qr(rand(m,n))
```

can be used to create a random $m \times m$ unitary matrix.

▷ **Exercise 365A** Try out qr. Check the result.

Function 365B The .M function hh, shown in Listing 18, computes the Householder decomposition. The command

```
#>  [Q R]=hh(A)
```

produces a unitary matrix Q and a matrix R in PREF so that the subsequent command

```
#>  Q*R,   A
```

produces two copies of the same matrix. Compare its performance with MINIMAT's built-in qr function.

12.4   Unitary Factorization

Reflections are to unitary matrices as elementary matrices are to invertible matrices. In particular, the following theorem should be compared with the Factorization Theorem 93B.

Unitary Factorization

Corollary 365C   *A real unitary matrix $P \in \mathbf{R}^{n \times n}$ is the product of reflection matrices. A complex unitary matrix $P \in \mathbf{C}^{n \times n}$ is a product of reflection matrices and unitary scales.*

Proof: By the Householder Decomposition, $P = QR$ with Q unitary and R in PREF. The proof of 362B actually showed that the matrix Q is a product of the appropriate form, since at each stage in the construction A was altered by multiplication either by a reflection matrix Q_N or by a unitary scale. (In the real case a unitary scale is a reflection.) Hence it is enough to show that $P = Q$, that is, that R is the identity.

Householder Decomposition in MINIMAT

```
function [Q, R] = hh(A)

    [m n] = size(A); r=0;
    Q=eye(m); R=A;
    for k=1:n
        if (r==m) return; end
        Y = R(:,k); Z=Y;
        c=sqrt(Y(r+1:m)'*Y(r+1:m));
        if c>1.0E-9
            r=r+1;
            Z(r)=c;
            Z(r+1:m) = zeros(m-r,1);
            P = flip(Y,Z);
            Q = Q*P'; R=P*R;
        end
    end
```

Listing 18

By the laws 198E for unitary multiplication, it follows that R is unitary. But its columns are then orthonormal. It follows (since R is triangular) that R must be diagonal, and hence, (since the diagonal entries are positive and the columns have norm one) that R is the identity matrix.   □

▷ **Question 366A** How many $n \times n$ diagonal real unitary matrices are there? What are the diagonal unitary matrices?

Using MINIMAT

▷ **Exercise 366B** Use the command

```
#> Q = qr(rand(3,3))
```

to generate a random 3×3 unitary matrix. Then write it as a product of reflection matrices.

12.5 Singular Values

Definition 367A Suppose $A \in \mathbf{F}^{m \times n}$ and $m \geq n$. A number σ is called **singular value** for a matrix A iff $\sigma \geq 0$, and there is a nonzero vector $X \in \mathbf{F}^{n \times 1}$ satisfying the condition

$$\langle AX, AY \rangle = \sigma^2 \langle X, Y \rangle$$

for all $Y \in \mathbf{F}^{n \times 1}$. Any X satisfying this condition is called a **singular vector** of A corresponding to the singular value σ. If $m < n$, the singular values of A are, by definition, the same as the singular values of A^*. (Exercise 368B explains the reason for this convention.)

Lemma 367B *The matrix A^*A is Hermitian and its eigenvalues are nonnegative.*

Proof: $(A^*A)^* = A^* (A^*)^* = A^*A$. If $A^*AX = \lambda X$ and $X \neq 0$, then

$$\lambda \langle X, X \rangle = \langle \lambda X, X \rangle = \langle A^*AX, X \rangle = \langle AX, AX \rangle \geq 0,$$

so $\lambda \geq 0$ since $\langle X, X \rangle > 0$. $\square$

Theorem 367C *Suppose that $A \in \mathbf{F}^{m \times n}$ with $m \geq n$. Then the singular values of A are precisely the nonnegative square roots of the eigenvalues of A^*A. The singular vectors of A corresponding to the singular value σ are the eigenvectors of A^*A corresponding to the eigenvalue σ^2.*

Proof: Assume that σ^2 is an eigenvalue of A^*A, and let X be a corresponding nonzero eigenvector. Then

$$A^*AX = \sigma^2 X,$$

so for any $Y \in \mathbf{F}^{n \times 1}$ we have

$$\langle AX, AY \rangle = \langle A^*A, X \rangle = \langle \sigma^2 X, Y \rangle = \sigma^2 \langle X, Y \rangle.$$

Conversely, assume
$$\langle AX, AY \rangle = \sigma^2 \langle X, Y \rangle$$

for all Y. Then $\langle (A^*AX - \sigma^2 X), Y \rangle = 0$ for all Y, which implies that the left factor is zero:
$$A^*AX = \sigma^2 X.$$

(To justify this step, you could take $Y = A^*AX - \sigma^2 X$.) Thus X is an eigenvector for the matrix A^*A with eigenvalue σ^2. $\square$

Corollary 368A *Any nonzero matrix has a nonzero singular value.*

Proof: The matrix A^*A is Hermitian, and hence, diagonalizable.   □

Exercise 368B Suppose that $A \in \mathbf{F}^{m \times n}$ with $m < n$. Show that

- 0 is an eigenvalue of A^*A;

- If A has maximal rank, then 0 is not a singular value of A;

- AA^* and A^*A have the same nonzero eigenvalues.

Exercise 368C Show that a matrix has maximal rank if and only if all its singular values are nonzero.

12.6   Singular Value Decomposition

Recall that two matrices $A, B \in \mathbf{F}^{m \times n}$ are said to be **unitarily biequivalent** iff there are unitary matrices $Q \in \mathbf{F}^{m \times n}$ and $P \in \mathbf{F}^{m \times n}$ with

$$A = QBP^{-1}.$$

The Biequivalence Decomposition Theorem 106D asserts that every matrix is biequivalent to a matrix in biequivalence normal form. The Singular Value Decomposition Theorem proved in this section is an analog where biequivalence is replaced by unitary biequivalence.

Lemma 368D *If the matrices A and B are unitarily biequivalent, then the corresponding Hermitian matrices A^*A and B^*B are unitarily similar.*

Proof: Suppose $A = QBP^{-1}$ with Q and P unitary:

$$Q^{-1} = Q^*, \qquad P^{-1} = P^*.$$

Then

$$
\begin{aligned}
A^*A &= (QBP^{-1})^*(QBP^{-1}) \\
&= (QBP^*)^*(QBP^{-1}) \\
&= (P^{**}B^*Q^*)QBP^{-1} \\
&= PB^*(Q^{-1}Q)BP^{-1} \\
&= P(B^*B)P^{-1}.
\end{aligned}
$$

□

Invariance of Singular Values

Corollary 369A *Unitarily biequivalent matrices have the same singular values.*

Singular Value Normal Form (SVNF)

Definition 369B A matrix $R \in \mathbf{F}^{m \times n}$ is said to be in **singular value normal form**, abbreviated **SVNF**, iff it has the form

$$D = \begin{bmatrix} \Delta & 0_{r \times (n-r)} \\ 0_{(m-r) \times r} & 0_{(m-r) \times (n-r)} \end{bmatrix}$$

where $\Delta \in \mathbf{F}^{r \times r}$ is a diagonal matrix with positive (real) entries on the diagonal arranged in descending order:

$$\Delta = \mathrm{diag}(\sigma_1, \sigma_2, \ldots, \sigma_r)$$

where $\sigma_1 \geq \sigma_2 \geq \cdots \geq \sigma_r > 0$.

Theorem 369C *The singular values of a matrix D in SVNF are its diagonal entries, that is, the diagonal entries of Δ, together with zero in case D is not of maximal rank (i.e. $r < n$ and $r < m$).*

Proof: Since Δ is diagonal with positive entries on the diagonal, we have that

$$\Delta^* \Delta = \Delta^2$$

is also diagonal, and hence,

$$D^* D = \begin{bmatrix} \Delta^2 & 0_{r \times (n-r)} \\ 0_{(n-r) \times r} & 0_{(n-r) \times (n-r)} \end{bmatrix}$$

is diagonal. Thus the eigenvalues of $D^* D$ are the diagonal entries of Δ^2, together with zero in case $r < n$. $\quad\square$

Lemma 370A *If D is in SVNF, so is D^*.*

Proof: For example,

$$D = \begin{bmatrix} \sigma_1 & 0 & 0 \\ 0 & \sigma_2 & 0 \end{bmatrix}, \qquad D^* = \begin{bmatrix} \sigma_1 & 0 \\ 0 & \sigma_2 \\ 0 & 0 \end{bmatrix}.$$

Lemma 370B *If A and B are unitarily biequivalent, so are A^* and B^**

Proof: $(QAP^*)^* = PA^*Q^*.$ □

Singular Value Decomposition

Theorem 370C *Any matrix is unitarily biequivalent to a unique matrix in SVNF (singular value normal form).*

Proof: In other words, any $A \in \mathbf{F}^{m \times n}$ may be written in the form

$$A = QDP^{-1}$$

where $Q \in \mathbf{F}^{m \times m}$ and $P \in \mathbf{F}^{n \times n}$ are unitary and $D \in \mathbf{R}^{m \times n}$ is in SVNF. (The theorem asserts that Q and P are real when A is real.) Moreover, the D in this decomposition is unique: if $QDP^{-1} = \widetilde{Q}\widetilde{D}\widetilde{P}^{-1}$, then $D = \widetilde{D}$.

The uniqueness is plausible. There is exactly one $m \times n$ matrix D in SVNF with prescribed nonzero entries $\sigma_1 > \sigma_2 > \cdots > \sigma_r > 0$. We have already proved that these are the nonzero singular values of D and hence of any matrix A unitarily biequivalent to D. Hence, two unitarily biequivalent matrices which are in SVNF (and have distinct singular values) must be equal. An additional argument, involving the idea of multiplicity, is required in case some of the σ's are repeated. See Remark 374A below.

Now we show existence. The matrix A^*A is Hermitian. By the Spectral Theorem 317C, there is a unitary matrix P such that P^*A^*AP is diagonal. Let $X_j = \mathrm{col}_j(P)$:

$$P = \begin{bmatrix} X_1 & X_2 & \cdots & X_n \end{bmatrix}.$$

The columns AX_j are pairwise orthogonal since

$$\langle AX_j, AX_k \rangle = \langle A^*AX_j, X_k \rangle$$

and X_j, X_k are eigenvectors of the Hermitian matrix A^*A. Suppose that $\sigma_j > 0$ for $j = 1, 2, \ldots, r$ and $\sigma_j = 0$ for $j = r+1, r+2, \ldots, n$. Let

$$\sigma_j = \|AX_j\|$$

for $j = 1, 2, \ldots, r$ so that

$$AX_j = \sigma_j Y_j \quad \text{for } j = 1, 2, \ldots, r$$
$$AX_j = 0 \quad \text{for } j = r+1, \ldots, n$$

(since $\sigma_j = 0$ for $j = r+1, r+2, \ldots, n$.) The columns

$$Y_j = \sigma_j^{-1} AX_j$$

are orthonormal. Extend $Y_1, \ldots, Y_r$ to an orthonormal basis for $\mathbf{F}^{m \times 1}$ and call the matrix with these columns Q:

$$Q = \begin{bmatrix} Y_1 & \cdots & Y_r & Y_{r+1} & \cdots & Y_m \end{bmatrix}.$$

Then

$$
\begin{aligned}
AP &= A \begin{bmatrix} X_1 & \cdots & X_r & X_{r+1} & \cdots & X_n \end{bmatrix} \\
&= \begin{bmatrix} AX_1 & \cdots & AX_r & AX_{r+1} & \cdots & AX_n \end{bmatrix} \\
&= \begin{bmatrix} \sigma_1 Y_1 & \cdots & \sigma_r Y_r & 0 & \cdots & 0 \end{bmatrix} \\
&= QD
\end{aligned}
$$

where D is as in Definition 369B. $\qquad \square$

▷ **Exercise 371A** The singular values of the matrix

$$A = \begin{bmatrix} 8.0 & 5.0 & 6.5 \\ 4.0 & 5.0 & 9.5 \\ 8.0 & -5.0 & 6.5 \\ 4.0 & -5.0 & 9.5 \end{bmatrix}$$

are 20, 10, 5. Find unitary matrices Q and P so that $Q^{-1}AP$ is in SVNF. (Do not use MINIMAT's built-in **svd** function.)

12.7   Invariants

One of the pleasant facts about the unitary theory is that the natural questions have neat answers. For example, according to the Normal Spectral

Theorem 320B, if we want to know whether a matrix A is unitarily diagonalizable, we simply check if the equation $AA^* = A^*A$ holds. Now we'll prove some more neat results.

Unitary Left Equivalence Test

Theorem 372A *Matrices A and B (of the same size) are unitarily left equivalent if and only if $A^*A = B^*B$.*

Proof: If $A = QB$ with Q unitary, then

$$A^*A = B^*Q^*QB = B^*B.$$

The proof of the converse is just a rehash of the proof of 370C. In that proof, the first step in finding the decomposition $A = QDP^{-1}$ was to construct P so that $P^*(A^*A)P$ is diagonal. Under the hypothesis that $A^*A = B^*B$, the same P works for B^*B as well. Now the proof produces unitary matrices Q_1 and Q_2 with Q_1^*AP and Q_2^*BP in SVNF. But the singular values are the square roots of the eigenvalues of $A^*A = B^*B$ and these, counted by geometric multiplicity, determine the SVNF. Hence,

$$Q_1^*AP = Q_2^*BP.$$

Hence, A and B are unitarily left equivalent (cancel P) as required.   □

Remark 372B Theorem 372A has the following geometrical interpretation. Let A_j denote the jth column of A:

$$A = \begin{bmatrix} A_1 & A_2 & \cdots & A_n \end{bmatrix},$$

so that

$$\text{entry}_{ij}(A^*A) = A_i^*A_j = \langle A_j, A_i \rangle.$$

Hence, the condition that $A^*A = B^*B$ says that *corresponding columns of A and B matrices have the same norm (take $i = j$), and corresponding pairs of columns make the same angle.* Theorem 372A says that these lengths and angles form a complete system of invariants for unitary left equivalence.

Definition 373A Suppose that $A \in \mathbf{F}^{m \times n}$. If $m \geq n$, **geometric multiplicity** of a singular value σ of A is the dimension

$$\dim \mathcal{E}_{\sigma^2}(A^*A) = \text{nullity}(\sigma^2 I - A^*A)$$

of the corresponding space of singular vectors. If $m < n$, the geometric multiplicity of σ for A is, by definition, the same as the geometric multiplicity of σ for A^*.

Remark 373B The definition is rigged so that the singular multiplicities are the same for A and A^*. When $m \geq n$, the multiplicity of 0 as a singular value of A is the same as the nullity of A, because

$$\mathcal{N}(A) = \mathcal{N}(A^*A).$$

When $m < n$, the nullity of A is $\nu + n - m$ where ν is the geometric multiplicity of the singular value 0. In either case, the nullity of A, the nullity of A^*A, and the geometric multiplicity of 0 as an *eigenvalue* of A^*A are all the same. In particular, a matrix has maximal rank if and only if all its singular values are nonzero.

Unitary Biequivalence Test

Corollary 373C *For two matrices A and B of the same size the following are equivalent:*

(1) *A and B are unitarily biequivalent*

(2) *A^*A and B^*B are unitarily similar.*

(3) *A and B have the same singular values each with the same multiplicity.*

Proof: The equivalence (2) $\Longleftrightarrow$ (3) is Theorem 321D. If $A = QBP^{-1}$ with Q, P unitary, then

$$A^*A = (P^{-1})^*B^*Q^*QBP^* = P(B^*B)P^{-1}.$$

This shows (1) $\Longrightarrow$ (2). For the converse, assume that $A^*A = P(B^*B)P^{-1}$ where P is unitary. As $P(B^*B)P^{-1} = (BP^*)^*(BP^*)$, it follows, from Theorem 372A, that BP^* and A are unitarily left equivalent, that is, that B and A are unitarily biequivalent. $\square$

Remark 374A For a matrix in SVNF, the geometric multiplicity of a nonzero singular value is the number of times it is repeated. Hence, two matrices in SVNF having the same singular values each with the same multiplicity must be equal. Hence, unitarily biequivalent matrices in SVNF are equal. This proves the uniqueness of D in the Singular Value Decomposition 370C.

12.8   More Exercises

▷ **Exercise 374B** What are the eigenvalues and eigenvectors of a reflection?

▷ **Exercise 374C** What are the singular vectors and singular values of a unitary matrix?

▷ **Exercise 374D** Derive the Biequivalence Decomposition 106D as a corollary of the Singular Value Decomposition.

▷ **Exercise 374E** The Householder Decomposition and the Gram Schmidt Decomposition 206C are the same when A is invertible, but, in general, they are different. In the Gram-Schmidt Decomposition, A and Q are the same size, and R is square; In the Householder Decomposition, A and R are the same size, and Q is square. Using the Gauss-Jordan Decomposition 105A, deduce the Householder Decomposition from the Gram-Schmidt Decomposition.

Real Unitary Equivalence

Recall the definitions of unitary left equivalence, unitary right equivalence, and unitary biequivalence from Definition 361A. They define A and B to be *equivalent* iff there exist unitary P and Q satisfying a certain matrix equation. We say that A and B are **unitarily equivalent over F** iff P and Q may be chosen from **F**. (Compare this with Section 7.3.)

Exercise 374F Show that *If two real square matrices $A, B \in \mathbf{R}^{m \times n}$ are unitarily left equivalent over* **C**, *then they are unitarily left equivalent over* **R**. Hint: Use Theorem 372A.

Exercise 374G Show that *If two real square matrices $A, B \in \mathbf{R}^{m \times n}$ are unitarily biequivalent over* **C**, *then they are unitarily biequivalent over* **R**. Hint: Use Theorem 373C and Corollary 323B.

Submultiplicative Norms

Exercise 375A (Maximum Singular Value) Let $\rho(A)$ be the largest singular value of the matrix $A \in \mathbf{F}^{m \times n}$. Prove that

$$\|AX\| \le \rho(A)\,\|X\|$$

for all $X \in \mathbf{F}^{n \times 1}$, but that this is false (for some X) if $\rho(A)$ is replaced by a smaller number. Show that the function $\rho(A)$ is a **submultiplicative norm**. This means that ρ satisfies the following four conditions:

(Positivity) $\rho(A) > 0$ if $A \ne 0$,

(Homogeneity) $\rho(bA) = |b|\rho(A)$,

(Triangle Inequality) $\rho(A_1 + A_2) \le \rho(A_1) + \rho(A_2)$,

(Submultiplicativity) $\rho(AB) \le \rho(A)\rho(B)$,

for $b \in \mathbf{F}$ $A, A_1, A_2 \in \mathbf{F}^{m \times n}$ and $B \in \mathbf{F}^{n \times p}$. The first three conditions say that $\rho(A)$ is a norm as defined in Section 6.10.

Exercise 375B Define $\gamma(A)$ by

$$\gamma(A) = \max_{1 \le i \le m} \sum_{j=1}^{n} |a_{ij}|$$

for $A \in \mathbf{F}^{m \times n}$ where $a_{ij} = \mathrm{entry}_{ij}(A)$. Show that $\gamma(A)$ is a submultiplicative norm.

Exercise 375C It is known that a matrix power series

$$f(A) = \sum_{p=0}^{\infty} c_p A^p$$

converges if

$$\sum_{p=0}^{\infty} \ell(c_p A^p) < \infty$$

for some norm $\ell(A)$. Use this to show that the **Neuman series**

$$(I - W)^{-1} = \sum_{p=0}^{\infty} W^p$$

converges if $\gamma(W) < 1$. (Any submultiplicative norm works here.)

Exercise 375D Give another proof of **Gershgorin's theorem** which was proved as Exercise 352H: If $D \in \mathbf{F}^{n \times n}$ is an invertible diagonal matrix, $W \in \mathbf{F}^{n \times n}$ has vanishing diagonal entries, and $\gamma(D^{-1}W) < 1$, then $D + W$ is invertible.

Polar Decomposition

Recall the notion of *positive semidefinite* matrix and *positive definite* from Section 10.7.

Exercise 376A (Square Roots) Show that a positive semidefinite matrix P has a unique positive semidefinite square root R: $P = R^2$.

Exercise 376B (Polar Decomposition) Show that any matrix $\Phi \in \mathbf{F}^{n \times \nu}$ with independent columns may be written uniquely in the form

$$\Phi = \Psi R$$

where $\Psi \in \mathbf{F}^{n \times \nu}$ has orthonormal columns and R is positive definite. Compare this with the Gram-Schmidt Decomposition 206C. Hint: Consider $\Phi^* \Phi$.

Exercise 376C By Exercise 376B, an invertible matrix A may be written uniquely in the form $A = QR$ where Q is unitary and R is positive definite. Show that A is normal if and only if $QR = RQ$.

Using MINIMAT

The Singular Value Decomposition is implemented as a built-in function svd in MINIMAT. The command

```
#>  [Q D P] = svd(A)
```

will assign to the variables Q, D, and P the factors in the Singular Value Decomposition. The one-output form

```
#>  L=svd(A)
```

assigns to L a row matrix containing the Singular Values.

▷ **Exercise 376D** Try out svd. Check the result.

Exercise 376E Generate a random 3×5 matrix A. Then use MINIMAT's built-in svd function to find the largest singular value $\sigma_{\max}$ of A and a nonzero column X such that $\|AX\| = \sigma_{\max}\|X\|$. Also confirm that $\|AX\|/\|X\| \leq \sigma_{\max}$ for a randomly chosen $X \in \mathbf{F}^{5 \times 1}$.

Exercise 376F The command

```
#> D=zeros(3,5); D(1,1)=7; D(2,2)=5; D(3,3)=2
#> A=qr(rand(3,3))*D*qr(rand(5,5))
```

will create a 3×5 matrix $A = QDP^{-1}$ whose singular values are $2, 5, 7$ without remembering the matrices Q and P. Using MINIMAT's built-in svd function find real unitary matrices Q and P with $A = QDP^{-1}$.

Exercise 377A Let $A \in \mathbf{F}^{3 \times 5}$ be the matrix constructed in the previous problem. Confirm that $\|AX\|/\|X\| \leq 7$ for a randomly chosen $X \in \mathbf{F}^{5 \times 1}$. Find an X with $\|AX\|/\|X\| = 7$.

Exercise 377B Using MINIMAT's built-in svd function, confirm the laws on the maximum singular value (Exercise 375A) by plugging in random matrices.

Exercise 377C The command

```
#> P=qr(rand(n,n)), D=diag(rand(1,n))
```

will assign to the variable P a random $n \times n$ real unitary matrix and to D a random $n \times n$ diagonal matrix. Execute these commands and confirm that A=P*D*P^(-1) is Hermitian and that P*P'=P'*P=eye(n).

Exercise 377D Confirm that $A^{*}A$ has positive eigenvalues by generating a random A and then computing the eigenvalues of $A^{*}A$ using MINIMAT's built-in eig function.

▷ **Exercise 377E** Find a Singular Value Decomposition for a random 5×3 matrix using MINIMAT's eig and qr functions but not the svd function.

Chapter Summary

Factorization Theorem

When $\mathbf{F} = \mathbf{R}$, any unitary matrix can be written as a product of reflection matrices. When $\mathbf{F} = \mathbf{C}$, any unitary matrix can be written as a product of reflection matrices and unitary scales.

Householder Decomposition

Every matrix is unitarily left equivalent to a unique matrix in PREF (positive row echelon form).

Singular Value Decomposition

Every matrix is unitarily biequivalent to a unique matrix in SVNF (singular value normal form).

Unitary Equivalence

- *Two matrices A and B are unitarily left equivalent iff*

$$A^*A = B^*B.$$

- *Two matrices A and B are unitarily biequivalent iff*

$$\nu_\lambda(A^*A) = \nu_\lambda(B^*B) \text{ for all } \lambda \geq 0.$$

Here
$$\nu_\lambda(C) = \text{nullity}(\lambda I - C)$$

is the geometric multiplicity. The eigenvalues of A^*A are the squares of the singular values of A. Hence, A and B are unitarily biequivalent iff they have the same singular values each with the same multiplicity.

13

BLOCK DIAGONALIZATION

Broadly speaking, we will learn two things in this chapter:

(1) *Most* matrices can be diagonalized, and

(2) *All* matrices can be block diagonalized.

Item (1) means that a randomly chosen matrix, that is, one which satisfies no unusual condition, will be diagonalizable; it will be similar to a diagonal matrix

$$D = \begin{bmatrix} \lambda_1 & & & \\ & \lambda_2 & & \\ & & \ddots & \\ & & & \lambda_n \end{bmatrix}$$

where the entries $\lambda_1, \lambda_2, \ldots, \lambda_n$ on the diagonal are the eigenvalues of A (and of D). Item (2) means that every matrix A is similar to a matrix Δ of form

$$\Delta = \begin{bmatrix} \Lambda_1 & & & \\ & \Lambda_2 & & \\ & & \ddots & \\ & & & \Lambda_m \end{bmatrix}$$

where the entries $\Lambda_1, \Lambda_2, \ldots, \Lambda_m$ on the diagonal are not necessarily numbers, but are themselves square matrices (of varying sizes). Each of the
blocks Λ_i has a single eigenvalue λ_i and for many purposes behaves as if it
were a number.

13.1   Generic Diagonalization

Not every square matrix is diagonalizable. However, a randomly chosen
(some people say **generic**) matrix *is* diagonalizable. You can confirm this
using MINIMAT with the following commands:

```
#> A=rand(5,5)
#> [D P]=eig(A)
#> A, P*D*P^(-1)
```

If you repeat this command sequence 20 times, you will discover that it
never fails to produce a diagonalization of the randomly chosen matrix A.
Our aim in this section is to explain why this is so.

We recall the theory of Section 8.3. There is a (nonlinear) function

$$\det : \mathbf{F}^{n \times n} \to \mathbf{F}$$

called the **determinant** with the following two crucial properties:

(I) A matrix $A \in \mathbf{F}^{n \times n}$ is invertible iff $\det(A) \neq 0$.

(II) The function f defined by

$$f(\xi) = \det(\xi I - A)$$

is a polynomial of degree n.

(Here $I = I_n$ is the $n \times n$ identity matrix.) The polynomial $f = f(\xi)$
defined in (II) is called the **characteristic polynomial** of A. The number
λ is an eigenvalue of $A \iff$ the matrix $\lambda I - A$ is *not* invertible $\iff$
$\det(\lambda I - A) = 0$. Thus condition (I) implies that

*the eigenvalues of a matrix are the roots of its characteristic
polynomial.*

This may expressed symbolically as

$$AX = \lambda X \text{ for some } X \neq 0 \iff f(\lambda) = 0.$$

It is plausible that a "generic" polynomial of degree n has n distinct (complex) roots and, hence, that a "generic" matrix has n distinct eigenvalues.

This could even be proven if we took the trouble to give a careful definition of "generic". In this section we show that an $n \times n$ matrix A with n distinct eigenvalues is diagonalizable. This justifies the idea that a generic matrix is diagonalizable when $\mathbf{F} = \mathbf{C}$. Incidentally, similar reasoning leads to the conclusion that a generic square matrix is invertible.

Remark 381A You may conclude from the previous discussion that most square matrices are both diagonalizable and invertible, but you should not conclude that nondiagonalizable matrices are unimportant in applications. There are two reasons for this. In the first place, a matrix that arises in an application will not usually be generic. It will be constructed in accordance with some theory that may impose special conditions on it. In the second place, one often needs parameterized families of matrices and such families will (generically) contain noninvertible and nondiagonalizable members. It may be that the interesting exceptional behavior occurs exactly when the matrix is noninvertible or nondiagonalizable.

Theorem 381B *Assume that $X_1, X_2, \ldots, X_k$ are nonzero eigenvectors of a square matrix $A \in \mathbf{F}^{n \times n}$:*

$$AX_j = \lambda_j X_j$$

for $j = 1, 2, \ldots, k$. If the eigenvalues $\lambda_1, \lambda_2, \ldots, \lambda_k$ are all distinct, then the sequence $(X_1, X_2, \ldots, X_k)$ is independent.

Proof: Let $I \in \mathbf{F}^{n \times n}$ denote the identity matrix and for each $i = 1, 2, \ldots, k$ define the matrix B_i to be the product of all the matrices $A - \lambda_j I$ for $j \neq i$:

$$B_i = (A - \lambda_1 I) \cdots (A - \lambda_{i-1} I)(A - \lambda_{i+1} I) \cdots (A - \lambda_k I).$$

Since $AX_j = \lambda_j$, we obtain

$$B_i X_j = (\lambda_j - \lambda_1) \cdots (\lambda_j - \lambda_{i-1})(\lambda_j - \lambda_{i+1}) \cdots (\lambda_j - \lambda_k) X_j.$$

Now if $i \neq j$, the coefficient on the right contains the factor $\lambda_j - \lambda_j = 0$ and hence vanishes:

$$B_i X_j = 0 \qquad \text{for } i \neq j.$$

On the other hand, for $i = j$, the coefficient is

$$\beta_i = (\lambda_i - \lambda_1) \cdots (\lambda_i - \lambda_{i-1})(\lambda_i - \lambda_{i+1}) \cdots (\lambda_i - \lambda_k),$$

which is *not* zero, since we are assuming that the eigenvalues are distinct: $\lambda_i - \lambda_j \neq 0$ for $j = 1, \ldots, i - 1, i + 1, \ldots, k$.

We show that $(X_1, X_2, \ldots, X_k)$ are independent. Assume that

$$c_1 X_1 + x_2 X_2 + \cdots c_k X_k = 0.$$

Multiply by B_i to obtain

$$c_i \beta_i X_i = B_i 0 = 0,$$

since the terms $c_j B_i X_j$ with $j \neq i$ drop out. But $\beta_i \neq 0$ and $X_i \neq 0$, so the only possibility is that $c_i = 0$ as required. For example, if $k = 3$ and

$$c_1 X_1 + c_2 X_2 + c_3 X_3 = 0,$$

then we would prove that $c_1 = 0$ by multiplying both sides by

$$B_1 = (A - \lambda_2 I)(A - \lambda_3 I)$$

to obtain $\beta_1 c_1 X_1 = 0$ where $\beta_1 = (\lambda_1 - \lambda_2)(\lambda_1 - \lambda_3) \neq 0$. $\square$

Generic Diagonalization

Corollary 382A *If an $n \times n$ matrix has n distinct eigenvalues, it is diagonalizable.*

Proof: By Theorem 381B (with $k = n$), the fact that a matrix is invertible iff its columns are independent (Corollary 175A), and the diagonalization Theorem 258A. $\square$

▷ **Question 382B** Is the converse of the corollary true?

13.2   Monotriangular Block Diagonal Form (MTBDF)

Definition 46A introduced the notation

$$D = \mathrm{diag}(\lambda_1, \lambda_2, \ldots, \lambda_n)$$

to denote the $n \times n$ diagonal matrix with diagonal entries $\lambda_1, \ldots, \lambda_n$:

$$D = \begin{bmatrix} \lambda_1 & & & \\ & \lambda_2 & & \\ & & \ddots & \\ & & & \lambda_n \end{bmatrix}.$$

(The white space in the matrix is filled with zeros.) We now introduce a more general notation where the entries λ_j are replaced by square matrices.

Suppose that $\Lambda_1, \Lambda_2, \ldots, \Lambda_m$ are square matrices of possibly different sizes: let Λ_k be of size $n_k \times n_k$. Let

$$n = n_1 + n_2 + \cdots + n_m.$$

Then the notation

$$\Delta = \mathrm{Diag}(\Lambda_1, \Lambda_2, \ldots, \Lambda_m)$$

means that Δ is the **block diagonal** matrix, of size $n \times n$, with blocks $\Lambda_1, \Lambda_2, \ldots, \Lambda_m$ on the diagonal:

$$\Delta = \begin{bmatrix} \Lambda_1 & & & \\ & \Lambda_2 & & \\ & & \ddots & \\ & & & \Lambda_m \end{bmatrix}.$$

The precise definition of the notation $\Delta = \mathrm{Diag}(\Lambda_1, \Lambda_2, \ldots, \Lambda_m)$ is

$$\mathrm{entry}_{ij}(\Delta) = \begin{cases} \mathrm{entry}_{xy}(\Lambda_k) & \text{if } s_{k-1} < i, j \le s_k, \\ 0 & \text{otherwise} \end{cases}$$

where Λ_k is $n_k \times n_k$, $s_k = n_1 + n_2 + \cdots + n_{k-1}$, and $x = i - s_{k-1}$, $y = j - s_{k-1}$. For example,

$$\mathrm{Diag}(\Lambda_1, \Lambda_2, \Lambda_3) = \begin{bmatrix} a_1 & b_1 & c_1 & 0 & 0 & 0 \\ d_1 & e_1 & f_1 & 0 & 0 & 0 \\ g_1 & h_1 & i_1 & 0 & 0 & 0 \\ 0 & 0 & 0 & a_2 & 0 & 0 \\ 0 & 0 & 0 & 0 & a_3 & b_3 \\ 0 & 0 & 0 & 0 & c_3 & d_3 \end{bmatrix},$$

where

$$\Lambda_1 = \begin{bmatrix} a_1 & b_1 & c_1 \\ d_1 & e_1 & f_1 \\ g_1 & h_1 & i_1 \end{bmatrix}, \qquad \Lambda_2 = \begin{bmatrix} a_2 \end{bmatrix}, \qquad \Lambda_3 = \begin{bmatrix} a_3 & b_3 \\ c_3 & d_3 \end{bmatrix}.$$

Definition 383A A square matrix $\Lambda \in \mathbf{F}^{n \times n}$ is called **monotriangular** iff it has the form

$$\Lambda = \lambda I + V$$

where I is the identity matrix and V is strictly triangular.

Remark 384A A matrix Λ is monotriangular iff it is triangular and all its diagonal entries are the same. Since the eigenvalues of a triangular matrix are precisely the diagonal entries (see Exercise 274E), the only eigenvalue of Λ is λ.

Monotriangular Block Diagonal Form (MTBDF)

Definition 384B A square matrix $\Delta \in \mathbf{F}^{n \times n}$ is said to be in **monotriangular block diagonal form**, abbreviated **MTBDF**, iff it has the block diagonal form

$$\Delta = \mathrm{Diag}(\Lambda_1, \Lambda_2, \ldots, \Lambda_m)$$

where each Λ_k is monotriangular.

Thus Λ_k has the form

$$\Lambda_k = \lambda_k I + V_k$$

where $I = I_{n_k}$ is the identity matrix, V_k is strictly triangular, and λ_k is the eigenvalue of Λ_k. Here Λ_k is of size $n_k \times n_k$, so that $n_1 + n_2 + \cdots + n_m = n$. For example, the matrix

$$\mathrm{Diag}(\Lambda_1, \Lambda_2, \Lambda_3) = \begin{bmatrix} \lambda_1 & * & * & & & \\ 0 & \lambda_1 & * & & & \\ 0 & 0 & \lambda_1 & & & \\ & & & \lambda_2 & & \\ & & & & \lambda_3 & * \\ & & & & 0 & \lambda_3 \end{bmatrix}$$

is in MTBDF. (The blank entries represent 0 and have been omitted to emphasize the block structure. The asterisks represent arbitrary numbers)

Block Diagonalization

Theorem 384C *Assume that* $\mathbf{F} = \mathbf{C}$. *Then every square matrix is similar to a matrix in MTBDF.*

Proof: In other words, any square matrix A may be written in the form

$$A = P\Delta P^{-1}$$

where P is invertible and Δ is in MTBDF.

By the Schur Triangularization Theorem 312A, we may assume that A is triangular. Also assume (see Remark 313C) that any repeated eigenvalues occur in adjacent positions on the diagonal. Let λ be the first eigenvalue and p be its (algebraic) multiplicity. Then A has the form

$$A = \begin{bmatrix} \Lambda & C \\ 0 & B \end{bmatrix}$$

where $\Lambda \in \mathbf{F}^{p \times p}$ is monotriangular, $B \in \mathbf{F}^{q \times q}$ is triangular, $q = n - p$, and $C \in \mathbf{F}^{p \times q}$. The matrix $B - \lambda I_q$ is invertible, since λ is not a diagonal entry of B. We will find an invertible matrix P such that $P^{-1}AP$ has a similar form but with $C = 0$. Then we proceed by induction.

We try

$$P = \begin{bmatrix} I_p & W \\ 0_{q \times p} & I_q \end{bmatrix}, \qquad P^{-1} = \begin{bmatrix} I_p & -W \\ 0_{q \times p} & I_q \end{bmatrix},$$

where $W \in \mathbf{F}^{p \times q}$. (The matrix P is automatically invertible, no matter what the choice of W; see Corollary 52B.) The block multiplication law 51A gives

$$P^{-1}AP = \begin{bmatrix} \Lambda & -WB + \Lambda W + C \\ 0_{p \times q} & B \end{bmatrix}.$$

We want to choose W so that

$$WB - \Lambda W - C = 0. \tag{\#}$$

Equation (#) is a linear system of pq equations (one for each entry of C) in pq unknowns (one for each entry of W). You can solve (#) with Gaussian elimination. Since the system (#) is inhomogeneous (C is the inhomogeneous term), we need to prove that there is a solution. To do this, we use the fact that $\Lambda = \lambda I + V$ where V is strictly triangular. Then we want to solve

$$W(B - \lambda I) - VW - C = 0.$$

We multiply on the right by $M = (B - \lambda I)^{-1}$ and rewrite this in the form

$$W = CM + VWM. \tag{*}$$

Since $V^p = 0$, we can write down an explicit solution:

$$W = CM + VCM + V^2CM^2 + \cdots V^{p-1}CM^{p-1}.$$

Now we proceed by induction. Assume that we have found Q so that $Q^{-1}BQ$ is in MTBDF. Then

$$\begin{bmatrix} I & 0 \\ 0 & Q^{-1} \end{bmatrix} \begin{bmatrix} \Lambda & 0 \\ 0 & B \end{bmatrix} \begin{bmatrix} I & 0 \\ 0 & Q \end{bmatrix} = \begin{bmatrix} \Lambda & 0 \\ 0 & Q^{-1}BQ \end{bmatrix}$$

is also. But we have just shown that this matrix is similar to A. $\square$

Exercise 386A Let

$$A = \begin{bmatrix} 3 & 6 & 5 & 5 & 7 \\ 0 & 3 & 4 & 3 & 9 \\ 0 & 0 & 1 & 4 & 8 \\ 0 & 0 & 0 & 2 & 7 \\ 0 & 0 & 0 & 0 & 2 \end{bmatrix}$$

Find an invertible matrix P such that $P^{-1}AP$ is in MTBDF.

Exercise 386B Equation $(*)$ on Page 385 is called a **fixed point equation**, because it has the form

$$W = F(W).$$

(Here $F(W) = CM + VWM$.) To solve a fixed point equation, it is natural to try iteration. This means we take $W_0 = 0$ and $W_{k+1} = F(W_k)$ and hope that the limit $\lim W_k$ exists. This limit would be a solution because

$$F\left(\lim_{k\to\infty} W_k\right) = \lim_{k\to\infty} F(W_k) = \lim_{k\to\infty} W_{k+1} = \lim_{k\to\infty} W_k.$$

Show that, in the case at hand, we have $W_p = W_{p-1}$. Conclude that

$$W_{p-1} = W_p = W_{p+1} = W_{p+2} = \cdots = \lim_{k\to\infty} W_k,$$

so passing to the limit is unnecessary.

13.3   Using MINIMAT

Exercise 386C Transform each of the following matrices A to MTBDF Delta

```
#> D=diag([1,2,3,4]), A=D+randsutr(4)
#> D=diag([1,1,3,4]), A=D+randsutr(4)
#> D=diag([1,1,1,4]), A=D+randsutr(4)
#> D=diag([1,1,2,2]), A=D+randsutr(4)
```

You are to find P so that A and P*Delta*P^(-1) are the same. Check your answer by computing the latter and comparing in to the original A.

13.4   Nilpotent Matrices

Definition 387A A matrix $N \in \mathbf{C}^{n \times n}$ is called **nilpotent** iff some positive power of it is zero:
$$N^p = 0.$$
The smallest such power p is called the **degree of nilpotence** of N.

Lemma 387B *Any matrix similar to a nilpotent matrix is itself nilpotent with the same degree of nilpotence.*

Proof: $(PNP^{-1})^p = PN^pP^{-1}.$ $\qquad\qquad\qquad\qquad\qquad\qquad\qquad$ □

Recall that a matrix N is called **strictly triangular** iff all its entries on or below the diagonal vanish:
$$\mathrm{entry}_{ij}(N) = 0$$
for $j \leq i$.

Theorem 387C *A strictly triangular matrix is nilpotent.*

Proof: Let $E_j = \mathrm{col}_j(I)$ the nth column of the identity matrix and assume that V is strictly upper triangular. Then
$$V E_j = \sum_{j=i+1}^{n} v_{ij} E_j$$
where $v_{ij} = \mathrm{entry}_{ij}(V)$. Then $V^{n-j+1}E_j = 0$. Hence,
$$\mathrm{col}_j(V^n) = V^n E_j = V^{j-1}V^{n-j+1}E_j = 0$$
for $j = 1, 2, \ldots, n$, i.e., $V^n = 0$. $\qquad\qquad\qquad\qquad\qquad\qquad$ □

Example 387D For $n = 4$ a strictly upper triangular matrix has form
$$V = \begin{bmatrix} 0 & u_{12} & u_{13} & u_{14} \\ 0 & 0 & u_{23} & u_{24} \\ 0 & 0 & 0 & u_{34} \\ 0 & 0 & 0 & 0 \end{bmatrix},$$

$$V^2 = \begin{bmatrix} 0 & 0 & v_{13} & v_{14} \\ 0 & 0 & 0 & v_{24} \\ 0 & 0 & 0 & 0 \\ 0 & 0 & 0 & 0 \end{bmatrix}, \qquad V^2 = \begin{bmatrix} 0 & 0 & 0 & w_{14} \\ 0 & 0 & 0 & 0 \\ 0 & 0 & 0 & 0 \\ 0 & 0 & 0 & 0 \end{bmatrix}.$$

where the coefficients v_{13}, v_{14}, v_{24}, and w_{14} are expressions in the u_{ij}. The proof of Theorem 387C yields

$$
\begin{aligned}
VE_1 &= u_{12}E_2 + u_{13}E_3 + u_{14}E_4 \\
VE_2 &= \phantom{u_{12}E_2 +} u_{23}E_3 + u_{24}E_4 \\
VE_3 &= \phantom{u_{12}E_2 + u_{23}E_3 +} u_{34}E_4 \\
VE_4 &= 0
\end{aligned}
$$

$$
\begin{aligned}
V^2E_1 &= v_{13}E_3 + v_{14}E_4 \\
V^2E_2 &= \phantom{v_{13}E_3 +} v_{24}E_4 \\
V^2E_3 &= 0
\end{aligned}
$$

$$
\begin{aligned}
V^3E_2 &= w_{14}E_4 \\
V^3E_3 &= 0
\end{aligned}
$$

$$
V^4E_4 = 0.
$$

Theorem 388A *A matrix is nilpotent if and only if zero is its only eigenvalue.*

Proof: Assume that zero is the only eigenvalue of N. By Schur's Triangularization Theorem 312A, N is similar to a strictly upper triangular matrix and is therefore nilpotent. Conversely, assume that $N^p = 0$. Assume that λ is an eigenvalue of N. Then $NX = \lambda X$ where $X \neq 0$. Hence, $0 = N^p X = \lambda^p X$, so $\lambda^p = 0$, so $\lambda = 0$ as required. $\square$

Exercise 388B Compute the coefficients v_{13}, v_{14}, v_{24}, and w_{14} in Example 387D.

Exercise 388C Do Exercise 37G.

13.5   Chevalley Decomposition

Chevalley Decomposition

Theorem 389A *A square matrix* $A \in \mathbf{F}^{n \times n}$ *may be written uniquely in the form*

$$A = S + N$$

where S is diagonalizable, N is nilpotent, and S and N commute. Moreover, if A is real, then so are S and N.

Proof: By the Block Diagonalization Theorem 384C, A is similar to a matrix Δ in MTBDF. Thus for some invertible matrix P:

$$\Delta = D + V, \qquad A = P\Delta P^{-1},$$

where D is diagonal and V is strictly triangular. Moreover, D and V have the block diagonal forms

$$
\begin{aligned}
D &= \mathrm{Diag}(\lambda_1 I_{n_1}, \lambda_2 I_{n_2}, \ldots, \lambda_m I_{n_m}), \\
V &= \mathrm{Diag}(V_1, V_2, \ldots, V_m),
\end{aligned}
$$

where I_{n_k} is the identity matrix of size $n_k \times n_k$ and V_k is strictly triangular and also of size $n_k \times n_k$. The matrices A, P, and Δ are $n \times n$ where $n = n_1 + n_2 + \cdots + n_m$.

From the block multiplication law 51A and the obvious formula

$$(\lambda_k I_{n_k})V_k = \lambda_k V_k = V_k(\lambda_k I_{n_k}),$$

we obtain $DV = VD$. The Chevalley Decomposition of A is given by

$$S = PDP^{-1}, \qquad N = PVP^{-1},$$

so $A = P(D + V)P^{-1} = PDP^{-1} + PVP^{-1} = S + N$. The matrix S is diagonalizable, since it is similar to the diagonal matrix D. The matrix N is nilpotent, since it is similar to the strictly upper triangular matrix V. The matrices S and N commute, since

$$
\begin{aligned}
SN &= (PDP^{-1})(PVP^{-1}) \\
&= PDVP^{-1} \\
&= PVDP^{-1} \\
&= (PVP^{-1})(PDP^{-1}) \\
&= NS.
\end{aligned}
$$

The proofs that that the matrices S and N are uniquely determined by A and that they are real when A is real are developed in the exercises in Section 13.6 on Page 397.

▷ **Question 390A** What is the Chevalley decomposition of Δ?

Remark 390B An explicit formula for the Chevalley Decomposition is

$$N = g(A), \qquad S = A - g(A),$$

where

$$g(\xi) = \prod_{k=1}^{m} (\xi - \lambda_k)$$

and the product is over the distinct eigenvalues of A.

Corollary 390C *If $A = S+N$ is the Chevalley Decomposition of a matrix A, then the matrix exponential is given by*

$$e^{tA} = e^{tS} e^{tN}.$$

Proof: $SN = NS$. See Exercise 294G.   □

Using MINIMAT

Exercise 390D Generate matrices B, P, A with the commands

```
#> B= [2*eye(3)+randsutr(3,3), zeros(3,2)
      zeros(2,3),              5*eye(2)+randsutr(2,2)]
#> P=rand(5,5),  A=P*B*inv(P)
```

Find a diagonalizable matrix S and a nilpotent matrix N such that A=S+N and S*N=N*S. Check your answer. (You may use the value of P.)

13.6   More Exercises

Exercise 390E Suppose that $A, B \in \mathbf{F}^{n \times n}$.

(1) Show that AB and BA have the same eigenvalues.

(2) Assume $n = 2$ and $ABAB = 0$. Show that $BABA = 0$.

(3) Assume $n = 3$. Find A, B with $ABAB = 0$ but $BABA \neq 0$.

Diagonalization

The exercises in this section prove Theorem 272A. Recall the key definitions. For a matrix A with distinct eigenvalue $\lambda_1, \ldots, \lambda_m$:

- the **algebraic multiplicity** of the eigenvalue λ is the highest power $\alpha(\lambda, A)$ of $(\xi - \lambda)$ dividing the characteristic polynomial

$$\det(\xi I - A) = \prod_{i=1}^{m} (\xi - \lambda_i)^{\alpha(\lambda_i, A)};$$

- the **geometric multiplicity** of the eigenvalue λ for the matrix A is the dimension $\nu(\lambda, A)$ of the eigenspace

$$\mathcal{E}_\lambda(A) = \mathcal{N}(\lambda I - A).$$

Assume that

$$\Delta = \text{Diag}(\Lambda_1, \Lambda_2, \ldots, \Lambda_m)$$

is in MTBDF, and let λ_i be the (unique) eigenvalue of Λ_i. (In particular, $\lambda_1, \lambda_2, \ldots, \lambda_m$ are assumed distinct.)

Exercise 391A Show that

(1) $\text{rank}(\Delta) = \sum_i \text{rank}(\Lambda_i)$;

(2) $\text{nullity}(\Delta) = \sum_i \text{nullity}(\Lambda_i)$; and

(3) $\det(\Delta) = \prod_i \det(\Lambda_i)$.

Exercise 391B Show that the characteristic polynomial of Δ is given by

$$\det(\xi I - \Delta) = \prod_{i=1}^{m} (\xi - \lambda)^{n_i}$$

where the size of Λ_i is $n_i \times n_i$. Conclude that the algebraic multiplicity is given by

$$\alpha(\lambda_i, \Delta) = n_i.$$

Exercise 391C Show that the geometric multiplicity is given by

$$\nu(\lambda_i, \Delta) = \text{nullity}(\lambda_i I - \Lambda_i).$$

Conclude that $\nu(\lambda_i, \Delta) \leq \alpha(\lambda_i, \Delta)$ with equality if and only if $\Lambda_i = \lambda_i I$.

Exercise 392A From the previous three exercises deduce that a matrix A is diagonalizable if and only if $\nu(\lambda_i, A) = \alpha(\lambda_i, A)$ for $i = 1, 2, \ldots, m$ where $\lambda_1, \lambda_2, \ldots, \lambda_m$ are the eigenvalues of A. (This is Theorem 272A.)

Exercise 392B Assume that Φ_i is a basis for the eigenspace $\mathcal{E}_{\lambda_i}(A)$. Let

$$P = \left[\begin{array}{cccc} \Phi_1 & \Phi_2 & \cdots & \Phi_m \end{array} \right].$$

Prove that

(1) $A\Phi_i = \lambda_i \Phi_i$;

(2) the columns of P are independent;

(3) P is square $\Longleftrightarrow$ A is diagonalizable;

(4) if P is square, $P^{-1}AP$ is diagonal.

This justifies Remark 272B.

Generalized Eigenspaces

Our aim in this section is to formulate the Block Diagonalization Theorem 384C in such a way as to make it look like Exercise 392B. Suppose that $A \in \mathbf{F}^{n \times n}$ is a square matrix. Recall from Definition 256A that the subspace

$$\mathcal{E}_\lambda(A) = \mathcal{N}(\lambda I - A)$$

is called the **eigenspace** of λ in A. The subspace

$$\mathcal{G}_\lambda(A) = \mathcal{N}\left((\lambda I - A)^n \right)$$

is called the **generalized eigenspace** of λ in A. Let $\lambda_1, \lambda_2, \ldots, \lambda_m$ be the eigenvalues of A listed without repetitions.

Exercise 392C Show that $X \in \mathcal{G}_\lambda(A)$ if and only if

$$(\lambda I - A)^p X = 0$$

for some p. (You must show that if the equation holds for some p, then it holds for $p = n$.)

Exercise 392D Show that each generalized eigenspace of a square matrix A satisfies:

$$A\mathcal{G}_\lambda(A) \subset \mathcal{G}_\lambda(A),$$

with equality if $\lambda \neq 0$. In the lingo of Definition 324H, this says that $\mathcal{G}_\lambda(A)$ is an A-invariant subspace.

Exercise 393A Let Φ_i be a basis for $\mathcal{G}_{\lambda_i}(A)$. Show that there are unique matrices Λ_i (for $i = 1, 2, \ldots, m$) such that

$$A\Phi_i = \Phi_i \Lambda_i.$$

(Hint: See Exercise 186B.) Compare with item (1) of Exercise 392B.

Exercise 393B Show that the matrix

$$P = \left[\begin{array}{cccc} \Phi_1 & \Phi_2 & \cdots & \Phi_m \end{array} \right]$$

is invertible and that

$$P^{-1}AP = \mathrm{Diag}(\Lambda_1, \Lambda_2, \ldots, \Lambda_m).$$

Compare with items (2) and (4) of Exercise 392B.

Exercise 393C Show that λ_i is the only eigenvalue of Λ_i. Conclude that $\Lambda_i - \lambda_i I$ is nilpotent.

Exercise 393D By definition the geometric multiplicity is given by

$$\nu(\lambda, A) = \dim \mathcal{E}_\lambda(A).$$

Show that the algebraic multiplicity is given by

$$\alpha(\lambda, A) = \dim \mathcal{G}_\lambda(A).$$

Exercise 393E Show that $\mathcal{E}_\lambda(A) \subset \mathcal{G}_\lambda(A)$ and that A is diagonalizable if and only if $\mathcal{E}_\lambda(A) = \mathcal{G}_\lambda(A)$ for all λ.

Exercise 393F Show that every $X \in \mathbf{F}^{n \times 1}$ can be written uniquely in the form

$$X = X_1 + X_2 + \cdots + X_m$$

where $X_i \in \mathcal{G}_{\lambda_i}(A)$. (One says that $\mathbf{F}^{n \times 1}$ is the **direct sum** of the subspaces $\mathcal{G}_\lambda(A)$.)

Remark 393G Once we know all the eigenvalues of A, here is how to block diagonalize A.

(1) Arrange the eigenvalues of A in a list $(\lambda_1, \lambda_2, \ldots, \lambda_m)$ without repetitions.

(2) For each i construct a basis Φ_i for the generalized eigenspace $\mathcal{G}_{\lambda_i}(A)$.

(3) Form the matrix
$$P = \begin{bmatrix} \Phi_1 & \Phi_2 & \cdots & \Phi_m \end{bmatrix}.$$

The matrix P is invertible.

(4) The matrix $P^{-1}AP$ is block diagonal.

Exercise 394A Use this algorithm to block diagonalize the matrix
$$A = \begin{bmatrix} 97 & 230 & 320 & 115 & 51 \\ -17 & -42 & -59 & -22 & -14 \\ -27 & -63 & -87 & -30 & -12 \\ 23 & 55 & 76 & 26 & 15 \\ 13 & 31 & 43 & 15 & 8 \end{bmatrix}.$$

The characteristic polynomial of A is
$$\det(\xi I - A) = (\xi - 1)^2 \xi^3.$$

Exercise 394B Generate a matrix A as follows.

```
#> B = diag([1 1 0 0 0]); B(1,2)=1; B(3,4)=1; B(4,5)=1;
#> P=intinv(5); A= P*B*P^(-1); P=[ ]
```

Find P so that $P^{-1}AP$ is block diagonal. (The variable P held an answer, but we have instructed MINIMAT to forget it.)

Matrix Exponential

▷ **Exercise 394C** A matrix is called **unipotent** iff it is the sum of a nilpotent matrix and the identity matrix. For example, A unitriangular matrix is unipotent. By Theorem 388A, a unipotent matrix is invertible, since 1 is its only eigenvalue. Find a formula for the inverse of the unipotent matrix $I - N$. (Hint: Factor $I - N^n$.)

Exercise 394D Show that the exponential of a nilpotent matrix is unipotent.

Exercise 394E Show that a unipotent matrix is the exponential of a nilpotent matrix. Hint: The natural logarithm satisfies
$$\ln(1 + \xi) = \sum_{k=1}^{\infty} \frac{(-\xi)^k}{k}.$$

Exercise 394F Prove that any invertible matrix A can be written in the form $A = SW$ where S is diagonalizable, W is unipotent, and $SW = WS$.

Exercise 395A Show that a complex square matrix B is an exponential (meaning that $B = e^A$ for some complex matrix A) if and only if B is invertible. (Exercise 294I treated the special case where B is diagonalizable.)

Exercise 395B Show that if N is nilpotent, there is a polynomial $p(t)$ such that
$$\|e^{tN}\| \le p(t)$$
for all complex numbers t. The submultiplicative inequality 225G and the triangle inequality from Theorem 193B will be helpful.

Exercise 395C Do Exercises 296A through 296D without the hypothesis that A is diagonalizable. Hint: If $a > 0$ and $p(t)$ is a polynomial, then
$$\lim_{t \to \infty} |e^{-at}p(t)| = \lim_{t \to -\infty} |e^{at}p(t)| = 0.$$

Minimal Polynomial

Exercise 395D Show that for any square matrix $A \in \mathbf{C}^{n \times n}$ there is a unique polynomial
$$f(\xi) = \xi^m + c_{m-1}\xi^{m-1} + \cdots + c_2\xi^2 + c_1\xi + c_0$$

with the following properties:

(I) the polynomial is **monic**, which means that the coefficient of its highest order term is 1 (as indicated in the formula);

(II) $f(A) = 0$;

(III) Any polynomial $g(\xi)$ which satisfies $g(A) = 0$ is divisible by $f(\xi)$.

This polynomial f is called the **minimal polynomial** of A. Hint: For $m \ge n^2$ the equation $\sum_{j=0}^{m} c_j A^j = 0$ has a nontrivial solution C. Find a solution of least degree so that $c_m \ne 0$. Prove (III) by long division of polynomials:
$$g(\xi) = q(\xi)f(\xi) + r(\xi)$$
where the remainder $r(\xi)$ has degree $< m$.

▷ **Exercise 395E** What is the minimal polynomial of a nilpotent matrix?

Exercise 395F Show that the degree of the minimal polynomial of any square matrix $A \in \mathbf{F}^{n \times n}$ is less than or equal to n. In particular, the degree of nilpotence of a nilpotent matrix $N \in \mathbf{F}^{n \times n}$ is less than or equal to n.

Exercise 396A Show that the minimal polynomial of a diagonal matrix

$$D = \text{Diag}(d_1, d_2, \cdots, d_n)$$

is

$$f(\xi) = \prod_{i=j}^{m} (\xi - \lambda_j)$$

where the λ_j are the d_i listed without repetitions, i.e.

$$\{d_1, d_2, \cdots, d_n\} = \{\lambda_1, \lambda_2, \cdots, \lambda_m\}$$

and $\lambda_i \neq \lambda_j$ for $i \neq j$.

Exercise 396B Show that similar matrices have the same minimal polynomial.

Exercise 396C Show that the minimal polynomial of a 2×2 matrix A is its characteristic polynomial $\det(\xi I - A)$ unless A has a single eigenvalue λ. In that case, the minimal polynomial is $\xi - \lambda$ if $A = \lambda I$, and $(\xi - \lambda)^2$ otherwise.

Exercise 396D Show that the eigenvalues of a matrix are the roots of its minimal polynomial.

Exercise 396E The command

```
#> X=A(:)
```

has the effect of assigning to X a single column formed from the entries of A. Use it and MINIMAT's `randsoln` function to find the minimal polynomial of a random 3×3 matrix directly from the definition.

Exercise 396F Prove the **Cayley-Hamilton Theorem**: *A square matrix satisfies its characteristic equation.* More precisely, if $A \in \mathbf{F}^{n \times n}$ and

$$p(\xi) = \det(\xi I - A)$$

is its characteristic polynomial, then

$$p(A) = 0.$$

▷ **Exercise 396G** Why is the following not a correct proof of the Cayley-Hamilton Theorem?

$$p(\xi) = \det(\tau I - A), \quad \text{so} \quad p(A) = \det(AI - A) = \det(0) = 0.$$

Chevalley Decomposition

Theorem 389A says that a square matrix $A \in \mathbf{F}^{n \times n}$ may be written uniquely in the form

$$A = S + N$$

where S is diagonalizable, N is nilpotent, and S and N commute. The existence was proved in Theorem 389A. Here we prove uniqueness. By the existence proof, A is similar to a matrix

$$\Delta = D + V, \qquad A = P\Delta P^{-1},$$

where D and V have the block diagonal form

$$
\begin{aligned}
D &= \operatorname{Diag}(\lambda_1 I_{n_1}, \lambda_2 I_{n_2}, \ldots, \lambda_m I_{n_m}) \\
V &= \operatorname{Diag}(V_1, V_2, \ldots, V_m)
\end{aligned}
$$

where I_{n_k} is the identity matrix of size $n_k \times n_k$ and V_k is strictly triangular and also of size $n_k \times n_k$. Thus $n_1 + n_2 + \cdots + n_m = n$ where Δ is $n \times n$. Throughout we'll assume that the eigenvalues $\lambda_1, \lambda_2, \ldots, \lambda_m$ are distinct. This actually follows from the construction given in the proof of Theorem 384C, but it can also be achieved by replacing Δ by $Q\Delta Q^{-1}$ where Q is a permutation matrix which rearranges the blocks so that any two blocks with the same eigenvalue are adjacent. (Then we rename the λ_k's to eliminate the repetitions.)

If the theorem is true for a matrix Δ, then it is true for any matrix A which is similar to Δ. Hence, we may reformulate the uniqueness of the Chevalley Decomposition as follows:

Lemma 397A *[Uniqueness]* *Suppose that $\Delta = D + V$ is in MTBDF as above, and also that $\Delta = S + N$ where S is diagonalizable, N is nilpotent, and S and N commute. Then $S = D$ and $N = V$.*

Exercise 397B Suppose $DC = CD$ with D as above. Show that C has block diagonal form, that is

$$C = \operatorname{Diag}(C_1, C_2, \ldots, C_m)$$

where C_k is $n_k \times n_k$. (This is where we use the fact that the numbers λ_k are distinct.) Hint: As a warmup assume $m = 2$ or $m = 3$ and that all the blocks are 1×1.

Exercise 397C Let Δ be as in 397A. Show that the minimal polynomial of Δ is the polynomial

$$f(\xi) = \prod_{p=1}^{m} (\xi - \lambda_k)^{p_k}$$

where p_k is the degree of nilpotence of V_k.

Exercise 398A Show that matrix is diagonalizable if and only if its minimal polynomial has no multiple roots, that is, has the form

$$g(\xi) = \prod_{p=1}^{m} (\xi - \lambda_k)$$

where the numbers λ_k are distinct.

Exercise 398B With $\Delta = D + V$ as in 397A, Show that

$$V = g(\Delta), \qquad D = \Delta - g(\Delta),$$

with g as in the last exercise. Conclude that, if $\Delta = S + N$ as in 397A, we have that $DS = SD$ and $DN = ND$.

Exercise 398C Let C have the block diagonal form

$$C = \text{Diag}(C_1, C_2, \ldots, C_m).$$

Show that C diagonalizable if and only if each component C_k is diagonalizable. Show that C is nilpotent if and only if each component C_k is nilpotent.

Combining the previous exercises, we obtain that S and N are block diagonal,

$$
\begin{aligned}
S &= \text{Diag}(S_1, S_2, \ldots, S_m) \\
N &= \text{Diag}(N_1, N_2, \ldots, N_m)
\end{aligned}
$$

where each S_k is diagonalizable and each N_k is nilpotent. Since $SN = NS$, we have that $S_k N_k = N_k S_k$. Since $\Delta = S + N$, we have that

$$\Delta = \text{Diag}(S_1 + N_1, \; S_2 + N_2, \; \ldots, \; S_m + N_m).$$

It only remains to prove the special case $m = 1$.

Exercise 398D Assume that $\lambda I + V = S + N$ where I is the identity matrix, S is diagonalizable, V and N are both nilpotent, and $SN = NS$. Show that $S = \lambda I$ and $V = N$. Hint: $S - \lambda I = V - N$ where $S - \lambda I$ is diagonalizable and $V - N$ is nilpotent.

Exercise 398E Suppose that $A = S + N$ is the Chevalley decomposition of a matrix A. Show that if A is real, then so are S and N. Hint: The complex roots of a real polynomial occur in conjugate pairs: if $f(\lambda) = 0$, then $f(\bar{\lambda}) = 0$.

Chapter Summary

Block Diagonalization

Every square matrix is similar to a block diagonal matrix

$$\Delta = \begin{bmatrix} \Lambda_1 & & & \\ & \Lambda_2 & & \\ & & \ddots & \\ & & & \Lambda_m \end{bmatrix}.$$

Here the blocks Λ_k on the diagonal have the form

$$\Lambda_k = \lambda_k I + V_k$$

where $I = I_{n_k}$ is the identity matrix and V_k is strictly triangular, so λ_k is the (only) eigenvalue of Λ_k.

Chevalley Decomposition

Every matrix A uniquely has the form

$$A = S + N$$

where

> *S is diagonalizable;*
>
> *N is nilpotent; and*
>
> *$SN = NS$.*

14

JORDAN NORMAL FORM

In this chapter we will find a complete system of invariants that characterize similarity. This means a collection of nonnegative integers $\rho_{\lambda,k}(A)$ – defined for each square matrix A, each positive integer k, and each complex number λ – such that for $A, B \in \mathbf{C}^{n \times n}$, we have that A and B are similar if and only if

$$\rho_{\lambda,k}(A) = \rho_{\lambda,k}(B) \qquad \text{for all } \lambda \in \mathbf{C} \text{ and all } k = 1, 2, \dots.$$

We will prove a normal form theorem for similarity called the *Jordan Normal Form Theorem*.

14.1 Similarity Invariants

Definition 401A Let $A \in \mathbf{C}^{n \times n}$, $\lambda \in \mathbf{C}$, and $k = 1, 2, 3, \dots$. Define

$$\rho_{\lambda,k}(A) = \operatorname{rank}(\lambda I - A)^k$$

where $I = I_n$ is the $n \times n$ identity matrix. The integer $\rho_{\lambda,k}(A)$ is called the kth **eigenrank** of A for the eigenvalue λ.

Remark 401B If λ is not an eigenvalue of A, then $\rho_{\lambda,k}(A) = n$. If $k \geq n$, $\rho_{\lambda,k}(A) = \rho_{\lambda,n}(A)$. (See Exercise 403B below.) Thus only finitely many of these numbers are of interest.

Definition 402A The **eigennullities** $\nu_{\lambda,k}(A)$ of the matrix A are defined by

$$\nu_{\lambda,k}(A) = \text{nullity}((\lambda I - A)^k) = \dim \mathcal{N}((\lambda I - A)^k)$$

From the Rank Nullity Relation 172A *(rank + nullity = n)*, we obtain

$$\rho_{\lambda,k}(A) + \nu_{\lambda,k}(A) = n \qquad\qquad (*)$$

for $A \in \mathbf{C}^{n \times n}$. Hence, the eigennullities and eigenranks contain the same information.

Remark 402B The eigennullity

$$\nu_{\lambda,1}(A) = \dim \mathcal{N}(\lambda I - A) = \dim \mathcal{E}_{\lambda}(A)$$

has been called the **geometric multiplicity** of the eigenvalue λ. It is the dimension of the eigenspace $\mathcal{E}_{\lambda}(A)$. In Exercise 393D, it was proved that the eigennullity

$$\nu_{\lambda,n}(A) = \dim \mathcal{N}(\lambda I - A)^n = \dim \mathcal{G}_{\lambda}(A)$$

is the **algebraic multiplicity** of λ for A. It is the dimension of the generalized eigenspace $\mathcal{G}_{\lambda}(A)$.

Invariance of the Eigenranks

Theorem 402C *Similar matrices have the same eigenranks.*

Proof: There are two key points: (1) Similar matrices are *a fortiori* biequivalent, for if $A = PBP^{-1}$, then $A = QBP^{-1}$ where $Q = P$. (2) Similar matrices have similar powers, for $(PBP^{-1})^k = PB^kP^{-1}$.

Now assume that A and B are similar. Then $A = PBP^{-1}$ where P is invertible. Choose $\lambda \in \mathbf{C}$. Then $\lambda I - A = P(\lambda I - B)P^{-1}$. Hence, $(\lambda I - A)^k = P(\lambda I - B^kP^{-1}$ for $k = 1, 2, \ldots$. By Theorem 171D, the matrices $(\lambda I - A)^k$ and $(\lambda I - B)^k$ have the same rank. By the definition of $\rho_{\lambda,k}$, we have $\rho_{\lambda,k}(A) = \rho_{\lambda,k}(B)$, as required. $\qquad\square$

Remark 402D Of course, by equation $(*)$ on page 402, similar matrices have the same eigennullities as well. Thus, by Remark 402B, Theorem 402C generalizes Theorem 271B. Below (Corollary 416A), we will prove the converse to Theorem 402C.

Exercise 403A Prove that a matrix $A \in \mathbf{C}^{n \times n}$ is diagonalizable if and only if $\rho_{\lambda,k}(A) = \rho_{\lambda,1}(A)$ for all eigenvalues λ of A and all $k = 1, 2, 3, \ldots$.

▷ **Exercise 403B** Prove that $\rho_{\lambda,k}(A) = \rho_{\lambda,n}(A)$ if $k \geq n$.

14.2 Jordan Normal Form

We can improve the monotriangular block diagonal form considerably by making further similarity transformations within each block. The resulting blocks will be almost diagonal except for a few nonzero entries above the diagonal. Here are the precise definitions.

The entries $\text{entry}_{ii}(A)$ of a matrix A are called the **diagonal entries**, and said to be *on the diagonal*. The entries $\text{entry}_{i,i+1}(A)$ are called the **superdiagonal** entries, and said to lie on the *on the superdiagonal*. The superdiagonal entries lie just above the diagonal. A **Jordan block** is a square matrix Λ having all its diagonal entries equal, zeros or ones on the superdiagonal, and zeros elsewhere. Thus Λ is a Jordan block iff

$$
\begin{aligned}
\text{entry}_{ii}(\Lambda) &= \lambda, \\
\text{entry}_{i,i+1}(\Lambda) &= 0 \text{ or } 1, \\
\text{entry}_{ij}(\Lambda) &= 0 \text{ if } j \neq i, i+1.
\end{aligned}
$$

Jordan Normal Form

A matrix J is in **Jordan normal form** iff it is in block diagonal form

$$J = \text{Diag}(\Lambda_1, \Lambda_2, \ldots, \Lambda_m)$$

where each Λ_k is a Jordan block.

In particular, each Jordan block is monotriangular (see Definition 383A), and so a matrix in Jordan normal form is automatically in monotriangular block diagonal form (see Definition 384B). For example, the 6×6 matrix

$$
J = \begin{bmatrix}
\lambda_1 & e_1 & 0 & & & \\
0 & \lambda_1 & e_2 & & & \\
0 & 0 & \lambda_1 & & & \\
& & & \lambda_2 & & \\
& & & & \lambda_3 & e_3 \\
& & & & 0 & \lambda_3
\end{bmatrix}
$$

is in Jordan normal form provided that each of the superdiagonal entries e_1, e_2, e_3 is either zero or one.

Jordan Normal Form Theorem

Theorem 404A *Every square matrix A is similar to a matrix J in Jordan normal form.*

In other words, any square matrix A may be written in the form

$$
A = PJP^{-1}
$$

where P is invertible and J is in Jordan normal form. By Theorem 402C and the Block Diagonalization Theorem 384C, we can assume that the matrix A is in MTBDF (monotriangular block diagonal form) as in Definition 384B. We can work a block at a time, so it is enough to prove the theorem for monotriangular matrices. As the matrices $\lambda I + V_1$ and $\lambda I + V_2$ are similar if and only if the matrices V_1 and V_2 are, it is enough to prove the theorem for nilpotent (in fact, strictly upper triangular) matrices. The proof will occupy most of the rest of this chapter.

14.3   Indecomposable Jordan Blocks

In this section we'll prove a special case of the Jordan Normal Form Theorem 404A as a warmup. The ideas in the general case are similar. We'll make a preliminary definition.

An **indecomposable Jordan block** is one where all the entries on the superdiagonal are one. It has the form $\lambda I + W$ where

$$
\text{entry}_{ij}(W) = \begin{cases} 1 & \text{if } j = i + 1, \\ 0 & \text{otherwise.} \end{cases}
$$

Notice that W is itself an indecomposable Jordan block (with eigenvalue zero). A Jordan block has form

$$\Lambda = \text{Diag}(\lambda I + W_1, \ \lambda I + W_2, \ \ldots, \ \lambda I + W_k)$$

where the matrices $\lambda I + W_1, \lambda I + W_2, \ldots, \lambda I + W_k$ are indecomposable Jordan blocks.[1] For example, the Jordan block

$$\Lambda = \begin{bmatrix} \lambda & 1 & 0 & & & \\ 0 & \lambda & 1 & & & \\ 0 & 0 & \lambda & & & \\ & & & \lambda & & \\ & & & & \lambda & 1 \\ & & & & 0 & \lambda \end{bmatrix}$$

has form

$$\Lambda = \text{Diag}(\lambda I + W_1, \ \lambda I + W_2, \ \lambda I + W_3)$$

where the constituent indecomposable Jordan blocks are

$$\lambda I + W_1 = \begin{bmatrix} \lambda & 1 & 0 \\ 0 & \lambda & 1 \\ 0 & 0 & \lambda \end{bmatrix}, \quad \lambda I + W_2 = \begin{bmatrix} \lambda \end{bmatrix}, \quad \lambda I + W_3 = \begin{bmatrix} \lambda & 1 \\ 0 & \lambda \end{bmatrix}.$$

▷ **Question 405A** What are the eigenranks of this last matrix Λ?

Theorem 405B *Let $N \in \mathbf{F}^{n \times n}$ be a matrix of size $n \times n$ and degree of nilpotence n, i.e. that $N^n = 0$ but $N^{n-1} \neq 0$. Then N is similar to the indecomposable $n \times n$ Jordan block W.*

Proof: Since $N^n = 0$ but $N^{n-1} \neq 0$, there is a vector $X \in \mathbf{F}^{n \times 1}$ such that $N^n X = 0$ but $N^{n-1} X \neq 0$. Form the matrix P whose jth column is $N^{n-j} X$. We will prove that

$$NP = PW.$$

Then we will show that P is invertible. Multiplying on the right by P^{-1} gives

$$N = PWP^{-1}.$$

We prove that $NP = PW$, i.e. that

$$\text{col}_j(NP) = \text{col}_j(PW)$$

---

[1] The terminology here is at slight variance with the general usage. Most authors call *Jordan block* what we have called *indecomposable Jordan block*.

for $j = 1, 2, \ldots, n$. By the definition of P,

$$P = \begin{bmatrix} N^{n-1}X & N^{n-2}X & \cdots & NX & X \end{bmatrix},$$

so

$$NP = \begin{bmatrix} 0 & N^{n-1}X & \cdots & N^2X & NX \end{bmatrix} = PW,$$

so

$$\begin{aligned}
\mathrm{col}_1(NP) &= 0, \\
\mathrm{col}_j(NP) &= \mathrm{col}_{j-1}(P) \text{ for } j = 2, 3, \ldots, n.
\end{aligned}$$

On the other hand, the first column of W is zero, and the jth column of W is the $(j-1)$st column of the identity matrix. Thus

$$\begin{aligned}
\mathrm{col}_1(PW) &= 0, \\
\mathrm{col}_j(PW) &= \mathrm{col}_{j-1}(P) \text{ for } j = 2, 3, \ldots, n.
\end{aligned}$$

This proves that $NP = PW$.

We prove that P is invertible. It is enough to show that its columns are independent. Suppose

$$0 = c_1 N^{n-1}X + c_2 N^{n-2}X + \cdots + c_{n-1}NX + c_n X.$$

Since $N^k = 0$ for $k \geq n$ we may apply N^{n-1} to both sides and obtain that $c_n N^{n-1}X = 0$. But $N^{n-1}X \neq 0$ so $c_n = 0$. Now apply N^{n-2} to both sides to prove that $c_{n-1} = 0$. Repeating in this way we obtain that $c_1 = c_2 = \cdots = c_n = 0$, as required.   □

14.4   Partitions

A little terminology from number theory is useful in describing the relations among the various eigennullities of a nilpotent matrix.

A **partition** of a positive integer n is a nonincreasing sequence π of positive integers which sum to n, that is,

$$\pi = (n_1, n_2, \ldots, n_m)$$

where

$$n_1 \geq n_2 \geq \cdots \geq n_m \geq 1$$

and

$$n_1 + n_2 + \cdots + n_m = n.$$

A partition $\pi = (n_1, n_2, \ldots, n_m)$ can be used to construct a diagram of stars called a **tableau**. The tableau consists of $n = n_1 + n_2 + \cdots + n_m$ stars arranged in m rows with the kth row having n_k stars. The stars in a row are left justified so that the jth columns align. The jth column of the tableau intersects the kth row exactly when $j \leq n_k$. The **dual partition** π^* of π is obtained by forming the transpose of this tableau. Thus $\pi^* = (\ell_1, \ell_2, \ldots, \ell_p)$ where ℓ_j is the number of indices k with $j \leq n_k$. For example, if

$$\pi = (5, 5, 4, 3, 3, 3, 1),$$

then the tableau is

$$
\begin{array}{ccccc}
\star & \star & \star & \star & \star \\
\star & \star & \star & \star & \star \\
\star & \star & \star & \star & \\
\star & \star & \star & & \\
\star & \star & \star & & \\
\star & \star & \star & & \\
\star & & & &
\end{array}
$$

and the dual partition

$$\pi^* = (7, 6, 6, 3, 2)$$

is obtained by counting the number of entries in successive columns. The dual of the dual is the original partition:

$$\pi^{**} = \pi.$$

14.5   Weyr Characteristic

Let $N \in \mathbf{F}^{n \times n}$ be a nilpotent matrix, and let p be the **degree of nilpotence** of N. This is the least integer for which $N^p = 0$:

$$N^p = 0, \qquad N^{p-1} \neq 0.$$

Recall that the kth **eigenrank** of N is the integer

$$\rho_k(N) = \mathrm{rank}(N^k) = \dim \mathcal{R}(N^k).$$

We have dropped the subscript λ since N is nilpotent: its only eigenvalue is zero. The sequence of integers $\omega = (\ell_1, \ell_2, \ldots, \ell_p)$ defined by

$$\ell_k = \rho_{k-1}(N) - \rho_k(N)$$

for $k = 1, 2, \ldots, p$ is called the **Weyr characteristic** of of the nilpotent matrix N.

Theorem 408A *The Weyr characteristic of a nilpotent matrix $N \in \mathbf{F}^{n \times n}$ is a partition of n.*

Proof: Successive terms ℓ_k and ℓ_{k+1} in the sum $\ell_1 + \cdots + \ell_p$ contain $\rho_k(N)$ with opposite signs. Hence, the sum "telescopes":

$$\ell_1 + \ell_2 + \cdots + \ell_p = \rho_0(N) - \rho_p(N) = n - 0 = n$$

as $N^0 = I$ and $N^p = 0$. To show that $\ell_k \geq \ell_{k+1}$, first note the obvious inclusion of ranges

$$\mathcal{R}(N^k) \subset \mathcal{R}(N^{k-1}).$$

This holds because $N^k X = N^{k-1}(NX)$. Let Φ be a basis for the subspace $\mathcal{R}(N^k)$, and extend it to a basis Ψ for $\mathcal{R}(N^{k-1})$ by adjoining additional columns Υ:

$$\Psi = \begin{bmatrix} \Phi & \Upsilon \end{bmatrix}.$$

Then Ψ has $\rho_{k-1}(N)$ columns, Φ has $\rho_k(N)$ columns, and Υ has ℓ_k columns. Now

$$\mathcal{R}(N^{k-1}) = \mathcal{R}(\Psi), \qquad \mathcal{R}(\mathbf{N}^k) = \mathcal{R}(\Phi),$$

so

$$\mathcal{R}(N^k) = \mathcal{R}(N\Psi), \qquad \mathcal{R}(N^{k+1}) = \mathcal{R}(N\Phi).$$

Discard some columns from $N\Phi$ to make a basis $\tilde{\Phi}$ for $\mathcal{R}(N^{k+1})$, and then discard some columns from

$$N\Psi = \begin{bmatrix} N\Phi & N\Upsilon \end{bmatrix},$$

so that

$$\tilde{\Psi} = \begin{bmatrix} \tilde{\Phi} & \tilde{\Upsilon} \end{bmatrix}$$

is a basis for $\mathcal{R}(N^k)$. Then $\tilde{\Upsilon}$ has ℓ_{k+1} columns. Since the discarded columns were taken from Υ, it follows that $\ell_{k+1} \leq \ell_k$, as required.   □

14.6   Segre Characteristic

For each k let $W_k \in \mathbf{F}^{k \times k}$ denote the $k \times k$ indecomposable Jordan block with eigenvalue zero:

$$\begin{aligned} \text{entry}_{ij}(W_k) &= 0 \quad \text{if } j \neq i+1 \\ \text{entry}_{i,i+1}(W_k) &= 1 \quad \text{for } i = 1, 2, \ldots, k-1. \end{aligned}$$

The subscript k indicates the size of the matrix W_k. For each partition

$$\pi = (n_1, n_2, \ldots, n_m)$$

denote by W_π the Jordan block given by the block diagonal matrix

$$W_\pi = \mathrm{Diag}(W_{n_1}, W_{n_2}, \ldots W_{n_m}).$$

A matrix of form W_π is called a **Segre matrix**.

A Segre matrix is in Jordan normal form. Conversely, any nilpotent matrix in Jordan normal form can be transformed to a Segre matrix by permuting the blocks so that their sizes decrease along the diagonal. (This can be accomplished by replacing J by PJP^{-1} for a suitable permutation matrix P.)

Now define the **Segre characteristic** of a nilpotent matrix to be the dual partition

$$\pi = \omega^*$$

of the Weyr characteristic ω. The key to understanding the Jordan Normal Form Theorem is the following

Theorem 409A *The Segre characteristic of the Segre matrix W_π is π.*

An example is more convincing than a general proof. Let

$$\pi = (3, 2, 2, 1), \qquad \omega = \pi^* = (4, 3, 1),$$

so

$$W = W_\pi = \mathrm{Diag}(W_3, W_2, W_2, W_1).$$

Written in full this is

$$W = \begin{bmatrix} 0 & 1 & 0 & & & & & \\ 0 & 0 & 1 & & & & & \\ 0 & 0 & 0 & & & & & \\ & & & 0 & 1 & & & \\ & & & 0 & 0 & & & \\ & & & & & 0 & 1 & \\ & & & & & 0 & 0 & \\ & & & & & & & 0 \end{bmatrix}.$$

(The blank entries represent 0; they have been omitted to make the block structure more evident.) In the notation of the definition

$$\pi = (n_1, n_2, n_3, n_4), \qquad \omega = (\ell_1, \ell_2, \ell_3),$$

where $n_1 = 3$, $n_2 = n_3 = 2$, $n_4 = 1$, $\ell_1 = 4$, $\ell_2 = 3$, $\ell_3 = 1$ and

$$n = n_1 + n_2 + n_3 + n_4 = \ell_1 + \ell_2 + \ell_3 = 8.$$

For $j = 1, 2, \ldots, 8$ let $E_j = \mathrm{col}_j(I_8)$ denote the jth column of the 8×8 identity matrix so that

$$W E_j = \begin{cases} 0 & \text{for } j = 1, 4, 6, 8; \\ E_{j-1} & \text{for } j = 2, 3, 5, 7. \end{cases}$$

Arrange these columns in a tableau

$$\begin{array}{ccc} E_1 & E_2 & E_3 \\ E_4 & E_5 & \\ E_6 & E_7 & \\ E_8 & & \end{array}$$

so that n_i is the number of entries in the ith row of the tableau and ℓ_j be the number of entries in the jth column. We can decorate the tableau with arrows to indicate the effect of applying W:

$$\begin{array}{ccccccc} 0 & \leftarrow & E_1 & \leftarrow & E_2 & \leftarrow & E_3 \\ 0 & \leftarrow & E_4 & \leftarrow & E_5 & & \\ 0 & \leftarrow & E_6 & \leftarrow & E_7 & & \\ 0 & \leftarrow & E_8 & & & & \end{array}$$

We now see a general principle:

> *Applying W^k to the tableau annihilates the elements in the first k columns and transforms the remaining elements into the columns of a basis for $\mathcal{R}(N^k)$.*

Thus the kth eigenrank is the number

$$\rho_k(W) = \ell_{k+1} + \ell_{k+2} + \cdots + \ell_p$$

of elements to the right of the kth column. This equation says precisely that $\omega = (\ell_1, \ell_2, \ldots, \ell_p)$ is the Weyr characteristic of $W = W_\pi$, as required.

14.7   Jordan-Segre Basis

Continue the notation of the last section. Let π be a partition of n and W_π be the corresponding Segre matrix. For $j = 1, 2, \cdots, n$ let

$$E_j = \mathrm{col}_j(I_n)$$

denote the jth column of the identity matrix I_n. Then $W_\pi E_j$ is either E_{j-1} or 0 depending on π. We'll use a double subscript notation to specify for which values of j the former alternative holds. Let

$$E_{1,1}, \ldots, E_{1,n_1}, E_{2,1}, \ldots, E_{2,n_2}, \ldots$$

denote the columns $E_1, E_2, \ldots, E_n$ in that order. Then

$$W_\pi E_{i,1} = 0 \qquad \text{for } i = 1, 2, \ldots, m,$$

$$W_\pi E_{i,j} = E_{i,j-1} \qquad \text{for } j = 2, 3, \ldots, n_i.$$

These relations say that the doubly indexed sequence E_{ij} forms a *Jordan-Segre Basis* for $(\mathbf{F}^{n\times 1}, W_\pi)$. Here's the definition.

Let $N \in \mathbf{F}^{n\times n}$ be a matrix and $\mathcal{V} \subset \mathbf{F}^{n\times 1}$ be a subspace. A **Jordan-Segre Basis** for $(\mathcal{V}, N)$ $N \in \mathbf{F}^{n\times n}$ is a doubly indexed sequence

$$X_{i,j} \in \mathcal{V}, \qquad (i = 1, 2, \ldots, m; \ \ j = 1, 2, \ldots, n_i)$$

of columns which forms a basis for $\mathcal{V}$ and satisfies

$$NX_{i,1} = 0 \qquad \text{for } i = 1, 2, \ldots, m,$$

$$NX_{i,j} = X_{i,j-1} \qquad \text{for } j = 2, 3, \ldots, n_i.$$

The sequence $\pi = (n_1, n_2, \ldots, n_m)$ is called the **associated partition** of the basis; it is a partition of the dimension of $\mathcal{V}$:

$$\dim(\mathcal{V}) = n_1 + n_2 + \cdots + n_m.$$

The condition that the elements $X_{i,j} \in \mathcal{V}$ form a basis for $\mathcal{V}$ means that every $X \in \mathcal{V}$ may be written uniquely as a linear combination of these $X_{i,j}$, in other words, that the inhomogeneous system

$$X = \sum_{i=1}^{m} \sum_{j=1}^{n_m} c_{ij} X_{ij}$$

(in which the $c_{i,j}$ are the unknowns) has a unique solution. Throughout most of this book we would have said instead that the matrix formed from these columns is a *basis* for $\mathcal{V}$, but the present terminology is more conventional. The matrix whose columns are

$$X_{1,1}, \ldots, X_{1,n_1}, X_{2,1}, \ldots, X_{2,n_2}, \ldots, X_{n,n_m}$$

(in that order) is called the **basis corresponding** to the Jordan-Segre basis. In case $\mathcal{V} = \mathbf{F}^{n\times 1}$, this is an invertible matrix.

Theorem 412A *Suppose that $P \in \mathbf{F}^{n \times n}$ is the basis (matrix) corresponding to a Jordan-Segre basis for $(\mathbf{F}^{n \times 1}, N)$. Then*

$$N = PW_\pi P^{-1}$$

where π is the associated partition.

Proof: Since P is invertible, the conclusion may be written as $NP = PW_\pi$. Let $E_{i,j}$ be the kth column of the identity matrix where $X_{i,j}$ be the kth column of P. Then

$$\mathrm{col}_k(NP) = N\,\mathrm{col}_k(P) = NX_{i,j} = \begin{cases} X_{i,j-1} = \mathrm{col}_{k-1}(P) & \text{if } j > 1, \\ 0 & \text{if } j = 1, \end{cases}$$

while

$$\mathrm{col}_k(PW_\pi) = P\,\mathrm{col}_k(I) = \begin{cases} PE_{i,j-1} = \mathrm{col}_{k-1}(P) & \text{if } j > 1, \\ 0 & \text{if } j = 1, \end{cases}$$

so

$$\mathrm{col}_k(NP) = \mathrm{col}_k(PW_\pi).$$

As k is arbitrary this shows that

$$NP = PW_\pi,$$

as required.   □

14.8   Improved Rank Nullity Relation

The Rank Nullity Relation 172A says that for $A \in \mathbf{F}^{m \times n}$ we have

$$\mathrm{rank}(A) + \mathrm{nullity}(A) = n.$$

For the proof of the Jordan Normal Form Theorem, we'll need a slight generalization.

The rank of A is the dimension of the range $\mathcal{R}(A) = A\mathbf{F}^{n \times 1}$ which is defined by

$$A\mathbf{F}^{n \times 1} = \{AX \in \mathbf{F}^{m \times 1} : X \in \mathbf{F}^{n \times 1}\}.$$

The nullity of A is the dimension of the null space $\mathcal{N}(A)$ which is defined by

$$\mathcal{N}(A) = \{X \in \mathbf{F}^{n \times 1} : AX = 0\}.$$

Thus the Rank Nullity Relation takes the form

$$\dim\left(A\mathbf{F}^{n \times 1}\right) + \dim\left(\mathcal{N}(A)\right) = n.$$

We can generalize by replacing $\mathbf{F}^{n \times 1}$ by a subspace.

Theorem 413A (Improved Rank Nullity Relation) *Suppose that* $\mathcal{V} \subset \mathbf{F}^{n \times 1}$ *is a subspace and that* $A \in \mathbf{F}^{m \times n}$. *Then*

$$\dim(A\mathcal{V}) + \dim(\mathcal{V} \cap \mathcal{N}(A)) = \dim(\mathcal{V})$$

where

$$A\mathcal{V} = \{AX \in \mathbf{F}^{m \times 1} : X \in \mathcal{V}\}$$

and

$$\mathcal{V} \cap \mathcal{N}(A) = \{X \in \mathcal{V} : AX = 0\}.$$

The theorem is an immediate consequence of the following lemma. It is an improved version of Lemma 179D.

Lemma 413B *Suppose that* $\Phi \in \mathbf{F}^{n \times r}$, $\Phi_1 \in \mathbf{F}^{n \times k}$, $\Phi_2 \in \mathbf{F}^{n \times (k+r)}$ *satisfy*

(1) $A\Phi_1$ *is a basis for* $A\mathcal{V}$;

(2) Φ_2 *is a basis for* $\mathcal{V} \cap \mathcal{N}(A)$;

(3) $\Phi = \begin{bmatrix} \Phi_1 & \Phi_2 \end{bmatrix}$.

Then Φ *is a basis for* $\mathcal{V}$.

Proof: Exercise.

14.9   Proof of the Jordan Normal Form Theorem

To prove the Jordan Normal Form Theorem 404A, it is enough to prove it for nilpotent matrices. For this, by Theorem 412A, it is enough to prove that if N is a nilpotent matrix, there is a Jordan-Segre basis for $(\mathbf{F}^{n \times 1}, N)$. We shall prove this inductively.

Let $N \in \mathbf{F}^{n \times n}$ be nilpotent, and let p be the degree of nilpotence of N. This means that

$$N^p = 0, \qquad N^{p-1} \neq 0.$$

Let $\mathcal{V}_k$ denote the range $\mathcal{R}(N^k)$ of N^k:

$$\mathcal{V}_k = N^k \mathbf{F}^{n \times 1} = N\mathcal{V}_{k-1}.$$

Clearly, $\mathcal{V}_k \subset \mathcal{V}_{k-1}$. (Proof: Choose $X \in \mathcal{V}_k$. Then $X = N^k Y$ for some Y, so $X = N^{k-1} Z$ where $Z = NY$, so $X \in \mathcal{V}_{k-1}$.) Hence, we have an increasing sequence

$$\{0\} = \mathcal{V}_k \subset \mathcal{V}_{p-1} \subset \cdots \subset \mathcal{V}_1 \subset \mathcal{V}_0 = \mathbf{F}^{n \times 1}$$

of subspaces of $\mathbf{F}^{n \times 1}$. The theorem follows by taking $k = 0$ in the following

Lemma 414A *There is a Jordan-Segre basis for* $(\mathcal{V}_k, N)$.

Proof: This is proved by reverse induction on k. This means that first we prove it for $k = p$, then for $k = p - 1$, then for $k = p - 2$, and so on. At the $(p - k)$th stage of the proof, we use the basis constructed for $\mathcal{V}_{k+1}$ to construct a basis for $\mathcal{V}_k$.

For $k = p$, the basis is empty, as $\mathcal{V}_k = \{0\}$. For $k = p - 1$, any basis for $\mathcal{V}_{p-1}$ is a Jordan-Segre basis, since $NX = 0$ for $X \in \mathcal{V}_{p-1}$. Now assume that we have constructed a Jordan-Segre basis

$$
\begin{array}{cccccc}
X_{1,1} & X_{1,2} & \cdots & \cdots & \cdots & X_{1,m_1} \\
& & \vdots & & & \\
X_{i,1} & X_{i,2} & \cdots & \cdots & X_{i,m_i} & \\
& & \vdots & & & \\
X_{h,1} & X_{h,2} & \cdots & X_{h,m_h} & &
\end{array}
$$

for $(\mathcal{V}_{k+1}, N)$. We shall extend it to a Jordan-Segre basis for $(\mathcal{V}_k, N)$ by adjoining an additional element to the end of every row and (possibly) some additional elements at the bottom of the first column.

As the elements of the basis lie in $\mathcal{V}_{k+1} = N\mathcal{V}_k$, each has the form NX for some $X \in \mathcal{V}_k$. In particular, this is true for these elements on the right edge of the tableau, so there are elements $X_{i,m_i+1} \in \mathcal{V}_k$ satisfying

$$X_{i,m_i} = NX_{i,m_i+1}.$$

We adjoin this element X_{i,m_i+1} to the right end of the ith row. The elements in the first column form a basis for $\mathcal{V}_{k+1} \cap \mathcal{N}(N)$. As $\mathcal{V}_{k+1} \subset \mathcal{V}_k$, these elements form an independent sequence in $\mathcal{V}_k \cap \mathcal{N}(N)$. Hence, we may extend to a basis

$$X_{1,1}, X_{2,1}, \ldots, X_{h,1}, X_{h+1,1}, \ldots, X_{g,1}$$

for $\mathcal{V}_k \cap \mathcal{N}(N)$.

We claim that this is a Jordan-Segre basis for $(\mathcal{V}_k, N)$. The elements $NX_{i,j}$ with $j > 1$ are precisely the elements of the Jordan-Segre basis for $\mathcal{V}_{k+1} = N\mathcal{V}_k$, while the elements $\mathcal{V}_{i,1}$ form a basis for $\mathcal{V}_k \cap \mathcal{N}(N)$ by construction. Thus by the Rank Nullity Lemma 413B, the elements $X_{i,j}$ $(i = 1, 2, \ldots, g, j \geq 1)$ form a basis for $\mathcal{V}_k$, as required. This completes the proof of the lemma and hence of the Jordan Normal Form Theorem 404A. □

Example 414B Suppose that the Segre characteristic of the nilpotent matrix N is the partition $\pi = (3, 2, 2, 1)$ of the example in the proof of

Theorem 409A. We follow the steps in the proof of 414A to construct a Jordan-Segre basis. Note that $N^3 = 0$.

- Let X_1 be a basis for $\mathcal{R}(N^2)$

- Extend to a basis $\begin{bmatrix} X_1 & X_2 & X_4 & X_6 \end{bmatrix}$ for $\mathcal{R}(N)$ by solving the inhomogeneous system $NX_2 = X_1$ for X_2 and extending $\begin{bmatrix} X_1 \end{bmatrix}$ to a basis $\begin{bmatrix} X_1 & X_4 & X_6 \end{bmatrix}$ of $\mathcal{R}(N) \cap \mathcal{N}(N)$.

- Extend to a basis

$$P = \begin{bmatrix} X_1 & X_2 & X_3 & X_4 & X_5 & X_6 & X_7 & X_8 \end{bmatrix}.$$

of $\mathbf{F}^{8 \times 1}$ by solving the inhomogeneous systems

$$NX_3 = X_2, \qquad NX_5 = X_4, \qquad NX_7 = X_6,$$

for X_3, X_5, and X_7, and extending $\begin{bmatrix} X_1 & X_4 & X_6 \end{bmatrix}$ to a basis $\begin{bmatrix} X_1 & X_4 & X_6 & X_8 \end{bmatrix}$ for $\mathcal{N}(N)$.

Theorem 415A *For two nilpotent matrices of the same size, the following conditions are equivalent:*

(1) *they are similar;*

(2) *they have the same eigenranks;*

(3) *they have the same eigennullities;*

(4) *they have the same Segre characteristic;*

(5) *they have the same Weyr characteristic.*

Proof: The eigennullities and the Weyr characteristic are related by the two equations

$$\begin{aligned} \nu_k(N) &= \ell_1 + \ell_2 + \cdots + \ell_k, \\ \ell_k &= \nu_k(N) - \nu_{k-1}(N), \end{aligned}$$

and so they determine one another. By the Rank Nullity Relation 172A,

$$\nu_k(N) + \rho_k(N) = n$$

the Weyr characteristic and the eigenranks determine one another. By d-uality, the Weyr characteristic and the Segre characteristic determine one

another. This shows that conditions (2) through (5) are equivalent. We
have seen that $(1) \implies (2)$ in Theorem 402C. We have proved that every nilpotent matrix is similar to some Segre matrix W_π (Theorems 412A
and 414A), and that the Segre characteristic of W_π is π (Theorem 409A).
Hence, $(4) \implies (1)$.   □

Characterization of Similarity

Corollary 416A *The eigenranks*

$$\rho_{\lambda,k}(A) = \operatorname{rank}(\lambda I - A)^k$$

*form a complete system of invariants for similarity. This means
that two square matrices $A, B \in \mathbf{F}^{n \times n}$ are similar if and only if*

$$\rho_{\lambda,k}(A) = \rho_{\lambda,k}(B)$$

for all $\lambda \in \mathbf{C}$ and all $k = 1, 2, \ldots$.

Proof: We have already proved "only if" as Theorem 402C. In the nilpotent case, "if" is Theorem 415A, just proved. The general case follows from
the nilpotent case as indicated in the discussion just after the statement of
Theorem 404A.

Remark 416B According to Exercise 274I, the similarity can be chosen
real when A and B are real.

14.10   More Exercises

▷ **Exercise 416C** Calculate the eigenranks $\rho_{\lambda,k}(A)$ where

$$A = \begin{bmatrix} 5 & 1 & 0 & & & \\ 0 & 5 & 1 & & & \\ 0 & 0 & 5 & & & \\ & & & 7 & 0 & 0 \\ & & & 0 & 7 & 1 \\ & & & 0 & 0 & 7 \end{bmatrix}.$$

▷ **Exercise 416D** A 24×24 matrix N satisfies $N^5 = 0$ and

$$\text{rank}(N^4) = 2, \quad \text{rank}(N^3) = 5, \quad \text{rank}(N^2) = 11, \quad \text{rank}(N) = 17.$$

Find its Segre characteristic.

Exercise 416E For a fixed eigenvalue λ there are 8 matrices in Jordan normal form of size 4×4 having λ as the only eigenvalue. (Each of the three entries on the superdiagonal can be either 0 or 1.) Which of these are similar? Hint: Compute the invariants $\rho_{\lambda,k}$.

Exercise 417A Show that a matrix and its transpose are similar.

Exercise 417B Suppose that N is nilpotent, that W is invertible, and that $WN = NW$. Show that N and NW are similar.

Exercise 417C Prove that if N is nilpotent, then $I+N$ and e^N are similar.

14.11   Using MINIMAT

Computer programs such as MINIMAT cannot compute the Jordan normal form of a matrix in full generality. The reason is that MINIMAT uses floating point arithmetic which can produce round-off error. But to compute the Jordan normal form we must do the arithmetic exactly with absolutely no round-off error. This is because the Jordan normal form of matrix does not depend continuously on the matrix. For example, the matrix

$$A_0 = \begin{bmatrix} 5.00 & 1.00 \\ 0.00 & 5.00 \end{bmatrix}$$

is already in Jordan normal form, whereas the nearby matrix

$$A = \begin{bmatrix} 5.00 & 1.00 \\ 0.00 & 5.01 \end{bmatrix}$$

has distinct eigenvalues 5.00 and 5.01 and therefore its Jordan normal form is diagonal:

$$J = \begin{bmatrix} 5.00 & 0.00 \\ 0.00 & 5.01 \end{bmatrix}.$$

Exercise 417D Compute the angle between the two eigenvectors of this last matrix A.

Exercise 417E Use MINIMAT's built-in `eig` function to diagonalize A_0 and A.

The MINIMAT command N=randsutr(n) (see 64F) assigns to the variable N a random strictly upper triangular matrix of size n ×n.

▷ **Exercise 417F** Confirm that a randomly chosen 5×5 strictly upper triangular matrix N is similar to the 5×5 indecomposable Jordan block W5.

Exercise 417G Generate a matrix N as follows.

```
#> W=zeros(8,6);
#> W(1,2)=1; W(2,3)=1; W(4,5)=1; W(6,7)=1;
#> P=intinv(8); N= P*W*P^(-1); P=[]
```

Find P so that $P^{-1}NP$ is in Jordan normal form. (The variable P held an answer, but we have instructed MINIMAT to forget it.) Hint: The associated partition is $\pi = (3, 2, 2, 1)$ as in Example 414B. You will have to express the range of one matrix as the null space of another. The formula

$$\mathcal{N}(A) \cap \mathcal{N}(B) = \mathcal{N}\left(\begin{bmatrix} A \\ B \end{bmatrix}\right)$$

also helps. You should use integer matrices to avoid round-off error.

Exercise 418A Write a .M function randnil so that the call

```
#> N = randnil(n)
```

produces a random $n \times n$ nilpotent matrix. This function should be likely to produce a matrix whose corresponding Jordan block W_π is not indecomposable. In other words, $N = PW_\pi P^{-1}$, where the only nonzero entries of W_π are on the super diagonal, and the entries on the super diagonal of W_π are equally likely to be zero or one. (Use the .M function intinv to generate P, so that the result N is not near any nilpotent matrix with a different Jordan normal form.)

Exercise 418B [Hard] Write a .M function jsnf which transforms the output of the previous .M function to Jordan normal form. In other words, the statements

```
#> N=randnil(n), [W P]=jsnf(N)
```

should find a matrix W in Jordan normal form and an invertible matrix P with $N = PWP^{-1}$.

Chapter Summary

Jordan Normal Form

Every matrix is similar to a matrix in Jordan normal form.

Eigennullities

Two matrices $A, B \in \mathbf{F}^{n \times n}$ are similar if and only if

$$\nu_{\lambda,k}(A) = \nu_{\lambda,k}(B)$$

for all $\lambda \in \mathbf{C}$ and all positive integers k. Here $\nu_{\lambda,k}(A)$ denotes the eigennullity defined by

$$\nu_{\lambda,k}(A) = \operatorname{nullity}(\lambda I - A)^k = \dim \mathcal{N}\big((\lambda I - A)^k\big).$$

The eigennullity $\nu_{\lambda,1}(A)$ is the geometric multiplicity; the eigennullity $\nu_{\lambda,n}(A)$ is the algebraic multiplicity.

15

DETERMINANTS-II

In this chapter we develop some more of the properties of determinants. This material can be studied immediately after Chapter 7. The **characteristic polynomial**, first defined in Section 8.4 is also studied in this chapter. We recall the definition:

$$p(\xi) = \det(\xi I - A)$$

for $A \in \mathbf{F}^{n \times n}$, where $I = I_n$ is the identity matrix.

15.1   Cofactors

Let $A \in \mathbf{F}^{n \times n}$ be a square matrix. There are certain formulas that are useful in evaluating the determinant of A. These formulas are especially easy to apply when A has a lot of zero entries.

Definition 421A For each pair (p, k) of indices between 1 and n, define the (p, k)-**minor** $M_{pk}(A)$ to be the determinant of the $(n - 1) \times (n - 1)$ submatrix that results from A by deleting the pth row and the kth column.

▷ **Question 421B** What are the minors of $A \in \mathbf{F}^{n \times n}$, if the rank of A is $n - 2$ or less?

Definition 422A For each pair (p, k) of indices between 1 and n, define the (p, k)-**cofactor** $C_{pk}(A)$ to be the signed minor:

$$C_{pk}(A) = (-1)^{p+k} M_{pk}(A).$$

Example 422B If

$$A = \begin{bmatrix} a_{11} & a_{12} & a_{13} & a_{14} \\ a_{21} & a_{22} & a_{23} & a_{24} \\ a_{31} & a_{32} & a_{33} & a_{34} \\ a_{41} & a_{42} & a_{43} & a_{44,} \end{bmatrix},$$

then

$$M_{23}(A) = \det\left(\begin{bmatrix} a_{11} & a_{12} & a_{14} \\ a_{31} & a_{32} & a_{34} \\ a_{41} & a_{42} & a_{44} \end{bmatrix}\right)$$

and

$$C_{23}(A) = -M_{23}(A).$$

Theorem 422C (Cofactor Expansion)[1]   *For any $p = 1, 2, \ldots, n$, the determinant $\det(A)$ may be computed by expansion by cofactors in the pth row. This means that the following formula holds:*

$$\det(A) = a_{p1}C_{p1} + a_{p2}C_{p2} + \cdots + a_{pn}C_{pn}.$$

Similarly, for any $q = 1, 2, \ldots, n$, the determinant $\det(A)$ may be computed by expansion by cofactors in the qth column. This means that the following formula holds:

$$\det(A) = a_{1q}C_{1q} + a_{2q}C_{2q} + \cdots + a_{nq}C_{nq}.$$

Proof: The argument is a bit cumbersome in full generality, so we do the special case where $n = 3$ and $p = 2$. (We'll use an argument that works more generally.) Let

$$A = \begin{bmatrix} a_1 & a_2 & a_3 \\ b_1 & b_2 & b_3 \\ c_1 & c_2 & c_3 \end{bmatrix}.$$

We must prove that

$$\det(A) = -b_1 \det(B_1) + b_2 \det(B_2) - b_3 \det(B_3)$$

---

[1] Also called the **Laplace Expansion**.

where

$$B_1 = \begin{bmatrix} a_2 & a_3 \\ c_2 & c_3 \end{bmatrix}, \qquad B_2 = \begin{bmatrix} a_1 & a_3 \\ c_1 & c_3 \end{bmatrix}, \qquad B_3 = \begin{bmatrix} a_1 & a_2 \\ c_1 & c_2 \end{bmatrix}.$$

By Theorem 241B, the determinant of a matrix is a linear function of each row, so

$$\det(A) = b_1 \det(A_1) + b_2 \det(A_2) + b_3 \det(A_3)$$

where

$$A_1 = \begin{bmatrix} a_1 & a_2 & a_3 \\ 1 & 0 & 0 \\ c_1 & c_2 & c_3 \end{bmatrix}, \qquad A_2 = \begin{bmatrix} a_1 & a_2 & a_3 \\ 0 & 1 & 0 \\ c_1 & c_2 & c_3 \end{bmatrix}, \qquad A_3 = \begin{bmatrix} a_1 & a_2 & a_3 \\ 0 & 0 & 1 \\ c_1 & c_2 & c_3 \end{bmatrix}.$$

To compute the determinants of the matrices A_j, first perform shears so that the jth column of A_j becomes the jth column of the identity matrix. This changes neither the determinant nor the entries not in the jth column. Then perform row swaps to transform A_j to a matrix whose first row is the first row of the identity matrix. Do this by swapping a row with an adjacent row so that the relative order of the other rows is maintained. In the case at hand, only $2 - 1 = 1$ swap(s) are required. Now swap columns so that the first column becomes the first column of the identity matrix. Again, only swap adjacent columns. For this $j - 1$ swaps are required. The result is that $\det(A_j) = (-1)^{2+j} \det(\widetilde{A}_j)$ where

$$\widetilde{A}_j = \begin{bmatrix} 1 & 0 \\ 0 & B_j \end{bmatrix}.$$

But $\det(\widetilde{A}_j) = \det(B_j)$ is immediate from Definition 239B, since the only elementary products that appear in the determinant are those corresponding to permutations σ with $\sigma(1) = 1$.

Expansion in columns follows from expansion in rows, since a matrix and its transpose have the same determinant (Theorem 243C). For example, in the notation of this proof,

$$\det(A) = a_1 \det(C_1) - b_1 \det(C_2) + c_1 \det(C_3)$$

where

$$C_1 = \begin{bmatrix} b_2 & b_3 \\ c_2 & c_3 \end{bmatrix}, \qquad C_2 = \begin{bmatrix} a_2 & a_3 \\ c_2 & c_3 \end{bmatrix}, \qquad C_3 = \begin{bmatrix} a_2 & a_3 \\ b_2 & b_3 \end{bmatrix}.$$

□

15.2   The Companion Matrix

In Theorem 301A, it was proved that the characteristic polynomial of the companion matrix of the monic polynomial is itself. Here we give another proof using expansion by cofactors.

Theorem 424A *If*

$$p(\xi) = \xi^n + p_1\xi^{n-1} + \cdots + p_{n-1}\xi + p_n$$

and

$$\xi I - C = \begin{bmatrix} \xi & -1 & \dots & 0 & 0 \\ 0 & \xi & \dots & 0 & 0 \\ & & \ddots & & \\ 0 & 0 & \dots & \xi & -1 \\ p_n & p_{n-1} & \dots & p_2 & (\xi + p_1) \end{bmatrix},$$

then $\det(\xi I - C) = p(\xi)$.

Proof: Expand $\det(\xi I - C)$ by cofactors on the bottom row. (See Theorem 422C.) The submatrix W_k, obtained from $\xi I - C$ by deleting the row and column containing the entry p_k (or the entry $\xi + p_1$ if $k = n$), has the form

$$W_k = \begin{bmatrix} U_k & 0_{(k-1)\times(n-k)} \\ 0_{(n-k)\times(k-1)} & V_k \end{bmatrix}$$

where $U_k \in \mathbf{F}^{(n-k)\times(n-k)}$ has ξ on the diagonal and -1 on the superdiagonal and $V_k \in \mathbf{F}^{(k-1)\times(k-1)}$ has -1 on the diagonal and ξ on the subdiagonal. Hence, the minor of the entry p_k is

$$\det(W_k) = \det(U_k)\det(V_k) = \xi^{n-k}(-1)^{k-1}$$

and the cofactor is

$$(-1)^{n+(n-k+1)}\det(W_k) = \xi^{n-k}$$

since p_k is in the $(n, n-k+1)$ position. The formula for expansion by cofactors gives

$$\begin{aligned} \det(\xi I - C) &= \\ &= p_n\xi^{n-n} + p_{n-1}\xi^{n-(n-1)} + \cdots + p_2\xi^{n-2} + (\xi + p_1)\xi^{n-1} \\ &= \xi^n + p_1\xi^{n-1} + p_2\xi^{n-2} \cdots + p_{n-1}\xi + p_n \\ &= p(\xi), \end{aligned}$$

as required.   □

Exercise 425A A square matrix A whose entries have the form

$$\text{entry}_{ij}(A) = x_j^{i-1}$$

is called a **Vandermonde's matrix**. Here $x_1, x_2, \ldots, x_n$ are any numbers. For example, if $n = 4$,

$$A = \begin{bmatrix} 1 & 1 & 1 & 1 \\ x_1 & x_2 & x_3 & x_4 \\ x_1^2 & x_2^2 & x_3^2 & x_4^2 \\ x_1^3 & x_2^3 & x_3^3 & x_4^3 \end{bmatrix}.$$

The determinant

$$V = V(x_1, x_2, \ldots, x_n) = \det(A)$$

is a function of n variables called **Vandermonde's determinant**. Show that it is given by

$$\det(A) = \prod_{i>j}(x_i - x_j).$$

Hint: Think of V as a polynomial in x_1. Evaluate for $x_1 = 0$ and for $x_1 = x_2, x_3, \ldots, x_n$.

15.3 Adjoint

We are going to give an explicit formula for the inverse of a matrix. The key step is the following

Definition 425B The **adjoint** $\text{adj}(A)$ of a square matrix A is the transposed matrix of cofactors:

$$\text{entry}_{ij}(\text{adj}(A)) = C_{ji}(A).$$

In other words,

$$\text{adj}(A) = \begin{bmatrix} C_{11}(A) & C_{21}(A) & \cdots & C_{n1}(A) \\ C_{12}(A) & C_{22}(A) & \cdots & C_{n2}(A) \\ & & \ddots & \\ C_{1n}(A) & C_{2n}(A) & \cdots & C_{nn}(A) \end{bmatrix}$$

for any square matrix $A \in \mathbf{F}^{n \times n}$.

Examples 426A When $n = 2$,

$$\text{adj}(A) = \begin{bmatrix} a_{22} & -a_{12} \\ -a_{21} & a_{11} \end{bmatrix}$$

for

$$A = \begin{bmatrix} a_{11} & a_{12} \\ a_{21} & a_{22} \end{bmatrix}.$$

When $n = 3$,

$$\text{adj}(A) =$$

$$= \begin{bmatrix} +(a_{22}a_{33} - a_{32}a_{21}) & -(a_{12}a_{33} - a_{32}a_{13}) & +(a_{12}a_{23} - a_{22}a_{13}) \\ -(a_{21}a_{33} - a_{31}a_{23}) & +(a_{11}a_{33} - a_{31}a_{13}) & -(a_{11}a_{23} - a_{21}a_{13}) \\ +(a_{21}a_{32} - a_{13}a_{22}) & -(a_{11}a_{32} - a_{31}a_{22}) & +(a_{11}a_{22} - a_{21}a_{12}) \end{bmatrix}$$

for

$$A = \begin{bmatrix} a_{11} & a_{12} & a_{13} \\ a_{21} & a_{22} & a_{23} \\ a_{31} & a_{32} & a_{33} \end{bmatrix}.$$

Theorem 426B *The product of a square matrix A with its adjoint matrix* $\text{adj}(A)$ *is the determinant of A times the identity matrix:*

$$A \, \text{adj}(A) = \text{adj}(A)A = \det(A)I.$$

Proof: The proof is simply an application of expansion by cofactors 422C. We prove $A \, \text{adj}(A) = \det(A)I$ and leave $\text{adj}(A)A = \det(A)I$ as an exercise. The (i, k)-entry of the product $A \, \text{adj}(A)$ is

$$\begin{aligned} \text{entry}_{ik}(A \, \text{adj}(A)) &= \sum_{j=1}^{n} \text{entry}_{ij}(A) \, \text{entry}_{jk}(\text{adj}(A)) \\ &= \sum_{j=1}^{n} a_{ij} C_{kj}(A) \\ &= a_{i1} C_{k1}(A) + a_{i2} C_{k2}(A) + \cdots + a_{in} C_{kn}(A) \end{aligned}$$

where $a_{ij} = \text{entry}_{ij}(A)$ is the (i, j)-entry of A. When $i = k$, this is the expansion of $\det(A)$ by cofactors on the ith row, so

$$\text{entry}_{ii}(A \, \text{adj}(A)) = \det(A). \tag{$*$}$$

We show that

$$\text{entry}_{ik}(A \, \text{adj}(A)) = 0 \tag{$\#$}$$

for $i \neq k$. To prove (#), form a matrix B from A by replacing the kth row of A by the ith row of A; that is, $\mathrm{row}_\ell(B) = \mathrm{row}_\ell(A)$ for $\ell \neq k$, whereas $\mathrm{row}_k(B) = \mathrm{row}_i(A)$. Since $C_{kj}(B)$ is computed by deleting the kth row (and the jth column) of B and then computing a sign and a determinant, and since B and A agree except in the kth row, we have that

$$C_{kj}(B) = C_{kj}(A).$$

Hence, expanding by cofactors of B in the ith row, we obtain

$$\det(B) = a_{i1}C_{k1}(A) + a_{i2}C_{k2}(A) + \cdots + a_{in}C_{kn}(A),$$

which is the (i,k)-entry of $A \operatorname{adj}(A)$. But $\det(B) = 0$, since B has a repeated row. This proves (#). Combining (*) and (#), we get

$$\mathrm{entry}_{ik}(A \operatorname{adj}(A)) = \det(A)\,\mathrm{entry}_{ik}(I)$$

for *any* $i, k = 1, 2, \ldots, n$ where $I = I_n$ is the $n \times n$ identity matrix. This shows that $A \operatorname{adj}(A) = \det(A)I$, as required. □

Corollary 427A *A square matrix A is invertible if and only if its determinant is not zero. When $\det(A) \neq 0$, the inverse A^{-1} of A is given by the formula*

$$A^{-1} = \det(A)^{-1} \operatorname{adj}(A).$$

Remark 427B When $n = 2$, we recover the formula 39D

$$\begin{bmatrix} a & b \\ c & d \end{bmatrix}^{-1} = (ad - bc)^{-1} \begin{bmatrix} d & -b \\ -c & a \end{bmatrix}$$

for the inverse of a 2×2 matrix.

▷ **Exercise 427C** Compute the adjoint matrix of $A = \begin{bmatrix} 1 & 2 & 3 \\ 4 & 5 & 6 \\ 7 & 8 & 9 \end{bmatrix}$. What is the rank of A? What is the rank of $\operatorname{adj}(A)$?

▷ **Exercise 427D** Is it always true that a matrix and its adjoint have the same rank? Ever true?

15.4   Cramer's Rule

When $A \in \mathbf{F}^{n \times n}$ is an invertible matrix, the unique solution of the inhomogeneous system $Y = AX$ is $X = A^{-1}Y$. With the help of the adjoint, this can be written in the form

$$X = \det(A)^{-1} \operatorname{adj}(A)Y.$$

Using the definition of the adjoint we obtain

$$\operatorname{entry}_j(\operatorname{adj}(A)Y) = C_{1j}(A)y_1 + C_{2j}(A)y_2 + \cdots + C_{nj}(A)y_n$$

where $y_i = \operatorname{entry}_i(Y)$. But this last formula is the expansion by cofactors along the jth column for the matrix B_j that results from A by replacing the jth column by Y:

$$\begin{aligned}
\operatorname{col}_j(B_j) &= Y, \\
\operatorname{col}_k(B_j) &= \operatorname{col}_k(A) \quad \text{for } k \neq j,
\end{aligned}$$

so

$$C_{ij}(B_j) = C_{ij}(A).$$

Hence, the formula for the jth entry $x_j = \operatorname{entry}_j(X)$ of the solution X is

$$x_j = \frac{\det(B_j)}{\det(A)}.$$

Expanding this out gives

Theorem 428A (Cramer's Rule)   *The solution of the inhomogeneous system*

$$\begin{aligned}
y_1 &= a_{11}x_1 + a_{12}x_2 + \cdots + a_{1n}x_n \\
y_2 &= a_{21}x_1 + a_{22}x_2 + \cdots + a_{2n}x_n \\
&\;\;\vdots \\
y_n &= a_{n1}x_1 + a_{n2}x_2 + \cdots + a_{nn}x_n
\end{aligned}$$

is given by

$$x_j = \frac{\det\left(\begin{bmatrix} a_{11} & \cdots & y_1 & \cdots & a_{1n} \\ a_{21} & \cdots & y_2 & \cdots & a_{2n} \\ & \cdots & & \cdots & \\ a_{n1} & \cdots & y_n & \cdots & a_{nn} \end{bmatrix}\right)}{\det\left(\begin{bmatrix} a_{11} & \cdots & a_{1j} & \cdots & a_{1n} \\ a_{21} & \cdots & a_{2j} & \cdots & a_{2n} \\ & \cdots & & \cdots & \\ a_{n1} & \cdots & a_{nj} & \cdots & a_{nn} \end{bmatrix}\right)}$$

provided that the matrix of coefficients is invertible.

Example 429A The solution of the system

$$\begin{aligned} y_1 &= ax_1 + bx_2 \\ y_2 &= cx_1 + dx_2 \end{aligned}$$

is

$$x_1 = \frac{\det\left(\begin{bmatrix} y_1 & b \\ y_2 & d \end{bmatrix}\right)}{\det\left(\begin{bmatrix} a & b \\ c & d \end{bmatrix}\right)}, \qquad x_2 = \frac{\det\left(\begin{bmatrix} a & y_1 \\ c & y_2 \end{bmatrix}\right)}{\det\left(\begin{bmatrix} a & b \\ c & d \end{bmatrix}\right)}.$$

Exercise 429B Redo Exercise 185D using Cramer's rule.

Using MINIMAT

Function 429C The .M function adj, shown in Listing 19, computes the adjoint of its input. For example, the commands

```
#> A=rand(3,3), B=adj(A)
#> B*A, det(A)*eye(3)
```

will produce the same multiple of the identity matrix.

Exercise 429D Write a .M function det3 that computes the determinant of its input recursively using the formula 422C for expansion by cofactors. Its general form should be

```
function d = det3(A)

[m n] = size(A);
if m==1
        d=A(1,1);
else
    .....
end
```

The dots indicate a patch of code that expresses d in terms of the expansion by cofactors in the first row. This code will contain a loop and expressions like det3(A([1:i-1, i+1:n], 2:n). WARNING: This function is very inefficient.

The Adjoint Matrix in MINIMAT

```
function B = adj(A)

    [m n] = size(A); B = zeros(n,n); si = 1;
    for i = 1:n
        si = -si;  sj = si;
        for j = 1:n
            sj = -sj;
            rows  = [1:i-1,i+1:n];
            cols  = [1:j-1,j+1:n];
            B(j,i) = sj*det(A(rows,cols));
        end % for j
    end % for i
```

Listing 19

15.5   Derivative of the Determinant

Theorem 430A *Suppose that $A(\xi) \in \mathbf{F}^{n \times n}$ is a differentiable matrix valued function of a real variable ξ. Let*

$$p(\xi) = \det(A(\xi))$$

denote its determinant. Then the derivative $\dot{p}(\xi)$ of $p(\xi)$ is given by

$$\dot{p}(\xi) = \mathrm{trace}(\dot{A}(\xi)A(\xi)^{-1}) \det(A(\xi))$$

for any value of ξ where $A(\xi)$ is invertible.

Proof: We first do a special case: we assume that $A(0) = I$, the identity matrix; we will verify the formula at $\xi = 0$, i.e. we will show that $\dot{p}(0) = \mathrm{trace}(\dot{A}(0))$. Since $\det(A(\xi))$ and $\det(A(0) + \xi\dot{A}(0))$ have the same derivative at $\xi = 0$, we may (by the chain rule) assume without loss of generality that

$$A(\xi) = I + \xi B, \qquad B = \dot{A}(0).$$

By Theorem 268B,

$$p(\xi) = \det(A(\xi)) = \prod_{i=1}^{n}(1 + \xi\lambda_i), \qquad \text{trace}(B) = \sum_{i=1}^{n}\lambda_i,$$

where $\lambda_1, \lambda_2, \ldots, \lambda_n$ are the eigenvalues of B. Differentiating with respect to ξ and evaluating at $\xi = 0$ gives

$$\frac{d}{d\xi}\prod_{j}(1 + \xi\lambda_j)\bigg|_{\xi=0} = \sum_{j}\lambda_j,$$

and hence, $\dot{p}(0) = \text{trace}(B) = \text{trace}(\dot{A}(0))$, as required.

Now we prove the general case. Fix ξ and define $C(h) = A(\xi+h)A(\xi)^{-1}$. Then $A(\xi + h) = C(h)A(\xi)$, so $p(\xi + h) = \det(C(h))\det(A(\xi))$. Differentiating with respect to h and evaluating at $h = 0$ gives the desired result. (The special case applies because $C(0) = I$.) $\qquad\square$

Corollary 431A *Let $A(\xi)$ and $p(\xi)$ be as in the Theorem 430A. Then*

$$\dot{p}(\xi) = \text{trace}(\dot{A}(\xi)\,\text{adj}(A(\xi)))$$

for all ξ.

Proof: At any value of ξ where $A(\xi)$ is invertible, we have

$$\begin{aligned}
\dot{p}(\xi) &= \text{trace}(\dot{A}(\xi)A(\xi)^{-1})\det(A(\xi)) \\
&= \text{trace}(\dot{A}(\xi)\det(A(\xi))^{-1}\,\text{adj}(A(\xi)))\det(A(\xi)) \\
&= \text{trace}(\dot{A}(\xi)\,\text{adj}(A(\xi))).
\end{aligned}$$

The formula must hold for all values of ξ by continuity. $\qquad\square$

Corollary 431B *Let $A \in \mathbf{F}^{n\times n}$ be a square matrix and*

$$p(\xi) = \det(\xi I - A)$$

be its characteristic polynomial. Then

$$\dot{p}(\xi) = \text{trace}(\text{adj}(\xi I - A)).$$

15.6   The Souriau-Frame Algorithm

The characteristic polynomial of a matrix is much harder to compute than the determinant. The next theorem gives an efficient algorithm for computing it.

Theorem 432A (Souriau-Frame Algorithm)   *Let*

$$p(\xi) = \det(\xi I - A) = \xi^n + p_1 \xi^{n-1} + \cdots p_{n-1} \xi + p_n$$

be the characteristic polynomial of $A \in \mathbf{F}^{n \times n}$. *Define* $n \times n$ *matrices* $B_0, B_1, \ldots, B_{n-1}$ *by*

$$\mathrm{adj}\,(\xi I - A) = B_0 \xi^{n-1} + B_1 \xi^{n-2} + \cdots B_{n-2} \xi + B_{n-1}. \qquad (1)$$

Then

$$B_0 = I; \qquad (2.0)$$

$$B_k = AB_{k-1} + p_k I \qquad \text{for } k = 1, 2, \ldots, n-1; \qquad (2.k)$$

$$0 = AB_{n-1} + p_n I; \qquad (2.n)$$

$$p_k = -\frac{1}{k} \, \mathrm{trace}(AB_{k-1}) \qquad \text{for } k = 1, 2, \ldots, n. \qquad (3)$$

where $I = I_n$ *is the* $n \times n$ *identity matrix.*

Proof: Reading $\xi I - A$ for A in Theorem 426B gives

$$(\xi I - A) \, \mathrm{adj}(\xi I - A) = p(\xi)I = \xi^n I + p_1 \xi^{n-1} I + \cdots p_{n-1} \xi I + p_n I,$$

while (1) gives

$$(\xi I - A) \, \mathrm{adj}(\xi I - A) =$$
$$= B_0 \xi^n + (B_1 - AB_0)\xi^{n-1} + \cdots (B_{n-1} - AB_{n-2})\xi - AB_{n-1}.$$

Equating coefficients gives

$$I = B_0,$$
$$p_1 I = B_1 - AB_0,$$
$$\vdots$$
$$p_{n-1} I = B_{n-1} - AB_{n-1},$$
$$p_n I = -AB_{n-1}.$$

This proves (2.k) for $k = 0, 1, \ldots, n$.

It remains to prove (3). Since

$$p(\xi) = \xi^n + \sum_{k=1}^{n} p_k \xi^{n-k},$$

we get

$$\dot{p}(\xi) = n\xi^{n-1} + \sum_{k=1}^{n-1} (n-k)p_k \xi^{n-k-1}.$$

By Corollary 431B,

$$\dot{p}(\xi) = \text{trace}(\text{adj}(\xi I - A)) = \sum_{k=0}^{n-1} \text{trace}(B_k)\xi^{n-k-1},$$

Equating coefficients gives

$$(n-k)p_k = \text{trace}(B_k).$$

By (2.k),

$$\begin{aligned}
\text{trace}(B_k) &= \text{trace}(AB_{k-1}) + p_k \text{trace}(I) \\
&= \text{trace}(AB_{k-1}) + np_k.
\end{aligned}$$

Combining these and subtracting np_k gives

$$-kp_k = \text{trace}(AB_{k-1}).$$

This proves (3) for $k < n$. The case $k = n$ follows from (2.n) by taking the trace.   □

The Souriau-Frame theorem provides an algorithm for computing the characteristic polynomial: Starting from $B_0 = I$ we compute p_1 using (3), and then B_1 using (2.1). Then we compute p_2 using (3), and then B_2 using (2.2). We proceed in this way till we have computed $B_0, B_1, \ldots, B_{n-1}$ and $p_1, p_2, \ldots, p_{n-1}$. Finally, we compute p_n using (3).

Exercise 433A Find $B_0, p_1, B_1, p_2, B_2, p_3$, if $A = \begin{bmatrix} a_{11} & a_{12} & a_{13} \\ a_{21} & a_{22} & a_{23} \\ a_{31} & a_{32} & a_{33} \end{bmatrix}$.

▷ **Exercise 433B** MINIMAT has two built-in functions sum and diag that can be used to compute the trace of a matrix. For a $1 \times n$ matrix L, the command

```
#>   s=sum(L)
```

assigns to s the sum of its entries. For a $n \times n$ matrix A, the command

```
#>   L=diag(A)
```

assigns to L a $1 \times n$ matrix whose entries are the diagonal elements of A. Compute the trace of a random 4×4 matrix and confirm your answer.

Function 434A The .M function sourfram, shown in Listing 20, computes the characteristic polynomial of its input using Souriau-Frame algorithm. More precisely, if A holds the $n \times n$ matrix A, the command

```
#> p = sourfram(A)
```

computes a $1 \times n$ matrix p whose entries are the coefficients of the characteristic polynomial

$$\det(\xi I - A) = \xi^n + p(1) * \xi^{n-1} + p(2) * \xi^{n-2} + \cdots + p(n-1) * \xi + p(n)$$

of the matrix in A as indicated.

▷ **Exercise 434B** Use the .M function sourfram to calculate a degree four polynomial whose roots are $1, 2, 3, 5$.

Characteristic Polynomial via Souriau-Frame

```
function p=sourfram(A)

        n = max(size(A));
        p = zeros(1,n);
        B = eye(n);
        for k = 1:n
                p(k) = -sum(diag(A*B))/k;
                B = A*B+p(k)*eye(n);
        end
```

Listing 20

Chapter Summary

An explicit formula for the inverse of a matrix can be determined from the formula

$$A \text{ adj}(A) = \det(A)I.$$

The adjoint matrix $\text{adj}(A)$ *is the transposed matrix of cofactors. A cofactor is a signed minor, and a minor is the determinant of a submatrix obtained by deleting a row and a column. The diagonal entries in this formula give a formula for computing the determinant called* expansion by cofactors *or* Laplace expansion.

If $A(\xi)$ *is a matrix valued function satisfying* $A(0) = I$, *then the derivative of the determinant is given by*

$$\frac{d}{d\xi} \det\big(A(\xi)\big)\bigg|_{\xi=0} = \text{trace}(\dot{A}(0)).$$

This is because

$$\frac{d}{d\xi} \prod_j (1 + \xi\lambda_j)\bigg|_{\xi=0} = \sum_j \lambda_j.$$

The Souriau-Frame algorithm gives an efficient method for calculating the determinant. It can be implemented in MINIMAT.

A

PROOFS

In this appendix we give some proofs that were omitted in the text. Although the arguments provide good practice in manipulating sigma notation, they are not very important. The abstract theory of linear algebra gives more conceptual proofs for the algebraic laws of matrix multiplication.

A.1   Matrix Algebra

Example 437A We prove the commutative law for addition. Suppose that $A, B \in \mathbf{F}^{m \times n}$. To prove that $A + B = B + A$ we must prove that

$$\text{entry}_{ij}(A + B) = \text{entry}_{ij}(B + A)$$

for $i = 1, 2, \ldots, m$ and $j = 1, 2, \ldots, n$. Choose i and j. Then

$$
\begin{aligned}
\text{entry}_{ij}(A + B) &= \text{entry}_{ij}(A) + \text{entry}_{ij}(B) \\
&= \text{entry}_{ij}(B) + \text{entry}_{ij}(A) \\
&= \text{entry}_{ij}(B + A)
\end{aligned}
$$

Note that the proof uses the commutative law for addition of numbers. $\square$

Example 438A We prove that $IA = A$ where $A \in \mathbf{F}^{m \times n}$ and $I = I_m \in \mathbf{F}^{m \times m}$ is the identity matrix. The matrices IA and A have the same size, so to show that they are equal we must show that they have the same entries. But

$$\text{entry}_{ij}(IA) = \sum_{k=1}^{m} \text{entry}_{ik}(I)\,\text{entry}_{kj}(A) = \text{entry}_{ij}(A)$$

since $\text{entry}_{ik}(I) = 0$ for $k \neq i$, whereas $\text{entry}_{ii}(I) = 1$. $\qquad\qquad$ □

Example 438B We prove the associative law for multiplication. Suppose $A \in \mathbf{F}^{m \times n}$, $B \in \mathbf{F}^{n \times p}$, and $C \in \mathbf{F}^{p \times q}$. Then $AB \in \mathbf{F}^{m \times p}$ and $BC \in \mathbf{F}^{n \times q}$. Both $(AB)C$ and $A(BC)$ have size $m \times q$. To show that $(AB)C = A(BC)$ we must prove that

$$\text{entry}_{il}((AB)C) = \text{entry}_{il}(A(BC))$$

for $i = 1, 2, \ldots, m$ and $l = 1, 2, \ldots, q$. Choose indices i and l. Then

$\text{entry}_{il}((AB)C) \quad =$

$$= \quad \sum_{k=1}^{p} \text{entry}_{ik}(AB)\,\text{entry}_{kl}(C) \tag{1}$$

$$= \quad \sum_{k=1}^{p} \left(\sum_{j=1}^{n} \text{entry}_{ij}(A)\,\text{entry}_{jk}(B) \right) \text{entry}_{kl}(C) \tag{2}$$

$$= \quad \sum_{k=1}^{p} \sum_{j=1}^{n} \text{entry}_{ij}(A)\,\text{entry}_{jk}(B)\,\text{entry}_{kl}(C) \tag{3}$$

$$= \quad \sum_{j=1}^{n} \sum_{k=1}^{p} \text{entry}_{ij}(A)\,\text{entry}_{jk}(B)\,\text{entry}_{kl}(C) \tag{4}$$

$$= \quad \sum_{j=1}^{n} \text{entry}_{ij}(A) \left(\sum_{k=1}^{p} \text{entry}_{jk}(B)\,\text{entry}_{kl}(C) \right) \tag{5}$$

$$= \quad \sum_{j=1}^{n} \text{entry}_{ij}(A)\,\text{entry}_{jl}(BC) \tag{6}$$

$$= \quad \text{entry}_{il}(A(BC)). \tag{7}$$

▷ **Exercise 438C** Following Example 437A, prove the scalar right distributive law: $(a+b)C = aC + bC$. Here a and b are numbers and C is a matrix.

▷ **Exercise 438D** Justify each of the steps in proof of 438B.

Exercise 439A Prove the scalar left distributive law, $c(A+B) = cA + cB$. Here c is a number and A and B are matrices of the same size.

▷ **Exercise 439B** Prove the right distributive law $(A + B)C = AC + AB$ for matrix multiplication.

Exercise 439C Prove the left distributive law $C(A + B) = CA + CB$ for matrix multiplication.

A.2   Block Multiplication

We prove the Block Multiplication Law. Refer to the notation in Theorem 51A. We must show that C_{wu} and $\sum_v A_{wv}B_{vu}$ have the same (y, z) entry for each $y = 1, 2, \ldots, m_w$ and $z = 1, 2, \ldots, p_u$. The hard part is keeping track of the subscripts. To do this, note that the labeling of entries in a submatrix is related to the labeling in the ambient matrix. For example, if

$$C = \left[\begin{array}{cc} C_{11} & C_{12} \\ C_{21} & C_{22} \end{array} \right]$$

with C_{wu} of size $m_w \times p_u$, then

$$\text{entry}_{yz}(C_{12}) = \text{entry}_{ik}(C)$$

for $y = 1, \ldots, m_1$, $z = 1, \ldots, p_2$, $i = y$, $k = p_1 + z$.

Suppose, as Theorem 51A, that A_{wv} of size $m_w \times n_v$, B_{vu} of size $n_v \times p_u$, C_{wu} is of size $m_w \times p_u$. Since the matrix C_{wu} is a submatrix of AB this entry is an entry of AB:

$$\text{entry}_{yz}(C_{wu}) = \text{entry}_{ik}(AB)$$

for certain indices (i, k):

$$
\begin{aligned}
i &= m_1 + m_2 + \cdots + m_{w-1} + y, \\
k &= p_1 + p_2 + \cdots + p_{u-1} + z.
\end{aligned}
$$

Since the blocks of A align, the i-th row of A appears in the row of blocks

$$\text{row}_i(A) = \left[\begin{array}{cccc} \text{row}_y(A_{w1}) & \text{row}_y(A_{w2}) & \cdots & \text{row}_y(A_{wr}) \end{array} \right]$$

and similarly the k-th column of B appears in the column of blocks

$$\text{col}_k(B) = \left[\begin{array}{c} \text{col}_z(B_{1u}) \\ \text{col}_z(B_{2u}) \\ \vdots \\ \text{col}_z(B_{ru}) \end{array} \right].$$

Thus

$$
\begin{aligned}
\mathrm{entry}_{yz}(C_{wu}) &= \mathrm{entry}_{ik}(AB) \\
&= \mathrm{row}_i(A)\,\mathrm{col}_k(B) \\
&= \sum_{v=1}^{r} \mathrm{row}_y(A_{wv})\,\mathrm{col}_z(B_{vu}) \\
&= \sum_{v=1}^{r} \mathrm{entry}_{yz}(A_{wv}B_{vu}) \\
&= \mathrm{entry}_{yz}\left(\sum_{v=1}^{r} A_{wv}B_{vu}\right). \qquad \square
\end{aligned}
$$

The **Row-Column Law** (33B)

$$\mathrm{entry}_{ik}(AB) = \mathrm{row}_i(A)\ \mathrm{col}_k(B).$$

can be contrasted with the following

Theorem 440A (Column-Row Law)   *We have*

$$AB = \sum_{j=1}^{n} \mathrm{col}_j(A)\,\mathrm{row}_j(B)$$

for $A \in \mathbf{F}^{m\times n}$ and $B \in \mathbf{F}^{n\times p}$.

Proof: $\mathrm{entry}_{ij}(A)\,\mathrm{entry}_{jk}(B) = \mathrm{entry}_{ik}(\mathrm{col}_j(A)\,\mathrm{row}_j(B))$. Now sum on j to get

$$\mathrm{entry}_{ik}(AB) = \mathrm{entry}_{ik}\left(\sum_{j=1}^{n} \mathrm{col}_j(A)\,\mathrm{row}_j(B)\right) \qquad \square$$

The following exercises are special cases of the block multiplication law. Test your understanding by finding direct proofs.

▷ **Exercise 440B** Prove that

$$\mathrm{row}_i(AB) = \mathrm{row}_i(A)B$$

for $A \in \mathbf{F}^{m\times n}$, $B \in \mathbf{F}^{n\times p}$, and $i = 1, 2\ldots, m$.

▷ **Exercise 440C** Prove that

$$\mathrm{col}_k(AB) = A\,\mathrm{col}_k(B)$$

for $A \in \mathbf{F}^{m\times n}$, $B \in \mathbf{F}^{n\times p}$, and $k = 1, 2\ldots, p$.

▷ **Exercise 441A** Prove that

$$AX = x_1 \operatorname{col}_1(A) + x_2 \operatorname{col}_2(A) + \cdots + x_n \operatorname{col}_n(A)$$

where $A \in \mathbf{F}^{m \times n}$, $X \in \mathbf{F}^{n \times 1}$, and $x_j = \operatorname{entry}_j(X)$ is the j-th entry of X.

Exercise 441B Prove that

$$HA = h_1 \operatorname{row}_1(A) + h_2 \operatorname{row}_2(A) + \cdots + h_m \operatorname{row}_m(A)$$

where $A \in \mathbf{F}^{m \times n}$, $H \in \mathbf{F}^{1 \times m}$, and $h_i = \operatorname{entry}_i(H)$ is the i-th entry of H.

A.3   The Fundamental Theorem

The proof of the Fundamental Theorem has three parts, one for each kind of elementary operation. In each part, we must show that $B = EA$ where B denotes the matrix which results from $A \in \mathbf{F}^{m \times n}$ by performing an elementary operation and $E \in \mathbf{F}^{m \times m}$ denotes the corresponding elementary matrix. We do this by proving that B and EA have the same rows, that is,

$$\operatorname{row}_i(B) = \operatorname{row}_i(EA)$$

for $i = 1, 2, \ldots, m$. This in turn is an easy consequence of the definitions and the equation

$$\operatorname{row}_i(EA) = \operatorname{row}_i(E)A$$

(See Exercise 440B). We develop the proof as a series of lemmas.

Lemma 441C *Assume* $\operatorname{row}_i(E) = \operatorname{row}_i(I)$ *where* $I = I_m$ *is the identity matrix. Then* $\operatorname{row}_i(EA) = \operatorname{row}_i(A)$.

Proof: If $\operatorname{row}_i(E) = \operatorname{row}_i(I)$, then

$$
\begin{aligned}
\operatorname{row}_i(EA) &= \operatorname{row}_i(E)A \\
&= \operatorname{row}_i(I)A \\
&= \operatorname{row}_i(IA) \\
&= \operatorname{row}_i(A). \qquad \square
\end{aligned}
$$

When an elementary row operation is applied to a matrix, only one (or two, in the case of a swap) of the rows of a matrix are changed. In each of the three following lemmas, Lemma 441C is used to to prove that $\operatorname{row}_i(EA) = \operatorname{row}_i(A)$ for these unchanged rows.

Lemma 442A $E = \text{Scale}(I_m, p, c) \implies EA = \text{Scale}(A, p, c)$.

Proof: Suppose $B = \text{Scale}(A, p, c)$, $E = \text{Scale}(I, p, c)$, $I = I_m$. We must show $B = EA$. By definition

$$\text{row}_i(B) = \begin{cases} \text{row}_i(A) & \text{if } i \neq p, \\ c \cdot \text{row}_i(A) & \text{if } i = p, \end{cases}$$

and

$$\text{row}_i(E) = \begin{cases} \text{row}_i(I) & \text{if } i \neq p, \\ c \cdot \text{row}_i(I) & \text{if } i = p. \end{cases}$$

For $i \neq p$, $\text{row}_i(EA) = \text{row}_i(A) = \text{row}_i(B)$ by Lemma 441C. For $i = p$:

$$\begin{aligned} \text{row}_p(EA) &= \text{row}_p(E)A \\ &= c \cdot \text{row}_p(I)A \\ &= c \cdot \text{row}_p(IA) \\ &= c \cdot \text{row}_p(A) \\ &= \text{row}_p(B). \end{aligned} \qquad \square$$

Lemma 442B $E = \text{Swap}(I_m, p, q) \implies EA = \text{Swap}(A, p, q)$.

Proof: Suppose $B = \text{Swap}(A, p, q)$, $E = \text{Swap}(I, p, q)$, $I = I_m$. We must show $B = EA$. By definition:

$$\text{row}_i(B) = \begin{cases} \text{row}_i(A) & \text{if } i \neq p, q, \\ \text{row}_q(A) & \text{if } i = p, \\ \text{row}_i(A) & \text{if } i = q, \end{cases}$$

and

$$\text{row}_i(E) = \begin{cases} \text{row}_i(I) & \text{if } i \neq p, q, \\ \text{row}_q(I) & \text{if } i = p, \\ \text{row}_p(I) & \text{if } i = q. \end{cases}$$

For $i \neq p, q$, $\text{row}_i(EA) = \text{row}_i(A) = \text{row}_i(B)$ by Lemma 441C. For $i = p$:

$$\begin{aligned} \text{row}_p(EA) &= \text{row}_p(E)A \\ &= \text{row}_q(I)A \\ &= \text{row}_q(IA) \\ &= \text{row}_q(A) \\ &= \text{row}_p(B). \end{aligned}$$

and for $i = q$:

$$\begin{aligned}
\text{row}_q(EA) &= \text{row}_q(E)A \\
&= \text{row}_p(I)A \\
&= \text{row}_i(IA) \\
&= \text{row}_i(A) \\
&= \text{row}_q(B).
\end{aligned}$$

$\square$

Lemma 443A $E = \text{Shear}(I_m, p, q, c) \implies EA = \text{Shear}(A, p, q, c)$.

Proof: Suppose $B = \text{Shear}(A, p, q, c)$, $E = \text{Shear}(I_m, p, q, c)$, $I = I_m$. We must show that $B = EA$. By definition:

$$\text{row}_i(B) = \begin{cases} \text{row}_i(A) & \text{if } i \neq p, \\ \text{row}_i(A) + c \cdot \text{row}_q(A) & \text{if } i = p, \end{cases}$$

and

$$\text{row}_i(E) = \begin{cases} \text{row}_i(I) & \text{if } i \neq p, \\ \text{row}_p(I) + c \cdot \text{row}_q(I) & \text{if } i = p. \end{cases}$$

For $i \neq p$, $\text{row}_i(EA) = \text{row}_i(A) = \text{row}_i(B)$ by Lemma 441C. For $i = p$:

$$\begin{aligned}
\text{row}_p(EA) &= \text{row}_p(E)A \\
&= \big(\text{row}_p(I) + c \cdot \text{row}_q(I)\big)A \\
&= \text{row}_i(IA) + c \cdot \text{row}_q(IA) \\
&= \text{row}_i(A) + c \cdot \text{row}_q(A) \\
&= \text{row}_i(B).
\end{aligned}$$

$\square$

B

MATHEMATICAL INDUCTION

Mathematical induction is a method for proving statements of form $P(n)$ *is true for all natural numbers* $n = 0, 1, 2, \ldots$. A proof by induction consist of two parts: a **basis step** where $P(0)$ is proved and an **induction step** where $P(n)$ is assumed and $P(n + 1)$ is proved. Then we reason as follows.

$$
\begin{array}{rll}
P(0) & \text{(Basis step)} \\
P(0) \implies P(1) & \text{(Induction step, } n = 0) \\
P(1) & \text{(Last two steps)} \\
P(1) \implies P(2) & \text{(Induction step, } n = 1) \\
P(2) & \text{(Last two steps)} \\
P(2) \implies P(3) & \text{(Induction step, } n = 2) \\
P(3) & \text{(Last two steps)}
\end{array}
$$

$\cdots$

and so on.

Example 445A Using mathematical induction, we prove the formula

$$P(n) \qquad\qquad \sum_{k=1}^{n} k = \frac{n(n+1)}{2}.$$

Basis Step: When $n = 1$, both sides evaluate to 1, so $P(1)$ is true.

Induction Step: Assume $P(n)$. Add $n + 1$ to both sides. The result is

$$\sum_{k=1}^{n+1} k = \frac{n(n+1)}{2} + (n+1) = \frac{(n+1)(n+2)}{2}.$$

This is $P(n+1)$. $\qquad\qquad\qquad\qquad\qquad\qquad\qquad\qquad\qquad$ □

Theorem 446A (Binomial Formula) *If a and b are numbers and n is a positive integer, then*

$$(a+b)^n = \sum_{p+q=n} \frac{n!}{p!q!} a^p b^q$$

where the sum is over all pairs (p, q) of nonnegative integers which satisfy $p + q = n$.

Proof: **Basis Step.** The only pair (p, q) of nonnegative integers with $p + q = 0$ is $(p, q) = (0, 0)$. Thus

$$(a+b)^0 = 1 = \frac{0!}{0!0!} a^0 b^0 = \sum_{p+q=0} \frac{n!}{p!q!} a^p b^q.$$

Induction Step. Assume

$$(a+b)^n = \sum_{p=0}^{n} \frac{n!}{p!(n-p)!} a^p b^{n-p}.$$

Multiply both sides by $(a + b)$:

$$(a+b)^{n+1} = \sum_{p=0}^{n} \frac{n!}{p!(n-p)!} a^{p+1} b^{n-p} + \sum_{p=0}^{n} \frac{n!}{p!(n-p)!} a^p b^{n+1-p}.$$

Let $k = p + 1$ in the first sum and let $k = p$ in the second:

$$(a+b)^{n+1} = \sum_{k=1}^{n+1} \frac{n!}{(k-1)!(n+1-k)!} a^k b^{n+1-k} + \sum_{k=0}^{n} \frac{n!}{k!(n-k)!} a^k b^{n+1-k}.$$

Now use the formula

$$\frac{n!}{(k-1)!(n+1-k)!} + \frac{n!}{k!(n-k)!} = \frac{(n+1)!}{k!(n+1-k)!}$$

for $1 \leq k \leq n$ and combine the two sums to get

$$(a+b)^{n+1} = \sum_{k=0}^{n+1} \frac{(n+1)!}{k!(n+1-k)!} a^k b^{n+1-k}$$

as required.   □

Recall from freshman calculus that the exponential formula has the infinite series

$$e^x = \sum_{n=1}^{\infty} \frac{x^n}{n!}. \qquad (*)$$

Using the binomial formula we can prove

Corollary 447A $e^{a+b} = e^a e^b$.

Proof:

$$
\begin{aligned}
e^{a+b} &= \sum_{n=0}^{\infty} \frac{(a+b)^n}{n!} \\
&= \sum_{n=0}^{\infty} \left(\sum_{p+q=n} \frac{a^p b^q}{p!\, q!} \right) \\
&= \left(\sum_{p=0}^{\infty} \frac{a^p}{p!} \right) \left(\sum_{q=0}^{\infty} \frac{b^q}{q!} \right) \\
&= e^a e^b.
\end{aligned}
$$

□

Exercise 447B Prove the formula

$$\sum_{k=1}^{n} k^2 = \frac{n(2n+1)(n+1)}{6}$$

using mathematical induction.

Exercise 448A If we take $x = i\theta$ in equation (∗) Page 447, we get

$$e^{i\theta} = \sum_{k=0}^{\infty} \frac{(-1)^k \theta^{2k}}{(2k)!} + i \sum_{k=0}^{\infty} \frac{(-1)^k \theta^{2k+1}}{(2k+1)!}.$$

Substitute $e^{i\theta} = \cos\theta + i\sin\theta$ and equate real and imaginary parts. We get the formulas

$$\cos\theta = \sum_{k=0}^{\infty} \frac{(-1)^k \theta^{2k}}{(2k)!}, \qquad \sin\theta = \sum_{k=0}^{\infty} \frac{(-1)^k \theta^{2k+1}}{(2k+1)!}.$$

(These formulas are usually derived directly from Taylor's formula in first year calculus.) Use a method similar to the proof of Corollary 447A to prove that $\sin(a+b) = \sin a \cos b + \cos a \sin b$.

C

SUMMARY OF MINIMAT

In this appendix we summarize the notations accepted by the computer program MINIMAT. MINIMAT's help file is fairly exhaustive and easy to use. Most of the features of MINIMAT were covered in Sections 1.8 and 2.9. This appendix supplements that material. It also provides page references to various portions of the text which discuss specific aspects of MINIMAT.

Almost everything in this book works with most versions of *Matlab* (including *PC-Matlab* and the *Student Edition of Matlab* published by Prentice-Hall) as well as with MINIMAT. (The only exceptions are some of the menu commands described in the MINIMAT Tutorial on Page 523.) These other versions may not have a user interface designed for use by students in an elementary matrix algebra course, but will have other compensating features.[1] If you are using some version of Matlab with this book, you will need to make the .M files supplied on the distribution diskette available to it.

---

[1] They will probably be faster, more accurate, implement more functions, and capable of handling larger matrices than MINIMAT. Many will have graphics.

C.1   Some Operations in MINIMAT

MINIMAT	Mathematics	English
C = A+B	$C = A + B$	Matrix Addition
C = A-B	$C = A - B$	Matrix Subtraction
C = a*B	$C = aB$	Scalar Multiplication
C = B/a	$C = a^{-1}B$	Scalar Division
C = A*B	$C = AB$	Matrix Multiplication
C = A^p	$C = A^p$	Power
C = A'	$C = A^*$	Conjugate Transpose
C = A.'	$C = A^\top$	Ordinary Transpose
C = zeros(m,n)	$C = 0_{m \times n}$	Zero Matrix
C = rand(m,n)		Random Matrix
I = eye(n)	$I = I_n$	Identity Matrix

These notations are used throughout the book. The operations described on this page may be combined using parentheses in usual way. For example,

```
#> A=[1 2 3]; B=[4 5 6]; C=[7 8 9];
#> C*(A+B)'
   ans = 172
```

C.2   Columnwise Operations

MINIMAT	Mathematics	English
B=all(A)	$B_i = 1$ if $(\forall j)A_{ij} \neq 0$ else $= 0$	For all
B=any(A)	$B_i = 1$ if $(\exists j)A_{ij} \neq 0$ else $= 0$	Exists
B=max(A)	$B_i = \max_j A_{ij}$	Maximum
[B I]=max(A)	$B_i = \max_j A_{ij}$	and where attained
B=min(A)	$B_i = \min_j A_{ij}$	Minimum
[B I]=min(A)	$B_i = \min_j A_{ij}$	and where attained
B=sum(A)	$B_i = \sum_j A_{ij}$	Sum

These operations return a row matrix with one entry for each column of the input matrix A if A is $m \times n$ with both m and n greater than 1. However, if A is a row matrix or a column matrix, the operation returns a scalar. If the input matrix A is complex, it is first replaced by its (elementwise) absolute value before the first four functions are performed. Thus if A is complex, max(A) and max(abs(A)) return the same result. The max and min have the additional ability to return the position of the extreme value. For example,

```
#> R = [1 2 3 6 4 2 5];    [m j] = max(R)
   m = 6
   j = 4
```

The appropriate way to test whether a matrix is zero (in a .M function) is as follows:

```
if max(max(abs(A))) < 1.0E-9  % then A is zero
```

If we know that A is a row or column, then one max is enough. The test

```
if max(max(abs(A))) == 0  % then A is zero
```

should be avoided unless we know that the entries of A are integers. (Round-off error can invalidate a test for equality.) The number 10^{-9} seems to work for the problems in this book.

C.3   Scalar Built-in Functions

MINIMAT	Mathematics	English
w=abs(z)	$w = \lvert z \rvert$	Absolute Value
w=acos(z)	$w = \arccos(z)$	Inverse Cosine
w=asin(z)	$w = \arcsin(z)$	Inverse Sine
w=atan(z)	$w = \arctan(z)$	Inverse Tangent
w=atan2(y,x)	$w = \arctan(y, x)$	Four Quadrant Arctan
w=conj(z)	$w = \bar{z}$	Complex Conjugate
w=cos(z)	$w = \cos(z)$	Cosine
w=exp(z)	$w = e^{z}$	Exponential
w=fix(z)	$w = \lfloor z \rfloor$	Integer Part
w=imag(z)	$w = \Im(z)$	Imaginary Part
w=log(z)	$w = \ln(z)$	Natural Logarithm
w=pi	$w = \pi$	3.14159...
w=real(z)	$w = \Re(z)$	Real Part
w=round(z)	$w = \lfloor z + \frac{1}{2} \rfloor$	Nearest Integer
w=sign(z)	$w = z/\lvert z \rvert$ ($= 0$ if $z = 0$)	Signum
w=sin(z)	$w = \sin(z)$	Sine
w=sqrt(z)	$w = \sqrt{z}$	Square Root
w=tan(z)	$w = \tan(z)$	Tangent

These functions usually operate elementwise. For example, the expression sin([a b c]) evaluates to [sin(a) sin(b) sin(c)]. This is a little different from the notations used in mathematics (see Page 290), but it's a handy notation.

Two handy built-in functions which operate elementwise are round, which rounds to the nearest integer and fix, which rounds toward zero. For example, the command

```
#> A = round(10*rand(3,5))
```

can be used to create a 3×5 matrix with random integer values from 0 to 10. (The entries of a matrix returned by the rand function are real numbers between 0 and 1.)

C.4   Matrix Built-in Functions

MINIMAT	Mathematics	English
C = inv(A)	$C = A^{-1}$	Inverse (38)
B=expm(A)	$B = \exp(A)$	Matrix Exponential (290)
d=det(A)	$d = \det(A)$	Determinant (248)
D=eig(A)	$AV = VD$	Eigenvalues (258)
[P D]=eig(A)	$A = PDP^{-1}$	and Eigenvectors
D=svd(A)	$A = UDV'$	Singular Values
[U D V]=svd(A)	$A = UDV'$	and Singular Vectors (370)
[Q R]=qr(A)	$A = QR$	Householder
Q=qr(A)	$A = QR$	Decomposition (362)
[L U]=lu(A)	$A = LU$	Gaussian
U=lu(A)	$A = LU$	Elimination (342)

These functions accept a matrix as input and produce one or more matrices as output. In general, they implement functions explained in the text, but the algorithms that MINIMAT uses are different from the algorithms described in the text for reasons of efficiency and accuracy. In the table, the numbers in parentheses indicate the page in the text where the function is explained.

C.5   Subscripts in MINIMAT

MINIMAT	Mathematics	English
c = A(i,j)	$c = \text{entry}_{ij}(A)$	(i,j)-entry
C = A(i,:)	$C = \text{row}_i(A)$	one row
C = A(:,j)	$C = \text{col}_j(A)$	one column
C = A(I,J)	$\text{entry}_{uv}(C) = \text{entry}_{ij}(A)$	submatrix
C = A(I,:)	$\text{row}_u(C) = \text{row}_i(A)$	rows from I
C = A(:,J)	$\text{col}_v(C) = \text{col}_j(A)$	columns from J
c = A(i)	$c = \text{entry}_i(A)$	(legal if $n = 1$ or $m = 1$)
C = A(I)	$\text{entry}_u(C) = \text{entry}_i(A)$	(legal if $n = 1$ or $m = 1$)
C = A(:)	$\text{entry}_{(j-1)m+i}(C) = \text{entry}_{ij}(A)$	one big column
L = p:q	$L = \begin{bmatrix} p & p+1 & \cdots & q \end{bmatrix}$	range

In this table, $A \in \mathbf{F}^{m \times n}$, I and J are rows, $i = \text{entry}_u(I)$, $j = \text{entry}_v(J)$.

The table shows the various ways of accessing the entries of a matrix in MINIMAT. The same notation A(i,j) which accesses an entry produces a submatrix when i and/or j are not scalars. In the table, the lower case letters i, j represent scalars; the upper case letters I, J represent rows. If the variable A has been assigned an $m \times n$ matrix A and MINIMAT evaluates an expression of form A(i,j) where i and j evaluate to scalars (that is, 1×1 matrices), then i must evaluate to an integer between 1 and m, and j must evaluate to an integer between 1 and n. Otherwise, MINIMAT will issue an error message ("Subscript out of range"). More generally, for a matrix expression of form A(I,J) where I and J are matrices, every entry of I must be an integer between 1 and m, and every entry of J must be an integer between 1 and n.

C.6   MINIMAT's Entrywise Operations

MINIMAT	Mathematics	English
C = A.*B	$C_{ij} = A_{ij}B_{ij}$	Entrywise Multiplication
C = A./B	$C_{ij} = A_{ij}/B_{ij}$	Entrywise Division
C = A.^B	$C_{ij} = A_{ij}^{B_{ij}}$	Entrywise Power
C = A.^b	$C_{ij} = A_{ij}^{b}$	Entrywise Power

These operations aren't to important, since entrywise operations (except in the case of diagonal matrices) are not used in elementary matrix algebra. Here are some examples:

```
#> A  = [1 2 3]; B = [5 3 2];

#> A.*B
   ans = [5 6 6]

#> A.^B
   ans = [1 8 9]

#> A.^3
   ans = [1 8 27]
```

C.7   Logical Operations

MINIMAT	Mathematics
C = (A<B)	$C_{ij} = 1$ if $A_{ij} < B_{ij}$ else 0
C = (A<=B)	$C_{ij} = 1$ if $A_{ij} \leq B_{ij}$ else 0
C = (A>B)	$C_{ij} = 1$ if $A_{ij} > B_{ij}$ else 0
C = (A>=B)	$C_{ij} = 1$ if $A_{ij} \geq B_{ij}$ else 0
C = (A==B)	$C_{ij} = 1$ if $A_{ij} = B_{ij}$ else 0
C = (A~=B)	$C_{ij} = 1$ if $A_{ij} \neq B_{ij}$ else 0
C = ~A	$C_{ij} = 1$ if $A_{ij} \neq 0$ else 0
C = (A & B)	$C_{ij} = 1$ if $A_{ij} \neq 0$ and $B_{ij} \neq 0$ else 0
C = (A | B)	$C_{ij} = 1$ if $A_{ij} \neq 0$ or $B_{ij} \neq 0$ else 0

The logical and relational operators are performed entrywise and yield the value one if the relation holds and zero otherwise. For example, if A and B are matrices of the same size the assignment

    $> C = (A<B)

produces C(i,j)=1 if A(i,j)<B(i,j) and C(i,j)=0 otherwise. The logical/relational operations include **negation, disjunction, conjunction** and the usual order relations. The **equality relation** is written A==B to avoid confusion with the assignment statement A=B. The notations

    ~A,   A & B,   A | B,

mean *not*-A, A *and* B, A *or* B respectively. These same notations are often used by mathematicians. The notations

    A~=B,   A<=B,   A>=B,

mean $A \neq B$, $A \leq B$, $A \geq B$, The symbols $\neq$, $\leq$ and $\geq$ don't exist on most computer keyboards.

MINIMAT does not distinguish between a numerical value and a Boolean value. For input to a Boolean operation, MINIMAT treats the numerical value zero as equivalent to *false* and any other value as equivalent to *true*. It always produces the particular non-zero value one when it wants to return the value *true* as indicated above. Hence the expression ~~A evaluates to a matrix with ones in positions occupied by the non-zero entries of A and zeros elsewhere.

C.8   Control Structures

MINIMAT offers the usual control structures (if-elseif-else, for, while) found
in most computer languages although their behavior may seem a bit strange
since there is no separate Boolean data type in MINIMAT: the only data
type in MINIMAT is that of a (doubly-indexed) array.

If, Elseif, Else

An **if statement** in MINIMAT is a compound statement of form

> **if** *expression*$_1$
> > *statement-list*$_1$
>
> **elseif** *expression*$_2$
> > *statement-list*$_2$
>
> $\cdots$
>
> **elseif** *expression*$_n$
> > *statement-list*$_n$
>
> **else**
> > *statement-list*$_{n+1}$
>
> **end**

There may be several **elseif** parts (or none) and the **else** part may also
be omitted. The expression is usually a boolean expression like $a == b$ or
$a < b$ but this is not required.

The *if statement* is evaluated as follows. First *expression*$_1$ is evaluated. If
it is true the statements in *statement-list*$_1$ are executed and the if-statement
terminates; otherwise the process is repeated with *expression*$_2$ (if present)
and so on up to *expression*$_n$. If all the expressions are false and the "else"
is present, then the statements in *statement-list*$_{n+1}$ are executed. Usually
the expressions are 1×1: in this context "true" means non-zero. If an
expression has more than one entry, "true" means that all its entries are
non-zero.

Example: The statement

```
$> if a>b,  c=1 elseif a<b, c=2 else c=3 end
```

results in c=2 after a=[1 2] and b=[3 4], but c=3 after a=[1 2] and b=[3
1].

For

A **for statement** in MINIMAT has form

$$\textbf{for } variable = expression$$
$$statement\text{-}list$$
$$\textbf{end}$$

The *expression* is evaluated and the statements in the *statement-list* are executed once for every entry[2] in the result; during this execution the value of the *variable* is that column (entry). Usually, the expression is of form m:n in which case the *variable* takes the values m, m+1, m+2, ..., n successively. If m > n, then the row m:n is empty and the statement-list is not executed at all.

Example: The statement

```
A=zeros(3,4);
for i = 1:3
    for   j=1:4
            A(i,j)=i+j;
        end
    end
```

sets A to the matrix $A = \begin{bmatrix} 2 & 3 & 4 & 5 \\ 3 & 4 & 5 & 6 \\ 4 & 5 & 6 & 7 \end{bmatrix}$. The purpose of the prelimi-

nary assignment statement A=zeros(3,4); is to assure that A has a value within the for-statement; without it, the subscripted assignment statement A(i,j)=i+j; would be illegal.

While

A **while statement** in MINIMAT has form

$$\textbf{while } expression$$
$$statement\text{-}list$$
$$\textbf{end}$$

The expression is evaluated and if it is true the statements in the statement-list are executed. This process is repeated until the expression is false. Usually the expression is a scalar (a 1×1 matrix) in which case "true" means non-zero. In case the expression has more than one entry, "true" means that *all* the entries are non-zero.

---

[2]Column, in classical MatLab.

Example: The statement

```
x=1,  while x<100, x = 2*x, end
```

causes the powers of 2 which are less than 100 to be printed.

Break

If the **break statement** is ever executed inside a for statement or a while statement the compound statement immediately terminates. (Usually the break will occur within an if statement.)

Example: The statement

```
for i=1:10
        if A(i)>7, break end
end
```

will give i the value of the first index where $A(i) > 7$ (or the value 10 if no such index exists).

Return

The **return statement** is used to terminate a .M function file. (It is an error to enter it from the keyboard.) It is not necessary to place a return statement at the end of a .M function.

Example: The .M function

```
function i=first7(A)
for i=1:10
        if A(i)>7, return end
end
```

would function the same if the **return** were replaced by **break**.

C.9   .M Functions Used in this Book

Syntax	Result	Page
N=randsutr(n)	Random Strictly Upper Trnglr	64
B=scale(A,p,c)	Elementary Operations	
B=swap(A,p,q)		
B=shear(A,p,q,c)		81
R=gj(A)	Reduced Row Echelon Form	90
[M,R]=gjm(A)	The same with Multiplier	92
A=intinv(n)	A, A^{-1} Integral random	98
[Q,D,P]=bed(A)	Biequivalence Decomposition	108
Phi=Nbasis1(A)	Basis for Null Space	140
Phi=Nbasis2(A)		146
Phi=Nbasis3(A)		165
Psi=Rbasis1(A)	Basis for the Range	140
Psi=Rbasis2(A)		146
Psi=Rbasis3(A)		149
Psi=Rbasis4(A)		166
X=randsoln(A)	Random Solution of $AX = 0$	149
X=randsoln(A,Y)	Random Solution of $AX = Y$	149
[Ps,R]=gs(Ph)	Gram-Schmidt	211
t=next(s)	Next Permutation	237
x=signum(s)	Sign of a Permutation	237
d=detm(s)	Determinant	248
C=compan(p)	Companion matrix	303
[P,R]=schur(A)	Schur Decomposition	315
[L,D,U]=rook(A)	Rook Decomposition	346
Q=flip(Y,Z)	Householder Reflection	360
[Q,R]=hh(A)	Householder Decomposition	365
B=adj(A)	Adjoint Matrix	429
p=sourfram(A)	Characteristic polynomial	434

The numbers in the table indicate the page on which the .M function
is explained; a listing of the .M function will be found nearby.[3] You can
always view the listing on line using MINIMAT's type command as in

        #> type gjm.m

---

[3]Exception: The function randsutr.m is not listed since it is rather prosaic.

Occasionally, the actual program will differ somewhat from the listing in the book: the actual program may incorporate error checking instructions which have little pedagogical value.

A command may invoke a function using fewer inputs and/or outputs than the header indicates. For example, either of the commands

```
#>  [M R] = gjm(R)
#>  M = gjm(A)
```

is correct. The general rule is that if the call has fewer outputs than the function heading, then the leftmost outputs in the heading are returned.

Within a .M function `nargin` evaluates to the number of inputs in the call and `nargout` evaluates to the number of outputs. This explains how the `randsoln` function (see Listing 9 Page 150) is able to respond differently when it is invoked with two inputs instead of one.

A .M function may invoke other .M functions. A .M function may even call itself recursively. For example, `randsoln` calls itself recursively: to find a random solution X of the inhomogeneous system $Y = AX$ it finds a random solution of the homogeneous system $AX - vY = 0$ and normalizes so $v = 1$.

C.10   Miscellaneous Functions

MINIMAT	English
[m, n]=size(A)	A is $m \times n$
x=input(msg)	Input number
disp(msg)	Display string
format e	Set format for output
pause	Wait for user to press a key

The format command was explained in Section 1.8; the other commands are useful in .M files. The command

> #>   [m, n]=size(A)

will assign to m the number of rows and to n the number of columns of A. (MINIMAT also allows the single output form t=size(A) which will assign a 2×1 matrix to the variable t.)

The disp and input commands display the string msg; the input command additionally returns the number the user types. The msg must be a character string enclosed within single quotes. Here's an example:

> n = input('Enter a number: ')
> if n>10, disp('too big'), return end

The pause command prints the prompt ''Press a key:'' and continues after the user responds.

C.11   Empty Matrices

It is convenient to define the space $\mathbf{F}^{p \times q}$ of all matrices with p rows and q columns even when $p = 0$ or $q = 0$. In this case we declare that $\mathbf{F}^{p \times q}$ consists of a single element denoted by 0:

$$\mathbf{F}^{0 \times q} = \{0_{0 \times q}\}, \qquad \mathbf{F}^{p \times 0} = \{0_{p \times 0}\}.$$

Thus $0_{0 \times q}$ denotes the unique empty matrix with q columns (and no rows) and $0_{p \times 0}$ denotes the unique empty matrix with p rows (and no columns). The matrix operations are defined as before: we can add matrices of the same size and multiply matrices if the number of columns in the first factor equals the number of rows in the second. The result of any matrix operation involving an empty matrix is always zero (or empty). For example, if

$A \in \mathbf{F}^{m \times n}$ and $B \in \mathbf{F}^{n \times p}$ the formula

$$\text{entry}_{ik}(AB) = \sum_{j=1}^{n} \text{entry}_{ij}(A)\,\text{entry}_{jk}(B)$$

gives $AB = 0_{m \times p}$ when $n = 0$. This is consistent with the mathematical convention that the empty sum be 0. We also get $AB = 0_{m \times p}$ when $m = 0$ or $n = 0$.

By this convention, the empty matrix $0_{0 \times n}$ with no rows and n columns has the set $\mathbf{F}^{n \times 1}$ as its null space and the empty matrix $0_{m \times 0}$ with no columns and m rows has the set $\{0_{m \times 1}\} \subset \mathbf{F}^{m \times 1}$ as its range. This means that the zero subspace $\{0_{m \times 1}\}$ has a the empty matrix (with no columns) as a basis:

$$\mathcal{R}(0_{m \times 0}) = \{0_{m \times 1}\}, \qquad \mathcal{N}(0_{m \times 0}) = \{0\},$$

and the space $\mathbf{F}^{n \times 1}$ has the empty matrix (with no rows) as a co-basis:

$$\mathcal{N}(0_{n \times 0}) = \{\mathbf{F}^{n \times 1}\}, \qquad \mathcal{R}(0_{n \times 0}) = \{0\}.$$

Empty matrices are handy in computer programming[4] since they often provide a handy way to initialize a loop. For a similar reason they are also handy in mathematics: they provide a way to begin and inductive argument. They provide a kind of 'identity element' for the operation of matrix catenation:

$$A = \begin{bmatrix} A & 0_{m \times 0} \end{bmatrix} = \begin{bmatrix} 0_{m \times 0} & A \end{bmatrix}.$$

MINIMAT (but not older versions of Matlab) does implement empty matrices so that we can use the zeros function to create empty matrices, and catenation works for empty matrices. Thus for *non-negative* integers which are not too big:

(1) the command

```
#> A = zeros(m,n)
```

produces an $m \times n$ matrix.

(2) The command

```
#> A = [A1 A2]
```

produces an $m \times n$ matrix A, if A1 is $m \times n_1$, A2 is $m \times n_2$, and $n = n_1 + n_2$.

---

[4]See Section 5.2.

(3) The command

```
#> A = [A1; A2]
```

produces an $m \times n$ matrix A, if A1 is $m_1 \times n$, A2 is $m_2 \times n$, and $m = m_1 + m_2$.

The reason that this feature is handy is that we often want to initialize a loop with the empty matrix. For example, consider the following two programs. Both will assign to Phi a basis for the null space of A. The first program works in MINIMAT but not in classical Matlab:

```
[m, n]=size(A)
Phi = zeros(n,0)
X = randsoln(A)
while max(max(abs(randsoln([Phi X])))) == 0,
     Phi = [Phi X]
     X = randsoln(A)
end
```

The second program is slightly more cumbersome but works in either version of Matlab:

```
if  max(max(abs(randsoln(A)))) > 0,
   Phi = randsoln(A)
   X = randsoln(A)
   while max(max(abs(randsoln([Phi X])))) == 0,
       Phi = [Phi X]
       X = randsoln(A)
   end
end
```

D

ANSWERS

8F $\{x^2 : -2 < x < 3\} = \{y : 0 \le y < 9\}$

8G $3 \in V_1$; $0, -1 \in V_2$; $0, -1, 7/9, 9/7 \in V_3$; $0, 3, 7/9, 9/7 \in V_4$. In all other cases $x \notin V_i$.

10C   It helps if you note that

$$V_1 = \{x \in \mathbf{R} : -2 < x < 13/3\}, \quad V_2 = \{-1, 0, 1, 2, 3, 4\},$$

$$V_3 = \{y \in \mathbf{R} : 13 < y < 20\}, \quad V_4 = \{y \in \mathbf{R} : 1 < y < 20\}.$$

Thus $V_2 \subset V_1$. $V_3, V_4 \not\subset V_1$ as $19 \in V_3, V_4$ but $19 \notin V_2$. Hence $V_3, V_4 \not\subset V_2$. $V_1 \not\subset V_2$ as $1/2 \in V_1$ but $1/2 \notin V_2$. $V_1 \not\subset V_3, V_4$ as $0 \in V_1$ but $0 \notin V_3, V_4$. $V_3 \subset V_4$ but $V_4 \not\subset V_3$ as $2 \in V_4$ but $2 \notin V_3$.

12B   We show $W \subset V$. Choose $n \in W$. If $n = 2$ then $n^2 + 7 = 11 < 12 = 6n$. If $n = 3$ then $n^2 + 7 = 16 < 18 = 6n$. If $n = 4$ then $n^2 + 7 = 23 < 24 = 6n$. In any case $n \in V$. This shows $W \subset V$. We show $V \subset W$. Choose $n \in V$. Then $n^2 + 7 < 6n$ and $n \in \mathbf{N}$. Hence $3 - \sqrt{2} < n < 3 + \sqrt{2}$. Since n is an integer, we must have $n \in W$. This shows $V \subset W$.

13A $|3 + 4i| = \sqrt{9 + 16} = 5, \ 1/(3 + 4i) = (3/25) - (4/25)i$.

13B Here's the proof that $\overline{zw} = \bar{z}\bar{w}$. Let $z = x + iy$ and $w = u + iv$ where $x, y, u, v \in \mathbf{R}$. Then

$$
\begin{aligned}
zw &= (x+iy)(u+iv) = (xu - yv) + i(xv + yu)\\
\overline{zw} &= (xu - yv) - i(xv + yu)\\
\bar{z}\bar{w} &= (x - iy)(u - iv) = (xu - yv) - i(xv + yu)
\end{aligned}
$$

13C That the answer to the exercise appears in this part of the book.

18A 1-a (intersecting lines, solution is $x_1 = 2$, $x_2 = 1$); 2-c (parallel lines); 3-b (same line).

18B 1-a (intersection is a line); 2-c (parallel planes); 3-b (same plane).

18C In each case subtract the first equation from the second and third. In (1) this gives a unique solution $x_3 = 3$, $x_2 = 2$ and $x_1 = -2$. In (2) the new third equation contradicts the old second equation. In (3) the new third equation coincides with the old second equation. Hence, the answer is 1-b, 2-c, 3-a.

20A The expression `real(z)` returns 3, `imag(z)` returns 4, `conj(z)` returns 3-i*4, `abs(z)` returns 5.

22B $e^{it} = \cos(t) + i\sin(t)$. See Exercise 290B if you haven't seen this before.

22C $e^{i(t+s)} = e^{it+is} = e^{it}e^{is}$.

30B No. A $1 \times n$ matrix and a $n \times 1$ matrix are of different sizes (unless $n = 1$.)

31D Since matrices of different size are never equal, we have $0_{2\times 3} \neq 0_{3\times 2}$ even though both might be denoted by the same symbol 0. (It is usually clear from the context what the size of a zero matrix is.)

32A $B = \begin{bmatrix} 2 & 0 & -2 \end{bmatrix}$.

34C The same size as A.

34D No. For $A \in \mathbf{F}^{m \times n}$, the product AA is defined only if $m = n$.

36A $AX = \begin{bmatrix} 6 \\ 15 \\ 24 \end{bmatrix}$, $\quad HA = \begin{bmatrix} 12 & 15 & 18 \end{bmatrix}$, $\quad HAX = 45$.

36B $\begin{bmatrix} a & b \\ c & d \end{bmatrix} \begin{bmatrix} d & -b \\ -c & a \end{bmatrix} = \begin{bmatrix} ad - bc & 0 \\ 0 & ad - bc \end{bmatrix} = (ad - bc) \begin{bmatrix} 1 & 0 \\ 0 & 1 \end{bmatrix}$.

36C The expression $X + Y$ is not meaningful since X has size 2×1 and Y has size 1×2 and matrices can be added only if they have the same size. The other two expressions are meaningful:

$$XY = \begin{bmatrix} x_1 y_1 & x_1 y_2 \\ x_2 y_1 & x_2 y_2 \end{bmatrix} \in \mathbf{F}^{2 \times 2}, \qquad YX = y_1 x_1 + y_2 x_2 \in \mathbf{F}^{1 \times 1} = \mathbf{F}.$$

36D $\text{entry}_{21}(A + B) = a_5 + a_4$, $\text{entry}_{21}(AB) = a_5^2 + a_2 a_4$.

36E $AX = \begin{bmatrix} x_1 a_1 + x_2 a_2 + x_3 a_3 \\ x_1 b_1 + x_2 b_2 + x_3 b_3 \end{bmatrix}$ and $AF = \begin{bmatrix} a_2 \\ b_2 \end{bmatrix}$. Note that

$$AX = x_1 \begin{bmatrix} a_1 \\ b_1 \end{bmatrix} + x_2 \begin{bmatrix} a_2 \\ b_2 \end{bmatrix} + x_3 \begin{bmatrix} a_3 \\ b_3 \end{bmatrix}.$$

F is the second column of I_3, and AF is the second column of A.

36F $D_1 D_2 = \begin{bmatrix} 1 & 0 \\ 0 & 1 \end{bmatrix}$ and $D_2 D_1 = \begin{bmatrix} 1 & 0 & 0 \\ 0 & 1 & 0 \\ 0 & 0 & 0 \end{bmatrix}$. Note that $D_1 D_2 = I_2$ but $D_2 D_1 \neq I_3$.

36G (1), (2), and (5) are true.

36I The A and B of (1) do not commute; the pairs A and B of (2) and (3) do commute.

37B A and B must have the same size and the number of rows in C must be the same as the number of columns in A (and B). In other words, there are integers m, n, p with $A, B \in \mathbf{F}^{m \times n}$ and $C \in \mathbf{F}^{n \times p}$.

37C $-A = (-1)A$, $\qquad A - B = A + (-1)B$.

37D In AD the columns are rescaled; in DA the rows are:

$$\begin{bmatrix} a & b & c \\ d & e & f \\ g & h & i \end{bmatrix} \begin{bmatrix} x & 0 & 0 \\ 0 & y & 0 \\ 0 & 0 & z \end{bmatrix} = \begin{bmatrix} ax & by & cz \\ dx & ey & fz \\ gx & hy & iz \end{bmatrix}$$

$$\begin{bmatrix} x & 0 & 0 \\ 0 & y & 0 \\ 0 & 0 & z \end{bmatrix} \begin{bmatrix} a & b & c \\ d & e & f \\ g & h & i \end{bmatrix} = \begin{bmatrix} xa & xb & xc \\ yd & ye & yf \\ zg & zh & zi \end{bmatrix}.$$

37E In AP the columns are permuted; in PA the rows are.

$$\begin{bmatrix} a & b & c \\ d & e & f \\ g & h & i \end{bmatrix} \begin{bmatrix} 0 & 1 & 0 \\ 0 & 0 & 1 \\ 1 & 0 & 0 \end{bmatrix} = \begin{bmatrix} c & a & b \\ f & d & e \\ i & g & h \end{bmatrix},$$

$$\begin{bmatrix} 0 & 1 & 0 \\ 0 & 0 & 1 \\ 1 & 0 & 0 \end{bmatrix} \begin{bmatrix} a & b & c \\ d & e & f \\ g & h & i \end{bmatrix} = \begin{bmatrix} d & e & f \\ g & h & i \\ a & b & c \end{bmatrix}.$$

37F
$$\begin{bmatrix} a_{11} & a_{12} & a_{13} \\ 0 & a_{22} & a_{23} \\ 0 & 0 & a_{33} \end{bmatrix} \begin{bmatrix} b_{11} & b_{12} & b_{13} \\ 0 & b_{22} & b_{23} \\ 0 & 0 & b_{33} \end{bmatrix} =$$

$$= \begin{bmatrix} a_{11}b_{11} & a_{11}b_{12} + a_{12}b_{22} & a_{11}b_{13} + a_{12}b_{23} + a_{13}b_{33} \\ 0 & a_{22}b_{22} & a_{22}b_{23} + a_{23}b_{33} \\ 0 & 0 & a_{33}b_{33} \end{bmatrix}$$

37G $AB = \begin{bmatrix} 0 & 0 & 0 & a_{12}b_{24} \\ 0 & 0 & 0 & 0 \\ 0 & 0 & 0 & 0 \\ 0 & 0 & 0 & 0 \end{bmatrix}.$

39B A matrix having a column of zeros is *never* invertible. For if A has a column of zeros, then so does BA for any matrix B:

$$\text{entry}_{ij}(BA) = \text{row}_i(B)\,\text{col}_j(A)$$

which shows that $\text{col}_j(BA) = 0$ if $\text{col}_j(A) = 0$. Hence, A cannot be invertible, for if it were we could take $B = A^{-1}$ and obtain $A^{-1}A = I$ which does *not* have a column of zeros.

39C The matrix $A = \begin{bmatrix} 1 & 1 \\ 1 & 1 \end{bmatrix}$ is not invertible (see Theorem 39D), but has only nonzero entries.

42B See the proof of part (3) of Theorem 40D.

42C $X = \begin{bmatrix} 2 & 1 \\ 1 & 1 \end{bmatrix} \begin{bmatrix} 3 \\ 5 \end{bmatrix} = \begin{bmatrix} 11 \\ 8 \end{bmatrix}$

42D Since $\det \begin{bmatrix} 3 & 1 \\ 1 & 1 \end{bmatrix} = 2$ the formula for the inverse of a 2×2 matrix gives

$$\begin{bmatrix} 3 & 1 \\ 1 & 1 \end{bmatrix}^{-1} = \begin{bmatrix} 1/2 & -1/2 \\ -1/2 & 3/2 \end{bmatrix}.$$

To check we compute

$$\begin{bmatrix} 3 & 1 \\ 1 & 1 \end{bmatrix} \begin{bmatrix} 1/2 & -1/2 \\ -1/2 & 3/2 \end{bmatrix} = \begin{bmatrix} 1/2 & -1/2 \\ -1/2 & 3/2 \end{bmatrix} \begin{bmatrix} 3 & 1 \\ 1 & 1 \end{bmatrix} = \begin{bmatrix} 1 & 0 \\ 0 & 1 \end{bmatrix}$$

42E If one row of a 2×2 matrix is a multiple of the other, then the determinant is zero:

$$\det \begin{bmatrix} a & b \\ ma & mb \end{bmatrix} = a(mb) - b(ma) = 0,$$

so the matrix is not invertible. For example, the matrix

$$A = \begin{bmatrix} 1 & 3 \\ 2 & 6 \end{bmatrix}$$

is not invertible.

42F Simply choose any integer matrix A with $\det(A) = \pm 1$. There are infinitely many examples. For example,

$$\begin{bmatrix} 2 & 1 \\ 1 & 1 \end{bmatrix}, \quad \begin{bmatrix} 4 & 3 \\ 3 & 2 \end{bmatrix}, \quad \begin{bmatrix} -4 & 3 \\ -3 & 2 \end{bmatrix}.$$

42G Assume A is invertible. If X satisfies $AX = 0$, then, by Remark 40B, the only solution of $AX = 0$ is $X = 0$.

Conversely, assume that A is not invertible; we must find a nonzero X with $AX = 0$. Since A is not invertible, $\det(A) = 0$. The proof of Theorem 39D shows how to find a nonzero X with $AX = 0$.

45E

$$A^\mathsf{T} = \begin{bmatrix} a & c & e \\ b & d & f \end{bmatrix}, \qquad H^\mathsf{T} = \begin{bmatrix} x \\ y \\ z \end{bmatrix},$$

$$HA = \begin{bmatrix} ax + cy + ez & bx + dy + fz \end{bmatrix},$$

$$(HA)^\mathsf{T} = A^\mathsf{T} H^\mathsf{T} = \begin{bmatrix} ax + cy + ez \\ bx + dy + fz \end{bmatrix},$$

$$\mathrm{row}_1(A^\mathsf{T}) = \mathrm{col}_1(A)^\mathsf{T} = \begin{bmatrix} a & c & e \end{bmatrix}.$$

48A See Exercise 37D.

48B

$$A = PCP^{-1} = \begin{bmatrix} 2 & 1 \\ 1 & 1 \end{bmatrix} \begin{bmatrix} 3 & 0 \\ 0 & 4 \end{bmatrix} \begin{bmatrix} 1 & -1 \\ -1 & 2 \end{bmatrix} = \begin{bmatrix} 2 & 2 \\ -1 & 5 \end{bmatrix}$$

$$B = PDP^{-1} = \begin{bmatrix} 2 & 1 \\ 1 & 1 \end{bmatrix} \begin{bmatrix} 2 & 0 \\ 0 & 1 \end{bmatrix} \begin{bmatrix} 1 & -1 \\ -1 & 2 \end{bmatrix} = \begin{bmatrix} 3 & -2 \\ 1 & 0 \end{bmatrix}$$

$$AB = BA = \begin{bmatrix} 8 & -4 \\ 2 & 2 \end{bmatrix}$$

48C If $AB = BA$, then

$$(PAP^{-1})(PBP^{-1}) = PABP^{-1} = PBAP^{-1} = (PBP^{-1})(PAP^{-1})$$

48D This follows from Theorem 47D by taking transposes.

54C $AE = \text{col}_2(A)$.

54D $H = \text{row}_2(I_4)$.

54E $AB = \begin{bmatrix} A_{11}B_{11} + A_{12}B_{21} & A_{11}B_{12} + A_{12}B_{22} \\ A_{21}B_{11} + A_{22}B_{21} & A_{21}B_{12} + A_{22}B_{22} \end{bmatrix}$ where

$$A_{11}B_{11} + A_{12}B_{21} \in \mathbf{F}^{3\times 7}, \qquad A_{11}B_{12} + A_{12}B_{22} \in \mathbf{F}^{3\times 6},$$

$$A_{21}B_{11} + A_{22}B_{21} \in \mathbf{F}^{2\times 7}, \qquad A_{21}B_{12} + A_{22}B_{22} \in \mathbf{F}^{2\times 6},$$

$$\text{row}_4(A) = \text{row}_1([\; A_{21} \quad A_{22} \;]), \qquad \text{col}_8(B) = \begin{bmatrix} \text{col}_1(B_{12}) \\ \text{col}_1(B_{22}) \end{bmatrix},$$

$$\text{entry}_{48}(AB) = \text{row}_1(A_{21})\,\text{col}_1(B_{12}) + \text{row}_1(A_{22})\,\text{col}_1(B_{22}).$$

56A $M = \begin{bmatrix} a_{11} & a_{12} & b_1 \\ a_{21} & a_{22} & b_2 \\ c_1 & c_2 & d \end{bmatrix}, \; Z = \begin{bmatrix} x_1 \\ x_2 \\ y \end{bmatrix}$. In this case, the block multiplication

law asserts that

$$MZ = \begin{bmatrix} AX + BY \\ CX + DY \end{bmatrix} = \begin{bmatrix} (a_{11}x_1 + a_{12}x_2) + b_1 y \\ (a_{21}x_1 + a_{22}x_2) + b_2 y \\ (c_1 x_1 + c_2 x_2) + dy \end{bmatrix}$$

in agreement with the answer obtained using the definition of matrix multiplication.

56D $HA = h_1\,\text{row}_1(A) + h_2\,\text{row}_2(A) + \cdots + h_m\,\text{row}_m(A)$ where $h_i = \text{entry}_i(H)$ is the ith entry of H. With $m = 3$ and $n = 2$,

$$[\; h_1 \quad h_2 \quad h_3 \;] \begin{bmatrix} a_{11} & a_{12} \\ a_{21} & a_{22} \\ a_{31} & a_{32} \end{bmatrix} =$$

$$= h_1 [\; a_{11} \quad a_{12} \;] + h_2 [\; a_{21} \quad a_{22} \;] + h_3 [\; a_{31} \quad a_{32} \;].$$

56F $X = \begin{bmatrix} 2 \\ 3 \\ 0 \\ -1 \end{bmatrix}$. A is not invertible since $AX = 0$ but $X \neq 0$.

57A $C = \begin{bmatrix} 1 & 0 & 1 \\ 0 & 1 & 1 \end{bmatrix}$.

57B Define B by

$$B = \begin{bmatrix} A_{11}^{-1} & -A_{11}^{-1}A_{12}A_{22}^{-1} \\ 0_{q \times p} & A_{22}^{-1} \end{bmatrix}.$$

We will show that $AB = BA = I_n$; then A is invertible and $B = A^{-1}$ by Definition 38A. By the block multiplication law 51A,

$$\begin{aligned} AB &= \begin{bmatrix} A_{11} & A_{12} \\ 0_{q \times p} & A_{22} \end{bmatrix} \begin{bmatrix} A_{11}^{-1} & -A_{11}^{-1}A_{12}A_{22}^{-1} \\ 0_{q \times p} & A_{22}^{-1} \end{bmatrix} \\ &= \begin{bmatrix} A_{11}A_{11}^{-1} & -A_{11}A_{11}^{-1}A_{12}A_{22}^{-1} + A_{12}A_{22}^{-1} \\ 0_{q \times p} & A_{22}A_{22}^{-1} \end{bmatrix} \\ &= \begin{bmatrix} I_p & 0_{p \times q} \\ 0_{q \times p} & I_q \end{bmatrix} \\ &= I_n \end{aligned}$$

and similarly, $BA = I_n$.

57D $A^{-1} = \begin{bmatrix} I_p & 0_{p \times q} \\ -C & I_q \end{bmatrix}$.

59A The last command differs from the one before only in that there is a space before the -1. MINIMAT thinks you are trying to construct a matrix with four entries in the first row but only three in the second. It doesn't allow this and produces an error message.

61A   #> I=eye(3), A=rand(3,4), B=rand(5,3)
      #> I*A, A,   B, B*I

62A MINIMAT rounds the answer before printing. If a format showing more decimal places were used, say **format short**, then the discrepancy would not occur.

62B We confirm the left distributive law.

      #> C=rand(2,3), A=rand(3,1), B=rand(3,1)
      #> AB = A+B,   CA = C*A,   CB = C*B
      #> C*AB,   CA+CB

(The last two answers agree.)

62C #>   A=rand(3,3), B=inv(A), A*B, B*A

63B   #> A=rand(3,3), B=rand(3,3), inv(A*B), inv(B)*inv(A)
      (The last two answers agree.)

63C   This is one example:

```
#> A = rand(4,4),
#> inv(inv(A)), A
#> A5=A^5, A3=A^3, B=A^{-2}
#> A5*B, A3
```

64A
```
#> D= diag(rand(1,3)), C= diag(rand(1,3))
#> D*C, C*D
```

64B
```
#> P= rand(4,4)
#> A=P*diag(rand(4,1))*P^(-1)
#> B=P*diag(rand(4,1))*P^(-1)
#> A*B, B*A
```

64C
```
#> L=rand(1,4), M=rand(1,4), LM=L.*M
#> DL=diag(L), DM=diag(M)
#> DL*DM, diag(LM)
```

64D
```
#> D=diag(rand(1,3))
#> A=rand(4,3)
#> AD=A*D
#> AD(:,2), D(2,2)*A(:,2)
```

(The last two answers agree.)

64E
```
#> L=[1 8 27 64], D=diag(L), C=diag(L.^(1/3))
#> C^3, D
```

produces a diagonal matrix C which is a "cube root" of the diagonal matrix D. The method

```
#> D = diag([1 8 27 64])  % note ([ and ])
#> C = D.^(1/3)
#> C^3, D
```

is equivalent since $0^{1/3} = 0$.

65A
```
#> N=randsutr(5)
#> N^2, N^3, N^4, N^5
```

(The last power is zero.)

65B   ```
#> N=randsutr(5), A=eye(5)-N
#> B=eye(5)+N+N^2+N^3+N^4
#> B*A, A*B
```

**65C**     ```
#> A=randsutr(5)+diag(rand(1,5)),
#> D=diag(diag(A)), C=diag(diag(A).^(-1)), D*C, C*D
#> U= C*A, A, D*U
#> N=eye(5)-U, V=eye(5)+N+N^2+N^3+N+4, U*V,V*U
#> B=V*C, A*B, B*A
```

65D   ```
#>   M*Z, A*X + B*Y, C*X + D*Y
```

**67A**   The entrywise product A.*B produces [1 16 81] while the ordinary product A*B produces an error message: *incompatible sizes for multiplication*. The expression A.^(1/3) evaluates to [1 2 3], but the expression A^(1/3) produces an error message.

**67B**   ```
#>      A=rand(2,3), B=rand(3,4), (A*B)', B'*A'
```

67C   The conjugate transpose is obtained from the transpose by taking the conjugate of every entry.

68A   ```
#>   A=rand(3,3), B=rand(3,3), inv(A*B), inv(B)*inv(A)
```

**68B**   E is the third column of the identity matrix and A*E is the third column of A. This illustrates formula 54A.

**68C**   Since $(B + B^\mathsf{T})^\mathsf{T} = B^\mathsf{T} + B^{\mathsf{T}\mathsf{T}} = B^\mathsf{T} + B = B + B^\mathsf{T}$, the MINIMAT command B=rand(n,n), A=B+B' generates a random symmetric matrix.

**68D**   The matrix $H$ has small entries, so $H^5$ has very small entries. But

$$(I - H)(I + H + H^2 + H^3 + H^4) = I - H^5$$

suggesting that $(I + H + H^2 + H^3 + H^4)$ is roughly $(I - H)^{-1}$.

**68E**     ```
#> H=0.1*rand(3,3), I=eye(3)
#> B=I+H+H2+H^3, inv(I-H)
```

For larger H we can take more terms:

```
#> H=0.5*rand(3,3), I=eye(3)
#> B=I+H+H^2+H^3+H^4+H^5, inv(I-H)
```

However for H which is too large this doesn't work at all:

```
#> H=5*rand(3,3), I=eye(3)
#> B=I+H+H^2+H^3+H^4+H^5, inv(I-H)
```

See Exercise 70C.

68F   `#> A = [ 1 2 3 -1; 4 5 6 -1; 7 8 9 1], X=randsoln(A), A*X`

68G   `#> A=rand(5,3), H=randsoln(A')',  H*A`

69A $C = BP_2 = AP_1^{-1}P_2$ so take $P = P_2^{-1}P_1$.

69B $A^\top = (B + B^\top)^\top = B^\top + (B^\top)^\top = B^\top + B = B + B^\top = A$.

69C   Let $x_i = \text{entry}_i(X)$. The equation $AX = 0$ is a system

$$
\begin{aligned}
x_1 + 2x_2 + 3x_3 &= 0 \\
4x_1 + 5x_2 + 6x_3 &= 0 \\
7x_1 + 8x_2 + 9x_3 &= 0
\end{aligned}
$$

Solving this system we see that the most general solution is $x_1 = x_3 = z$ and $x_2 = -2z$ where z is arbitrary. Since (for $z \neq 0$) $AX = 0$ but $X \neq 0$, we may conclude that A is not invertible by the Noninvertibility Remark 40B.

69E $\begin{bmatrix} 1 & a \\ 0 & 1 \end{bmatrix}\begin{bmatrix} 1 & b \\ 0 & 1 \end{bmatrix} = \begin{bmatrix} 1 & a+b \\ 0 & 1 \end{bmatrix}$, so $\begin{bmatrix} 1 & 1 \\ 0 & 1 \end{bmatrix}^n = \begin{bmatrix} 1 & n \\ 0 & 1 \end{bmatrix}$.

78B

$$
\begin{bmatrix} c & 0 & 0 & 0 \\ 0 & 1 & 0 & 0 \\ 0 & 0 & 1 & 0 \\ 0 & 0 & 0 & 1 \end{bmatrix}
\begin{bmatrix} c^{-1} & 0 & 0 & 0 \\ 0 & 1 & 0 & 0 \\ 0 & 0 & 1 & 0 \\ 0 & 0 & 0 & 1 \end{bmatrix}
=
\begin{bmatrix} 1 & 0 & 0 & 0 \\ 0 & 1 & 0 & 0 \\ 0 & 0 & 1 & 0 \\ 0 & 0 & 0 & 1 \end{bmatrix}
$$

$$
\begin{bmatrix} c^{-1} & 0 & 0 & 0 \\ 0 & 1 & 0 & 0 \\ 0 & 0 & 1 & 0 \\ 0 & 0 & 0 & 1 \end{bmatrix}
\begin{bmatrix} c & 0 & 0 & 0 \\ 0 & 1 & 0 & 0 \\ 0 & 0 & 1 & 0 \\ 0 & 0 & 0 & 1 \end{bmatrix}
=
\begin{bmatrix} 1 & 0 & 0 & 0 \\ 0 & 1 & 0 & 0 \\ 0 & 0 & 1 & 0 \\ 0 & 0 & 0 & 1 \end{bmatrix}
$$

$$
\begin{bmatrix} 1 & 0 & 0 & 0 \\ 0 & 0 & 1 & 0 \\ 0 & 1 & 0 & 0 \\ 0 & 0 & 0 & 1 \end{bmatrix}
\begin{bmatrix} 1 & 0 & 0 & 0 \\ 0 & 0 & 1 & 0 \\ 0 & 1 & 0 & 0 \\ 0 & 0 & 0 & 1 \end{bmatrix}
=
\begin{bmatrix} 1 & 0 & 0 & 0 \\ 0 & 1 & 0 & 0 \\ 0 & 0 & 1 & 0 \\ 0 & 0 & 0 & 1 \end{bmatrix}
$$

$$
\begin{bmatrix} 1 & 0 & 0 & 0 \\ 0 & 1 & c & 0 \\ 0 & 0 & 1 & 0 \\ 0 & 0 & 0 & 1 \end{bmatrix}
\begin{bmatrix} 1 & 0 & 0 & 0 \\ 0 & 1 & -c & 0 \\ 0 & 0 & 1 & 0 \\ 0 & 0 & 0 & 1 \end{bmatrix}
=
\begin{bmatrix} 1 & 0 & 0 & 0 \\ 0 & 1 & 0 & 0 \\ 0 & 0 & 1 & 0 \\ 0 & 0 & 0 & 1 \end{bmatrix}
$$

$$
\begin{bmatrix} 1 & 0 & 0 & 0 \\ 0 & 1 & -c & 0 \\ 0 & 0 & 1 & 0 \\ 0 & 0 & 0 & 1 \end{bmatrix}
\begin{bmatrix} 1 & 0 & 0 & 0 \\ 0 & 1 & c & 0 \\ 0 & 0 & 1 & 0 \\ 0 & 0 & 0 & 1 \end{bmatrix}
=
\begin{bmatrix} 1 & 0 & 0 & 0 \\ 0 & 1 & 0 & 0 \\ 0 & 0 & 1 & 0 \\ 0 & 0 & 0 & 1 \end{bmatrix}
$$

78C

$$E_2 E_3 = \begin{bmatrix} 0 & 1 & 0 \\ 0 & 0 & 1 \\ 1 & 0 & 0 \end{bmatrix}, \quad (E_2 E_3)^{-1} = E_3 E_2 = \begin{bmatrix} 0 & 0 & 1 \\ 1 & 0 & 0 \\ 0 & 1 & 0 \end{bmatrix},$$

$$E_1 E_2 E_3 = \begin{bmatrix} 0 & 0 & 1 \\ 1 & c & 0 \\ 0 & 1 & 0 \end{bmatrix}, \quad (E_1 E_2 E_3)^{-1} = \begin{bmatrix} 0 & 1 & -c \\ 0 & 0 & 1 \\ 1 & 0 & 0 \end{bmatrix},$$

$$E_1 E_2 E_1 = \begin{bmatrix} 0 & 1 & c \\ 1 & 0 & c \\ 0 & 0 & 1 \end{bmatrix}, \quad (E_1 E_2 E_1)^{-1} = \begin{bmatrix} 0 & 1 & -c \\ 1 & 0 & -c \\ 0 & 0 & 1 \end{bmatrix}.$$

78D $M^{-1} = \begin{bmatrix} 1 & 0 & -3 \\ 0 & 1 & 0 \\ 0 & 0 & 1 \end{bmatrix} \begin{bmatrix} 1 & 0 & 0 \\ 0 & 1 & -4 \\ 0 & 0 & 1 \end{bmatrix} \begin{bmatrix} 1 & 0 & 0 \\ -5 & 1 & 0 \\ 0 & 0 & 1 \end{bmatrix} \begin{bmatrix} 1/3 & 0 & 0 \\ 0 & 1 & 0 \\ 0 & 0 & 1 \end{bmatrix}$

80A
```
#>   A0=rand(3,4)
#>   E=eye(3); c=7; E(1,:)=c*E(1,:)
#>   A=A0; A(1,:)=c*A(1,:), E*A0

#>   A0=rand(3,4)
#>   E=eye(3); E([2 3],:)=E([3 2],:)
#>   A=A0; A([2 3],:)=A([3 2],:), E*A0

#>   A0=rand(3,4)
#>   E=eye(3); c=7; E(2,:)=E(2,:)+c*E(3,:)
#>   A=A0,  A(2,:)=A(2,:)+c*A(3,:), E*A0
```

In each of these the final values of A and E*A0 should be the same.

80B
```
#> I=eye(3), E=I, F=I, c=rand(1,1)
#> E(1,:)=c*E(1,:),  F(1,:)=(1/c)*F(1,:)
#> E*F, F*E

#> I=eye(3), E=I
#> E([2 3],:)=E([3 2],:), F=E
#> E*F, F*E

#> I=eye(3), E=I, F=I, c=rand(1,1)
#> E(2,:)=E(2,:)+c*E(3,:),  F(2,:)=F(2,:)-c*F(3,:)
#> E*F, F*E
```

84B If $\text{row}_k(R) = 0$, then $\text{row}_k(RB) = 0$ no matter what B is. Hence, we can never have $RB = I$ since $\text{row}_k(I) \neq 0$.

88A $E_1 = \begin{bmatrix} 0 & 1 \\ 1 & 0 \end{bmatrix}$, $E_2 = \begin{bmatrix} 1/2 & 0 \\ 0 & 1 \end{bmatrix}$, $E_3 = \begin{bmatrix} 1 & 0 \\ 0 & 1/2 \end{bmatrix}$, $E_4 = \begin{bmatrix} 1 & -3 \\ 0 & 1 \end{bmatrix}$.

$M = E_4 E_3 E_2 E_1 = \begin{bmatrix} -3/2 & 1/2 \\ 1/2 & 0 \end{bmatrix}$.

90B Generate a 3×5 matrix A by the command

```
#> A =  rand(3,5); A(1,1)=0, R=A
#> R([1 2],:) = R([2 1],:)
```

Now perform steps as in Example 90A.

94C

$$\text{start:} \quad \begin{bmatrix} 1 & 2 & | & 1 & 0 \\ 3 & 4 & | & 0 & 1 \end{bmatrix}$$

$$\text{shear:} \quad \begin{bmatrix} 1 & 2 & | & 1 & 0 \\ 0 & -2 & | & -3 & 1 \end{bmatrix}$$

$$\text{shear:} \quad \begin{bmatrix} 1 & 0 & | & -2 & 1 \\ 0 & -2 & | & -3 & 1 \end{bmatrix}$$

$$\text{scale:} \quad \begin{bmatrix} 1 & 0 & | & -2.0 & 1.0 \\ 0 & 1 & | & 1.5 & -0.5 \end{bmatrix}$$

$$\text{so:} \quad \begin{bmatrix} 1 & 2 \\ 3 & 4 \end{bmatrix}^{-1} = \begin{bmatrix} -2.0 & 1.0 \\ 1.5 & -0.5 \end{bmatrix}.$$

94D With $A_0 = A$:

$$E_1 = \begin{bmatrix} 1/2 & 0 \\ 0 & 1 \end{bmatrix} \qquad A_1 = E_1 A_0 = \begin{bmatrix} 1 & 1/2 \\ 3 & 1 \end{bmatrix}$$

$$E_2 = \begin{bmatrix} 1 & 0 \\ -3 & 1 \end{bmatrix} \qquad A_2 = E_2 A_1 = \begin{bmatrix} 1 & 1/2 \\ 0 & -1/2 \end{bmatrix}$$

$$E_3 = \begin{bmatrix} 1 & 0 \\ 0 & -2 \end{bmatrix} \qquad A_3 = E_3 A_2 = \begin{bmatrix} 1 & 1/2 \\ 0 & 1 \end{bmatrix}$$

$$E_4 = \begin{bmatrix} 1 & -1/2 \\ 0 & 1 \end{bmatrix} \qquad A_4 = E_4 A_3 = \begin{bmatrix} 1 & 0 \\ 0 & 1 \end{bmatrix}.$$

In other words

$$E_4 E_3 E_2 E_1 A = A_4 = I.$$

Thus $A^{-1} = E_4 E_3 E_2 E_1$ so $A = E_1^{-1} E_2^{-1} E_3^{-1} E_4^{-1}$:

$$\begin{bmatrix} 2 & 1 \\ 3 & 1 \end{bmatrix} = \begin{bmatrix} 2 & 0 \\ 0 & 1 \end{bmatrix} \begin{bmatrix} 1 & 0 \\ 3 & 1 \end{bmatrix} \begin{bmatrix} 1 & 0 \\ 0 & -1/2 \end{bmatrix} \begin{bmatrix} 1 & 1/2 \\ 0 & 1 \end{bmatrix}.$$

94E We put A in RREF:

$$MA = R = \begin{bmatrix} 1 & 0 & 1 \\ 0 & 1 & 1 \\ 0 & 0 & 0 \end{bmatrix}$$

from which we see that

$$X = \begin{bmatrix} 1 \\ 1 \\ -1 \end{bmatrix}$$

satisfies $AX = 0$ but $X \neq 0$. Hence, by Corollary 94A, A is not invertible.

95A

(1) Not invertible: $AX = 0$ for $X = \begin{bmatrix} -3 \\ -6 \\ 1 \end{bmatrix}$.

(2) Invertible: $A^{-1} = A = I$ (the identity matrix).

(3) Not invertible: $AX = 0$ for $X = \begin{bmatrix} 1 \\ -2 \\ 1 \end{bmatrix}$.

(4) Invertible: $A^{-1} = \begin{bmatrix} 14 & -2 & -3 \\ -2 & 1 & 0 \\ -3 & 0 & 1 \end{bmatrix}$.

(5) Not invertible: $AX = 0$ for $X = \begin{bmatrix} 1 \\ -2 \\ 1 \end{bmatrix}$.

(6) Not invertible: $AX = 0$ for $X = \begin{bmatrix} 0 \\ 0 \\ 1 \end{bmatrix}$.

95C Not invertible. If $X = \begin{bmatrix} 1 \\ 1 \\ -1 \end{bmatrix}$, then $AX = 0$ but $X \neq 0$.

95D By Exercise 95C, $HA^\top = 0$ where $H = \begin{bmatrix} 1 & 1 & -1 \end{bmatrix}$. Let $Y = \begin{bmatrix} 1 \\ 0 \\ 0 \end{bmatrix}$.

Then $HY \neq 0$. But if there were a solution to $A^\top X = Y$, we would have $0 = HA^\top X = HY \neq 0$, a contradiction. Hence, there is no such solution X.

95F $A^{-1} = \begin{bmatrix} I_p & 0 \\ -C & I_q \end{bmatrix}$ This formula comes from transposing the block multiplication law 52A. If $p = q = 1$ then A^{-1} is a shear. In general, A^{-1} is a product of pq shears.

96C $M = \begin{bmatrix} I_r & 0_{r \times (m-r)} \\ -C & I_{(m-r)} \end{bmatrix}$. At most $r(m-r)$ shears are required, one for each entry of C.

96D If all the diagonal entries are nonzero, the matrix can be transformed to the identity matrix by elementary row operations and is hence invertible. If some diagonal entry is zero, the matrix can be transformed to a matrix with a zero row by shears. To see this, let k be the largest subscript such that the kth diagonal entry is zero. By scales, transform the given triangular matrix to a matrix B with with $\text{entry}_{ii}(B) = 1$ for $i > k$. Then, for each $i > k$, shear an appropriate multiple of the ith row from the kth row to achieve a matrix where the kth row vanishes.

97D
```
#> A =  rand(4,4)
#> [B, I] = gjm(A)
#> B*A, A*B, I
#> B, inv(A)
```

97E
```
function B = inverse(A)
     B = gjm(A);
```

98A
```
#> A=eye(3)
#> A(2,:)=A(2,:)+2*A(1,:)
#> A(3,:)=A(3,:)-3*A(2,:)
#> A=A*A'
#> B=A^(-1), A*B, B*A
```

98C
```
#> A=intinv(4), B=A^(-1)
#> A*B, eye(4)
```

03D $Y = \begin{bmatrix} y \\ x \\ w \\ z \\ v \\ u \end{bmatrix}$. The answer $Y = \begin{bmatrix} -b \\ a \\ 0 \\ 0 \\ 0 \\ 0 \end{bmatrix}$ is wrong: it might be zero.

06B When it is the identity matrix: $m = n = r$.

07A From $Q_1 A_1 = Q_2 A_3$ we conclude $A_3 = Q_2^{-1} Q_1 A_1$. Hence, $A_4 = Q_3 A_3 = Q_3 Q_2^{-1} Q_1 A_1 = Q A_1$ where $Q = Q_3 Q_2^{-1} Q_1$.

07D (1) is easy: since A is invertible, we may take $Q = A$ and $D = P = I$ the identity. We do (3). Take $Q = \begin{bmatrix} 1 & 2 \\ 3 & 7 \end{bmatrix}$, so $Q^{-1} = \begin{bmatrix} 7 & -2 \\ -3 & 1 \end{bmatrix}$. Since Q consists of the first two columns of $A = \begin{bmatrix} 1 & 2 & a \\ 3 & 7 & b \end{bmatrix}$ we get

$$Q^{-1} A = \begin{bmatrix} 1 & 0 & c \\ 0 & 1 & d \end{bmatrix} = \begin{bmatrix} 1 & 0 & 0 \\ 0 & 1 & 0 \end{bmatrix} \begin{bmatrix} 1 & 0 & c \\ 0 & 1 & d \\ 0 & 0 & 1 \end{bmatrix} = DP^{-1}.$$

(In the actual example, $a = 2$, $b = 6$, $c = 2$, $d = 0$.)

07E In all seven cases Q^{-1} is the same, namely the matrix M of Exercise 88E. The matrices P and $D = Q^{-1} A P$ are:

(1) $$P = \begin{bmatrix} 1 & 0 & 0 & -1 \\ 0 & 1 & 0 & -2 \\ 0 & 0 & 1 & -3 \\ 0 & 0 & 0 & 1 \end{bmatrix}, \qquad D = \begin{bmatrix} 1 & 0 & 0 & 0 \\ 0 & 1 & 0 & 0 \\ 0 & 0 & 1 & 0 \end{bmatrix}.$$

(2) $$P = \begin{bmatrix} 1 & 0 & -1 & -3 \\ 0 & 1 & -2 & -4 \\ 0 & 0 & 1 & 0 \\ 0 & 0 & 0 & 1 \end{bmatrix}, \qquad D = \begin{bmatrix} 1 & 0 & 0 & 0 \\ 0 & 1 & 0 & 0 \\ 0 & 0 & 0 & 0 \end{bmatrix}.$$

(3) $$P = \begin{bmatrix} 0 & 0 & 1 & 0 \\ 1 & 0 & 0 & -2 \\ 0 & 0 & 0 & 1 \\ 0 & 1 & 0 & 0 \end{bmatrix}, \qquad D = \begin{bmatrix} 1 & 0 & 0 & 0 \\ 0 & 1 & 0 & 0 \\ 0 & 0 & 0 & 0 \end{bmatrix}.$$

(4) $$P = \begin{bmatrix} 1 & 0 & 0 & -1 \\ 0 & 1 & 0 & -2 \\ 0 & 0 & 0 & 1 \\ 0 & 0 & 1 & 0 \end{bmatrix}, \qquad D = \begin{bmatrix} 1 & 0 & 0 & 0 \\ 0 & 1 & 0 & 0 \\ 0 & 0 & 1 & 0 \end{bmatrix}.$$

(5) $$P = I_2, \qquad D = R.$$

(6) $$P = \begin{bmatrix} 1 & -2 \\ 0 & 1 \end{bmatrix}, \qquad D = \begin{bmatrix} 1 & 0 \\ 0 & 0 \\ 0 & 0 \end{bmatrix}.$$

(7) $$P = \begin{bmatrix} 0 & 1 \\ 1 & 0 \end{bmatrix}, \qquad D = \begin{bmatrix} 1 & 0 \\ 0 & 0 \\ 0 & 0 \end{bmatrix}.$$

107F
$$R = \begin{bmatrix} 1 & c_{11} & 0 & c_{12} & 0 & c_{13} & c_{14} \\ 0 & 0 & 1 & c_{22} & 0 & c_{23} & c_{24} \\ 0 & 0 & 0 & 0 & 1 & c_{33} & c_{34} \\ 0 & 0 & 0 & 0 & 0 & 0 & 0 \end{bmatrix}$$

Let $R_j = \mathrm{col}_j(R)$, $E_j = \mathrm{col}_j(I_7)$, $H_i = \mathrm{row}_i(I_7)$, and define a permutation matrix T by

$$T = \begin{bmatrix} E_1 & E_3 & E_5 & E_2 & E_4 & E_6 & E_7 \end{bmatrix}$$

$$= \begin{bmatrix} 1 & 0 & 0 & 0 & 0 & 0 & 0 \\ 0 & 0 & 0 & 1 & 0 & 0 & 0 \\ 0 & 1 & 0 & 0 & 0 & 0 & 0 \\ 0 & 0 & 0 & 0 & 1 & 0 & 0 \\ 0 & 0 & 1 & 0 & 0 & 0 & 0 \\ 0 & 0 & 0 & 0 & 0 & 1 & 0 \\ 0 & 0 & 0 & 0 & 0 & 0 & 1 \end{bmatrix} = \begin{bmatrix} H_1 \\ H_4 \\ H_2 \\ H_5 \\ H_3 \\ H_6 \\ H_7 \end{bmatrix}$$

By the Column Permutation Theorem 101B,

$$RT = \begin{bmatrix} R_1 & R_3 & R_5 & R_2 & R_4 & R_6 & R_7 \end{bmatrix}$$

where

$$R = \begin{bmatrix} R_1 & R_2 & R_3 & R_4 & R_5 & R_6 & R_7 \end{bmatrix};$$

that is,

$$RT = \begin{bmatrix} 1 & 0 & 0 & c_{11} & c_{12} & c_{13} & c_{14} \\ 0 & 1 & 0 & 0 & c_{22} & c_{23} & c_{24} \\ 0 & 0 & 1 & 0 & 0 & c_{33} & c_{34} \\ 0 & 0 & 0 & 0 & 0 & 0 & 0 \end{bmatrix}.$$

We want $P = TP_1$ where

$$
P_1 = \begin{bmatrix}
1 & 0 & 0 & -c_{11} & -c_{12} & -c_{13} & -c_{14} \\
0 & 1 & 0 & 0 & -c_{22} & -c_{23} & -c_{24} \\
0 & 0 & 1 & 0 & 0 & -c_{33} & -c_{34} \\
0 & 0 & 0 & 1 & 0 & 0 & 0 \\
0 & 0 & 0 & 0 & 1 & 0 & 0 \\
0 & 0 & 0 & 0 & 0 & 1 & 0 \\
0 & 0 & 0 & 0 & 0 & 0 & 1
\end{bmatrix}.
$$

By the Row Permutation Theorem 101A,

$$
P = TP_1 = \begin{bmatrix}
1 & 0 & 0 & -c_{11} & -c_{12} & -c_{13} & -c_{14} \\
0 & 0 & 0 & 1 & 0 & 0 & 0 \\
0 & 1 & 0 & 0 & -c_{22} & -c_{23} & -c_{24} \\
0 & 0 & 0 & 0 & 1 & 0 & 0 \\
0 & 0 & 1 & 0 & 0 & -c_{33} & -c_{34} \\
0 & 0 & 0 & 0 & 0 & 1 & 0 \\
0 & 0 & 0 & 0 & 0 & 0 & 1
\end{bmatrix}.
$$

08A $D_1 = \begin{bmatrix} I_3 & 0_{3 \times 2} \end{bmatrix}$, $D_2 = \begin{bmatrix} I_3 \\ 0_{2 \times 3} \end{bmatrix}$, $D_3 = \begin{bmatrix} I_3 & 0_{3 \times 3} \\ 0_{1 \times 3} & 0_{1 \times 3} \end{bmatrix}$, $D_4 = B = D_5$.

08B Zero. A and B are left equivalent so their RREF's are equal by the uniqueness asserted in the Gauss-Jordan Decomposition Theorem 105A.

22A $MY = \begin{bmatrix} 1 \\ 2 \\ 0 \end{bmatrix}$ so $X = \begin{bmatrix} 1 \\ 2 \\ 0 \end{bmatrix}$ solves $MY = MAX$. The most general solution

is $X = \begin{bmatrix} 1 - 2x_3 \\ 2 - 3x_3 \\ x_3 \end{bmatrix}$ where x_3 is arbitrary. If $\tilde{Y} = \begin{bmatrix} -2 \\ 5 \\ 0 \end{bmatrix}$, then $M\tilde{Y} = $

$\begin{bmatrix} -7 \\ -2 \\ -10 \end{bmatrix}$ and the system $M\tilde{Y} = MAX$ has no solution since the last row

of MA is zero and $-10 \neq 0$.

22C $MY_1 = \begin{bmatrix} 1 \\ 2 \\ 0 \end{bmatrix}$, $MY_2 = \begin{bmatrix} -7 \\ -2 \\ -10 \end{bmatrix}$, $MY_3 = \begin{bmatrix} 8 \\ 8 \\ 0 \end{bmatrix}$, where M as in

Exercise 88E. For matrices (1) and (4) all the rows of the RREF R are nonzero, so there is a solution X of $MY_i = RX$. For matrices (2), (3), and (5) the last row of R is zero, so there is a solution of $MY_1 = RX$ and $MY_3 = RX$ but no solution of $MY_2 = RX$. For matrices (5) and (7) the last two rows of R are zero, so the is no solution of $MY_i = RX$.

122D $x_1 = 46$, $x_2 = -39$, $x_3 = 3$, $x_4 = 4$. In other words, $X = 3X_1 + 4X_2$.

122E $Y_1 = \begin{bmatrix} 1 \\ 1 \\ 2 \end{bmatrix}$; $Y_2 = \begin{bmatrix} 1 \\ 1 \\ 1 \end{bmatrix}$; no.

122F The system can be transformed to the system $RX = MY$ given by

$$
\begin{aligned}
x_1 - 6x_3 - 7x_4 &= 2y_1 - y_2 \\
x_2 + 5x_3 + 6x_4 &= -y_1 + y_2 \\
0 &= 3y_1 - 2y_2 - y_3
\end{aligned}
$$

The matrices M, A, and R are given by

$$
M = \begin{bmatrix} 2 & -1 & 0 \\ -1 & 1 & 0 \\ 3 & -2 & -1 \end{bmatrix}, \; A = \begin{bmatrix} 1 & 1 & -1 & -1 \\ 1 & 2 & 4 & 5 \\ 1 & -1 & -11 & -13 \end{bmatrix}, \; R = \begin{bmatrix} 1 & 0 & -6 & -7 \\ 0 & 1 & 5 & 6 \\ 0 & 0 & 0 & 0 \end{bmatrix}.
$$

The system has a solution only if $3y_1 - 2y_2 - y_3 = 0$. When solutions exist they are not unique: the most general solution is given by $X = X_0 + x_3 X_1 + x_4 X_2$ where x_3 and x_4 are arbitrary and

$$
X_0 = \begin{bmatrix} 2y_1 - y_2 \\ -y_1 + y_2 \\ 0 \\ 0 \end{bmatrix}, \quad X_1 = \begin{bmatrix} 6 \\ -5 \\ 1 \\ 0 \end{bmatrix}, \quad X_2 = \begin{bmatrix} 7 \\ -6 \\ 0 \\ 1 \end{bmatrix}.
$$

124B $\mathcal{N}(0_{m \times n}) = \mathbf{F}^{n \times 1}$ and $\mathcal{N}(I_n) = \{0\}$.

125A $\mathcal{R}(0_{m \times n}) = \{0\}$ and $\mathcal{R}(I_n) = \mathbf{F}^{n \times 1}$.

125E False. The null space is never empty since it contains 0. What is true is: *If the null space of a square matrix not $\{0\}$, then the matrix is not invertible.* Do not confuse the empty set $\emptyset$, the zero matrix 0, and the singleton $\{0\}$.

126B $X = \begin{bmatrix} 1 \\ -1 \\ 0 \\ 3 \end{bmatrix}$.

126C No. Yes. Yes.

126D $Y_i \in \mathcal{R}(A)$ if and only if the system $Y_i = AX$ has a solution X. This exercise has the same answer as Exercise 122C.

126E   The block multiplication law 54A says that

$$\text{col}_j(A) = AE_j$$

where $E_j = \text{col}_j(I_n)$ is the jth column of the identity matrix. Thus the system $\text{col}_j(A) = AX$ has a solution X (namely $X = E_j$), so $\text{col}_j(A) \in \mathcal{R}(A)$.

126F   Firstly, $X_j \notin \mathcal{R}(A)$, $X_j \notin \mathcal{R}(R)$ $Y_j \notin \mathcal{N}(A)$, and $Y_j \notin \mathcal{N}(R)$, because the vectors are the wrong size. Secondly,

$$X_1 \in \mathcal{N}(A), \qquad X_2 \in \mathcal{N}(A), \qquad X_3 \in \mathcal{N}(A), \qquad X_4 \notin \mathcal{N}(A),$$

$$X_1 \in \mathcal{N}(R), \qquad X_2 \in \mathcal{N}(R), \qquad X_3 \in \mathcal{N}(R), \qquad X_4 \notin \mathcal{N}(R).$$

Finally,

$$Y_1 \notin \mathcal{R}(A), \qquad Y_2 \in \mathcal{R}(A), \qquad Y_3 \in \mathcal{R}(A), \qquad Y_4 \notin \mathcal{R}(A),$$

$$Y_1 \in \mathcal{R}(R), \qquad Y_2 \in \mathcal{R}(R), \qquad Y_3 \notin \mathcal{R}(R), \qquad Y_4 \notin \mathcal{R}(R).$$

For the null space the answers are the same for A and R (this is illustrates Theorem 128A), whereas for the range they are not. (The matrices M, A, R of this example are the same as those in Exercise 122F.)

127A   Since $Y \in \mathcal{R}(A) \iff MY \in \mathcal{R}(R)$ we get

$$MY_1 \notin \mathcal{R}(R), \quad MY_2 \in \mathcal{R}(R), \quad MY_3 \in \mathcal{R}(R), \quad MY_4 \notin \mathcal{R}(R).$$

Since $Y \in \mathcal{R}(R) \iff M^{-1}Y \in \mathcal{R}(A)$ we get

$$M^{-1}Y_1 \in \mathcal{R}(A), \ M^{-1}Y_2 \in \mathcal{R}(A), \ M^{-1}Y_3 \notin \mathcal{R}(A), \ M^{-1}Y_4 \notin \mathcal{R}(A).$$

129A $A_1 \in V$, $A_1 \in W$, $B_1 \in W$, but $B_1 \notin V$. $W \not\subset V$ since $B_1 \in W$ but $B_1 \notin V$. $V \not\subset W$ although you can't use the first column A_1 of A to prove this. However, the second column of A lies in V but not in W.

129B   The null space $\mathcal{N}(A)$ is a line through the origin and B is a point on this line. To show that $\mathcal{N}(A) = \mathcal{R}(B)$ we must show that each is a subset of the other.

Assume that $X \in \mathcal{R}(B)$. Then $X = Bz$ for some $z \in \mathbf{F} = \mathbf{F}^{1\times1}$. Since $AB = 0$ we have that $AX = ABz = 0$ so $X \in \mathcal{N}(A)$. This shows that $\mathcal{R}(B) \subset \mathcal{N}(A)$.

Conversely, choose $X = \begin{bmatrix} x_1 \\ x_2 \end{bmatrix} \in \mathcal{N}(A)$. Then $a_1x_1 + a_2x_2 = AX = 0$.

If $a_1 \neq 0$, let $z = x_2/a_1$; otherwise let $z = -x_1/a_2$. In either case, $X = Bz$ so $X \in \mathcal{R}(A)$. This shows $\mathcal{N}(A) \subset \mathcal{R}(B)$. Hence, $\mathcal{N}(A) = \mathcal{R}(B)$, as required.

129C   Choose $X \in \mathcal{R}(C)$. Then $X = CW$ for some $W \in \mathbf{F}^{k \times 1}$. Form $Z \in \mathbf{F}^{(k+1) \times 1}$ by inserting a zero in the ith row and shifting the lower entries down. Then $X = CW = BZ$ so $X \in \mathcal{R}(B)$. This shows $\mathcal{R}(C) \subset \mathcal{R}(B)$. Here's an example with $k = i = 2$.

$$B = \begin{bmatrix} a_1 & b_1 & c_1 \\ a_2 & b_2 & c_2 \\ a_3 & b_3 & c_3 \end{bmatrix}, \quad C = \begin{bmatrix} a_1 & c_1 \\ a_2 & c_2 \\ a_3 & c_3 \end{bmatrix}, \quad W = \begin{bmatrix} u \\ v \end{bmatrix}, \quad Z = \begin{bmatrix} u \\ 0 \\ v \end{bmatrix}.$$

129D   Notice that B is constructed so that each of its columns X is a solution of $AX = 0$. Thus $AB = 0$ and hence $A(BZ) = 0$ for every Z. This shows that $\mathcal{R}(B) \subset \mathcal{N}(A)$. Now suppose (say) that $a_3 \neq 0$. We will show that $\mathcal{R}(B_3) = \mathcal{N}(A)$. Choose $X \in \mathcal{N}(A)$. Then

$$a_1 x_1 + a_2 x_2 + a_3 x_3 = 0,$$

so $x_3 = -(a_1 x_1 + a_2 x_2)/a_3$ so

$$X = \begin{bmatrix} x_1 \\ x_2 \\ x_3 \end{bmatrix} = \begin{bmatrix} 0 & a_3 \\ -a_3 & 0 \\ a_2 & -a_1 \end{bmatrix} \begin{bmatrix} -x_2/a_3 \\ x_1/a_3 \end{bmatrix}$$

which shows that $X = B_3 Z$ for some Z and hence that $X \in \mathcal{R}(B_3)$. Thus (under the hypothesis that $a_3 \neq 0$) we have shown that $\mathcal{R}(B_3) = \mathcal{N}(A)$. Since $\mathcal{R}(B_3) \subset \mathcal{R}(B)$, this shows that $\mathcal{R}(B) = \mathcal{N}(A)$. Similar reasoning shows that $\mathcal{R}(B_2) = \mathcal{N}(A)$ if $a_2 \neq 0$ and that $\mathcal{R}(B_1) = \mathcal{N}(A)$ if $a_1 \neq 0$. Since $A \neq 0$, at least one of these three cases applies.

129E   Not necessarily. The simplest example is $B = \begin{bmatrix} 1 \\ 0 \end{bmatrix}$ and $A = QB$ where Q is any 2×2 invertible matrix whose first column is not a multiple of B.

129F   Choose $X \in \mathcal{N}(B)$. Then $BX = 0$. Hence $AX = QBX = 0$. Therefore $X \in \mathcal{N}(A)$. This shows that $\mathcal{N}(B) \subset \mathcal{N}(A)$. If $Q = 0$ and $B \neq 0$, then $A = 0$, so $\mathcal{N}(A) = \mathbf{F}^{n \times 1}$, but $\mathcal{N}(B) \neq \mathbf{F}^{n \times 1}$.

129H   Yes. If $\mathcal{R}(B) \subset \mathcal{N}(A)$, then $AB = 0$. Here's the proof. Assume $\mathcal{R}(B) \subset \mathcal{N}(A)$. The kth column $B_k = \text{col}_k(B)$ is an element of $\mathcal{R}(B)$. Hence $B_k \in \mathcal{N}(A)$. Then, by the block multiplication law 54B, we obtain $\text{col}_k(AB) = A \, \text{col}_k(B) = 0$. In other words, every column of AB is zero. Hence $AB = 0$.

131B   The commands

```
#> A=rand(5,3), B=A*rand(3,3)
#> Y=A*rand(3,1), X=randsoln(B, Y)
#> Y, B*X % (equal)
```

confirm that $\mathcal{R}(A) \subset \mathcal{R}(B)$. Reversing the roles of A and B in the last two line confirms the reverse inclusion.

131C

```
#> Y = B*rand(2,1), X=randsoln(A,Y)
#> Y, A*X % (equal)
#> Y = A*rand(2,1), X=randsoln(B,Y)
#> Y, B*X % (not equal)
```

133B $\mathcal{R}(A_4) \subset \mathcal{R}(A_1) = \mathcal{R}(A_2)$; all other inclusions are false.

133C   Take $\mathrm{col}_1(B) = \mathrm{col}_1(A)$, $\mathrm{col}_2(B) = \mathrm{col}_1(A)$, $\mathrm{col}_3(B) = \mathrm{col}_1(A) + \mathrm{col}_2(A)$.

133F $A(\mathbf{F}^{n \times 1}) = \mathcal{R}(A)$, $A\{0\} = \{0\}$, $A^{-1}\mathbf{F}^{m \times 1} = \mathbf{F}^{n \times 1}$, $A^{-1}(\{0\}) = \mathcal{N}(A)$.

134E $X \in PV \iff P^{-1}X \in V$. Since $P^{-1}X = \begin{bmatrix} x_1 - x_2 \\ -x_1 + 2x_2 \end{bmatrix}$ the conditions is
$x_1 - x_2 = -x_1 + 2x_2$ or $2x_1 = 3x_2$. Therefore only $X_4 \in PV$.

135C   Not if $Y \neq 0$. A subspace must contain 0.

135D   The ambiguity is only apparent – if the two interpretations really disagreed, mathematicians would have invented different notations. Assume A is invertible and let V_1 denote the preimage of W by A and V_2 denote the image of W by A^{-1}. Then $X \in V_1$ if and only if $AX \in W$ and $X \in V_2$ if and only if $X = A^{-1}Y$ for some $Y \in W$. But

$$X = A^{-1}Y \iff AX = Y$$

so

$$\begin{aligned} X \in V_2 & \iff X = A^{-1}Y \text{ for some } Y \in W \\ & \iff AX \in W \\ & \iff X \in V_1 \end{aligned}$$

so $V_2 = V_1$ as claimed.

135F   When its null space is $\{0\}$.

139B   It can be formed from the two columns X_1 and X_2 defined in Example 120A:

$$\Phi = \begin{bmatrix} X_1 & X_2 \end{bmatrix} = \begin{bmatrix} -C \\ I_2 \end{bmatrix} = \begin{bmatrix} 6 & 7 \\ -5 & -6 \\ 1 & 0 \\ 0 & 1 \end{bmatrix}.$$

A column X lies in the null space if and only if it has the form

$$X = \Phi Z = x_3 X_1 + x_4 X_2, \qquad Z = \begin{bmatrix} x_3 \\ x_4 \end{bmatrix}.$$

139C $\mathcal{N}(A) = \{0\}$ for parts (1), (2), (4), (6). Matrices (3) and (7) have the same

null space and $\begin{bmatrix} -2 \\ 0 \\ 1 \end{bmatrix}$ is a basis for it. $\begin{bmatrix} -2 \\ 1 \\ 0 \end{bmatrix}$ is a basis for the null

space of (5) , and $\begin{bmatrix} -1 & 1 \\ -1 & -1 \\ 1 & 0 \\ 0 & 1 \end{bmatrix}$ is a basis for the null space of (8).

139D Only (5) has zero null space. The following gives a possible answer in each of the other cases although any matrix that is right equivalent to the given answer is also correct.

$$(1)\ \begin{bmatrix} -1 \\ -2 \\ -3 \\ 1 \end{bmatrix}. \qquad (2)\ \begin{bmatrix} -1 & -3 \\ -2 & -4 \\ 1 & 0 \\ 0 & 1 \end{bmatrix}. \qquad (3)\ \begin{bmatrix} 1 & 0 \\ 0 & -2 \\ 0 & 1 \\ 0 & 0 \end{bmatrix}.$$

$$(4)\ \begin{bmatrix} -1 \\ -2 \\ 1 \\ 0 \end{bmatrix}. \qquad (6)\ \begin{bmatrix} -2 \\ 1 \end{bmatrix}. \qquad (7)\ \begin{bmatrix} 1 \\ 0 \end{bmatrix}.$$

145A (1) $\widetilde{\Phi}$ = last two columns of I_5, Φ = last two columns of P. (2) $\widetilde{\Phi} = \Phi = []$ (the empty matrix). (3) $\widetilde{\Phi}$ = last three columns of I_5, Φ = last three columns of P. (4) $\widetilde{\Phi}$ = last column of I_3, Φ = last column of P.

145B (1) $\widetilde{\Psi} = I_3$, $\Psi = Q$. (2) $\widetilde{\Psi} = D=$ first three columns of I_5. Ψ = first three columns of Q. (3) $\widetilde{\Psi}$ = first two columns of I_3, Ψ = first two columns of Q. (4) $\widetilde{\Psi}$ = first two columns of I_5, Ψ = first two columns of Q.

145C Upon multiplying B by a suitable permutation matrix Q, the matrices QB and $Q\Psi$ will have the forms to which the original proof applies. From $A = BP$ follows $QA = QBP$ so (by the original proof) $Q\Psi$ is a basis for $\mathcal{R}(QA)$. Hence, Ψ is a basis for $\mathcal{R}(A)$ by Theorem 142B.

145D $\Psi = \begin{bmatrix} y_1 & y_2 & y_3 & y_4 \\ x_1 & x_2 & x_3 & x_4 \\ w_1 & w_2 & w_3 & w_4 \\ z_1 & z_2 & z_3 & z_4 \\ v_1 & v_2 & v_3 & v_4 \\ u_1 & u_2 & u_3 & u_4 \end{bmatrix}.$

148A $\Phi = \begin{bmatrix} \mathrm{col}_2(A) & \mathrm{col}_4(A) \end{bmatrix}$

148B

$$
(1) \begin{bmatrix} 4 & -3 & -2 \\ -5 & 5 & 2 \\ -2 & 2 & 1 \end{bmatrix}, \quad
(2) \begin{bmatrix} 4 & -3 \\ -5 & 5 \\ -2 & 2 \end{bmatrix}, \quad
(3) \begin{bmatrix} 4 & -3 \\ -5 & 5 \\ -2 & 2 \end{bmatrix},
$$

$$
(4) \begin{bmatrix} 4 & -3 \\ -5 & 5 \\ -2 & 2 \end{bmatrix}, \quad
(5) \begin{bmatrix} 4 & -3 \\ -5 & 5 \\ -2 & 2 \end{bmatrix}, \quad
(6) \begin{bmatrix} 4 \\ -5 \\ -2 \end{bmatrix}, \quad
(7) \begin{bmatrix} 4 \\ -5 \\ -2 \end{bmatrix}.
$$

Any answer obtained from the give answer by right multiplication by an invertible matrix is also correct. For example, any 3×3 invertible matrix is a correct answer for (1).

148C For (1), (2), (3), and (5) the range is $\mathbf{F}^{2\times 1}$, so I_2 (or any other invertible 2×2 matrix) is a basis. Matrices (4) and (8) have the same range and (4) is a basis for it. Matrix (6) is a basis for its own range. The first two columns of (8) form a basis for its range.

148E $X = \begin{bmatrix} 0 \\ z_1 \\ 0 \\ z_2 \end{bmatrix}.$

148F Let $R_j = \mathrm{col}_j(R)$. Then

$$
R_1 = c_{11} R_2 + c_{21} R_4, \qquad R_3 = c_{12} R_2 + c_{22} R_4.
$$

Multiply by M^{-1} and use the block multiplication law 54B:

$$
A_1 = c_{11} A_2 + c_{21} A_4, \qquad A_3 = c_{12} A_2 + c_{22} A_4.
$$

Substitute:

$$
\begin{aligned}
Y &= x_1 A_1 + x_2 A_2 + x_3 A_3 + x_4 A_4 \\
  &= (x_1 c_{11} + x_2 + x_3 c_{12}) A_2 + (x_1 c_{21} + x_3 c_{22} + x_4) A_4 \\
  &= \Psi Z
\end{aligned}
$$

where

$$
\Psi = \begin{bmatrix} A_1 & A_4 \end{bmatrix}, \qquad
Z = \begin{bmatrix} x_1 c_{11} + x_2 + x_3 c_{12} \\ x_1 c_{21} + x_3 c_{22} + x_4 \end{bmatrix}.
$$

148G First assume the leading columns are at the left so that

$$
R = \begin{bmatrix} I & C \\ 0 & 0 \end{bmatrix}, \qquad
\widetilde{\Psi} = \begin{bmatrix} I \\ 0 \end{bmatrix}
$$

Then for $X \in \mathbf{F}^{n \times 1}$ we have

$$RX = \left[\begin{array}{c} W + CZ \\ 0 \end{array} \right] = \widetilde{\Psi}(W + CZ), \qquad X = \left[\begin{array}{c} W \\ Z \end{array} \right]$$

where $W \in \mathbf{F}^{r \times 1}$ and $Z \in \mathbf{F}^{(n-r) \times 1}$. This shows $\mathcal{R}(R) = \mathcal{R}(\widetilde{\Psi})$. Since $\widetilde{\Psi} W = \left[\begin{array}{c} W \\ 0 \end{array} \right]$ it follows that $\mathcal{N}(\widetilde{\Psi}) = \{0\}$. As any RREF is right equivalent to one where the leading columns are at the left the general case follows by Theorem 128B.

148H This is the content of Theorem 147A. There will be other (right equivalent) bases whose columns are not columns of A by Theorem 136A.

151A When $\mathcal{R}(A) = \mathbf{F}^{m \times 1}$.

152B The matrix $\Theta = \left[\begin{array}{cc} I_r & 0_{(n-r) \times r} \end{array} \right]$ is a co-basis for the null space $\mathcal{N}(D)$; the matrix $\Lambda = \left[\begin{array}{cc} 0_{(m-r) \times r} & I_{m-r} \end{array} \right]$ is a co-basis for the range $\mathcal{R}(D)$ of D.

153C Delete the zero rows.

156C NO!!!!!!

159A $Y_i \in \mathcal{R}(\Psi) = \mathcal{R}(\Phi) = \{\Phi Z : Z \in \mathbf{F}^{\nu \times 1}\}$ so $Y_i = \Phi Z_i$ for some Z_i. Define $b_{i1}, b_{i2}, \ldots, b_{i\nu}$ by

$$Z_i = \left[\begin{array}{c} b_{i1} \\ b_{i2} \\ \vdots \\ b_{i\nu} \end{array} \right].$$

Then (a) is the block multiplication law 53B. For (b) note that

$$Z_i = \mathrm{row}_i(B)^{\top} = \mathrm{col}_i(B^{\top})$$

and

$$\Phi Z_i = Y_i = \mathrm{col}_i(\Psi) = \mathrm{col}_i(\Phi T) = \Phi \, \mathrm{col}_i(T)$$

so

$$\mathrm{col}_i(B^{\top}) = Z_i = \mathrm{col}_i(T).$$

Hence $B^{\top} = T$.

159C Yes.

160A We have $\Phi_1 = \Phi_2 T_1$, $\Phi_4 = \Phi_3 T_2$ where $T_1 = \left[\begin{array}{cc} 0 & 1 \\ 1 & 0 \end{array} \right]$, $T_2 = \left[\begin{array}{cc} 2 & 1 \\ 1 & 1 \end{array} \right]$ so $\mathcal{V}_1 = \mathcal{V}_2$ and $\mathcal{V}_3 = \mathcal{V}_4$. (If $\Phi_k = \Phi_j T$ then $\Phi_k T^{-1} = \Phi_j$.) If $X_1 = \mathrm{col}_1(\Phi_1)$ and $X_2 = \mathrm{col}_3(\Phi_3)$ then $X_1 \in \mathcal{V}_1$, $X_1 \notin \mathcal{V}_3$, $X_2 \notin \mathcal{V}_1$, $X_2 \in \mathcal{V}_3$. This answers 12 of the 30 questions. Also $\Phi_5 = A$ and $\Phi_6 = B$ where A and B are as in Exercise 129A.

161A   Assume that Ph has three columns:

```
#> X1=Ps(:,1); T1=randsoln(Ph,X1);
#> X2=Ps(:,2); T2=randsoln(Ph,X2);
#> X3=Ps(:,3); T3=randsoln(Ph,X3);
#> T=[T1 T2 T3], Ps, Ph*T
```

The solutions T1, T2, T3, are not really random; they are uniquely determined.

165A   The commands

```
#> randsoln(Ph1), randsoln(Ph2), randsoln(Ph3)
```

reveal that Ph1 and Ph2 have null space zero but Ph3 does not. After the commands

```
#> X=randsoln(A)
#> Z1 = randsoln(Ph1, Y)
#> Z2 = randsoln(Ph2, Y)
#> Z3 = randsoln(Ph3, Y)
```

the command

```
#> X, Ph1*Z1, Ph2*Z2, Ph3*Z3
```

shows that X, Ph2*Z2 and Ph3*Z3 are the same but X and Ph1*Z1 are different. We conclude that Ph2 and Ph3 have range equal to the null space of A, but not Ph1. The matrix Ph2 is a basis for the null space of A.

165B   (1) $\mathcal{N}(\Phi_k) = \{0\}$ for $k < 4$.

   (2) $\mathcal{R}(\Phi_k) \subset \mathcal{N}(A)$ for all k.

   (3) $\mathcal{R}(\Phi_k) = \mathcal{N}(A)$ for $k = 3, 4$.

   (4) $\mathcal{R}(\Phi_k) \subset \mathcal{R}(\Phi_{k+1})$ for all k.

   (5) $\mathcal{R}(\Phi_3) = \mathcal{R}(\Phi_4)$.

   (Item (2) is true because $A\Phi_k = 0$.)

173A   rank$(A) = 2$ and nullity$(A) = 3$.

173B   Equation (1) implies that $\mathcal{R}(\Phi) \subset \mathcal{N}(A)$. Hence, equation (2) implies that $\mathcal{R}(\Phi) = \mathcal{N}(A)$(see Exercise 170C). Finally, equation (3) asserts that $\mathcal{N}(A) = \{0\}$.

173C Both commands gj(rand(3,5)) and gj(rand(5,3)) produce a matrix with 3 nonzero rows.

173D Since A is a random 3×5 matrix its rank is 3 (the maximum possible rank for a 3×5 matrix) and hence its nullity is 2. Hence applying the proof of Theorem 163A produces a basis with two columns.

173F #> gj(A), gj(A'). (Both answers will have the same number of nonzero rows.)

176D If $AX = Y$, then $X = IX = \Gamma AX = \Gamma Y$ so $X = \Gamma Y$ is the only solution.

176E Since $A(\Omega Y) = (A\Omega)Y = IY = Y$ so $X = \Omega Y$ is a solution.

176F In the former we are given a left inverse Γ to A which enables us to find *the* (there's at most one) solution X of $Y = AX$ knowing Y but not A. In the latter we are given a right inverse Ω to A which enables us to find *a* solution X (but not all solutions) of $Y = AX$ knowing Y but not A.

176G

    (1) Assume that $\Gamma A = I$. Choose $X \in \mathcal{N}(A)$. Then $AX = 0$. Hence, $X = \Gamma AX = \Gamma 0 = 0$. This shows that $\mathcal{N}(A) \subset \{0\}$.

    (2) Assume that $A\Omega = I$. Choose $Y \in \mathbf{F}^{m \times 1}$. Let $X = \Omega Y$. Then $AX = A\Omega Y = Y$. Therefore $Y \in \mathcal{R}(A)$. This shows that $\mathbf{F}^{m \times 1} \subset \mathcal{N}(A)$.

176H $SD = I \implies (PSQ^{-1})A = (PSQ^{-1})(QDP^{-1}) = PSDP^{-1} = PP^{-1} = I.$
$DS = I \implies A(PSQ^{-1}) = (QDP^{-1})(PSQ^{-1}) = QDSQ^{-1} = QQ^{-1} = I.$

177A The transpose of D is given by $D^{\top} = D_{n,m,r}$ and is also in biequivalence normal form. Hence, both assertions amount to an easy application of the block multiplication law:

$$[\, I \quad 0 \,] \begin{bmatrix} I \\ 0 \end{bmatrix} = I. \qquad \qquad \square$$

177B We must show that a matrix $A \in \mathbf{F}^{m \times n}$

    (1) has a left inverse if and only if $\mathrm{rank}(A) = n$.

    (2) has has a right inverse if and only if $\mathrm{rank}(A) = m$.

    (1) By Theorem 106D, A is biequivalent to $D = D_{m,n,r}$ where $r = \mathrm{rank}(A)$. By 176H, A has a left inverse if and only if D does. By 177A, D has a left inverse if and only if $r = n$. If A has a left inverse, then $\mathcal{N}(A) = \{0\}$ by 176G, so $r = n$ by 172B.

(2)] By Theorem 106D, A is biequivalent to $D = D_{m,n,r}$ where $r = \text{rank}(A)$. By 176H, A has a right inverse if and only if D does. By 177A, D has a right inverse if and only if $r = m$. If A has a right inverse, then $\mathcal{R}(A) = \mathbf{F}^{m \times 1}$ by 176G, so $r = m$ by 172B.

177C $\Gamma = \Gamma I = \Gamma(A\Omega) = (\Gamma A)\Omega = I\Omega = \Omega$. By Exercise 177B, $m = n$. Now use the definition of A^{-1}.

177D By Exercise 177B, a square matrix which has a left inverse also has a right inverse and conversely. The rest follows from Exercise 177C.

177E Matrix (1) has no left inverse since

$$\begin{bmatrix} 1 & 2 & 3 \\ 2 & 4 & 6 \end{bmatrix} \begin{bmatrix} 2 \\ -1 \\ 0 \end{bmatrix} = \begin{bmatrix} 0 \\ 0 \end{bmatrix}.$$

Its transpose is matrix (2) which also has no left inverse as

$$\begin{bmatrix} 1 & 2 \\ 2 & 4 \\ 3 & 6 \end{bmatrix} \begin{bmatrix} 2 \\ -1 \end{bmatrix} = \begin{bmatrix} 0 \\ 0 \\ 0 \end{bmatrix}.$$

Matrix (3) has no left inverse as

$$\begin{bmatrix} 2 & 1 & 3 \\ 1 & 1 & 2 \end{bmatrix} \begin{bmatrix} 1 \\ 1 \\ -1 \end{bmatrix} = \begin{bmatrix} 0 \\ 0 \end{bmatrix}.$$

Its transpose is matrix (4) which does have a left inverse. To find it note that the first two rows form an invertible 2×2 matrix so we can simply write down a left inverse:

$$\begin{bmatrix} 1 & -1 & 0 \\ -1 & 2 & 0 \end{bmatrix} \begin{bmatrix} 2 & 1 \\ 1 & 1 \\ 3 & 2 \end{bmatrix} = \begin{bmatrix} 1 & 0 \\ 0 & 1 \end{bmatrix}.$$

(Of course, there are other left inverses.)

177F These problems can all be done the same way: (1) Find an invertible M so that $R = MA$ is in reduced form. (2) If R has a row of zeros choose $\tilde{Y}$ having one as its last entry: then $\tilde{Y} \neq RX$ for all X so we may take $Y = M^{-1}\tilde{Y}$ to get $Y \neq AX$ for all X. (3) If MA has no row of zeros, we may find a left inverse $\tilde{\Gamma}$ for R; then $\Gamma = \tilde{\Gamma}M$ is a lefy inverse for $A = M^{-1}R$. For example, for the matrix A of part (1) we have

$$MA = \begin{bmatrix} 1 & 2 & 3 \\ 0 & 0 & 0 \end{bmatrix}$$

where
$$M = \begin{bmatrix} 1 & 0 \\ -2 & 1 \end{bmatrix}.$$

Clearly
$$\begin{bmatrix} 0 \\ 1 \end{bmatrix} \neq MAX$$

for all X (since the last row of MA is zero) so
$$M^{-1} \begin{bmatrix} 0 \\ 1 \end{bmatrix} \neq AX$$

for all X. For the matrix A of part (3) we have
$$\begin{bmatrix} 1 & -1 \\ -1 & 2 \end{bmatrix} \begin{bmatrix} 2 & 1 & 3 \\ 1 & 1 & 2 \end{bmatrix} = \begin{bmatrix} 1 & 0 & 1 \\ 0 & 1 & 1 \end{bmatrix}$$

so a right inverse is
$$\begin{bmatrix} 1 & 0 \\ 0 & 1 \\ 0 & 0 \end{bmatrix} \begin{bmatrix} 1 & -1 \\ -1 & 2 \end{bmatrix} = \begin{bmatrix} 1 & -1 \\ -1 & 2 \\ 0 & 0 \end{bmatrix}$$

178A $A\Omega = I \implies I = I^\top = (A\Omega)^\top = \Omega^\top A^\top$.

178B The matrix $B(AB)^{-1}$ is a right inverse for A, and the matrix $(AB)^{-1}A$ is a left inverse for B.

178C If Γ is a left inverse for A, then $\Gamma A = I$, so A is a right inverse for Γ.

178E A square matrix is invertible $\iff$ its columns are independent (see Corollary 175A). Thus the columns of A are independent $\iff$ A is invertible $\iff$ $A^\top$ is invertible $\iff$ the columns of $A^\top$ are independent $\iff$ the rows of A are independent.

178I With $\Gamma = (CA)^{-1}C$ we have
$$\Gamma A = ((CA)^{-1})C)A = (CA)^{-1})(CA) = I.$$

178J In fact, for most $n \times m$ matrices C, the matrix $\Gamma = (CA)^{-1}C$ will be a left inverse to A.

```
#> A=rand(3,2)
#> C=rand(2,3), G=inv(C*A)*C, G*A
```

Repeat the last line to produce a different left inverse G for the same A.

178L

```
function B = inv1(A)
%
[Q D P] = bed(A);
B = P*D'*Q^(-1);
```

182A

(1) $\text{rank}(A) = \text{rank}(D) = 3$, $\text{rank}(Q) = \text{rank}(Q^{-1}) = 3$, $\text{rank}(P) = \text{rank}(P^{-1}) = 5$.

(2) $\text{rank}(A) = \text{rank}(D) = 3$, $\text{rank}(Q) = \text{rank}(Q^{-1}) = 5$, $\text{rank}(P) = \text{rank}(P^{-1}) = 3$.

(3) $\text{rank}(A) = \text{rank}(D) = 2$, $\text{rank}(Q) = \text{rank}(Q^{-1}) = 3$, $\text{rank}(P) = \text{rank}(P^{-1}) = 5$.

(4) $\text{rank}(A) = \text{rank}(D) = 2$, $\text{rank}(Q) = \text{rank}(Q^{-1}) = 5$, $\text{rank}(P) = \text{rank}(P^{-1}) = 3$.

184A Let $E_j = \text{col}_j(I_5)$. $MX = E_1$ and $M(X - Y) = E_2$. Hence, $R = MA = \begin{bmatrix} 0 & E_1 & 3E_1 & E_2 & E_1 - E_2 & E_1 + 3E_2 \end{bmatrix}$.

194A The positivity law holds because the square of a nonzero number is positive and

$$\langle X, X \rangle = \sum_{j=1}^{n} x_j^2$$

if $x_j = \text{entry}_j(X)$. The transpose law is proved as as follows:

$$\langle AX, Y \rangle = (AX)^{\mathsf{T}}Y = X^{\mathsf{T}}A^{\mathsf{T}}Y = \langle X, A^{\mathsf{T}}Y \rangle.$$

194C

$$
\begin{aligned}
\|X + Y\|^2 &= \langle X + Y, X + Y \rangle \\
&= \langle X, X \rangle + \langle Y, Y \rangle + \langle X, Y \rangle + \langle Y, X \rangle \\
&= \langle X, X \rangle + \langle Y, Y \rangle + \langle X, Y \rangle + \overline{\langle X, Y \rangle} \\
&= \langle X, X \rangle + \langle Y, Y \rangle + 2\Re(\langle X, Y \rangle) \\
&= \|X\|^2 + \|Y\|^2 + 2\Re(\langle X, Y \rangle) \\
&= \|X\|^2 + \|Y\|^2 + 2\langle X, Y \rangle \text{ if } X, Y \text{ are real.}
\end{aligned}
$$

Here $\Re(z)$ denotes the real part of the complex number z.

198A The Schwarz inequality

$$\|X\|\,\|Y\|\,|\cos\theta| = |\langle X,Y\rangle| \leq \|X\|\,\|Y\|$$

says that the cosine of any angle has absolute value at most one.

198B $\|X+Y\|$ and $\|X-Y\|$ are the diagonals a parallelogram with edges $\|X\|$ and $\|Y\|$.

199B A unitary matrix is (by definition) a square matrix. Either of the two equations $QQ^* = I$ or $Q^*Q = I$ says that Q has a one-sided inverse and is hence invertible. See Exercise 177D.

200A

$$
\begin{aligned}
R(\theta)R(\phi) &= \begin{bmatrix} \cos\theta & \sin\theta \\ -\sin\theta & \cos\theta \end{bmatrix}\begin{bmatrix} \cos\phi & \sin\phi \\ -\sin\phi & \cos\phi \end{bmatrix} \\[4pt]
&= \begin{bmatrix} \cos\theta\cos\phi - \sin\theta\cos\phi & \cos\theta\sin\phi + \sin\theta\cos\phi \\ -\cos\theta\sin\phi - \sin\theta\cos\phi & \cos\theta\cos\phi - \sin\theta\cos\phi \end{bmatrix} \\[4pt]
&= \begin{bmatrix} \cos(\theta+\phi) & \sin(\theta+\phi) \\ -\sin(\theta+\phi) & \cos(\theta+\phi) \end{bmatrix} \\[4pt]
&= R(\theta+\phi)
\end{aligned}
$$

and

$$R(\theta)^\mathsf{T} = \begin{bmatrix} \cos\theta & \sin\theta \\ -\sin\theta & \cos\theta \end{bmatrix} = R(-\theta)$$

so $R(\theta)R(\theta)^\mathsf{T} = R(\theta)R(-\theta) = R(\theta - \theta) = R(0) = I$ as $\cos 0 = 1$ and $\sin 0 = 0$. The matrix

$$Q = \begin{bmatrix} 1 & 0 \\ 0 & -1 \end{bmatrix}$$

is unitary, but is not of form $Q = R(\theta)$.

200B

$$
\begin{aligned}
QQ^\mathsf{T} &= \begin{bmatrix} q_{11} & q_{12} \\ q_{21} & q_{22} \end{bmatrix}\begin{bmatrix} q_{11} & q_{21} \\ q_{12} & q_{22} \end{bmatrix} \\[4pt]
&= \begin{bmatrix} q_{11}^2 + q_{12}^2 & q_{11}q_{21} + q_{12}q_{22} \\ q_{21}q_{11} + q_{22}q_{12} & q_{21}^2 + q_{22}^2 \end{bmatrix}.
\end{aligned}
$$

The condition that QQ^T be the identity matrix is equivalent to the three equations in the problem.

202A The conjugate transpose Q^* of a unitary matrix Q is again unitary since

$$(Q^*)^{-1} = (Q^{-1})^{-1} = Q = (Q^*)^*$$

and, as the rows of Q^* are the conjugate transposes of the columns of Q, these rows must also be orthonormal.

203B If (and only if) all the elements of the sequence are nonzero.

203C By Theorem 202E,

$$X = \sum_{k=1}^{\nu} \langle X, U_k \rangle U_k, \qquad Y = \sum_{j=1}^{\nu} \langle Y, U_j \rangle U_j.$$

Hence

$$
\begin{aligned}
\langle X, Y \rangle &= \sum_{k=1}^{\nu} \sum_{j=1}^{\nu} \langle X, U_k \rangle \overline{\langle Y, U_j \rangle} \langle U_k, U_j \rangle \\
&= \sum_{k=1}^{\nu} \langle X, U_k \rangle \overline{\langle Y, U_k \rangle} \\
&= \sum_{k=1}^{\nu} \langle X, U_k \rangle \langle U_k, Y \rangle.
\end{aligned}
$$

203D $\text{entry}_{ij}(AA^*) = \langle \text{row}_i(A), \text{row}_j(A) \rangle.$

204B A triangular unitary matrix must be diagonal since its columns are pairwise orthogonal. A diagonal matrix is unitary if and only if its diagonal entries have absolute value one. In particular, there are 2^n $n \times n$ real diagonal unitary matrices.

209A Two vectors make an acute angle if their inner product is positive. The Gram-Schmidt process chose U_j so that

$$\langle V_j, U_j \rangle = z_{jj} > 0.$$

211A

(1)
$$
\begin{bmatrix} 1 & 7 \\ 1 & 7 \\ 1 & 3 \\ 1 & 3 \end{bmatrix}
=
\begin{bmatrix} 1/2 & 1/2 \\ 1/2 & 1/2 \\ 1/2 & -1/2 \\ 1/2 & -1/2 \end{bmatrix}
\begin{bmatrix} 2 & 10 \\ 0 & 4 \end{bmatrix}.
$$

(2)
$$
\begin{bmatrix} 1 & 7 & 17 \\ 1 & 7 & 11 \\ 1 & 3 & 5 \\ 1 & 3 & -1 \end{bmatrix}
=
\begin{bmatrix} 1/2 & 1/2 & 1/2 \\ 1/2 & 1/2 & -1/2 \\ 1/2 & -1/2 & 1/2 \\ 1/2 & -1/2 & -1/2 \end{bmatrix}
\begin{bmatrix} 2 & 10 & 16 \\ 0 & 4 & 12 \\ 0 & 0 & 6 \end{bmatrix}.
$$

(3)
$$
\begin{bmatrix} 1 & 7 & 17 & 20 \\ 1 & 7 & 11 & -2 \\ 1 & 3 & 5 & -6 \\ 1 & 3 & -1 & -12 \end{bmatrix}
=
\begin{bmatrix} 1/2 & 1/2 & 1/2 & 1/2 \\ 1/2 & 1/2 & -1/2 & -1/2 \\ 1/2 & -1/2 & 1/2 & -1/2 \\ 1/2 & -1/2 & -1/2 & 1/2 \end{bmatrix}
\begin{bmatrix} 2 & 10 & 16 & 0 \\ 0 & 4 & 12 & 18 \\ 0 & 0 & 6 & 14 \\ 0 & 0 & 0 & 8 \end{bmatrix}.
$$

211B
```
#> V=rand(5,3), U=zeros(5,3), T=zeros(3,3)
%
#> W=V(:,1)
#> T(1,1)= sqrt(W'*W)
#> U(:,1)=W/T(1,1)      % column 1
%
#> T(1,2)= U(:,1)'*V(:,2),
#> W= V(:,2)-T(1,2)*U(:,1)
#> T(2,2)= sqrt(W'*W)
#> U(:,2)= W/T(2,2)     % column 2
%
#> T(1,3)= U(:,1)'*V(:,3),   T(2,3)= U(:,2)'*V(:,3)
#> W= V(:,3)-T(1,3)*U(:,1)-T(2,3)*U(:,2)
#> T(3,3)= sqrt(W'*W)
#> U(:,3)= W/T(3,3)     % column 3
%
#> X, U*T                % check
```

211C
```
#> A=rand(2,5), Psi=Nbasis1(A)
#> W = Psi(:,1);
#> U1= W1/sqrt(W'*W);
#> W = Psi(:,2)-(U1'*Psi(:,2))*U1
#> U2= W/sqrt(W'*W)
#> W = Phi(:,3)-(U1'*Phi(:,3))*U1-(U2*Phi(:,3)*U2,
#> U3= W/sqrt(W'*W)
#> Phi=[U1 U2 U3]
#> Phi'*Phi % Phi has orthonormal columns
#> A*Phi    % basis for null space
#> T=Phi'*Psi, Psi, Phi*T % check Gram-Schmidt
```

211E
```
#> Psi = rand(5,3), [Phi T]=gs(Psi)
#> Phi'*Phi, eye(3)
#> Psi, Phi*R
```

217A Assume that $Y_1, Y_2 \in V^\perp$ and $a_1, a_2 \in \mathbf{F}$. For any $X \in V$ we have

$$\langle a_1 Y_1 + a_2 Y_2, X \rangle = a_1 \langle Y_1, X \rangle + a_2 \langle Y_2, X \rangle = 0.$$

Thus $a_1 Y_1 + a_2 Y_2 \in V^\perp$.

219A $A^* A = I_n$ so that $\Gamma = A^*$ is a left inverse to A. Hence, $A^* A = \Gamma A$ is the orthogonal projection matrix on $\mathcal{R}(A)$.

219C $\Pi = \frac{1}{42} \begin{bmatrix} 10 & -8 & 16 \\ -8 & 40 & 4 \\ 16 & 4 & 34 \end{bmatrix}$

220A Read A^* for A in the Left Inverse Theorem 218C. You must also prove that $\mathcal{N}(A^*)^\perp = \mathcal{R}(A)$ to complete the proof.

220D Because $(A^*)^* = A$ and $(\mathcal{V}^\perp)^\perp = \mathcal{V}$.

220F
```
#> Ph=rand(5,3), Ps=rand(5,2), Q =[Ph Ps]
#> Pi=[Ph zeros(5,2)]*inv(Q)
#> Pi*Pi-Pi, Pi*Ph-Ph, Pi*Ps
```

220G
```
#> Ph = rand(5,3),  Pi = Ph*inv(Ph'*Ph)*Ph'
#> Pi*Pi-Pi, Pi*Ph-Ph, Pi*Ps, Pi'-Pi
```

225G Let $c_{ij} = \text{entry}_{ik}(AB)$, $A_i = \text{row}_i(A)$, $B_j = \text{col}_j(B)$. Then

$$
\begin{aligned}
\|AB\|^2 &= \sum_{ij} |c_{ij}|^2 \\
&= \sum_{ij} |A_i B_j|^2 \\
&= \sum_{ij} |\langle A_i, B_j^* \rangle|^2 \\
&\leq \sum_{ij} \|A_i\|^2 \|B_j\|^2 \\
&= \left(\sum_i \|A_i\|^2 \right) \left(\sum_j \|B_j\|^2 \right) \\
&= \|A\|^2 \|B\|^2.
\end{aligned}
$$

233B There are no crossings so the sign is $(-1)^0 = 1$.

238A
```
function t = perminv(s)
%
[m n] = size(s);
t = zeros(1,n);
t(s) = 1:n     % bombs here if m~=1
```

244B Swapping columns in A corresponds to swapping rows in the transpose $A^\top$. Hence, $\det(B) = \det(B^\top) = -\det(A^\top) = -\det(A)$ by 243C and 242C.

248A $\det(A) = 5$.

248B By Theorems 240C and 246B.

248D
```
#> A=rand(5,5), B=rand(5,5), I=eye(5)
#> det(I), 1
#> det(A^(-1)), 1/det(A)
#> det(A*B), det(A)*det(B)
#> det(A'), det(A)
#> det(A([5 4 2 3 1],:)), -det(A)
```

250B For part (3) let

$$Z = X \wedge Y = \begin{bmatrix} z_1 \\ z_2 \\ z_3 \end{bmatrix}.$$

Then the matrix M is invertible. But its determinant is

$$\det(M) = (x_2 y_3 - x_3 y_2)z_1 + (x_3 y_1 - x_1 y_3)z_2 + (x_1 y_2 - x_2 y_1)z_3$$
$$= z_1^2 + z_2^2 + z_3^2$$

which is positive if $Z = X \wedge Y \neq 0$.

250D The equation $aX + bY = 0$ is equivalent to the three matrix equations

$$\begin{bmatrix} x_2 & y_2 \\ x_3 & y_3 \end{bmatrix} \begin{bmatrix} a \\ b \end{bmatrix} = \begin{bmatrix} 0 \\ 0 \end{bmatrix},$$

$$\begin{bmatrix} x_1 & y_1 \\ x_3 & y_3 \end{bmatrix} \begin{bmatrix} a \\ b \end{bmatrix} = \begin{bmatrix} 0 \\ 0 \end{bmatrix},$$

$$\begin{bmatrix} x_1 & y_1 \\ x_2 & y_2 \end{bmatrix} \begin{bmatrix} a \\ b \end{bmatrix} = \begin{bmatrix} 0 \\ 0 \end{bmatrix}$$

though of course these latter equations contain some redundancy. The
determinants of the matrices

$$\begin{bmatrix} x_2 & x_3 \\ y_2 & y_3 \end{bmatrix}, \quad \begin{bmatrix} x_1 & x_3 \\ y_1 & y_3 \end{bmatrix}, \quad \begin{bmatrix} x_1 & x_2 \\ y_1 & y_2 \end{bmatrix}$$

are the entries in $X \wedge Y$. Hence, if $X \wedge Y \neq 0$ then at least one of these
matrices is invertible so that the only solution of $aX + bY = 0$ is $a = b = 0$.

250G If $\mathbf{F} = \mathbf{C}$ and

$$X = \begin{bmatrix} 1 \\ i \\ 0 \end{bmatrix}, \quad Y = \begin{bmatrix} 1 \\ i \\ 1 \end{bmatrix},$$

then

$$X \wedge Y = \begin{bmatrix} i \\ -1 \\ 0 \end{bmatrix} \neq 0$$

but $(X, Y, X \wedge Y)$ is dependent since $iX = X \wedge Y$.

251C If $A = A_0 + iA_1$ and $B = B_0 + iB_1$ and P is real then

$$AP = B \iff A_0 P = B_0 \text{ and } A_1 P = B_1.$$

The theory of real equivalence of complex matrices is subsumed under the more general (and very difficult) theory of simultaneous equivalence.

254B Matrices A and B are biequivalent if $A = QBP^{-1}$; similarity requires $Q = P$.

254C $PIP^{-1} = I$ so the only matrix similar to the identity matrix I is itself. A matrix is biequivalent to the identity if and only if it is invertible.

254E Yes (take P to be the identity).

255C

$$\begin{aligned}
A_4 &= P_4 A_2 P_4^{-1} \\
&= P_4 P_2 A_1 P_2^{-1} P_4^{-1} \\
&= P_4 P_2 P_3^{-1} A_3 P_3 P_2^{-1} P_4^{-1} \\
&= P_4 P_2 P_3^{-1} A_3 (P_4 P_2 P_3^{-1})^{-1}
\end{aligned}$$

so take $P = P_4 P_2 P_3^{-1}$.

256B The only eigenvalue of the identity matrix I_n is 1 and every column $X \in \mathbf{F}^{n \times 1}$ is an eigenvector. The only eigenvalue of the zero matrix $0_{n \times n}$ is 0 and every column $X \in \mathbf{F}^{n \times 1}$ is an eigenvector.

259A The eigenvalues of D and of A are 3 and 7. The identity matrix is an invertible matrix whose columns are eigenvectors of D. The matrix P is an invertible matrix whose columns are eigenvectors of A.

259B The matrix equation is equivalent to the homogeneous system $(\lambda_1 - \lambda_2)x_1 = 0$ and $(\lambda_3 - \lambda_2)x_3 = 0$. When $\lambda_1, \lambda_2, \lambda_3$ are distinct the most general solution is $x_1 = x_3 = 0$ and x_2 arbitrary. However if (for example) $\lambda_1 = \lambda_2 \neq \lambda_3$ there are more solutions: the most general solution is $x_3 = 0$ with x_1, x_2 arbitrary.

259D $AX = X$, $AY = 2Y$, $AZ = 3Z$.

259E $AX = 2X$ and $AZ = -Z$ but AY is not a multiple of Y.

260A
```
#> A=[13 6 30; 1 2 3; -4 -2 -9], I=eye(3)
#> X1=solve(A-1*I), X2=solve(A-2*I), X3=solve(A-3*I)
#> P=[X1 X2 X3], D=diag([1 2 3])
#> A*P, P*D
```

260C Let
$$D = \mathrm{diag}(\lambda_1, \lambda_2, \ldots, \lambda_n)$$
be the diagonal matrix with entries $\lambda_1, \lambda_2, \ldots, \lambda_n$ on the diagonal and
$$E_j = \mathrm{col}_j(I_n)$$
be the jth column of the identity matrix. Then
$$DE_j = \lambda_j E_j$$
which says that λ_j is an eigenvalue with eigenvector E_j. Conversely, for $X \in \mathbf{F}^{n \times 1}$, we have
$$\mathrm{entry}_j(DX) = \lambda_j \, \mathrm{entry}_j(X),$$
so that if $DX = \lambda X$ where the jth entry of X is not zero, then $\lambda = \lambda_j$. $\square$

263B It is important to remember that there is no single correct answer here for we can take the eigenvalues in any order and a nonzero multiple of an eigenvector is again an eigenvector. The following answers are correct in that $A = PDP^{-1}$. You can generate other correct answers from these by rearranging the diagonal entries of D (provided that the columns of P are similarly rearranged) and by replacing each column of P by a nonzero multiple of itself.

(1) $D = \begin{bmatrix} 2 & 0 \\ 0 & 3 \end{bmatrix}$, $\quad P = \begin{bmatrix} 2 & 1 \\ 1 & 1 \end{bmatrix}$.

(2) $D = \begin{bmatrix} \frac{3+\sqrt{5}}{2} & 0 \\ 0 & \frac{3-\sqrt{5}}{2} \end{bmatrix}$, $\quad P = \begin{bmatrix} -1 & -1 \\ \frac{1-\sqrt{5}}{2} & \frac{1+\sqrt{5}}{2} \end{bmatrix}$.

(3) $D = \begin{bmatrix} -1 & 0 \\ 0 & 1 \end{bmatrix}$, $\quad P = \begin{bmatrix} 1 & 1 \\ -1 & 1 \end{bmatrix}$.

(4) $D = \begin{bmatrix} 3+4i & 0 \\ 0 & 3-4i \end{bmatrix}$ $\quad P = \begin{bmatrix} 1 & 1 \\ -i & i \end{bmatrix}$.

(5) $D = \begin{bmatrix} 1 & 0 \\ 0 & 3 \end{bmatrix}$, $\quad P = \begin{bmatrix} 1 & 1 \\ 0 & 1 \end{bmatrix}$.

(6) $D = \begin{bmatrix} 4 & 0 \\ 0 & 6 \end{bmatrix}$, $\quad P = \begin{bmatrix} 2 & 0 \\ -5 & 1 \end{bmatrix}$.

264B

(1) $D = \begin{bmatrix} 2 & 0 & 0 & 0 \\ 0 & 3 & 0 & 0 \\ 0 & 0 & 1 & 0 \\ 0 & 0 & 0 & 4 \end{bmatrix}$, $\quad P = \begin{bmatrix} 2 & 1 & 0 & 0 \\ 1 & 1 & 0 & 0 \\ 0 & 0 & 3 & 1 \\ 0 & 0 & 2 & 1 \end{bmatrix}$.

(2) $D = \begin{bmatrix} 1 & 0 & 0 & 0 \\ 0 & 4 & 0 & 0 \\ 0 & 0 & 2 & 0 \\ 0 & 0 & 0 & 3 \end{bmatrix}$, $\quad P = \begin{bmatrix} 3 & 1 & 0 & 0 \\ 2 & 1 & 0 & 0 \\ 0 & 0 & 2 & 1 \\ 0 & 0 & 1 & 1 \end{bmatrix}$.

(3) $D = \begin{bmatrix} 1 & 0 & 0 & 0 \\ 0 & 4 & 0 & 0 \\ 0 & 0 & 1 & 0 \\ 0 & 0 & 0 & 4 \end{bmatrix}$, $\quad P = \begin{bmatrix} 2 & 1 & 0 & 0 \\ 1 & 1 & 0 & 0 \\ 0 & 0 & 2 & 1 \\ 0 & 0 & 1 & 1 \end{bmatrix}$.

(4) $D = \begin{bmatrix} 1 & 0 & 0 & 0 \\ 0 & 4 & 0 & 0 \\ 0 & 0 & 1 & 0 \\ 0 & 0 & 0 & 4 \end{bmatrix}$, $\quad P = \begin{bmatrix} 1 & 0 & 0 & 0 \\ 0 & 1 & 0 & 0 \\ 0 & 0 & 2 & 1 \\ 0 & 0 & 1 & 1 \end{bmatrix}$.

(5) $D = \begin{bmatrix} 4 & 0 & 0 \\ 0 & 1 & 0 \\ 0 & 0 & 4 \end{bmatrix}$, $\quad P = \begin{bmatrix} 1 & 0 & 0 \\ 0 & 2 & 1 \\ 0 & 1 & 1 \end{bmatrix}$.

(6) $D = \begin{bmatrix} 1 & 0 & 0 \\ 0 & 4 & 0 \\ 0 & 0 & 4 \end{bmatrix}$, $\quad P = \begin{bmatrix} 2 & 1 & 0 \\ 1 & 1 & 0 \\ 0 & 0 & 1 \end{bmatrix}$.

265D

```
#> X=P(:,1), A*X, 1*X
#> X=P(:,2), A*X, 3*X
#> X=P(:,3), A*X, 7*X
```

266A

```
#> I= eye(3)
#> X1 = randsoln(A-3*I)
#> X2 = randsoln(A-5*I)
#> X3 = randsoln(A-0*I)
#> P=[X1 X2 X3]
#> A, P*D*inv(P) % (These agree)
```

270C Let $f(\xi) = \det(\xi I - A)$. Then $\overline{f(\xi)} = f(\bar{\xi})$ because the coefficients are real. Hence, $f(\lambda) = 0 \implies f(\bar{\lambda}) = \overline{f(\lambda)} = 0$.

270D   #> A=rand(5,5), sum(eig(A)), sum(diag(A))

270E   #> A=rand(5,5),   exp(sum(log(eig(A)))), det(A)

271A   The number of times it appears on the diagonal.

271C   No. $\nu(4, A) = 2 \neq 1 = \nu(4, B)$.

273A   (1), (5), (7), and (8) are true, the others are false.

273C $P = \begin{bmatrix} E_2 & E_4 & E_1 & E_3 \end{bmatrix}$ where $E_j = \mathrm{col}_j(I)$.

274A   We want $P = \begin{bmatrix} X_1 & X_2 & \cdots & X_n \end{bmatrix}$ where $AX_j = \lambda_j X_j$. (Then $AP = PD$.)
Let $E_k = \mathrm{col}_k(I)$. $AE_k = \lambda_{\sigma(k)} E_k$ so we want $X_j = E_k$ where $\sigma(k) = j$.
In the last problem,

$$\sigma = (3, 1, 4, 2), \qquad P = \begin{bmatrix} E_2 & E_4 & E_1 & E_3 \end{bmatrix}.$$

274C   "Only if" is Theorem 271B. To prove "if" it is enough to prove it for diagonal matrices. The geometric multiplicity of any eigenvalue of a diagonal
matrix is the number of times it appears on the diagonal. If D and $\tilde{D}$ are
diagonal matrices with same diagonal entries, counting repetitions, then
$\tilde{D} = PDP^{-1}$ for a suitable permutation matrix P. (See Exercise 274A.)

274D   Assume that $AX = \lambda X$, λ is the only eigenvalue, and $A \neq \lambda I$. Let P be
invertible with $\mathrm{col}_1(P) = X$. Then $P^{-1}AP = \begin{bmatrix} \lambda & c \\ 0 & \lambda \end{bmatrix}$. Take

$$B = \begin{bmatrix} 1 & 0 \\ 0 & c \end{bmatrix} \begin{bmatrix} \lambda & c \\ 0 & \lambda \end{bmatrix} \begin{bmatrix} 1 & 0 \\ 0 & c^{-1} \end{bmatrix}.$$

274E   By Theorem 240C, the determinant

$$\det(\xi I - A) = \prod_{j=1}^{n} (\xi - a_{jj})$$

is the product of the diagonal entries. It vanishes if and only if $\xi = a_{jj}$ for
some j.

274F   Suppose that $\lambda \neq 0$ is an eigenvalue for BA. Then there is a nonzero eigenvec-
tor X such that $BAX = \lambda X$. Hence $(AB)(AX) = A(BAX) = A(\lambda X) = \lambda(AX)$. Now $AX \neq 0$ (otherwise we would have $0 = BAX = \lambda X$ contradicting $X \neq 0$ and $\lambda \neq 0$.) Thus λ is an eigenvalue for AB with eigenvector
AX.

274G $AB = \begin{bmatrix} 1 & -2 \\ 4 & 7 \end{bmatrix}$ so its eigenvalues are 3 and 5. Since A has more columns than rows there is at least one nonzero $X \in \mathbf{R}^{3 \times 1}$ with $AX = 0$ and hence $BAX = 0X$ so that 0 is an eigenvalue of BA, Hence, the eigenvalues of BA are 3, 5 and 0.

274H
```
#> A=rand(3,5), B=rand(5,3)
#> eig(A*B), eig(B*A)
```

282A $u_1(t) = e^{7t} - e^{3t}$, $u_2(t) = -e^{7t} + 2e^{3t}$.

286A There are p different solutions c_i of $c_i^p = d_i$ (if $d_i \neq 0$) and so the method will construct p^n different matrices D provided that A has n distinct nonzero eigenvalues.

286D $D^2 = D$, $A^2 = A$.

286F Assume that $AX = \lambda X$. Multiply by A:

$$A^2 X = A(\lambda X) = \lambda A X = \lambda^2 X.$$

Multiply by A:
$$A^3 X = A^2(\lambda X) = \lambda A^2 X = \lambda^3 X.$$

Repeat p times to get $A^p X = \lambda^p X$. (This is essentially a proof by induction.)

287A
```
#> B=rand(3,3), P=rand(3,3), A=P^(-1)*B*P
#> A^4, P^(-1)*B^4*P
```

287B $D^2 = D$ and $A^2 = A$.

287C Enter the matrix. Use MINIMAT's *eig* function to diagonalize A:

```
#> [P D]=eig(A)
```

and confirm that $A = PDP^{-1}$

```
#> P*D*P^(-1), A
```

Solve $C^3 = D$:

```
#> C=diag(diag(D).^(1/3))
#> C^3, D
```

and transform back to solve $B^3 = A$:

```
#> B=P*C*P^(-1)
#> B^3, A
```

287D See Exercise 98A and Remark 255A.

290A *(geometric)* $c_j = 1$.

(exponential) $c_j = 1/j!$.

(sine) $c_j = 0$ if j is even, $c_j = 1/j!$ for $j = 1, 5, 9, \ldots$, and $c_j = -1/j!$ for $j = 3, 7, 11, \ldots$.

(cosine) $c_j = 0$ if j is odd, $c_j = 1/j!$ for $j = 0, 4, 8, \ldots$, and $c_j = -1/j!$ for $j = 2, 6, 10, \ldots$.

290B The Taylor series are

$$e^\xi = \sum_{n=0}^\infty \frac{\xi^n}{n!}, \qquad \cos\theta = \sum_{k=0}^\infty \frac{\theta^{2k}}{(2k)!}, \qquad \sin\theta = \sum_{k=0}^\infty \frac{\theta^{2k+1}}{(2k+1)!}.$$

Substitute $\xi = i\theta$ and use $i^2 = -1$. The even powers form the cosine series, the odd powers form the sine series.

290C Since $e^{\rho + i\theta} = e^\rho e^{i\theta}$ the equation $e^w = z$ takes the form

$$x = r\cos\theta, \qquad y = r\sin\theta, \qquad e^\rho = r.$$

where $r = |z| = \sqrt{x^2 + y^2}$. If $z = 0$ there is no solution, otherwise take $\rho = \ln r$ and

$$\theta = \begin{cases} \arctan(y/x) & \text{if } x > 0; \\ \arctan(y/x) + \pi & \text{if } x < 0; \\ \pi/2 & \text{if } x = 0 \text{ and } y > 0; \\ -\pi/2 & \text{if } x = 0 \text{ and } y < 0. \end{cases}$$

Of course, replacing θ by $\theta + 2n\pi$ gives another solution.

290D $\sinh(\xi) = \sum_{k=0}^\infty \frac{\xi^{2k+1}}{(2k+1)!}, \qquad \cosh(\xi) = \sum_{k=0}^\infty \frac{\xi^{2k}}{(2k)!}$.

293B $U(t) = e^{tA} U(0) = \begin{bmatrix} 4e^{-t} + 5e^{2t} \\ 2e^{-t} + 5e2t \end{bmatrix}$

293C $A = PDP^{-1}$ where $D = \begin{bmatrix} 2 & 0 \\ 0 & 3 \end{bmatrix}$, $\quad P = \begin{bmatrix} 2 & 1 \\ 1 & 1 \end{bmatrix}$. so

$$e^{tA} = Pe^{tD}P^{-1} = \begin{bmatrix} 2 & 1 \\ 1 & 1 \end{bmatrix} \begin{bmatrix} e^{2t} & 0 \\ 0 & e^{3t} \end{bmatrix} \begin{bmatrix} 1 & -1 \\ -1 & 2 \end{bmatrix}.$$

Hence

$$e^{tA} = \begin{bmatrix} 2e^{2t} - 3e^{3t} & -2e^{2t} + 2e^{3t} \\ e^{2t} - e^{3t} & -e^{2t} + 2e^{3t} \end{bmatrix}.$$

293F e^t, e^{4t}, e^{9t},

293G Since $N^k = 0$ for $k > 1$, only two terms appear in the series.

294A $J^2 = -I$, $J^3 = -J$, $J^4 = I$, $J^5 = J$ and in general $J^{4q+r} = J^r$. The series

$$e^{tJ} = \sum_{j=0}^{\infty} \frac{t^j J^j}{j!}$$

receives contributions on the diagonal from the even terms and off the diagonal from the odd terms:

$$e^{tJ} = \sum_{k=0}^{\infty} \frac{(-1)^k t^{2k}}{(2k)!} I + \sum_{k=0}^{\infty} \frac{(-1)^k t^{2k+1}}{(2k+1)!} J.$$

Hence $e^{tJ} = (\cos t)I + (\sin t)J$.

294B $K^2 = I$ so

$$e^{tK} = \sum_{k=0}^{\infty} \frac{t^{2k}}{(2k)!} I + \sum_{k=0}^{\infty} \frac{t^{2k+1}}{(2k+1)!} K.$$

Hence $e^{tK} = (\cosh t)I + (\sinh t)K$.

294C It is the matrix $U(t)$ of Exercise 69H.

294D $(A+B)^2 = A^2 + AB + BA + B^2$. If $AB = BA$, then $(A+B)^2 = A^2 + 2AB + B^2$ as in high school algebra.

294E $(A+B)^3 = A^3 + A^2B + ABA + BA^2 + AB^2 + BAB + B^2A + B^3$. If $AB = BA$, then $(A+B)^3 = A^3 + 3A^2B + 3AB^2 + B^3$ as in high school algebra.

294F See Theorem 446A.

294G See Corollary 447A.

295E $\dot{C}(t) = \dot{A}(t)B(t) + A(t)\dot{B}(t)$.

295F $\dot{C}(0) = AB + BA$.

295G $\dot{C}(t)(A + tB) + C(t)B = 0$ so $\dot{C}(0)A + A^{-1}B = 0$ so $\dot{C}(0) = -A^{-1}BA^{-1}$.

295H $\dot{C}(0) = pA^{p-1}B$. (This works for negative integers p also.)

295J If $D = \begin{bmatrix} a & 0 \\ 0 & b \end{bmatrix}$, then the matrix equation equation $D^2 = -D$ is equivalent to the two scalar equations $a^2 = -a$ and $b^2 = -b$. The solutions are $a = 0, -1$ and $b = 0, -1$. There are thus four diagonal 2×2 matrices D satisfying $D^2 = -D$:

$$D = \begin{bmatrix} -1 & 0 \\ 0 & -1 \end{bmatrix}, \begin{bmatrix} -1 & 0 \\ 0 & 0 \end{bmatrix}, \begin{bmatrix} 0 & 0 \\ 0 & -1 \end{bmatrix}, \begin{bmatrix} 0 & 0 \\ 0 & 0 \end{bmatrix}.$$

If P is invertible, then $A = PDP^{-1}$ solves $A^2 = -A$. For example,

$$P = \begin{bmatrix} 2 & 1 \\ 1 & 1 \end{bmatrix}, \quad D = \begin{bmatrix} 1 & 0 \\ 0 & 0 \end{bmatrix}, \quad A = \begin{bmatrix} 2 & -2 \\ 1 & -1 \end{bmatrix}.$$

295K $e^{tA} = I + A - e^{-t}A.$

298A W and expm(D) should be equal.

298B W is the 3×3 identity matrix but F has all its entries equal to one.

298C L=rand(1,4), diag(exp(L)), expm(diag(L))

298D
```
#> A=rand(4,4), [P D]=eig(A),
#> L=diag(D), W=diag(exp(L))
#> P*D*inv(P), A
#> P*W*inv(P), expm(A)
```

298E
```
#> A=0.1*rand(3,3)
#> expm(A), eye(3)+A+(1/2)*A^2+(1/6)*A^3
```

298F
```
#> A=rand(3,3), B=rand(3,3)
#> expm(A+B), expm(A)*expm(B)
```

(The last two are different.)

298I
```
#> A=rand(3,3), t=rand(1,1), h=0.01
#> ( expm( (t+h)*A )-expm( t*A ) )/h
#> A*expm(t*A)
```

The last two results should be roughly equal. If not, repeat using a smaller value of h.

298J
```
#> J=[0 1; -1 0], t=rand(1,1),
#> expm(t*J), cos(t), sin(t)
```

299A Here are the commands:
```
#> P=rand(3,3)
#> A=P*diag(rand(1,3)/10)*P^(-1)
#> B=P*diag(rand(1,3)/10)*P^(-1)
#> A*B, B*A
#> expm(A+B), expm(A)*expm(B)
```

(The last two answers should be the same. The reason for dividing by 10 is that otherwise round-off error would cause the answers to disagree.)

299B
```
#> A=rand(3,3), I=eye(3), expm(A)
#> (I+(1/5)*A)^5, (I+(1/10)*A)^10, (I+(1/50)*A)^50
```

300C $\det(\xi I - C) = \xi^2 + b\xi + c.$

300E $\det(\xi I - C) = \xi^3 + a\xi^2 + b\xi + c.$

303B
```
#> p=rand(1,3), r=eig(compan(p))
#> t=r(1), t^3+p(1)*t^2+p(2)*t+p(3)
#> t=r(2), t^3+p(1)*t^2+p(2)*t+p(3)
#> t=r(3), t^3+p(1)*t^2+p(2)*t+p(3)
```

Each of the last three lines should produce a zero. To save typing you can use MINIMAT's elementwise operations

```
#> r.^3+ p.*r.^2+p.*r+p
```

306B $A = \begin{bmatrix} 0 & i \\ -i & 0 \end{bmatrix}$ and $B = \begin{bmatrix} 0 & i \\ i & 0 \end{bmatrix}.$

307A $AE_1 = \mathrm{col}_1(A), \ AE_2 = \mathrm{col}_2(A), \ \langle AE_1, E_2 \rangle = \mathrm{entry}_{21}(A) = a_{21},$ and
$\langle E_1, AE_2 \rangle = \overline{\mathrm{entry}_{12}(A)} = \bar{a}_{12}.$

307B $A^* = (B + B^*)^* = B^* + (B^*)^* = B^* + B = A.$

310B No, a diagonal matrix is automatically unitarily diagonalizable but not need not have distinct eigenvalues. For example, a scalar multiple of the identity matrix has only one eigenvalue. It is true however that a matrix which is unitarily similar to a real diagonal matrix is Hermitian.

310C

(1) $P = \frac{1}{\sqrt{2}} \begin{bmatrix} 1 & 1 \\ -1 & 1 \end{bmatrix}$ $\qquad D = \begin{bmatrix} 5 & 0 \\ 0 & 3 \end{bmatrix}$

(2) $P = \begin{bmatrix} 0.6 & 0.8 \\ -0.8 & 0.6 \end{bmatrix}$ $\qquad D = \begin{bmatrix} 5 & 0 \\ 0 & 10 \end{bmatrix}$

(3) $P = \begin{bmatrix} 0.6 & 0.8 & 0 \\ -0.8 & 0.6 & 0 \\ 0 & 0 & 1 \end{bmatrix}$ $\quad D = \begin{bmatrix} 5 & 0 & 0 \\ 0 & 10 & 0 \\ 0 & 01 \end{bmatrix}$

310E The proof does not show that the matrix P is square, that is, that the sum of the geometric multiplicities dim $\mathcal{N}(\lambda_i I - A)$ is n. This *is* true for a Hermitian matrix (by the Spectral Theorem), but false in general.

310H `#> B=rand(5,5), A=B+B'.`

310I The diagonal matrices D and D1 will have the same entries on the diagonal but the order might be different. The matrices Q and P will be different, even after rearranging the columns, as the `eig` function does not normalize its columns. Since the matrix Q is unitary, the expressions `Q'*Q` and `Q*Q'` evaluate to the identity matrix. Since the columns of P are orthogonal, `P'*P` evaluates to a diagonal matrix; however `P*P'` does not, for the rows of P are not orthogonal.

311A True, the command `[P D]=eig(A)` need not produce a unitary matrix P, but by Theorem 309B, the columns of P will be pairwise orthogonal as they are eigenvectors of A. If we normalize the columns of P we will obtain a unitary matrix whose columns are eigenvectors of A.

```
#> B=rand(3,3), A=B+B', [P D]=eig(A)
#> P(:,1)=P(:,1)/sqrt(P(:,1)'*P(:,1))
#> P(:,2)=P(:,2)/sqrt(P(:,2)'*P(:,2))
#> P(:,3)=P(:,3)/sqrt(P(:,3)'*P(:,3))
#> P'*P, A, P*D*P'
```

314B No. If $AX = \lambda Z$ and $Z = X + iY$ where A, λ, X, Y are real, then $AX = \lambda X$ and $AY = \lambda Y$. If Z is nonzero, at least one of X and Y must be nonzero. It follows that a real matrix with a real eigenvalue has a corresponding real eigenvector.

315A
$$
P = \begin{bmatrix} 0.5 & 0.5 & 0.5 & 0.5 \\ 0.5 & 0.5 & -0.5 & -0.5 \\ 0.5 & -0.5 & 0.5 & -0.5 \\ 0.5 & -0.5 & -0.5 & 0.5 \end{bmatrix}, \quad
R = \begin{bmatrix} 3.0 & 6.0 & 5.0 & 5.0 \\ 0.0 & 3.0 & 4.0 & 3.0 \\ 0.0 & 0.0 & 1.0 & 4.0 \\ 0.0 & 0.0 & 0.0 & 2.0 \end{bmatrix}.
$$

315D By the Gram-Schmidt Decomposition, $P = QR$ where Q is orthogonal and R is triangular. Let $B = P^{-1}AP$. Then $Q^{-1}AQ = R^{-1}BR$. But B is triangular; hence so is $R^{-1}BR$.

320D Yes when $\mathbf{F} = \mathbf{C}$. (It is normal.)

325F The subspace V_k consists of all columns of form
$$
X = \begin{bmatrix} Z \\ 0_{n-k} \end{bmatrix}, \quad Z \in \mathbf{F}^{k \times 1}.
$$

For such X we have
$$
AX = \begin{bmatrix} A_{11}Z \\ A_{21}Z \end{bmatrix}.
$$

The latter lies in $\mathcal{V}_k$ iff $A_{21}Z = 0$. Hence,

$$
\begin{aligned}
A\mathcal{V} \subset \mathcal{V}_k \quad &\Longleftrightarrow \quad AX \in \mathcal{V}_k \ \forall X \in \mathcal{V}_k \\
&\Longleftrightarrow \quad A_{21}Z = 0 \forall Z \in \mathbf{F}^{k \times 1} \\
&\Longleftrightarrow \quad A_{21} = 0.
\end{aligned}
$$

325G $B = A_{11}$.

326B Choose $Y \in \mathcal{V}^{\perp}$; we must show that $AY \in \mathcal{V}^{\perp}$. Choose $X \in \mathcal{V}$; we must show that $\langle X, AY \rangle = 0$. But $AX \in \mathcal{V}$, since $\mathcal{V}$ is A-invariant so

$$
\langle X, AY \rangle = \langle AX, Y \rangle = 0
$$

as required.

335C No. $\begin{bmatrix} 1 & 0 \\ 0 & 2 \end{bmatrix} \begin{bmatrix} 0 & 1 \\ 1 & 0 \end{bmatrix} = \begin{bmatrix} 0 & 1 \\ 2 & 0 \end{bmatrix}$ is a product of two Hermitian matrices, but it is not Hermitian.

35D Mimic the proof for unitary matrices (Theorem 198E).

335F The set of invertible diagonal matrices.

336C For the general case see Exercise 394C. Here's the 3×3 case. The product

$$
\begin{bmatrix} 1 & a_1 & b_1 \\ 0 & c_1 & 0 \\ 0 & 0 & 1 \end{bmatrix} \begin{bmatrix} 1 & a_2 & b_2 \\ 1 & c_2 & 0 \\ 0 & 0 & 1 \end{bmatrix} = \begin{bmatrix} 1 & a_1 + a_2 & b_1 + b_1 + a_1 c_2 \\ 0 & c_1 + c_2 & 0 \\ 0 & 0 & 1 \end{bmatrix}
$$

of two unitriangular matrices is unitriangular. The inverse

$$
\begin{bmatrix} 1 & a & b \\ 0 & c & 0 \\ 0 & 0 & 1 \end{bmatrix}^{-1} = \begin{bmatrix} 1 & -a & ac - b \\ 0 & -c & 0 \\ 0 & 0 & 1 \end{bmatrix}
$$

of a unitriangular matrix is unitriangular.

41B

$$
L = \begin{bmatrix} 1 & 0 & 0 & 0 \\ 9 & 1 & 0 & 0 \\ 5 & 2 & 1 & 0 \\ 3 & 1 & 3 & 1 \end{bmatrix}, \quad D = \begin{bmatrix} 0 & 1 & 0 & 0 & 0 \\ 1 & 0 & 0 & 0 & 0 \\ 0 & 0 & 0 & 0 & 0 \\ 0 & 0 & 0 & 1 & 0 \end{bmatrix},
$$

$$
U^{-1} = \begin{bmatrix} 1 & -6 & 19 & -63 & 336 \\ 0 & 1 & -4 & 13 & -68 \\ 0 & 0 & 1 & -4 & 20 \\ 0 & 0 & 0 & 1 & -7 \\ 0 & 0 & 0 & 0 & 1 \end{bmatrix}.
$$

341D $R = DU^{-1}$.

345B $\delta_{1q} = \delta_{2q} = \delta_{3q} = 0$ for $q = 1, 2$; $\delta_{1q} = \delta_{2q} = 1$ for $q = 3, 4, 5$; $\delta_{3q} = 0$ for $q = 1, 2$; $\delta_{33} = 1$; $\delta_{3q} = 2$ for $q = 4, 5$ $\delta_{4q} = 1$ for $q = 1, 2$; $\delta_{4,3} = 2$; $\delta_{4q} = 3$ for $q = 4, 5$

345C $D = \begin{bmatrix} 0 & 1 & 0 \\ 1 & 0 & 0 \end{bmatrix}$.

345D $R = \begin{bmatrix} 0 & 0 & 1 & -a \\ 1 & -b & 0 & -c \end{bmatrix}$.

346D
```
#> DO=randrook(m,n,r)
#> A=randlow(m)*RO*randuni(n)
#> [L D U]=rook(A)
#> A, L*DO*U^(-1), RO, D
```

The reason for requiring that `randlow` and `randuni` return integer matrices is that the algorithm for transforming a matrix to RNF is quite sensitive to round-off error.

348A If $y_j = \text{entry}_j(Y)$ and $x_i = \text{entry}_i(X)$, then

$$\begin{aligned}
x_4 &= y_4/2 \\
x_3 &= (y_3 - a_{34}x_4)/3 \\
x_2 &= (y_2 - a_{23}x_3 - a_{24}x_4)/4 \\
x_1 &= (y_1 - a_{12}x_2 - a_{13}x_3 - a_{14}x_4)/5
\end{aligned}$$

348B Let $y_j = \text{entry}_j(Y)$ and $x_i = \text{entry}_i(X)$. Then

$$x_i = \frac{y_i - a_{i,i+1}x_{i+1} - a_{i,i+2}x_{i+2} - \cdots - a_{i,n}x_n}{a_{ii}}.$$

Use this to find $x_n, x_{n-1}, x_{n-2}, \ldots, x_1$ (in that order).

348C $X = \begin{bmatrix} x_1 \\ x_2 \\ 1 \\ 0 \end{bmatrix}$ where $x_2 = -a_{23}/4$ and $x_1 = -(a_{12}x_2 + a_{13})/5$.

348D
```
#> x2=-A(2,3)/A(2,2), x1=-(A(1,2)*x2+A(1,3)/A(1,1)
#> X = [x1; x2; 1; 0], A*X
```

348E Let $a_{ij} = \text{entry}_{ij}(A)$. Let the first zero entry on the diagonal be a_{pp}, that is, $a_{pp} = 0$ but $a_{ii} \neq 0$ for $i < p$. There is a unique $X \in \mathbf{F}^{n \times 1}$ with $AX = 0$, $x_j = 0$ for $j > p$ and $x_p = 1$. Here $x_j = \text{entry}_j(X)$. The entries $x_{p-1}, x_{p-2}, \ldots, x_1$ may be defined successively by

$$x_i = -\frac{a_{i,i+1}x_{i+1} + a_{i,i+2}x_{i+2} + \cdots + a_{i,n}x_n}{a_{ii}}.$$

348F Exercises 348B and 348E show that the null space of a triangular matrix is the zero space $\iff$ the diagonal entries of that matrix are nonzero. By the theory in Chapter 5 (see Corollary 175A) a square matrix is invertible $\iff$ its null space is zero.

351A One can modify the argument in the text to handle the new situation but it is more instructive to use a trick. Apply the first form of the Bruhat Decomposition with A replaced by WA. (Here W is the reversal matrix of Exercise 350B.) Then $WA = LPU^{-1}$ where L is lower triangular. Now take $V = WLW$ and $Q = WP$.

351D This is the same argument as the Alternate Bruhat Decomposition; the only point is that if D is in RNF, then so is WD.

352A First, as in the proof of the Rook Decomposition 340A we may transform R and $\tilde{R}$ to RNF, that is, $R = DU^{-1}$ and $\tilde{R} = \tilde{D}\tilde{U}^{-1}$ where D and $\tilde{D}$ are in RNF and U and $\tilde{U}$ are unitriangular. Then $LRU^{-1} = \tilde{L}\tilde{R}\tilde{U}^{-1}$ so by the uniqueness of the RNF, $D = \tilde{D}$. But, if V is unitriangular, then DV and D have the same positions of leading entries and zero rows. This is because of the formula

$$\text{row}_i(DV) = \text{row}_i(D)V$$

and the fact that $\text{row}_i(D) = \text{row}_j(I)$, if the leading entry in the ith row of D occurs in the jth column. Apply this with $V = U^{-1}$ and $V = \tilde{U}^{-1}$ to prove the lemma. $\square$

352B This follows from the formula

$$\text{row}_i(\tilde{L})\,\text{col}_j(\tilde{R}) = \text{entry}_{ij}(\tilde{L}\tilde{R}) = \text{entry}_{ij}(LR) = \text{row}_i(L)\,\text{col}_j(R)$$

for matrix multiplication. The entries in the kth row of R do not affect the answer since they vanish. For example, suppose that

$$L = \begin{bmatrix} b_{11} & 0 & 0 \\ b_{21} & b_{22} & 0 \\ b_{31} & b_{32} & b_{33} \end{bmatrix}, \qquad R = \begin{bmatrix} c_1 \\ 0 \\ c_3 \end{bmatrix},$$

$$\tilde{L} = \begin{bmatrix} \tilde{b}_{11} & 0 & 0 \\ \tilde{b}_{21} & \tilde{b}_{22} & 0 \\ \tilde{b}_{31} & \tilde{b}_{32} & \tilde{b}_{33} \end{bmatrix}, \qquad \tilde{R} = \begin{bmatrix} \tilde{c}_1 \\ 0 \\ \tilde{c}_3 \end{bmatrix},$$

so that

$$M = \begin{bmatrix} b_{11} & 0 \\ b_{31} & b_{33} \end{bmatrix}, \quad S = \begin{bmatrix} c_1 \\ c_3 \end{bmatrix}, \quad \widetilde{M} = \begin{bmatrix} \tilde{b}_{11} & 0 \\ \tilde{b}_{31} & \tilde{b}_{33} \end{bmatrix}, \quad \tilde{S} = \begin{bmatrix} \tilde{c}_1 \\ \tilde{c}_3 \end{bmatrix}.$$

The equations $LR = \tilde{L}\tilde{R}$ and $MS = \widetilde{M}\tilde{S}$ carry the same information. □

352C This is a special case of 352D as a unitriangular matrix is in LENF. It is quite easy since the equation $\tilde{L}^{-1}L = \tilde{U}U^{-1}$ asserts the equality of a lower triangular matrix and a unitriangular matrix. This can happen only if both are the identity matrix. □

352D By 352A and 352B, we may assume that R and $\tilde{R}$ have no zero row. Multiplication on the right by a permutation matrix permutes the columns so we may choose a permutation matrix P so that

$$RP = \begin{bmatrix} U & C \end{bmatrix}, \quad \tilde{R}P = \begin{bmatrix} \tilde{U} & \tilde{C} \end{bmatrix}, \tag{a}$$

where U and $\tilde{U}$ are unitriangular. The same P works for both R and $\tilde{R}$ by 352A. Both U and $\tilde{U}$ are unitriangular by the definition of LENF which requires that the entries below a leading entry vanish. Multiply the hypothesis that $LR = \tilde{L}\tilde{R}$ by P to obtain $LRP = \tilde{L}\tilde{R}P$ which, by block multiplication, becomes

$$LU = \tilde{L}\tilde{U}, \qquad LC = \tilde{L}\tilde{C}. \tag{b}$$

From 352C and equation (a) we obtain $L = \tilde{L}$ and $U = \tilde{U}$. By the invertibility of L and equation (b) we obtain $C = \tilde{C}$. Hence, $RP = \tilde{R}P$ and so (as P is invertible) $R = \tilde{R}$ as required. □

359D

(1) For $X \in \mathbf{F}^{n \times 1}$:

$$\begin{aligned} Q_N X &= (I - 2NN^*)X \\ &= X - 2N(N^*X) \\ &= X - 2(N^*X)N \\ &= X - 2\langle X, N \rangle N \end{aligned}$$

(2)

$$\begin{aligned} Q_N N &= (I - 2NN^*)N \\ &= N - 2(N^*N)N \\ &= -N. \end{aligned}$$

since $N^*N = \|N\|^2 = 1$.

(3) If $\langle X, N \rangle = 0$, then

$$\begin{aligned} Q_N X &= (I - 2NN^*)X \\ &= X - 2\langle X, N \rangle N \\ &= X. \end{aligned}$$

(4) The matrix Q_N is Hermitian:

$$\begin{aligned} Q_N^* &= (I - 2NN^*)^* \\ &= I - 2N^{**}N^* \\ &= I - 2NN^* \\ &= Q_N \end{aligned}$$

(5) The matrix Q_N is a square root of the identity matrix

$$\begin{aligned} Q_N^2 &= (I - 2NN^*)(I - 2NN^*) \\ &= I - 4NN^* + 4(NN^*)(NN^*) \\ &= I - 4NN^* + 4(N^*N)NN^* \\ &= I \end{aligned}$$

as $N^*N = \|N\|^2 = 1$.

(6) The matrix Q_N is unitary since $Q_N^{-1} = Q_N$ by part (3) and $Q_N = Q_N^*$ by (2).

(7) If the vector N is real, then so is the matrix Q_N since $N^* = N^{\mathsf{T}}$ is also real.

359E $\quad Q_N = \begin{bmatrix} 1 - 2a^2 & -2ab \\ -2ab & 1 - 2b^2 \end{bmatrix}$

360A

```
#> N=rand(4,1), N=N/sqrt(N'*N)
#> Q=eye(4)-2*N*N'
#> Q*N, -N
#> X=randsoln(N), Q*X
#> Q*Q', Q*Q
```

360B
```
#> Y=rand(3,1), Z=rand(3,1),
#> Z=Z/sqrt(Z'*Z), Z=sqrt(Y'*Y)*Z
#> Z'*Z, Y'*Y            % these are the same
#> N= Y-Z, N=N/sqrt(N'*N)
#> Q= eye(3)-2*N*N'
#> Q*Y, Z
#> Q*Z, Y
```

360D
```
#> Y = rand(4,1), Y = Y=Y/sqrt(Y'*Y)
#> Z = rand(4,1), Z = Z/sqrt(Z'*Z)
#> Q*Q, Q*Q' % two copies of identity
#> Z, Q*Y    % two copies of Z
#> Y, Q*Z    % two copies of Y
```

364B Two matrices A and B are **REF equivalent** iff $A = QBP^{-1}$ where Q is unitary and P is invertible diagonal.

364C The following commands create a random 3×5 matrix A, a (real) unitary matrix 3×3 matrix Q, and an upper triangular 3×5 matrix R satisfying A=Q*R:

```
#> A = rand(3,5), R=A
#> Y = A(:,1),   r = sqrt(Y'*Y), Z = [r;0;0]
#> N = (Y-Z), N = N/sqrt(N'*N),
#> Q1 = eye(3)-2*N*N', R = Q1*R
#> Y = A(2:3,2),  r = sqrt(Y'*Y), Z = [r;0]
#> N = (Y-Z), N = N/sqrt(N'*N), N = [ 0; N]
#> Q2 = eye(3)-2*N*N', R = Q2*R
#> Q=Q1*Q2,  Q'*Q, Q*R, A
```

Since Q is unitary, Q'*Q will give the 3×3 identity matrix. Evaluating the expression Q*R should be the same result as A. The matrix R should have the form

$$R = \begin{bmatrix} r_{11} & r_{12} & r_{13} & r_{14} & r_{15} \\ 0 & r_{22} & r_{23} & r_{24} & r_{25} \\ 0 & 0 & r_{33} & r_{34} & r_{35} \end{bmatrix}$$

364D The following commands create a random 4×5 matrix A, a (real) unitary matrix 4×4 matrix Q, and an upper triangular 4×5 matrix R satisfying A=Q*R:

```
#> A = rand(4,5), R=A
#> Y = A(:,1),   r = sqrt(Y'*Y), Z = [r;0;0;0]
```

```
#> N = (Y-Z), N = N/sqrt(N'*N),
#> Q1 = eye(4)-2*N*N', R = Q1*R
#> Y = A(2:4,2),  r = sqrt(Y'*Y), Z = [r;0;0]
#> N = (Y-Z), N = N/sqrt(N'*N), N = [0; N]
#> Q2 = eye(4)-2*N*N', R = Q2*R
#> Y = A(3:4,3),  r = sqrt(Y'*Y), Z = [r;0]
#> N = (Y-Z), N = N/sqrt(N'*N), N = [0;0; N]
#> Q3 = eye(4)-2*N*N', R = Q3*R
#> Q=Q1*Q2*Q3,  Q'*Q, Q*R, A
```

Since Q is unitary, Q'*Q will give the 4×4 identity matrix. Evaluating the expression Q*R should be the same result as A. The matrix R should have the form

$$R = \begin{bmatrix} r_{11} & r_{12} & r_{13} & r_{14} & r_{15} \\ 0 & r_{22} & r_{23} & r_{24} & r_{25} \\ 0 & 0 & r_{33} & r_{34} & r_{35} \\ 0 & 0 & 0 & r_{44} & r_{45} \end{bmatrix}$$

364E The following commands create a random 5×3 matrix A, a (real) unitary matrix 5×5 matrix Q, and an upper triangular 5×3 matrix R satisfying A=Q*R:

```
#> A = rand(5,3), R=A
#> Y = A(:,1),  r = sqrt(Y'*Y), Z = [r;0;0;0;0]
#> N = (Y-Z), N = N/sqrt(N'*N),
#> Q1 = eye(5)-2*N*N', R = Q1*R
#> Y = A(2:5,2),  r = sqrt(Y'*Y), Z = [r;0;0;0]
#> N = (Y-Z), N = N/sqrt(N'*N), N = [ 0; N]
#> Q2 = eye(5)-2*N*N', R = Q2*R
#> Y = A(3:5,3),  r = sqrt(Y'*Y), Z = [r;0;0]
#> N = (Y-Z), N = N/sqrt(N'*N), N = [0;0; N]
#> Q3 = eye(5)-2*N*N', R = Q3*R
#> Q=Q1*Q2*Q3,  Q'*Q, Q*R, A
```

Since Q is unitary, Q'*Q will give the 5×5 identity matrix. Evaluating the expression Q*R should be the same result as A. The matrix R should have the form

$$R = \begin{bmatrix} r_{11} & r_{12} & r_{13} \\ 0 & r_{22} & r_{23} \\ 0 & 0 & r_{33} \\ 0 & 0 & 0 \\ 0 & 0 & 0 \end{bmatrix}$$

365A
```
#> A=rand(3,5)
#> [Q R]= qr(A)
#> Q, Q', Q*Q' % check that Q is unitary
#> A, Q*R      % check that A=Q*R
```

366A Suppose that
$$D = \text{diag}(d_1, d_2, \ldots, d_n)$$
is diagonal and unitary. As
$$D^* = \text{diag}(\bar{d}_1, \bar{d}_2, \ldots, \bar{d}_n)$$
we obtain
$$DD^* = \text{diag}(\|d_1\|^2, \|d_2\|^2, \ldots, \|d_n\|^2)$$
so that D is unitary if and only if each of its diagonal entries has absolute value 1:
$$\|d_k\| = 1 \text{ for } k = 1, 2, \ldots, n.$$
In the real case this means that each $d_k = \pm 1$ so there are 2^n different possibilities (one of which is the identity matrix). In the complex case each entry must have form
$$d_k = \exp(i\theta_k) = \cos\theta_k + i\sin\theta_k$$
where θ_k is a real number between 0 and 2π. Here we have used the so-called **polar decomposition** of a complex number: if $z = z + iy$ where x and y are real, then
$$z = re^{i\theta} = r\cos\theta + i\sin\theta$$
where
$$r = \|z\| = \sqrt{z\bar{z}} = \sqrt{x^2 + y^2}$$
and
$$\tan\theta = y/x.$$

366B
```
#> Q = qr(rand(3,5)), R = Q
#> Y = Q(:,1),   r = sqrt(Y'*Y), Z = [r;0;0]
#> N = (Y-Z), N = N/sqrt(N'*N),
#> Q1 = eye(3)-2*N*N', R = Q1*R
#> Y = Q(2:3,2),  r = sqrt(Y'*Y), Z = [r;0]
#> N = (Y-Z), N = N/sqrt(N'*N), N = [ 0; N]
#> Q2 = eye(3)-2*N*N', R = Q2*R
#> Q3 = R, Q1*Q2*Q3, Q
```

371A
$$Q = \begin{bmatrix} 1/2 & 1/2 & 1/2 & 1/2 \\ 1/2 & 1/2 & -1/2 & -1/2 \\ 1/2 & -1/2 & 1/2 & -1/2 \\ 1/2 & -1/2 & -1/2 & 1/2 \end{bmatrix},$$

$$P = \begin{bmatrix} 3/5 & 0 & 4/5 \\ 0 & 1 & 0 \\ 4/5 & 0 & -3/5 \end{bmatrix}, \qquad D = \begin{bmatrix} 20 & 0 & 0 \\ 0 & 10 & 0 \\ 0 & 0 & 5 \\ 0 & 0 & 0 \end{bmatrix}.$$

374B -1 with multiplicity 1 and 1 with multiplicity $n - 1$. Since $Q_N N = -N$ the axis of the reflection (and any multiple of it) is a eigenvector for eigenvalue -1. The orthogonal complement to the axis consists of eigenvectors for the eigen value 1.

374C By Theorem 199A, a unitary matrix Q has the property that $\langle QX, QV \rangle = \langle X, V \rangle$ so that $\langle QX, QV \rangle = 0$ whenever $\langle X, V \rangle = 0$. Thus every vector X is a singular vector for Q. Since $\|QX\| = \|X\|$, it follows that the only singular value is $\sigma = 1$.

374D If D is SVNF, there is an invertible diagonal matrix U (a product of elementary scales) so that UD is in biequivalence normal form. Hence, if $A = QDP^{-1}$ is the singular value decomposition, then $A = (QU^{-1})(UD)P^{-1}$ is a biequivalence decomposition.

374E By Gauss-Jordan, $A = MW$ where W is in RREF and M is invertible. By Gram-Schmidt, $M = QB$ where Q is unitary and B is triangular with positive diagonal entries. Now $R = BW$ is in PREF.

376D
```
#> A=rand(3,5)
#> [Q D P]= svd(A)
#> Q^(-1), Q'   % check that Q is unitary
#> P^(-1), P'   %    "       "  P   "
#> A, Q*D*P'    % chexk that A=Q*D*P'
```

377E From the Singular Value Decomposition, $A = QDP^*$, we obtain $A^*A = P(D^*D)P^*$ we can find D^*D and P using the eig function and the method of the previous problem. Since D^*D is diagonal we find D with MINIMAT's sqrt function. (It works elementwise, but that's okay for diagonal matrices.) From this we can obtain the first three columns W of Q using the equation $AP = QD$. Now the command Q=qr([W rand(5,2)]) will produce the correct answer up to a possible change of the signs of the first three columns.

382B No. For example, a diagonal matrix is automatically diagonalizable, but it need not have distinct diagonal entries.

390A $\Delta = D + V$.

394C The equation

$$I - N^n = (I - N)(I + N + N^2 + \cdots + N^{n-1})$$

is familiar from high school algebra where it is used to sum a geometric series. Here's the proof:

$$
\begin{aligned}
&(I - N)(I + N + N^2 + \cdots + N^{n-1}) \\
&= \quad I(I + N + N^2 + \cdots + N^{n-1}) \\
&\quad\; -N(I + N + N^2 + \cdots + N^{n-1}) \\
&= \quad I + N + N^2 + \cdots + N^{n-1} \\
&\quad\; -N - N^2 - N^3 - \cdots - N^n \\
&= \quad I - N^n.
\end{aligned}
$$

When $N^n = 0$ the formula shows that $I - N$ is invertible and

$$(I - N)^{-1} = (I + N + N^2 + \cdots + N^{n-1}).$$

395E ξ^p where p is the degree of nilpotence.

396G There is no justification for interchanging the order of the two operations of evaluating the determinant and substituting A for ξ. Indeed, if this method of proof were correct we would also prove that $g(A) = 0$ where $g(\xi) = \text{trace}(\xi I - A)$. But for $A = \begin{bmatrix} a & b \\ c & d \end{bmatrix}$ we have $g(\xi) = 2\xi - a - d$ and $g(A) = \begin{bmatrix} a - d & 2b \\ 2c & d - a \end{bmatrix}$ which is generally *not* zero.

403B By the block diagonalization Theorem 384C, we may assume that $A \in \mathbf{C}^{n \times n}$ is in MTBDF:

$$A = \text{Diag}(\Lambda_1, \Lambda_2, \ldots, \Lambda_m)$$

where each Λ_j is monotriangular with eigenvalue λ_j. The matrix $\lambda I - \Lambda_j$ is invertible if $\lambda \neq \lambda_j$ and nilpotent if $\lambda = \lambda_j$. Hence,

$$(\lambda I - A)^k = \text{Diag}(C_1, C_2, \cdots, C_m)$$

where $C_j = (\lambda I - \Lambda_j)^k$ is invertible for $\lambda \neq \lambda_j$ and zero for $\lambda = \lambda_j$ and $k \geq n_j$ where Λ_j has size $n_j \times n_j$. Thus $\rho_{\lambda,k}(A) = n$ for $\lambda \neq \lambda_j$ and $\rho_{\lambda,k}(A) = n - n_j$ for $\lambda = \lambda_j$ and $k \geq n_j$.

405A $\rho_{\mu,k}(\Lambda) = 6$ for $\mu \neq \lambda$, $\rho_{\lambda,1}(\Lambda) = 3$, $\rho_{\lambda,2}(\Lambda) = 1$, and $\rho_{\lambda,k}(\Lambda) = 0$ for $k > 2$.

416C $\rho_{5,1}(A) = 5$, $\rho_{5,2}(A) = 4$, $\rho_{5,k}(A) = 3$ for $k \geq 3$, $\rho_{7,2}(A) = 4$, $\rho_{7,k}(A) = 3$ for $k \geq 2$, and $\rho_{\lambda,k}(A) = 6$ for $\lambda \neq ,5,7$.

416D $\pi = (5,5,4,3,3,3,1)$.

417F
```
#> N=randsutr(5), X=rand(5,1)
#> P=[N^4*X  N^3*X  N^2*X  N*X  X]
#> W5=P^(-1)*N*P
```

421B They are all zero since no $(n-2) \times (n-2)$ submatrix of A is invertible.

427C $\mathrm{adj}(A) = \begin{bmatrix} -3 & 6 & -3 \\ 6 & -12 & 6 \\ -3 & 6 & -3 \end{bmatrix}$. The original matrix A has rank 2 and $\mathrm{adj}(A)$ has rank 1.

427D Suppose that A is $n \times n$. If A has rank n, so does $\mathrm{adj}(A)$. If A has rank $n-1$, then $\mathrm{adj}(A)$ has rank 1, since every column of A is in the (one-dimensional) null-space of A. If A has rank $\leq n-2$, then $\mathrm{adj}(A) = 0$ since every $(n-1) \times (n-1)$ submatrix has determinant zero. Thus A and $\mathrm{adj}(A)$ always have the same rank for $n = 1, 2$, while for $n \geq 3$ this is true iff A is invertible or $A = 0$.

433B
```
#> A=rand(4,4)
#> sum(diag(A)), A(1,1)+A(2,2)+A(3,3)+A(4,4)
```

434B Of course, the polynomial is

$$p(\xi) = (\xi - 1)(\xi - 2)(\xi - 3)(\xi - 5),$$

but MINIMAT doesn't do symbolic calculation. However, the command

```
#> p = sourfram(diag([1 2 3 5]))
```

will compute the coefficients. Check this using

```
#> 1^4+p(1)*1^3+p(2)*1^2+p(3)*1+p(4)
#> 2^4+p(1)*2^3+p(2)*2^2+p(3)*2+p(4)
#> 3^4+p(1)*3^3+p(2)*3^2+p(3)*3+p(4)
#> 5^4+p(1)*5^3+p(2)*5^2+p(3)*5+p(4)
```

438C
$$
\begin{aligned}
\mathrm{entry}_{ij}((a+b)C) &= (a+b)\,\mathrm{entry}_{ij}(C) \\
&= a\,\mathrm{entry}_{ij}(C) + b\,\mathrm{entry}_{ij}(C) \\
&= \mathrm{entry}_{ij}(aC) + \mathrm{entry}_{ij}(bC) \\
&= \mathrm{entry}_{ij}(aC + bC).
\end{aligned}
$$

438D   Steps (1), (2), (6), (7) use the definition of matrix multiplication. Step (3) uses the right distributive law for multiplication (of numbers)

$$\left(\sum_{j=1}^{n} a_{ij} b_{jk} \right) c_{kl} = \sum_{j=1}^{n} a_{ij} b_{jk} c_{kl}.$$

The absence of parentheses on the right is really a tacit use of the associative law (for numbers). Step (5) similarly uses the left distributive law. Step (4) is a special case of the law

$$\sum_{k=1}^{p} \sum_{j=1}^{n} x_{jk} = \sum_{j=1}^{n} \sum_{k=1}^{p} x_{jk}$$

which comes from the commutative and associative laws for addition of numbers: numbers may be added in any order. Of course, all the steps use the principle that equals may be substituted for equals.

439B   Assume $A, B \in \mathbf{F}^{m \times n}$ and $C \in \mathbf{F}^{n \times p}$. Then $(A+B)C \in \mathbf{F}^{m \times p}$ and $AC+BC \in \mathbf{F}^{m \times p}$. Choose integers i and k with $1 \le i \le m$ and $1 \le k \le p$.

$$\text{entry}_{ik}((A+B)C) =$$

$$= \sum_{j=1}^{n} \text{entry}_{ij}(A+B)\,\text{entry}_{jk}(C)$$

$$= \sum_{j=1}^{n} \left(\text{entry}_{ij}(A)\,\text{entry}_{jk}(C) + \text{entry}\,ij(B)\,\text{entry}_{jk}(C) \right)$$

$$= \sum_{j=1}^{n} \text{entry}_{ij}(A)\,\text{entry}_{jk}(C) + \sum_{j=1}^{n} \text{entry}_{ij}(B)\,\text{entry}_{jk}(C)$$

$$= \text{entry}_{ik}(AC+BC).$$

440B   Both sides have size $1 \times p$. The k-th entry in the i-th row of a matrix C is the (i, k)-entry:

$$\text{entry}_k(\text{row}_i(C)) = \text{entry}_{ik}(C).$$

Hence, by Theorem 33B,

$$\begin{aligned}
\text{entry}_k(\text{row}_i(AB)) &= \text{entry}_{ik}(AB) \\
&= \text{row}_i(A)\,\text{col}_k(B) \\
&= \text{entry}_k(\text{row}_i(A)B).
\end{aligned}$$

This show that the two rows $\text{row}_i(AB)$ and $\text{row}_i(A)B$ have the same k-th entry for $k = 1, 2 \ldots, p$ and are hence equal.

440C   The same sort of argument as was used for Exercise 440B works. Alternatively, once you have learned about taking the transpose, you can derive this as a corollary of the Exercise 440B. Here's the argument:

$$\mathrm{col}_k(AB)^\top = \mathrm{row}_k(B^\top A^\top) = \mathrm{row}_k(B^\top)A^\top = \left(A\,\mathrm{col}_k(B)\right)^\top$$

where the middle step is by the last exercise. Now take transposes. This equation is Theorem 54B (in different notation).

441A   By the definition of matrix multiplication, the i-th entry of the column vector AX is

$$\mathrm{entry}_i(AX) = \mathrm{entry}_{i1}(A)x_1 + \mathrm{entry}_{i2}(A)x_2 + \cdots + \mathrm{entry}_{in}(A)x_n.$$

Also $\mathrm{entry}_{ij}(A)$ is the i-th entry of $\mathrm{col}_j(A)$. The equation $AE_j = \mathrm{col}_j(A)$ is the special case is where $x_j = 1$ and $x_k = 0$ for $k \neq j$.

E

MINIMAT Tutorial (PC Version)

The DOS version of the computer program MINIMAT (for MINImal MAT-lab) is provided free with this book.[1] This program implements a subset of the popular programming language *Matlab* and is upward compatible with commercial versions such as *PC-Matlab* published by the Mathworks and *Student Edition of Matlab* published by Prentice-Hall. MINIMAT contains only those features which are useful in teaching an elementary course. These include a scrolling transcript which enables the user to review earlier output, the capability of saving the transcript at any time, a built-in editor for writing and listing small .M files, and carefully designed menus and help files. All the programs listed in the text are duplicated in .M files supplied on the distribution diskette. These .M files work with most versions of Matlab as well. In particular, they have been tested with *The Student Edition of Matlab*.

This tutorial should give you a feeling for how MINIMAT interacts with the user and what it can do. The tutorial is not intended to be a definitive account of anything. MINIMAT's commands are explained more carefully in Appendix C. Most users will find that the information in the text together with the online information accessible through MINIMAT's *help*

---

[1] Versions which run on other platforms (such as the Macintosh or Windows) can be obtained directly from the author for a nominal charge.

facility (explained below) is more than adequate for their needs. To use this tutorial simply type the indicated commands at MINIMAT and compare the output with the text.

That portion of the text in this tutorial which is actually produced by the computer is set in typewriter type

```
Like this.
```

All text typed into the computer by the user appears in typewriter type on a line beginning with the MINIMAT's prompt as in

```
#>  A = [1 2; 3 4]
```

All other text in typewriter type represents output produced by MINIMAT as in

```
A = [ 1.000    2.000
      3.000    4.000 ]
```

E.1   Before You Begin

Before you use MINIMAT you should make a backup copy using the `copy` command from the Operating System. If you are completely new to computers you will have to learn a few things first, like (1) how to turn the computer on and off; (2) how to boot the operating system; (3) how to change the default disk/directory; (4) how to execute a program from the operating system; (5) how to copy files from one disk to another. These details will be peculiar to the particular computer system you are using and will not be explained here.

The easiest way to use MINIMAT is to put all the files on the distribution diskette in one directory (if you have a hard disk) or diskette (if you do not) and to make that disk/directory the default. We shall assume you have done this.

E.2   Starting Up

We begin the session by typing `minimat` (followed by [Enter]) at the DOS prompt; this causes DOS to start the MINIMAT program. First MINIMAT displays a sign-on message. Pressing any key will cause MINIMAT to continue; MINIMAT will continue in any case after three seconds.

MINIMAT finds the file `mmlogin.m` on the default directory and executes the commands in that file. (If MINIMAT does not find this file it will simply print the prompt.) If you press a key while MINIMAT is executing a command, MINIMAT asks you if you want to interrupt the calculation. If you respond affirmatively, then MINIMAT prints the prompt. The commands in `mmlogin.m` cause MINIMAT to print the following message.

```
At the  #> prompt, the following may be entered:

help                   (to view the file minimat.hlp)
type filename.ext      (to view the file filename.ext)
edit filename.ext      (to edit the file filename.ext)
mmdemo                 (a demonstration)
Press a key:
```

Notice that message lines at the top and bottom of the secreen sgow which function keys MINIMAT will respond to. We press a key and MINIMAT continues by clearing the screen and printing the prompt.

E.3 The Prompt

When MINIMAT is waiting for input it issues a prompt:

```
#>
```

If at some point in the session MINIMAT has not issued the prompt when you think it should, it may be that you have not completed typing the last line (perhaps there is an open parenthesis). In this case (if you do not know how to complete the statement correctly) you may press [Esc] (the escape key) to recover the prompt. If you press [Enter] before you have typed a complete statement, the message line at the top of the screen shows what symbol is needed to complete that statement.

E.4 Sample Session

We evaluate some simple MINIMAT expressions and assign them to variables. Note how the variable **ans** contains the result of the last expression evaluated which was not explicitly assigned to some variable.

```
#> 3*4-1

   ans = 11.000

#> t=rand(1,1), x=cos(t), y=sin(t), x^2+y^2
```

```
t    =    0.557
x    =    0.849
y    =    0.528
ans  =    1.000
```

MINIMAT's notations are similar to those used in mathematics and most programming languages like BASIC. It uses the asterisk * to denote multiplication (if we write AB when we mean $A*B$, then MINIMAT will think of AB as a symbol by itself having nothing to do with either A or B). Note also that several statements may appear the same line if they are separated by commas.

At this point we ask for help.

```
#> help
```

```
**********     MINIMAT HELP FILE.  *************
  Here are some MINIMAT statements:

  a=[1 2 3; 4 5 6; 7 8 9]
  b=ref(a)
  x=[1 -2 1]'
  a*x,    b*x
  [m n] = size(a)
  b=a'+a
  c=a*b/2
  c=a+2*c


*************************************************
*  Press the  PgDn  key  to see more.      *
*  Type Alt-x to return to MINIMAT         *
*************************************************
```

We show only one screenful of the help facility here. We type [Alt-X], as indicated on the message line, to return to MINIMAT.

At this point we use MINIMAT's convenient notations to perform Gaussian elimination. First we enter a simple matrix:

```
#> A = [  1 2 3; 4 5 6; 7 8 9  ]

    A = [ 1.000    2.000    3.000
          4.000    5.000    6.000
          7.000    8.000    9.000 ]
```

Next we subtract an appropriate multiple of the first row from the second
to obtain a zero in the $(2, 1)$-position:

```
#> A(2,:) = A(2,:)-A(2,1)*A(1,:)

    A =  [ 1.000    2.000    3.000
           0.000   -3.000   -6.000
           7.000    8.000    9.000 ]
```

We do the same for the third row to produce a zero in the $(3, 1)$-position:

```
#> A(3,:) = A(3,:)-A(3,1)*A(1,:)

    A =  [ 1.000    2.000    3.000
           0.000   -3.000   -6.000
           0.000   -6.000  -12.000  ]
```

We scale the second row to produce a one in the $(2, 2)$-position:

```
#> A(2,:) = A(2,:)/A(2,2)

    A =  [ 1.000    2.000    3.000
           0.000    1.000    2.000
           0.000   -6.000  -12.000  ]
```

We subtract a multiple of the second row from the third to produce a zero
in the $(3, 2)$-position.

```
#> A(3,:) = A(3,:)-A(3,2)*A(2,:)

    A =  [ 1.000    2.000    3.000
           0.000    1.000    2.000
           0.000    0.000    0.000  ]
```

Finally we produce a zero in the $(1, 2)$-position.

```
#> A(1,:) = A(1,:)-A(1,2)*A(2,:)

    A = [ 1.000    0.000   -1.000
          0.000    1.000    2.000
          0.000    0.000    0.000  ]
```

We use MINIMAT's what command to find out which .M functions
MINIMAT can find.

```
#> what
```

```
Directory of C:\MM\*.M
ADJ.M           BED.M          COMPAN.M       DETM.M       FLIP.M
GJ.M            GJM.M          GS.M           HH.M         INTINV.M
MMDEMO.M        NBASIS1.M      NBASIS2.M      NBASIS3.M    NEXT.M
MMLOGIN.M       RANDSOLN.M     RANDSUTR.M     RBASIS1.M    MMTEST.M
RBASIS2.M       RBASIS3.M      RBASIS4.M      ROOK.M       SCALE.M
SCHURTRI.M      SHEAR.M        SIGNUM.M       SOURFRAM.M   SWAP.M
(30 files.)
```

We decide to find out what the .M file gjm.m does.

```
#> type gjm
```

```
function [M, R] = gjm(A)
    % M*A=R the reduced row echelon form of A
    [m,n] = size(A);
    RaM = gj([A eye(m)]);
    R = RaM(:,1:n);
    M = RaM(:,n+1:n+m);
```

The type command listed the file jgm.m on the screen. The comments at the beginning of the listing indicate that the function jgm accepts a matrix A as input and produces two matrices R and M as output: R is the reduced row echelon form of A and M is the matrix which pre-multiplies A to put A into reduced row echelon form.

We try it out. We press [Alt-X] to exit the type facility (just as we exited the help facility). Then we enter the same matrix we entered before.

```
#> A=[1 2 3; 4 5 6; 7 8 9]
```

```
A = [ 1.000      2.000      3.000
      4.000      5.000      6.000
      7.000      8.000      9.000  ]
```

Next we invoke the .M function using the form suggested by the heading (the first line of the file, following the keyword function):

```
#> [R M]=jgm(A)
```

```
R = [ 1.000    0.000   -1.000
      0.000    1.000    2.000
      0.000    0.000    0.000 ]
```

```
M = [  0.000  -2.667   1.667
       0.000   2.333  -1.333
       1.000  -2.000   1.000  ]
```

We used the same letters – R, M, A – in calling jgm that are used in the text
of the file jgm.m, but this is not necessary. Notice that the value for R is
the same as the value for A we obtained before. We check the claim made
in the listing of jgm.m above:

```
#> M*A
```

```
ans =[ 1.000   0.000  -1.000
       0.000   1.000   2.000
      -0.000   0.000   0.000  ]
```

It works! (M*A=R.)
   At this point we could terminate MINIMAT and return to DOS by typing

```
#> quit
```

Don't do this yet, however. The next few sections show how to use various
aspects of MINIMAT's user interface, and you might as well try them out
now.

E.5   Function Keys and Menus

If you press the function key [F1] at the MINIMAT prompt, you will will
bring up the **Help menu**. If you press the function key [F10] you will
get the **Main menu**. (This is indicated on the menu line at the bottom
of the screen, so you don't have to remember which key to press.) You
select menu items by either using the up/down arrow keys to highlight
the item you want and then selecting that item by pressing the [Enter]
(Return) key or else by typing the letter enclosed in the brackets. The
Menu facility works by passing an appropriate text line to MINIMAT's
command processor (this line will then appear on the screen) so you don't
need to use the Menu system if you can remember the commands and like
to type.

E.6   Snow and Color

If MINIMAT creates "snow" on your monitor when it writes to the screen,
execute the **snow** command from the **change defaults** option on the

main menu ([F10]). This will eliminate the snow but slow down the screen response. If you are using a monochrome monitor and the screen is hard to read, try changing colors using the **colors** command from the defaults menu. In either case MINIMAT will print the command line which causes the change. You can put a copy of this line(s) in your mmlogin.m file. If you cannot see the menus or the menu bar on your monitor type

    #> colors bland

to change the colors to black and white. (After this command, the menu bar and menus should be visible on *any* monitor.)

E.7   Transcript

MINIMAT remembers everything it prints to the screen. In effect, it keeps a **transcript** of what has transpired. You can recall what has scrolled off the screen by using the cursor keys on the numeric keypad. Using the cursor keys by themselves will move the cursor around the screen. Pressing the control key [Ctrl] and a cursor key simultaneously will scroll the screen or move to the begining or end of your transcript. The **saveas** command described below can be used to write the transcript to a file on the disk for subsequent viewing, editing, or printing.[2]

WARNING
If MINIMAT runs out of memory, it may be forced to discard some of the transcript. It does this from the front, so that the output created first is discarded first.

If you type input when the cursor is before the prompt, the cursor will immediately move to the prompt. This makes it impossible for you to alter the transcript from MINIMAT[3] except by adding new text at the end.

---

[2]If you know that you will want to save the transcript, it is best to use the diary command (described below) rather than the **saveas** command. The diary command saves the output as it is being produced so that it will not be lost in the event something goes wrong with the computer.

[3]Of course, you can always edit any diary file which MINIMAT may have produced.

E.8   Recall

If you press the **recall key** [F3] at the prompt, the last statement you
entered will be copied to the prompt where you can edit it and re-execute
it if you like. Use the [End] key to move the cursor to the end of the line
or the [Ctrl-End] key to move it to the end of the text (these will be the
same if the cursor is on the last line of the text) before pressing [Enter].

If the cursor is before the active prompt, the recall key [F3] behaves
differently depending on what kind of text the cursor is at. The message
line at the top of the screen indicates what the recall key will do at the
current cursor position. Try it out with the following:

1. Enter a statement like `A=rsnd(3,4)` with a typing error in it. Press
   [F3] and correct it to `A=rand(3,4)`,

2. Move the cursor to the line containing the error message produced
   by your erroneous statement and press [F3].

3. Move the cursor to a line containing an earlier statement you typed
   and press [F3].

4. Move the cursor to a matrix which MINIMAT has printed and press
   [F3].

In each case observe how the message line at the top of the screen indicates
what will happen when you press [F3].

E.9   Diary

MINIMAT provides the capability of keeping a record of all or part a session
on the disk. If you type

```
#> diary  a:\homework.1
```

MINIMAT will create a file called `homework.1` in the root directory on drive
`a:` (this is the floppy disk drive) and echo all subsequent screen output to
this file. You can substitute any filename you like for `a:\homework.1`.
If the drive specification `a:` or the path specification (this is `\` in the
example) is absent then the default is assumed so that the command

```
#> diary  homework.1
```

will create the file `homework.1` in the default directory on the default disk
(which is probably `c:` if your computer has a hard drive.) Finally if the
file specification is missing altogether, as in

```
#> diary
```

MINIMAT with create a file named `diary`. in the default directory of the default disk. If a file matching the file specification already exists, MINIMAT will ask you whether you want to Overwrite the existing file, Append to it, or Cancel the `diary` command altogether. [4]

To stop echoing to the diary file you simply type

```
#> diary off
```

at the prompt. This closes the diary file. The diary file will also be automatically closed when you quit MINIMAT.

E.10   SaveAs

If you want to save the entire transcript the command

```
#> saveas  myfile
```

will save it in a disk file `myfile`. This is useful if you forgot to open a diary file. If appropriate, MINIMAT gives you the Overwrite, Append, Cancel option as for the diary file.

The `saveas` command operates independently of the `diary` command. Thus it is possible to create two (or more) copies of the transcript. The difference between the two commands is that the `diary` command saves the transcript as it is produced, while the `saveas` command saves it all at once. Remember that if you have produced a very long transcript, MINIMAT may have forgotten the earliest portion of it.

E.11   Viewing the Diary

If you want to look at the diary file from MINIMAT you can *type* it like any other text file. For example, the command

```
#> type a:\homework.1
```

will display that file on the screen. You can even type a diary file that is currently being created.

If you want to study your diary file away from the computer you will have to print it. If you have a printer, you can print the diary file using a suitable print program like the DOS `copy` program. This can be invoked

---

[4]WARNING: Other versions of Matlab may not do this.

from MINIMAT using a **shell escape**. It works as follows. You enter a command line to MINIMAT exactly as you would enter it in DOS, but you precede it with an exclamation point. DOS will process the command and return control to MINIMAT. For example, to send the file a:\homework.1 to the printer you type[5]

```
#> !copy a:\homework.1 prn
```

WARNING

Never execute a "Terminate but Stay Resident" program (that is, one which continues to work after other programs have been invoked) as a shell escape. The DOS PRINT program is such a program (since after it starts printing but before it is finished you can invoke other programs). The MINIMAT shell escape !print filename may cause the computer to freeze. This caveat applies to any program which invokes other programs, not just MINIMAT.

E.12   Comments

MINIMAT will ignore any line which begins with a percent sign %. This allows programmers to insert comments in .M files, but can also be used to insert text into diary files. For example, if you enter

```
#> %  This is
#> %  a comment.
```

MINIMAT will not respond but will copy the lines into the diary file if one is open. (If the percent sign % were not present, MINIMAT would print an error message.)

E.13   Homework

The diary file and % comments provide a handy way for keeping a record of your homework assignments. You can get a blank diskette[6] (a dou-

---

[5]If nothing happens make sure the printer is turned on and is connected properly.

[6]You will have to *format* the diskette before using it. This is done by executing a DOS (Disk Operating System) program called FORMAT.

ble density 360K diskette ought to be more than adequate for the entire semester) and store all your homework assignments on it, say in files named homework.1, homework.2, ... You can insert % comments in the diary file to keep a record of what problem you are solving and to answer any questions which may be part of the problem. As an example, suppose you have been assigned the following:

Exercise. Type the following command into MINIMAT:

```
#> a=4,  a1=sqrt(a*a)
#> b=-3, b1=sqrt(b*b)
```

(sqrt(c) is MINIMAT's name for $\sqrt{c}$.) Why does $a = a_1$ but $b \neq b_1$?

Let's solve this simple problem to illustrate the proper format.

1. Open the diary file (MINIMAT responds):

```
#> diary a:\hw.0

* Diary file: A:\hw.0
* Opened at  Date/Time: 88-07-09 14:19:40
* (Type   DIARY OFF   to close the diary.)
```

2. Identify yourself and the problem:

```
#> % Sally Student.  Homework #1 due September 8.
#> % Exercise A
```

3. Do the computation:

```
#> a=4, a1=sqrt(a*a)

  a  =    4.000
  a1 =    4.000

#> b=-3, b1=sqrt(b*b)

  b  =   -3.000
  b1 =    3.000
```

4. Answer the question:

```
#> sqrt(9)

    ans =  3.000

#> % sqrt is the square root function.
#> % it computes the positive square root.
```

E.14   Editing and Shell Escape

MINIMAT has a built-in editor which you can use for writing .M functions or editing the transcript. You can invoke it from the main menu ([F10]) or by typing the edit command as in

```
#> edit myfunc.m
```

(Here myfunc.m can be replaced by any other legal filename.) If the file myfunc.m exists it is loaded for editing; otherwise, a new empty file by that name is created.

MINIMAT's built-in editor is rather primitive (though it should be adequate for the simple tasks required in the book). If you have another editor which you prefer, you can invoke it from MINIMAT with the **shell escape** as in

```
#> !qedit myfunc.m
```

Indeed, any command line preceeded by a ! will be passed to DOS and executed as if you had typed it at the DOS prompt. This is faster than exiting MINIMAT and does not discard any variables you have defined. On returning from a shell escape (or MINIMAT's edit command) MINIMAT will check the time stamps on any .M functions in its memory. If it finds a nonmatch, it will clear all the .M functions from its memory so that subsequent calls invoke the newer versions.

Command	Key Stroke	Command	Key Stroke
Page up	[PgUp]	Page down	[PgDn]
Cursor left	[←]	Cursor right	[→]
Cursor up	[↑]	Cursor down	[↓]
Start of line	[Home]	End of line	[End]
Top of file	[Ctrl-Home]	End of file	[Ctrl-End]
New line	[Enter]	To end & new line	[Shift-Enter]
Scroll up	[Ctrl-PgUp]	Scroll down	[Ctrl-PgDn]
Scroll right	[Ctrl-←]	Scroll left	[Ctrl-→]
Delete under	[Del]	Delete before	[Backspace]
Delete line	[Ctrl-Y]	Cancel statement	[Esc]
Help menu	[F1]	Local help	[F2]
Recall	[F3]	Main menu	[F10]

(The local help key [F2] usually displays this table)

MINIMAT's Function Keys

F

INDEX

MINIMAT Symbols[2]

(See also Appendix C pages 449-464.)

---

[2]Math symbols inside front cover.